Lecture Notes in Computer Science   13929

Founding Editors

Gerhard Goos
Juris Hartmanis

The series Lecture Notes in Computer Science (LNCS), including its subseries Lecture Notes in Artificial Intelligence (LNAI) and Lecture Notes in Bioinformatics (LNBI), has established itself as a medium for the publication of new developments in computer science and information technology research, teaching, and education.

LNCS enjoys close cooperation with the computer science R & D community, the series counts many renowned academics among its volume editors and paper authors, and collaborates with prestigious societies. Its mission is to serve this international community by providing an invaluable service, mainly focused on the publication of conference and workshop proceedings and postproceedings. LNCS commenced publication in 1973.

Luis Gomes · Robert Lorenz

Editors

Application and Theory of Petri Nets and Concurrency

44th International Conference, PETRI NETS 2023
Lisbon, Portugal, June 25–30, 2023
Proceedings

 Springer

Editors
Luis Gomes (iD)
NOVA University Lisbon
Caparica, Portugal

Robert Lorenz (iD)
University of Augsburg
Augsburg, Germany

ISSN 0302-9743   ISSN 1611-3349 (electronic)
Lecture Notes in Computer Science
ISBN 978-3-031-33619-5   ISBN 978-3-031-33620-1 (eBook)
https://doi.org/10.1007/978-3-031-33620-1

This Springer imprint is published by the registered company Springer Nature Switzerland AG
The registered company address is: Gewerbestrasse 11, 6330 Cham, Switzerland

Preface

This volume contains the proceedings of the 44th International Conference on Application and Theory of Petri Nets and Concurrency (Petri Nets 2023). The aim of this series of conferences is to create an annual opportunity to discuss and disseminate the latest results in the field of Petri nets and related models of concurrency, including their tools, applications, and theoretical progress.

The 44th conference and affiliated events were organized by the R&D Group on Reconfigurable and Embedded Systems (GRES) at School of Science and Technology of NOVA University Lisbon (Campus of Caparica), during June 25–30, 2023. The conference was organized for the third time in Portugal, twenty-five years after the first visit, also organized at Costa da Caparica, in Lisbon region.

This year, 47 papers were submitted to Petri Nets 2023. Each paper was single-blind reviewed by at least four reviewers. The discussion phase and final selection process by the Program Committee (PC) were supported by the EasyChair conference system. From 38 regular papers and 9 tool papers, the PC selected 21 papers for presentation: 17 regular papers and 4 tool papers. After the conference, some of these authors were invited to submit an extended version of their contribution for consideration in a special issue of a journal.

We thank the PC members and other reviewers for their careful and timely evaluation of the submissions and the fruitful constructive discussions that resulted in the final selection of papers. The Springer LNCS team provided excellent and welcome support in the preparation of this volume.

The keynote presentations were given by

- Stefanie Rinderle-Ma, Technical University of Munich, on "Process Mining and Process Automation in Manufacturing and Transportation",
- Valeriy Vyatkin, Aalto University and Lulea University of Technology, on "Formal Modelling, Analysis, and Synthesis of Modular Industrial Systems inspired by Net Condition/Event Systems", and
- Boudewijn van Dongen, Eindhoven University of Technology, on "Challenges in Conformance Checking: Where Process Mining meets Petri Net Theory".

The conference series is coordinated by a steering committee with the following members: W. van der Aalst (Germany), G. Ciardo (USA), J. Desel (Germany), S. Donatelli (Italy), S. Haddad (France), K. Hiraishi (Japan), J. Kleijn (The Netherlands), F. Kordon (France), M. Koutny (UK) (chair), L. M. Kristensen (Norway), C. Lin (China), W. Penczek (Poland), L. Pomello (Italy), W. Reisig (Germany), G. Rozenberg (The Netherlands), A. Valmari (Finland), and A. Yakovlev (UK).

Alongside Petri Nets 2023, the following workshops took place:

- Algorithms and Theories for the Analysis of Event Data (ATAED 2023),
- International Workshop on Petri Nets and Software Engineering (PNSE 2023),

- International Workshop on Petri Nets for Twin Transition (PN4TT 2023), and
- Petri Net games, examples, and quizzes for education, contest, and fun (PENGE 2023).

Other colocated events included the Petri Net Course and Tutorials, coordinated by Jörg Desel and Jetty Kleijn, as well as a Tool Exhibition, coordinated by Filipe Moutinho and Fernando Pereira.

We greatly appreciate the efforts of all members of the Local Organizing Committee, chaired by Anikó Costa and Isabel Sofia Brito, and including Filipe Moutinho, Fernando Pereira, Carolina Lagartinho-Oliveira, José Ribeiro, and Rogério Campos-Rebelo, for their time spent in the organization of this event.

We hope you enjoy reading the contributions in this LNCS volume.

June 2023   Luis Gomes
                                                                                              Robert Lorenz

Organisation

Program Committee

Elvio Gilberto Amparore	University of Turin, Italy
Abel Armas Cervantes	The University of Melbourne, Australia
Paolo Baldan	Università di Padova, Italy
Joao Paulo Barros	Instituto Politécnico de Beja, Portugal
Beatrice Berard	LIP6, Sorbonne Université & CNRS, France
Luca Bernardinello	Università degli studi di Milano-Bicocca, Italy
Didier Buchs	University of Geneva, Switzerland
Jörg Desel	Fernuniversität in Hagen, Germany
Raymond Devillers	ULB, Belgium
Susanna Donatelli	Università di Torino, Italy
Javier Esparza	Technical University of Munich, Germany
João M. Fernandes	University of Minho, Portugal
David Frutos Escrig	Universidad Complutense de Madrid, Spain
Luis Gomes (Co-chair)	NOVA University Lisbon, Portugal
Stefan Haar	Inria, ENS Paris-Saclay, France
Xudong He	Florida International University, USA
Loic Helouet	Inria, France
Ryszard Janicki	McMaster University, Canada
Anna Kalenkova	University of Adelaide, Australia
Jörg Keller	Fernuniversität in Hagen, Germany
Ekkart Kindler	Technical University of Denmark, Denmark
Michael Köhler-Bußmeier	University of Applied Science Hamburg, Germany
Irina Lomazova	National Research University Higher School of Economics, Russia
Robert Lorenz (Co-chair)	University of Augsburg, Germany
Lukasz Mikulski	Nicolaus Copernicus University, Poland
Andrew Miner	Iowa State University, USA
Marco Montali	Free University of Bozen-Bolzano, Italy
Laure Petrucci	Université Paris 13, France
Artem Polyvyanyy	University of Melbourne, Australia
Pierre-Alain Reynier	Aix-Marseille Université, France
Arnaud Sangnier	IRIF, Univ. Paris Diderot, CNRS, France
Natalia Sidorova	Technische Universiteit Eindhoven, The Netherlands

Jaco van de Pol   Aarhus University, Denmark
Boudewijn Van Dongen   Eindhoven University of Technology,
                                   The Netherlands
Remigiusz Wisniewski   University of Zielona Gora, Poland
Alex Yakovlev   Newcastle University, UK

Additional Reviewers

Adobbati, Federica	Alkhammash, Hanan
Aubel, Adrián Puerto	Balasubramanian, A. R.
Barylska, Kamila	Bergenthum, Robin
Bozorgi, Zahra Dasht	Coet, Aurélien
Guillou, Lucie	Habermehl, Peter
Helfrich, Martin	Kaniecki, Mariusz
Kim, Yan	Lime, Didier
Morard, Damien	Nesterov, Roman
Pomello, Lucia	Racordon, Dimi
Reisig, Wolfgang	Remi, Morin
Rivkin, Andrey	Rosa-Velardo, Fernando
Rubio, Rubén	Rykaczewski, Krzysztof
Shershakov, Sergey	Sommers, Dominique
Su, Zihang	Tour, Andrei
Verbeek, Eric	Weininger, Maximilian
Winkler, Sarah	Zaman, Eshita

Challenges in Conformance Checking: Where Process Mining Meets Petri Net Theory (Extended Abstract)

Boudewijn van Dongen ⓘ

Department of Mathematics and Computer Science, Eindhoven University
of Technology, Eindhoven, The Netherlands
b.f.v.dongen@tue.nl

1 Conformance Checking

Over the past 20 years, process mining has developed as a research area focusing on the analysis of data to create insights into processes. Processes are typically expressed in the form of control-flow models using languages such as Petri nets and data is available in the form of collections of events referring to discrete state changes of objects in the environment.

In practice, all processes share the property that their day to day operations differ from what is described in models and conformance checking [6] has become a significant field in process mining dealing with the question how process models, data and reality relate to each other.

Various types of conformance checking techniques exist for control flow only [1, 4, 7, 13]. More advanced techniques also consider data and resources [3, 11, 12] In this keynote, we focus on the techniques based on *synchronous product nets*, as first introduced for this purpose by Adriansyah et al. [2]. These synchronous products were developed in a setting where the Petri net is a workflow net and the data consists of sequences of events for a specific instance of that workflow (the case). Conformance is then determined using an A* based search strategy on the statespace of a synchronous product net [1, 7], or by means of logic programming [5] or planning [10]. Furthermore, techniques exist to compute representations of classes of alignments [8] while other techniques focus on approximations of alignments [9, 16]. However, all these techniques have one fundamental property in common, namely that in the end, they all reason over the synchronous product (although the techniques are not always instantiating this net).

2 Synchronous Products

In Petri net terms, a synchronous product Petri net is a low-level Petri net combining a process model with a sequence of events. The question: "how well does this sequence fit the given (process) model?" can be answered by solving a reachability question in this synchronous product, with the additional challenge to minimize a cost function over the transition firings. The witness solution for this question provides the so-called alignment and since reachability is typically guaranteed by the properties of the model, the challenge is to find the witness minimizing the cost, not so much to establish reachability itself.

More formally, let $PN = ((P, T, F), m_i, m_f)$ be a marked Petri net, with an initial and final marking and let $c : T \rightarrow \mathfrak{R}^+$ be a cost function assigning non-negative cost to each transition T. If PN is a synchronous product of a model M and a sequence σ, then an alignment $\gamma \in T*$ is a firing sequence in PN, such that $m_i \xrightarrow{\gamma} m_f$, i.e. m_f is reached by firing γ from m_i. An optimal alignment γ^{opt} is an alignment such that there does not exist another alignment γ with $c(\gamma) < c(\gamma^{opt})$, where c is lifted to sequences by simply summing over the transitions in the sequence.

It is easy to see that finding an alignment is at least as complex as reachability as the question for a given Petri net $PN = (P, T, F)$ whether m_f is reachable from m_i is the same as looking for an alignment in $PN = ((P, T, F), m_i, m_f)$ with for all $t \in T$ holds that $c(t) = 0$, making every alignment optimal by definition.

The problem of finding an alignment with specific cost C can also be translated to reachability, provided that the cost function assigns integer (or rational) costs to each transition. The intuition here is that you let each transition produce a number of tokens in a cost place p and you try to reach the marking $m_f \cup [p^C]$. If this is possible, the witness sequence is an alignment with cost C. An optimal alignment can then be found by first finding an alignment with a cost function assigning 0 cost to all transitions. The witness of this reachability problem provides an upperbound for C and hence a binary search can be conducted to find C^{min}, i.e. the minimal cost with which $m_f \cup [p^C]$ is reachable.

While alignments for specific, isolated cases in a process are interesting, the more challenging problem is to find alignments for entire event logs, taking into account multiple perspectives, such as data and resources, as well as inter-case dependencies introduced by these perspectives or by process properties such as batching.

In this keynote, we introduce the relation between conformance checking and reachability and we see look at the scenario where the process model becomes a system model, explicitly modeling inter-case dependencies in the form of v-nets [14, 15]. We consider the case where the event data is no longer a sequence representing a single case and how this impacts the complexity of the conformance checking problem. We conclude with a challenge for the Petri net community: To develop reachability techniques tailored towards the problem of finding alignments between event data and system models.

References

1. van der Aalst, W.M.P., Adriansyah, A., van Dongen, B.F.: Replaying history on process models for conformance checking and performance analysis. Rev. Data Min. Knowl. Discov. 2(2), 182–192 (2012). https://doi.org/10.1002/widm.1045. Wiley Interdiscip.
2. Adriansyah, A., Sidorova, N., van Dongen, B.F.: Cost-based fitness in conformance checking. In: Caillaud, B., Carmona, J., Hiraishi, K. (eds.) 11th International Conference on Application of Concurrency to System Design, ACSD 2011, Newcastle Upon Tyne, UK, 20–24 June, 2011, pp. 57–66. IEEE Computer Society (2011). https://doi.org/10.1109/ACSD.2011.19

3. Alizadeh, M., Lu, X., Fahland, D., Zannone, N., van der Aalst, W.M.P.: Linking data and process perspectives for conformance analysis. Comput. Secur. **73**, 172–193 (2018). https://doi.org/10.1016/j.cose.2017.10.010

4. Berti, A., van der Aalst, W.M.P.: A novel token-based replay technique to speed up conformance checking and process enhancement. Trans. Petri Nets Other Model. Concurr. **15**, 1–26 (2021). https://doi.org/10.1007/978-3-662-63079-2_1

5. Boltenhagen, M., Chatain, T., Carmona, J.: Optimized SAT encoding of conformance checking artefacts. Computing **103**(1), 29–50 (2021). https://doi.org/10.1007/s00607-020-00831-8

6. Carmona, J., van Dongen, B.F., Weidlich, M.: Conformance checking: foundations, milestones and challenges. In: van der Aalst, W.M.P., Carmona, J. (eds.) Process Mining Handbook, LNBIP, vol. 448, pp. 155–190. Springer, Cham (2022). https://doi.org/10.1007/978-3-031-08848-3_5

7. van Dongen, B.F.: Efficiently computing alignments - using the extended marking equation. In: Weske, M., Montali, M., Weber, I., vom Brocke, J. (eds.) BPM 2018. LNCS, vol. 11080, pp. 197–214. Springer (2018). https://doi.org/10.1007/978-3-319-98648-7_12

8. Garca-Bañuelos, L., van Beest, N., Dumas, M., Rosa, M.L., Mertens, W.: Complete and interpretable conformance checking of business processes. IEEE Trans. Softw. Eng. **44**(3), 262--290 (2018). https://doi.org/10.1109/TSE.2017.2668418

9. Lee, W.L.J., Verbeek, H.M.W., Munoz-Gama, J., van der Aalst, W.M.P., Sepúlveda, M.: Replay using recomposition: alignment-based conformance checking in the large. In: Clarisó, R., et al. (eds.) Proceedings of the BPM Demo Track and BPM Dissertation Award co-located with 15th International Conference on Business Process Modeling (BPM 2017), Barcelona, Spain, 13 September 2017. CEUR Workshop Proceedings, vol. 1920. CEUR-WS.org (2017). http://ceur-ws.org/Vol-1920/BPM_2017_paper_157.pdf

10. de Leoni, M., Lanciano, G., Marrella, A.: Aligning partially-ordered process-execution traces and models using automated planning. In: de Weerdt, M., Koenig, S., Röger, G., Spaan, M.T.J. (eds.) Proceedings of the Twenty-Eighth International Conference on Automated Planning and Scheduling, ICAPS 2018, Delft, The Netherlands, 24–29 June 2018, pp. 321–329. AAAI Press (2018). https://aaai.org/ocs/index.php/ICAPS/ICAPS18/paper/view/17739

11. Mannhardt, F., de Leoni, M., Reijers, H.A., van der Aalst, W.M.P.: Balanced multi-perspective checking of process conformance. Computing **98**(4), 407–437 (2016). https://doi.org/10.1007/s00607-015-0441-1

12. Mehr, A.S.M., de Carvalho, R.M., van Dongen, B.F.: Detecting privacy, data and control-flow deviations in business processes. In: Nurcan, S., Korthaus, A. (eds.) Intelligent Information Systems - CAiSE Forum 2021, Melbourne, VIC, Australia, 28 June – 2 July 2021, Proceedings. LNBIP, vol. 424, pp. 82–91. Springer, Cham (2021). https://doi.org/10.1007/978-3-030-79108-7_10

13. Rozinat, A., van der Aalst, W.M.P.: Conformance checking of processes based on monitoring real behavior. Inf. Syst. **33**(1), 64–95 (2008). https://doi.org/10.1016/j.is.2007.07.001

14. Sommers, D., Sidorova, N., van Dongen, B.F.: Aligning event logs to resource-constrained ν-petri nets. In: Bernardinello, L., Petrucci, L. (eds.) PETRI NETS

2022. LNCS, vol. 13288, pp. 325–345. Springer, Cham (2022). https://doi.org/10.1007/978-3-031-06653-5_17

15. Sommers, D., Sidorova, N., van Dongen, B.F.: Exact and approximated log alignments for processes with inter-case dependencies. In: Bernardinello, L., Petrucci, L. (eds.) PETRI NETS 2023. LNCS. Springer (2023)

16. Taymouri, F., Carmona, J.: An evolutionary technique to approximate multiple optimal alignments. In: Weske, M., Montali, M., Weber, I., vom Brocke, J. (eds.) BPM 2018. LNCS, vol. 11080, pp. 215–232. Springer (2018). https://doi.org/10.1007/978-3-319-98648-7_13

Contents

Timed Models

Model Transformation

Invited Papers

From Process-Agnostic to Process-Aware Automation, Mining, and Prediction

Stefanie Rinderle-Ma[(✉)] [ID], Janik-Vasily Benzin[ID], and Juergen Mangler[ID]

Technical University of Munich,
TUM School of Computation, Information and Technology,
Boltzmannstrasse 3, 85748 Garching, Germany
{stefanie.rinderle-ma,janik.benzin,juergen.mangler}@tum.de

Abstract. The entire research area of (business) process management has experienced a tremendous push with the advent of process mining, robotic process automation, and predictive process monitoring. While this development is highly appreciated, the current process-agnostic pipelines for process mining, robotic process automation, and predictive process monitoring have several limitations. Taking a system perspective, this keynote elaborates the limitations of process-agnostic automation. Then, it shows how a shift towards process-aware automation and predictive compliance monitoring can be achieved and how process-aware pipelines contribute to overcome the limitations of process-agnostic automation. Finally, research implications with a focus on Petri nets are derived.

Keywords: Process Automation · Process Mining · Predictive Process Monitoring · Predictive Compliance Monitoring

1 Introduction

Process mining and robotic process automation are two mega trends. *"The global process mining software market is projected to grow from $933.1 million in 2022 to $15,546.4 million by 2029, at a CAGR of 49.5% in the forecast period."*[1]. The combination of both technologies is expected to even increase their market penetration [6].

Process mining comprises a set of techniques for the discovery and analysis of process models and their executions based on process event logs [1] and the expectations in practice are high [35]. Robotic process automation refers to the automation of single process tasks by replacing human-task interaction with a software bot [2]. The currently applied *mine and automate* pipeline (e.g., [14]) is depicted in Fig. 1a). Process mining is applied to discover process models, and within these models tasks with the potential for automation are detected. As an intermediate step between process mining and the automation of tasks, [14]

---

[1] https://www.fortunebusinessinsights.com/process-mining-software-market-104792.

L. Gomes and R. Lorenz (Eds.): PETRI NETS 2023, LNCS 13929, pp. 3–15, 2023.
https://doi.org/10.1007/978-3-031-33620-1_1

advocate to standardize the process models by removing variations in the process that might result due to, e.g., product variants. Once tasks are automated, process mining can be used to continuously monitor their performance.

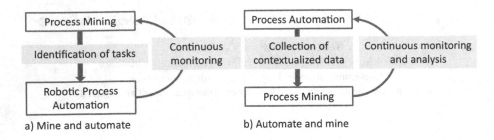

Fig. 1. Pipelines: a) Mine and automate; b) Automate and mine

However, the *mine and automate* pipeline as depicted in Fig. 1a) has several limitations:

1. *Task-oriented automation:* Robot process automation aims at the automation of single, often simple and repetitive interactions of humans with software. However, a process is a task-overarching, orchestrating concept. Real performance gains and analysis insights can only be achieved by taking an orchestration point of view for process automation.
2. *Data acquisition and preparation:* Process mining relies on process event logs emitted or extracted from information systems, e.g., ERP systems. If the underlying system is not process-aware or a black box (e.g., legacy systems), mechanisms for extraction and preparation of data are to be defined and employed. Moreover, if the data is spread over multiple and possibly heterogeneous information systems [18,27], mechanisms for integrating the data are to be defined and employed. Existing commercial systems support a range of adaptors to different systems and data sources, e.g., data connections as supported in Celonis[2]. Using an object-centric approach offers the opportunity to capture objects and their life cycles in the process event logs [8] and can be used even if no case id is available or can be extracted form the underlying data. However, data connections are not robust towards changes in the data structures, i.e., data structure changes possibly require the adaptation of one, several, or all of the established data connections.
3. *Ex-post point of view:* Most of the process mining analysis tasks are conducted in an ex-post manner, i.e., based on process event logs that reflect already finished process executions. This holds true for all three pillars of process mining, i.e., process discovery, conformance checking, and process enhancement. However, the monitoring and analysis of process executions during runtime (online) based on process event streams provides current insights into the

---

[2] https://docs.celonis.com/en/data-connections.html.

process, e.g., detecting exceptions when they are actually happening, and hence enabling a quicker reaction to potential problems such as compliance violations [9,19,36]. Moreover, in practice, many analysis questions refer to the monitoring of the process perspectives time, resources, and data, e.g., a temperature sensor exceeding a certain threshold or temporal deviations that *"are mostly caused by humans, e.g., someone stepping into the safety area of a machine causing a delay, and hint to problems with work organization"* [34]. Even predictive process monitoring, though suggesting to be applied in an online manner due to the term "monitoring", is mostly applied in an ex-post way. More precisely, a process event log is split into training and test data. One or several prediction models are learned based on the training data. These prediction models are then applied to the test data, i.e., prefixes from the test data are used to reflect a process event stream. Prediction goals comprise, for example, the remaining time of cases, the next activity, and the outcome of a process [13].

4. *Dealing with uncertainty and concept drift:* Ex-post mining allows to obtain a picture of the past. However, an ever changing process environment and uncertainties force processes to adapt constantly [5,9]. In the manufacturing domain, for example, if new processes are set up, several adaptation cycles are necessary until a process runs in a robust way. In health care, due to unforeseen situations, ad-hoc changes of process instances can be frequently required, e.g., the blood pressure exceeds a threshold such that the surgery has to be delayed. *"This uncertainty often manifests itself in significant changes in the executed processes"* [5]. Process changes, in turn, manifest as *concept drifts* in process event logs [5] and as *unseen behavior* in process event streams [23]. A selection of use cases for process changes from different domains can be found in [17].

Limitations 1. and 2. refer to the system and data perspective and Limitations 3. and 4. to the mining and analysis perspective of a process. In order to address Limitations 1. and 2., we advocate an inversion of the *mine and automate* pipeline into an *automate and mine* pipeline as depicted in Fig. 1b). The *automate and mine* pipeline starts with automated and orchestrated processes, driven and managed by process engines or process-aware information systems. These systems can be exploited to collect data in an integrated, orchestrated, and contextualized manner at arbitrary granularity which, in turn, offers novel process mining insights [29], for example, the combined analysis process event logs/streams and sensors streams [11,37].

Limitations 3. and 4. emphasize the need to move towards approaches applied during runtime when mining and monitoring processes. Most promising here are approaches for online process mining such as [9] and predictive process monitoring (cf., e.g., survey in [13]). One of the most crucial (business) goals of predictive process monitoring is the prediction of possible compliance violations [26]. For this, in existing approaches, the compliance constraint of interest, for example, service level agreement "90% of the orders must be processed within 2 h", is encoded as prediction goal in a prediction model (referred to as *predicate pre-diction* [21]). Predicate prediction is illustrated through the *comply and predict* pipeline depicted in Fig. 2a): compliance constraints are encoded as prediction goals (comply) into a prediction model each, based on which violations of the constraint are predicted (predict). The *comply and predict* pipeline for predicate prediction comes with the following limitations (ctd.):

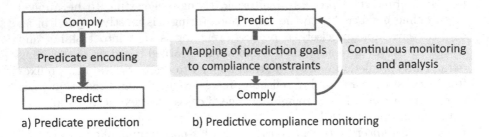

Fig. 2. Pipelines: a) Predicate prediction; b) Predictive compliance monitoring

5. *Performance:* In literature, predicate prediction, is mostly applied in the context of simple scenarios. Simple here refers to i) compliance constraints of limited complexity such as service level agreements and ii) a limited number of predicates. The reason is that the encoding of i) is more manageable for simple compliance constraints and ii) keeps the number of prediction models limited that are necessary for predicate prediction (recall that for n compliance constraints, n prediction models are to be created). However, real-world scenarios can look very different [32]: contrary to i), compliance constraints that stem from regulatory documents such as the GDPR are complex and refer to multiple process perspectives. Contrary to ii) there might be several hundred compliance constraints that are imposed on one process [28]. Supporting predicate prediction for full-blown real-world scenarios would possibly lead to a large number of complex prediction models, resulting in performance issues.

6. *Transparency:* Predicate prediction yields a binary answer, i.e., either "the predicate is violated" (possibly with a counterexample) or "the predicate is not violated". Though this constitutes an essential information, in particular in the case of violations, often some sort of reaction is required. At least, it should become transparent why a violation occurred and for which instance(s) (root cause). Without this information, it is difficult for users to decide on remedy actions.

7. *Maintainability:* In predicate and compliance prediction in general, two sources of change might occur. First of all, changes of the process and its instances might become necessary, reflected by concept drift in the process event log. Secondly, changes in the set of compliance constraints might be performed by adding, deleting, and updating compliance constraints.

Compliance constraint changes can occur frequently, e.g.: *"Bank regulations change about every 12 min"*[3]. For predicate prediction, a compliance constraint change requires the adaptation of the associated prediction model, i.e., m compliance constraint changes result in the adaptation of m prediction models.

In order to tackle Limitations 5.– 7., again, we advocate an inversion of the *comply and predict* pipeline shown in Fig. 2a). Instead of encoding compliance constraints and predicting their violations afterwards, we suggest the *predict and comply* pipeline denoted as *predictive compliance monitoring* [32], depicted in Fig. 2b): at first, predicting takes places through process monitoring approaches with different prediction goals such as next activity, remaining time, outcome, and other key performance indicators are applied (predict), followed by a mapping to the set of compliance constraints (comply).

In the following, we will contrast the different pipelines and approaches. For this, we take the perspective of a holistic system and generalize the pipelines into process-agnostic and process-aware automation (cf. Sect. 2). Finally, research implications with a focus on Petri nets will be provided in Sect. 3.

2   Process-Agnostic and Process-Aware Automation

In the introduction, pipelines for process automation and mining as well as prediction and compliance are shown, i.e., the current *mine and automate* and the inverted *automate and mine* pipeline as well as the current *comply and predict* and the inverted *predict and comply* pipeline. From a system perspective, the two pipelines are not separated from each other, i.e., a holistic system can support both. Figure 3 shows the system perspective realizing the *mine and automate* and *comply and predict* pipelines on the left side and the system perspective realizing the *automate and mine* and *predict and comply* pipelines on the right side. Due to the fact, that the system perspective on the right side takes an explicit process-aware point of view by employing a process engine or process-aware information system, we refer to it as *PAWA: process-aware automation*. Symmetrically, we refer to the system on the left side, where automation is restricted to single tasks, as *PAGA: process-agnostic automation*.

In current PAGA systems, the event and data stream is extracted by ETL pipelines from logs of the machine, ERP systems, and further systems as depicted in Fig. 3. A multi-perspective process model is mined through process discovery, conformance checking, and enriching the process model with additional perspectives using further mining methods, e.g., decision and organizational mining [12]. The machine, ERP systems and further systems are then *enhanced* through process analysts, domain experts, and/or developers as a result of insights gained from analysing the multi-perspective process model. Enhancing refines robotic process automation as shown for the *mine and automate* pipeline depicted in

---

[3] https://thefinanser.com/2017/01/bank-regulations-change-every-12-minutes (last accessed 2023-04-03).

Fig. 1 by additionally optimizing existing automatic activities in the process model, ad-hoc activities to mitigate possible problems that conformance checking has unveiled, or circumventing bottlenecks by assigning further resources to an activity.

The current state is dominated by the relational perspective of ERP systems that comes with major drawbacks. First, directly connecting to ERP and further relational information systems necessitates sophisticated ETL pipelines that emphasize ex-post over ex-ante views. Second, the lack of the process perspective in relational systems nudges the analysis to choose the traditional mine and automate line of action (cf. Sect. 1 and Fig. 1) such that the corresponding disadvantages apply.

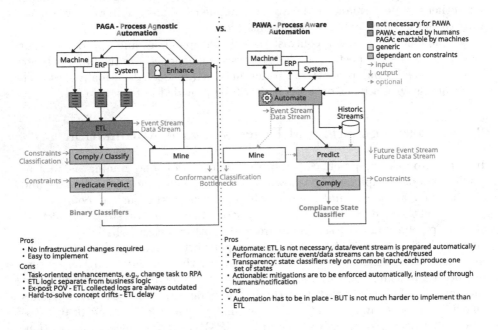

Fig. 3. System View Comparison

PAWA systems serve as an orchestration and automation environment that integrates the machine, ERP, and the system views (cf. Fig. 3). This enables the implementation and execution of arbitrary processes ($\rightarrow$ Limitation 1). To illustrate this, in [29], we provide a classification of process automation scenarios in manufacturing along the two dimensions of "human involvement" and "green field – brown field". This results in four automation classes that we have found and realized across 16 real-world process scenarios. More precisely, the process scenarios were modeled, implemented, and hence automated using the cloud process execution engine cpee.org [25]. The scenarios comprise i) a robotic process automation scenario (low human involvement, brown field), ii) fully automated process orchestration (low human involvement, green field), iii) process-oriented user support (high human involvement, brown field), and iv) interactive process automation (high human involvement, green field). i) was chosen to automate a task due to a black box application system to be invoked. ii) orchestrates the tasks of a robot, a machine, and measurement equipment. A video of the execution of the process orchestration can be found here[4]. iii) includes the automatically generated instructions to be shown at work stations for staff in a process with more than 20.000 variants. iv) features the on the fly creation and routing in process models based on interactions between human users and physical devices such as machines [22] or other utilities, e.g., in the care domain [33]. Such process scenarios are not only prevalent for the manufacturing domain, but also for other domains such as health care and logistics which integrate "physical" aspects (machines, vehicles) and human work. The variety of scenarios underpins that robotic process automation can be supported by a PAWA system, but is only one piece. PAWA systems are able to support any process orchestration and integrate different systems, human work, and physical devices along the process logic.

Moreover, PAWA systems can be employed to collect data in a systematic, integrated, and contextualized manner ($\rightarrow$ Limitation 2), i.e., they log every event emitted during process execution and on top of that, PAWA systems can collect and log process context data, e.g., IoT data in domains such as production, health care, and logistics. The combined collection of process and IoT data has gained interest lately, resulting in an extension of the process event log standard eXtensible event stream (XES)[5], i.e., the XES Sensor Stream extension [24]. This way, process engines and process-aware information systems serve as systems for the process-oriented and contextualized collection of process data at an arbitrary granularity (as defined in the process models) and a trusted, high quality level (****(*) star level according to the L* data quality model for process mining [3]) [30]. Using, for example, cpee.org as process collection system, we collected and published three real-world process event logs with additional context data[6]. Two data sets comprise data from public transport, augmented with context data on weather, traffic, etc. and one data set stems from the production domain on producing a chess piece.

In addition, PAWA systems collect and log data at an time, i.e., in an ex-post manner as process event logs and during runtime as process event streams ($\rightarrow$ Limitation 3). This also includes the runtime collection of context data such as sensor streams. In particular, the online collection of event streams facilitates the early detection of concept drifts [35] ($\rightarrow$ Limitation 4).

Up to this point, we discussed how PAGA and PAWA systems realize the *mine and automate* and *automate and mine* pipelines shown in Fig. 1. PAGA and PAWA systems can also realize the *comply and predict* and *predict and*

---

[4] https://lehre.bpm.in.tum.de/~mangler/.Slides/media/media1.mp4, last accessed 2023-04-04.

[5] www.xes-standard.org.

[6] https://zenodo.org/communities/processmining.

comply pipelines shown in Fig. 2, i.e., on the PAGA side by predicate prediction and on the PAWA side as predictive compliance monitoring components.

We conducted a comprehensive literature review covering the research areas of predictive process monitoring and compliance monitoring (see, e.g., [20]) with respect to functionalities required for building a predictive compliance monitoring system [31]. A system that supports predictive compliance monitoring employs the *predict and comply* pipeline (cf. Fig. 2b) to predict the future progress of the monitored system and to monitor compliance on top of the predictions and interprets them from a systems perspective. An abstract view on how to integrate predictive compliance monitoring into a PAWA system is depicted in Fig. 3, contrasted by the current state of how predicate prediction is conceptualized and implemented in a PAGA system.

In PAGA systems, due to the current lack of the process perspective in the monitored system and in the prediction models, the results of predicting compliance violations have to be manually transformed into actions that can be executed on an ERP system by notifying the respective employee (*enhance*).

In PAWA systems, the goal of predictive compliance monitoring centers around the process perspective (cf. Fig. 3). By automating existing ERP systems or substituting existing systems through a PAWA system, ETL pipelines are replaced by a simple connection to the logging service of PAWA system. The optional *mine* and the compulsory *predict* separately consume the event and data stream from the logging services. While *mine* is concerned with discovering process models, analysing structural and behavioral properties of process models and checking conformance, *predict* focuses on a single prediction model trained to predict the future event and data stream of the overall process, i.e., the prediction goal is a stream prediction ($\rightarrow$ Limitations 3. and 4.). The prediction model can additionally take the mined process model as input such that the prediction of the event and data stream is based on the respective execution states of running instances in the process model. Overall, the prediction goal consists of future events and, in particular, data attributes. If required for very important compliance constraints, the inverted, specialized *comply and predict* pipeline (predicate prediction) can be added to the predictive compliance monitoring system such that an independent prediction model for the very important compliance constraint is trained. Stream predictions of the process are the input to *comply*, while predicted violations of independent prediction models can directly trigger mitigation actions in the monitored system. Given a stream prediction, *comply* checks compliance of the compliance constraints resulting in various compliance states. Due to the process perspective inherent in PAWA systems, predicted compliance states can automatically trigger mitigation actions, e.g., by adding ad-hoc activities to an ongoing process instance.

Note that predictive compliance monitoring could also be integrated into the PAGA system, inheriting its limitations due to enhance and the data collection. More importantly, note that the distinction into predicate prediction and predictive compliance monitoring does not only apply to the domain of process mining and automation, but also to the more general area of *event prediction* [7].

At this point, we have to say that there is no solution for predictive compliance monitoring yet and the "sweet spot" between predicate prediction and predictive compliance monitoring w.r.t. prediction quality and limitations has to be investigated [32].

Due to its process centricity, the PAWA system comes with the following advantages regarding predictive compliance monitoring:

- *Performance and maintainability of the prediction model* (→ *Limitations 5. and 7.):* If the set of compliance constraints is updated, no retraining of the prediction model is necessary due to the clear separation of the prediction model and the compliance checking. Furthermore, no new prediction models have to be trained for new or updated constraints.
- *Transparency and explainability of the predictve process monitoring system* (→ *Limitation 6.):* As the prediction model predicts the future event and data streams, violations of compliance states can be pinpointed to their respective events or data attributes in the stream. Hence, the predicted violation is transparent and explainable.
- *Actionable mitigations:* Due to the process centricity of the PAGA system, compliance states can directly trigger actions in the process engine, e.g., through adding ad-hoc activities, or spawning instances of specialized mitigation processes.

3   Implications on Research

In the introduction, we raise seven limitations with current *mine and automate* as well as *comply and predict* pipelines which are integrated and analyzed through the systems perspective (PAGA vs. PAWA in Fig. 3). In the following, we will derive research implications with a focus on Petri net based research.

Soundness Verification for Automatic Changes to Automation. The system view comparison in Fig. 3 shows the two extreme sides of a continuous automation scale supported by process mining. A company on the move to process-aware automation can exhibit both automation systems, i.e., PAGA and PAWA, at the same time, as not all parts of the company are yet shifted to PAWA. During the transition, companies can benefit from support on how to shift from the manual *enhance* to the machine-enactable *automate* (cf. Fig. 3).

Petri Nets for Process-Aware Automation. Although Petri nets have been proposed and applied for process execution in the past (cmp. FUNSOFT Nets [10] in 1998), it remains not fully clear which Petri net class is sufficient to be used as execution model in PAWA. Recent candidates include object-centric Petri nets [4], Petri nets with identifiers [38], and colored Petri nets [16]. The main question is to keep the balance between expressive power to model all process perspectives and preventing problems such as checking soundness from becoming undecidable. Hence, research on Petri nets classes such as object-centric Petri nets or Petri nets with identifiers is ongoing.

Conformance Checking on Petri Net Process Models of Collaborative Systems. Conformance checking techniques for object-centric Petri nets and Petri nets with identifiers comparable to alignment-based conformance checking for sound workflow nets are missing. Also, replay-based techniques are yet missing, as the only existing object-centric Petri net implementation in PM4PY[7] does not feature replay.

Rescheduling Processes Execution - Checking and Balancing Resource Utilization. Whenever automatic changes are made, resource utilization may be affected. As multiple processes may share the same resources, optimization regarding resource utilization leads to better throughput. Scheduling of resource allocation with timed Petri nets (cmp. [15]) based on process models, can allow for simple, automatic and explainable solutions.

Instance and Process Spanning Constraints. Research on predicting and checking compliance has focused on intra-instance constraints so far. Predicting compliance states for instance and process spanning constraints remains an open research problem [31].

Provision of Mitigation Actions. Automatically providing mitigation actions for compliance violations, in particular at different granularity levels, and analyzing and visualizing their effects is relevant for both, predicate prediction and predictive process monitoring, but yet to be solved [31].

Visualization and explanation of predictions and violations. Visualization approaches for prediction results and future compliance violations are mostly missing. Moreover, root cause analysis has to be extended in order to deal with predicting violations of real-world compliance constraints [31].

Online Predictive Process Monitoring and Updating Compliance States. Since predictive process monitoring predicts future event and data streams given current event and data streams, prediction methods such as deep learning cannot be applied for cases with frequent process adaptations. It is not clear for which process environments existing prediction methods are capable of updating the prediction model after each incoming event with data or which batching methods are required such that existing prediction methods exhibit a sufficient performance. Continuous update of prediction models and predictions also results in continuous update of compliance states. It is open which granularity levels for compliance states, i.e., event-level, instance-level, process-level, multi-process-level, and multi-organisation-level, are supporting the users in understanding the current system state. Moreover, it is unclear how compliance states can be transformed between different granularity levels [31].

---

[7] https://pm4py.fit.fraunhofer.de/.

Data Properties and Quality. Exploiting data properties and quality is an emerging research topic. Considering data quality, data values of low quality may point to a compliance violation, e.g., redundant sensors fail quickly after each other. The relation of data quality with compliance violation that goes beyond merely removing low quality data points or imputating data values may reveal further insights.

References

1. van der Aalst, W.M.P.: Process Mining - Data Science in Action. 2nd edn. Springer, Heidelberg (2016). https://doi.org/10.1007/978-3-662-49851-4
2. van der Aalst, W.M.P., Bichler, M., Heinzl, A.: Robotic process automation. Bus. Inf. Syst. Eng. **60**(4), 269–272 (2018). https://doi.org/10.1007/s12599-018-0542-4
3. van der Aalst, W.M.P., et al.: Process mining manifesto. In: Business Process Management Workshops, pp. 169–194 (2011). https://doi.org/10.1007/978-3-642-28108-2_19
4. van der Aalst, W.M., Berti, A.: Discovering object-centric Petri nets. Fundamenta informaticae **175**(1–4), 1–40 (2020)
5. Adams, J.N., van Zelst, S.J., Rose, T., van der Aalst, W.M.P.: Explainable concept drift in process mining. Inf. Syst. **114**, 102177 (2023). https://doi.org/10.1016/j.is.2023.102177
6. Badakhshan, P., Wurm, B., Grisold, T., Geyer-Klingeberg, J., Mendling, J., vom Brocke, J.: Creating business value with process mining. J. Strateg. Inf. Syst. **31**(4), 101745 (2022). https://doi.org/10.1016/j.jsis.2022.101745
7. Benzin, J.V., Rinderle-Ma, S.: A survey on event prediction methods from a systems perspective: bringing together disparate research areas, February 2023. http://arxiv.org/abs/2302.04018
8. Berti, A., van der Aalst, W.M.P.: OC-PM: analyzing object-centric event logs and process models. Int. J. Softw. Tools Technol. Transf. **25**(1), 1–17 (2023). https://doi.org/10.1007/s10009-022-00668-w
9. Ceravolo, P., Tavares, G.M., Junior, S.B., Damiani, E.: Evaluation goals for online process mining: a concept drift perspective. IEEE Trans. Serv. Comput. **15**(4), 2473–2489 (2022). https://doi.org/10.1109/TSC.2020.3004532
10. Deiters, W., Gruhn, V.: Process management in practice applying the FUNSOFT net approach to large-scale processes. Autom. Softw. Eng. **5**(1), 7–25 (1998). https://doi.org/10.1023/A:1008654224389
11. Ehrendorfer, M., Mangler, J., Rinderle-Ma, S.: Assessing the impact of context data on process outcomes during runtime. In: Service-Oriented Computing, pp. 3–18 (2021). https://doi.org/10.1007/978-3-030-91431-8_1
12. Fahland, D.: Multi-dimensional process analysis. In: Business Process Management, pp. 27–33 (2022). https://doi.org/10.1007/978-3-031-16103-2_3
13. Francescomarino, C.D., Ghidini, C., Maggi, F.M., Milani, F.: Predictive process monitoring methods: which one suits me best? In: Business Process Management, pp. 462–479 (2018). https://doi.org/10.1007/978-3-319-98648-7_27
14. Geyer-Klingeberg, J., Nakladal, J., Baldauf, F., Veit, F.: Process mining and robotic process automation: a perfect match. In: Dissertation Award, Demonstration, and Industrial Track at BPM, pp. 124–131. CEUR-WS.org (2018)
15. Huang, B., Zhou, M., Lu, X.S., Abusorrah, A.: Scheduling of resource allocation systems with timed petri nets: a survey. ACM Comput. Surv. **55**(11), 230:1–230:27 (2023). https://doi.org/10.1145/3570326

16. Jensen, K.: Coloured Petri Nets: Basic Concepts, Analysis Methods and Practical Use, vol. 1. Springer, Heidelberg (1996). https://doi.org/10.1007/978-3-662-03241-1

17. Kaes, G., Rinderle-Ma, S., Vigne, R., Mangler, J.: Flexibility requirements in real-world process scenarios and prototypical realization in the care domain. In: Meersman, R., et al. (eds.) OTM 2014. LNCS, vol. 8842, pp. 55–64. Springer, Heidelberg (2014). https://doi.org/10.1007/978-3-662-45550-0_8

18. Koenig, P., Mangler, J., Rinderle-Ma, S.: Compliance monitoring on process event streams from multiple sources. In: Process Mining, pp. 113–120 (2019). https://doi.org/10.1109/ICPM.2019.00026

19. Lee, W.L.J., Burattin, A., Munoz-Gama, J., Sepúlveda, M.: Orientation and conformance: a hmm-based approach to online conformance checking. Inf. Syst. **102**, 101674 (2021). https://doi.org/10.1016/j.is.2020.101674

20. Ly, L.T., Maggi, F.M., Montali, M., Rinderle-Ma, S., van der Aalst, W.M.P.: Compliance monitoring in business processes: functionalities, application, and tool-support. Inf. Syst. **54**, 209–234 (2015). https://doi.org/10.1016/j.is.2015.02.007

21. Maggi, F.M., Francescomarino, C.D., Dumas, M., Ghidini, C.: Predictive monitoring of business processes. In: Advanced Information Systems Engineering, vol. 8484, pp. 457–472 (2014). https://doi.org/10.1007/978-3-319-07881-6_31

22. Mangat, A.S., Mangler, J., Rinderle-Ma, S.: Interactive process automation based on lightweight object detection in manufacturing processes. Comput. Ind. **130**, 103482 (2021). https://doi.org/10.1016/j.compind.2021.103482

23. Mangat, A.S., Rinderle-Ma, S.: Next-activity prediction for non-stationary processes with unseen data variability. In: Enterprise Design, Operations, and Computing, pp. 145–161 (2022). https://doi.org/10.1007/978-3-031-17604-3_9

24. Mangler, J., et al.: Datastream XES extension: embedding IoT sensor data into extensible event stream logs. Future Internet **15**(3) (2023). https://doi.org/10.3390/fi15030109

25. Mangler, J., Rinderle-Ma, S.: CPEE - cloud process execution engine. In: BPM Demo Sessions, p. 51 (2014). http://ceur-ws.org/Vol-1295/paper22.pdf

26. Márquez-Chamorro, A.E., Resinas, M., Ruiz-Cortés, A.: Predictive monitoring of business processes: a survey. IEEE Trans. Serv. Comput. **11**(6), 962–977 (2018). https://doi.org/10.1109/TSC.2017.2772256

27. Oberdorf, F., Schaschek, M., Weinzierl, S., Stein, N., Matzner, M., Flath, C.M.: Predictive end-to-end enterprise process network monitoring. Bus. Inf. Syst. Eng. **65**(1), 49–64 (2023). https://doi.org/10.1007/s12599-022-00778-4

28. Rinderle-Ma, S., Kabicher-Fuchs, S.: An indexing technique for compliance checking and maintenance in large process and rule repositories. Enterp. Model. Inf. Syst. Archit. Int. J. Concept. Model. **11**, 2:1–2:24 (2016). https://doi.org/10.18417/emisa.11.2

29. Rinderle-Ma, S., Mangler, J.: Process automation and process mining in manufacturing. In: Business Process Management, pp. 3–14 (2021). https://doi.org/10.1007/978-3-030-85469-0_1

30. Rinderle-Ma, S., Stertz, F., Mangler, J., Pauker, F.: Process Mining-Discovery, Conformance, and Enhancement of Manufacturing Processes, pp. 363–383. Springer, Heidelberg (2023). https://doi.org/10.1007/978-3-662-65004-2_15

31. Rinderle-Ma, S., Winter, K., Benzin, J.V.: Predictive compliance monitoring in process-aware information systems: state of the art, functionalities, research directions, March 2023. http://arxiv.org/abs/2205.05446, accepted in Information Systems

32. Rinderle-Ma, S., Karolin Winter, J.V.B.: Predictive compliance monitoring in process-aware information systems: state of the art, functionalities, research directions. Inf. Syst. (2023). https://doi.org/10.1016/j.is.2023.102210
33. Stertz, F., Mangler, J., Rinderle-Ma, S.: Balancing patient care and paperwork automatic task enactment and comprehensive documentation in treatment processes. Enterp. Model. Inf. Syst. Archit. Int. J. Concept. Model. **15**, 11:1–11:28 (2020). https://doi.org/10.18417/emisa.15.11
34. Stertz, F., Mangler, J., Rinderle-Ma, S.: The role of time and data: online conformance checking in the manufacturing domain. arXiv:2105.01454 (2021)
35. Stertz, F., Mangler, J., Scheibel, B., Rinderle-Ma, S.: Expectations vs. experiences - process mining in small and medium sized manufacturing companies. In: Business Process Management Forum, pp. 195–211 (2021). https://doi.org/10.1007/978-3-030-85440-9_12
36. Stertz, F., Rinderle-Ma, S.: Process histories - detecting and representing concept drifts based on event streams. In: On the Move to Meaningful Internet Systems, pp. 318–335 (2018). https://doi.org/10.1007/978-3-030-02610-3_18
37. Stertz, F., Rinderle-Ma, S., Mangler, J.: Analyzing process concept drifts based on sensor event streams during runtime. In: Business Process Management, pp. 202–219 (2020). https://doi.org/10.1007/978-3-030-58666-9_12
38. van der Werf, J.M.E.M., Rivkin, A., Montali, M., Polyvyanyy, A.: Correctness notions for Petri nets with identifiers, December 2022. http://arxiv.org/abs/2212.07363, arXiv:2212.07363 [cs]

Formal Modelling, Analysis, and Synthesis of Modular Industrial Systems Inspired by Net Condition/Event Systems

Midhun Xavier[1], Sandeep Patil[1], Victor Dubinin[2], and Valeriy Vyatkin[1,3]($\boxtimes$)

[1] Department of Computer Science, Electrical and Space Engineering, Luleå Tekniska Universitet, Luleå, Sweden
{midhun.xavier,sandeep.patil}@ltu.se
[2] Independent Researcher, Penza, Russia
[3] Department of Electrical Engineering and Automation, Aalto University, Espoo, Finland
vyatkin@ieee.org

Abstract. This paper summarises recent developments in the application of modular formalisms to model-based verification of industrial automation systems. The paper is a tribute to the legacy of Professor Hans-Michael Hanisch who invented Net Condition/Event Systems (NCES) and passionately promoted the closed-loop modelling approach to modelling and analysis of automation systems. The paper surveys the related works and highlights the impact NCES has made on the current progress of modular automation systems verification.

1 Introduction

Modularity is a fundamental feature of technical systems, in particular in industrial automation and cyber-physical systems. On the other hand, modular systems is a good example of distributed systems. Petri nets (PN) have been known as a formal language specifically focused on modelling of distributed state systems. That suggests a clear overlap and the need to address modularity in formal modelling. Petri nets inspired an uncountable number of derivatives.

Modularity in the context of PN has been discussed for a long time. According to [6], the concept of Modular Petri Nets has been through four generations of development.

On the other hand, the concept of Condition/Event Systems (C/ES) [31] was invented to model modular systems composed of communicating modules and study their generic properties.

Net Condition/Event Systems (NCES) [29] is a particular case of C/ES where modules are defined as (extended) Petri nets. It was proposed to model more efficiently distributed systems that are modular.

One should note that computational analysis of NCES is in general undecidable as shown by Starke and Hanisch [32]. Nevertheless, the formalism fits very well with the emerging engineering concepts for CPS such as service-oriented architecture (SOA) and IEC 61499 function blocks due to the properly addressed event-driven semantics. The initial effort of NCES application to IEC 61499 modelling is summarised in [14].

In this paper, we attempt to observe the developments related to the modelling and analysis of distributed modular industrial automation systems from the particular perspective of how the modular derivatives of Petri nets influence them.

The rest of the paper is structured as follows. In Sect. 2 the necessary definitions of NCES are provided. It is followed by a brief illustration of some features NCES provide for modelling distributed modular systems in Sect. 3. Section 4 contains some observations of similarities between IEC 61499 and NCES. Section 5 attempts to overview the related research works on the modelling of modular systems. The recently developed modelling framework for modular systems based on IEC 61499 and influenced by the Condition/Event Systems paradigm is described in Sect. 6. The paper is concluded with a short summary and outlook in Sect. 7 and acknowledgements.

2   Some Definitions

Net Condition/Event Systems (NCES) is a finite state formalism that preserves the graphical notation and the non-interleaving semantics of Petri nets [27], and extends them with a clear and concise notion of signal inputs and outputs. The formalism was introduced in [28] in 1995 and has been used in dozens of applications, especially in embedded industrial automation systems.

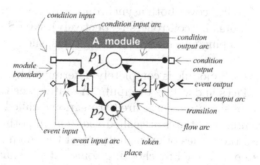

Fig. 1. Graphical notation of an NCES module.

Given a place/transition net $N = (P, T, F, m_0)$, the Net Condition/Event System (NCES) is defined as a tuple $\mathcal{N} = (N, \theta_N, \Psi_N, Gr)$, where θ_N is an internal structure of signal arcs, Ψ_N is an input/output structure, and $Gr \subseteq T$ is a set of so-called "obliged" transitions that fires as soon as it is enabled. Figure 1 illustrates an example of NCES module. The structure Ψ_N consists of condition and event inputs and outputs (ci, ei, eo, co). The structure θ_N is formed from two types of *signal* arcs. Condition arcs lead from places and condition inputs to transitions and condition outputs. They provide additional enableness conditions of the recipient transitions. Event arcs from transitions and event inputs to transitions and event outputs provide one-sided synchronization of the recipient transitions: firing of the source transition forces firing of the recipient if the latter is enabled by the marking and conditions.

The NCES modules can be interconnected by the condition and event arcs, forming thus distributed and hierarchical models as illustrated in Fig. 2. NCES having no inputs can be analyzed without any additional information about its external environment.

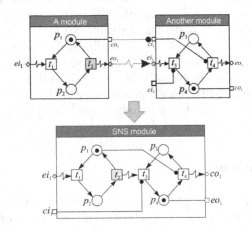

Fig. 2. Composition of NCES modules.

The semantics of NCES cover both asynchronous and synchronous behaviour (required to model plants and controllers respectively). NCES are supported by a family of model-developing and model-checking tools, such as a graphic editor, SESA and ViVe [2].

The state of an NCES module is completely determined by the current marking $m : P \to N_0$ of places and values of inputs. A state transition is determined by the subset $\tau \subseteq T$ of simultaneously fired transitions, called *step*. The transitions having no incoming event arcs are called *spontaneous*, otherwise *forced*. The step fully determines the values of event outputs of the module. In the original NCES version the step is formed by choosing some[1] of the enabled spontaneous transitions, and all the enabled transitions forced by the transitions already included in the step.

A state of NCES is fully described by the marking of all its places (in the timed version also by clocks). A transition step specifies a state transition. When used for system analysis, a set of all reachable states (complete or partial) of NCES model is generated and then analyzed.

---

[1] This means the step in NCES is non-deterministic.

For describing the execution model of function blocks we use a deterministic dialect of NCES and the modeling approach that guarantee certain properties of the models as follows:

1. In the chosen dialect a step is formed from all enabled spontaneous transitions and all forced transitions;
2. The models are designed so that there is no conflicts (i.e., deficient marking in some places) leading to non-deterministic choice of some of the enabled transitions;
3. The models also guarantee bounded marking in all places.

3 Modelling Distributed Systems with NCES

To illustrate the key features of NCES modelling for distributed systems, let us consider an example of a simple distributed control system. In the system of two cylinders in Fig. 3 each cylinder pushes a workpiece to the destination hole. The process starts when the workpiece appears in front of the corresponding cylinder as indicated by sensors P1 and P2 respectively. As it is clear from the Figure, cylinders can collide in the middle point, therefore the goal of controller design is to avoid such a situation.

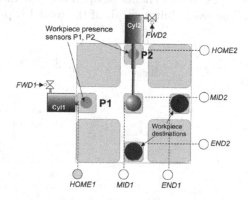

Fig. 3. Two cylinders example of a distributed system.

There are many possible ways to achieve the desired behaviour, which can be done by designing a "central" controller of both cylinders, or a protocol ensuring that distributed controllers collaborate correctly. Distributed control is of interest for many practical reasons, for example, for the case when control logic is "embedded" in each cylinder, so they can start working as soon as powered on.

NCES model of the two cylinders system with distributed control is presented in Fig. 5.

An abstract model of two processes interacting with each other with the help of buffer is presented in Fig. 4. Here Process 1 adds a token to the Buffer, and Process 2 sees it and removes it from the buffer.

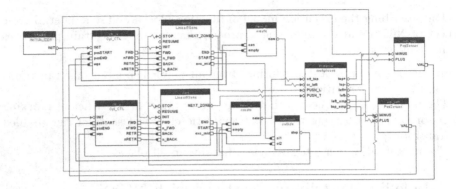

Fig. 4. NCES model of the two cylinders system.

A more sophisticated synchronous communication mechanism between clock-driven processes through a rendezvous channel is modelled by means of NCES formalism in Fig. 6. The example represents a part of the previously considered system (Fig. 4), where the Position_Control is a component inside the Robot, and the input channel Position connects it to the Coordinator block.

To verify the correctness of the channel's operation the model-checking tool ViVe can be applied. The reachability graph of the model is presented in Fig. 4.

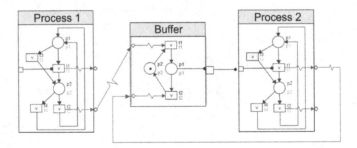

Fig. 5. NCES model of interprocess communication.

4   IEC 61499 Based Modular Engineering of Automation Systems

The IEC 61499 architecture [1] is getting increasingly recognised as a powerful mechanism for engineering cyber-physical systems. In IEC 61499, the basic design construct is called function block (FB). Each FB consists of a graphical event-data interface and a set of executable functional specifications (algorithms), represented as a state machine (in basic FB), as a network of other FB

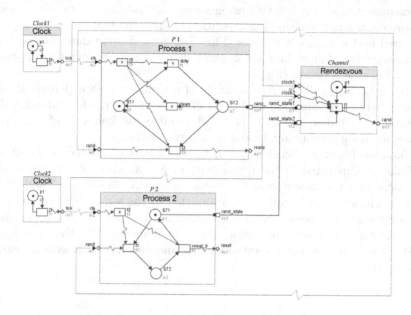

Fig. 6. NCES model of the process synchronisation.

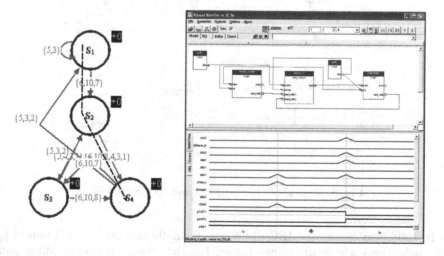

Fig. 7. Reachability graph of the model (left) and the behaviour along the $S1{\rightarrow}S2{\rightarrow}S4$ trace (right), where the rendezvous occurs at the state transition $S2{\rightarrow}S4$.

instances (composite FB), or as a set of services (service interface FB). FBs can be interconnected into a network using event and data connections to specify the entire control application. The execution of an individual FB in the network is triggered by the events it receives. This well-defined event-data interface and the encapsulation of local data and control algorithms make each FB a reusable functional unit of software (Fig. 7).

A basic Function Block (FB) consists of a signal interface (left-hand side) and an Execution Control Chart (ECC) state machine (right-hand side). The algorithms executed in the ECC states determine the behavior of the FB in response to changes in its inputs and its internal state.

A function block application is a network of FBs connected by event and data links as illustrated in the upper part of Fig. 8, which illustrates models of the same one pneumatic cylinder system with IEC 61499 (top) and NCES (bottom). The structural similarity is supported by the semantic similarity since both modelling languages are event-based. Connections between modules in both modelling languages are passing events and data. This simplifies the modelling of IEC 61499 with NCES and several modelling and analysis tools were developed to explore it.

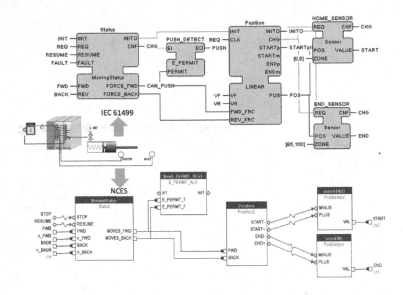

Fig. 8. Similarity of IEC 61499 and NCES models.

In 1998, way before the IEC 61499 was formally accepted as a standard by IEC, using an early draft, Hans-Michael Hanisch observed this stunning similarity and wrote a research proposal together with Peter Starke, supported by the German Research Council (DFG), on formal verification of IEC 61499 applications by means of NCES. That gave rise to a number of developments summarised in [14].

In particular, in 2001, Vyatkin and Hanisch developed a software package called "Verification Environment for Distributed Applications" (VEDA) for model-based simulation and verification [33]. NCES is used for modelling and IEC 61499 function blocks are automatically converted with the help of VEDA for efficient simulation and verification.

But, surprisingly, the NCES-IEC 61499 similarity helped develop a modelling approach in which IEC 61499 itself was directly used as a modelling language as it will be illustrated in Sect. 6.

5   Survey of Works on Modular Engineering and Modelling

To put the above-referenced developments on NCES and IEC 61499 to the broader context, in this section we present a brief survey of other related works on formal modelling and analysis of modular automation systems.

5.1   Modelling of Flexible Reconfigurable Systems

Reconfigurable Manufacturing Systems (RMS) are flexible and adaptable to manufacture various products to meet changing market demands. Meng et al. explain how complex RMS can be hierarchically modularized for modelling reconfigurability using coloured Object Oriented Petri nets [16]. The RMS model is developed with the help of the macro-level Petri net and the changes in RMS drive the change in Petri net.

Later, Wu et al. introduced Intelligent Token Petri Net (ITPN) for modelling reconfigurable Automated Manufacturing Systems (AMS) [35]. The ITPN model captures dynamic changes in the system and the deadlock-free policy makes the model always deadlock-free and reversible. The change in configuration modifies only changed part of the current model and the deadlock-free policy remains the same.

In real-time systems temporal constants are inevitable and these systems need to be modelled to ensure that it satisfies functional and non-functional requirements. Recently, Kaid et al. developed Intelligent Colored Token Petri Net (ICTPN) and it models dynamic changes in a modular manner and produces a compact model which ensures PN behavioural properties like boundness, liveness and reversibility but the ITCPN model lacks a conversion method to industrial control languages.

Reconfigurable Discrete Event System (RDES) such as reconfigurable manufacturing systems (RMS) has the ability to change the configuration of the system to adapt to changes in conditions and requirements. Reconfigurable discrete event control systems (RDECS) are an important part of RDESs. Reconfiguration done at the run time is called Dynamic reconfiguration and it should occur without influencing the working environment and with no deadlock. Zhang et al. introduced the reconfiguration based on the Timed Net Condition/Event system (R-TNCES) and it is a formalism for the modelling and verification of RDECSs. SESA model checker does the layer-by-layer verification of R-TNCES [43].

Modern manufacturing systems switch energy-intensive machines between working and idle mode with the help of dynamic reconfiguration to save energy. The later works of Zhang *et al.* developed how formal modelling and verification of reconfigurable and energy-efficient manufacturing systems can be done using R-TNCES formalism and SESA tool is applied to check functional, temporal and energy efficient properties [42,45].

System reconfiguration in run-time is inevitable and a discrete event system with dynamic reconfigurability is called (DRDES). NCES is widely applied in DRDESs in the past decade. NCES are a modular extension of PN and it is used for modelling, analysis and control of DRDES. Many researchers worked on the modelling, analysis and verification of reconfigurable RMS.

The system reconfiguration should be completed before the permissible reconfiguration delay. Whenever a reconfiguration event is triggered then DRDES should be able to go to the target state within the permissible reconfiguration delay otherwise it should reject the reconfiguration requirement. Zhang *et al.* developed to compute a shortest legal firing sequence (SLFS) of an NCES using Integer Linear Programming (ILP) under a given maximum permissible reconfiguration delay [44].

Interpreted time Petri net (ITPN) is used to model real-time systems, which helps to increase the modelling power and expressiveness compared to (Timed Petri net) TPN's. Hadjidj *et al.* proposed RT-studio (Real-time studio) for modelling, simulation and automatic verification. [13]. RT-studio tries to tighten the gap with the UPPAAL model checker by modularizing the ITPN model.

Dehnert *et al.* introduced a new probabilistic model checker [7,15] called Storm that can analyze both discrete- and continuous-time variants of Markov chains and Markov decision processes (MDPs), using the Prism and JANI modelling languages, probabilistic programs, dynamic fault trees and generalized stochastic Petri nets. It has a flexible design that allows for easy exchange of solvers and symbolic engines, and it offers a Python API for rapid prototyping. Benchmark experiments have shown that Storm has competitive performance.

5.2   Modelling of IEC 61499

Another approach to verify the application of IEC 61499 was presented by Schnakenhourg *et al.* , who explained the method to verify IEC 61499 function blocks by converting to the SIGNAL model [30]. The specification also converts to a SIGNAL model and verifies using SILDEX from the TNI society.

In order to formally model function blocks in IEC 61499, it is necessary to first define their complete execution semantics. The semantic ambiguities in IEC 61499, can lead to different interpretations of function blocks. To address this, the Sequential Hypothesis can be used, which defines a more intuitive and clear sequential execution model of function blocks. Pang *et al.* [21], developed IEC 61499 basic function blocks using the sequential hypothesis, which assumes that blocks within a network are activated sequentially. They used NCES and verified the behaviour of the model using model-checking tools such as iMATCH and SESA. They later proposed a model generator [22] that automatically translates IEC 61499 function blocks into the NCES formal model for the purpose of verification. The function blocks developed using the FBDK (Function Block Development Kit) are translated into functionally and semantically equivalent NCES models following the sequential execution model. This NCES model can be opened in ViEd and properties are verified using the ViVe tool.

Cengic *et al.* [5] introduced a new runtime environment called Fuber, which uses a formal execution model to make the behaviour of IEC 61499 applications deterministic and predictable. They developed a tool to translate IEC 61499 function blocks into a set of finite-state automata and used the Supremica tool for supervisor verification and synthesis. After that, they introduced a software tool to automatically generate formal closed-loop system models between control code and process models expressed as IEC 61499 function blocks, using extended finite automata (EFA) and Supremica for formal verification [3]. They further extended this by defining the buffered sequential execution model (BSEM) and its formal verification using Supremica by analyzing the EFA model [4]. In another study, Yoong *et al.* developed a tool to translate IEC 61499 function blocks to Esterel for verification [41]. Existing verification tools for Esterel help to analyze the safety properties of IEC 61499 function block programs.

Formal verification of embedded control systems using closed-loop plant-controller models is becoming more popular. However, the use of non-determinism in the model of the plant can lead to the complexity explosion in the model-checking process and make it difficult to verify the correctness of the plant model itself before it can be used in the closed-loop verification process. The paper [23] describes the integration of modelling principles into the Veridemo toolchain, and it also explains the implementation of controlled non-determinism in NCES systems. The controlled non-determinism limits the state space and eventually results in better verification times. This approach provides better model-checking performance with ViVe and SESA compared to NuSMV and UPPAAL model checkers with fully deterministic state machines. Later they introduced [26] a framework for model checking and counter-example playback in simulation models used to verify the system. The control logic and dynamics of the plant are modelled using Net Condition/Event Systems formalism and ViVe/SESA toolchain is used for model checking. The counter examples for failures during model checking are played back in simulation models for a better understanding of the failures.

The IEC 61499 standard is used for the development of distributed control systems, but it has limited support for reconfigurable architectures. To address this limitation, Guellouz *et al.* proposed a new model called reconfigurable function blocks (RFBs) in their study [11]. They use GR-TNCES, a derivative of NCES, to model the system and applied the proposed approach to a medical platform called BROS. Further studies [10,12] proposed translating RFBs to GR-TNCES in order to verify their correctness and alleviate state space explosion in model checking. Additionally, the latter paper aimed to analyze probabilistic properties and used a smart-grid system as a case study to demonstrate the approach. The study also developed a visual environment called ZiZo v3 for modelling reconfigurable distributed systems.

The formal verification technique is a reliable approach to ensure the correctness of instrumentation and control (I&C) systems. It mentions that model-checking is widely used in avionics, the automotive industry, and nuclear power plants but has some difficulties in locating errors in the model. The Oeritte tool, presented in the first study of Ovsiannikova et al. [19], is a solution for assisting analysts in the debugging process of formal models of instrumentation and control systems. It uses a method for automatic visual counterexample explanation and includes reasoning for both the falsified LTL formula and the NuSMV function block diagram of the formal model of the system. The tool addresses the challenges of counterexample visualization, LTL formulae, and counterexample explanation by providing methods, visual elements, and user interface. The second study, [20], presents the development of a model-checking plugin for IEC 61499 systems in the FBME graphical development environment. The plugin automates the process of converting the system to a formal model, model-checking, and providing a visual explanation of counterexamples.

The next step to verification is the formal synthesis of correct-by-design systems with ensured safe operation. Missal and Hanisch [17,18] present a modular synthesis approach. It is based on the modular backward search in order to avoid the complexity of generating all states and state transitions of the plant model. It uses modular backward steps that describe the trajectories leading to forbidden states. The generation of these trajectories is stopped as soon as a preventable step is found. From this information, the models of the controllers are generated. Each controller has decision functions and communications functions. Together they establish a network of local, interacting controllers with communication. It is assumed that the plant is completely observable, i.e. the local controllers have complete information about the local states of the partial plants they are supposed to control. The paper also contributes with the definition of the behaviour of the plant without its complete composition. This means that the behaviour can be studied by means of modular steps within the modules and their interaction across module boundaries.

Dubinin et al. in [9] demonstrate safety controller synthesis using the description of the plant and forbidden behaviour, proposing a method of synthesis of adaptive safety controller models for distributed control systems based on reverse safe Net Condition/Event Systems (RsNCES). The method allows for the generation of prohibiting rules to prevent the movement of closed-loop systems to forbidden states. The method is based on a backward search in the state space of the model.

6 Use IEC 61499 for Condition/Event Modelling: A Comprehensive Tool Chain

The works on formal modelling and verification of IEC 61499 systems by means of NCES and its analysis tools have confirmed the benefits of exploring their structural and semantic similarities. On the other hand, applying the verification to systems of industrial scale has raised several questions:

- Model-checking tools for NCES require constant support and improvement, which was lacking. A bridge to industrially supported powerful tools was desirable.
- Verification should be a part of the regular engineering and testing process that includes testing by simulation, and analysis of results.

Towards the first goal, Patil *et al.* [25] introduce a method for formally modelling and verifying IEC 61499 function blocks, a component model used in distributed embedded control system design, using the Abstract State Machines (ASM) as an intermediate model and the SMV model checker. The ASM model is translated into the input format of the SMV model checker, which is used to formally verify the properties. The proposed verification framework enables the formal verification of the IEC 61499 control systems, and also highlights other uses of verification such as the portability of IEC 61499-based control applications across different implementation platforms compliant with the IEC 61499 standard. Their other work [24] proposes a general approach for neutralizing semantic ambiguities in the IEC 61499 standard by the formal description of the standard in ASM.

Another study [25], highlights the importance of formally verifying function block applications in different execution semantics and the benefits of verifying the portability of component-based control applications across different platforms compliant with the IEC 61499 standard. The paper applies the formal model to an example IEC 61499 application and compares the verification results of the two-stage synchronous execution model with the earlier cyclic execution model, to verify the portability of the IEC 61499 applications across different platforms.

After that, they addressed the SMV modelling of the IEC 61499 specific timer function block types, particularly in hierarchical function block systems with timers located at different levels of hierarchy [8]. The paper also introduces plant abstraction techniques to reduce the complexity of cyber-physical systems models using discrete-timed state machine models implemented in UPPAAL. The approach is demonstrated with an example of formal verification of a modular mechatronic automated system and is shown to extend the abilities in the validation of real cyber-physical automation systems. A toolchain was developed to support the described modelling method, including an automatic FB-to-SMV converter for the transformation of IEC 61499 FB applications to the control part of SMV models. This approach can be used for the verification of newly developing industrial safety-critical systems such as smart grids.

Addressing the second goal, the road map on the creation of a tool-chain connecting engineering with verification seamlessly was outlined in [34]. A problem-oriented notation within the IEC 61499 syntax for creating formal closed-loop models of cyber-physical automation systems [40] is proposed. The notation enables the creation of a comprehensive toolchain for the design, simulation, formal verification, and distributed deployment of automation software.

The toolchain includes an IEC 61499-compliant engineering environment, a converter for functions blocks to SMV code, the NuSMV model-checker and utilities for interpreting counterexamples. The proposed method aims to overcome the hurdle of verifying and analyzing function blocks implemented in IEC 61499 standard by providing a toolchain for continuous development and testing of distributed control systems.

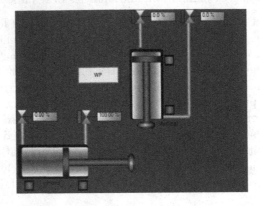

Fig. 9. Visualisation of the Two Cylinder system produced by the model of the plant implemented in IEC 61499.

The two-cylinder system consists of two orthogonal pneumatic cylinders controlled by a switch button shown in Fig. 9. It is built using five basic function blocks, including a controller function block (Button FB) that triggers the movement of the cylinders when pressed, plant function blocks (HorCyl and VerCyl FBs) that model the physical device of each cylinder, and controller function blocks (HorCTL FB and VerCTL FB) that control the plant by analyzing sensor signals and triggering actuator signals. These blocks receive information from the switch FB and send orders to the plant FB.

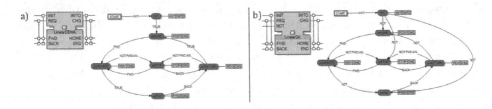

Fig. 10. a) Deterministic discrete state linear motion process model without NDT, b) Discrete state linear motion process model with NDT.

To implement the closed-loop approach to system modelling, the model of the plant needs also to be modelled using function blocks. A discrete state linear motion of a cylinder for a linear motion, for example, a linear axis, can be represented by a LinearDDtrA function block with two states (sHOME and sEND) that transition between them based on input signals (BACK or FWD) Fig. 10, a. However, this minimal approach may not capture all possible errors that can occur during transitions between states.

By using NDT (Non-deterministic transition), a more comprehensive model can be created by adding two dynamic states (ddMOVETO and ddRETURN) to capture potential errors during transitions Fig. 10,b.

The axis moves from the stHOME state to the stEND state via the motion state ddMOVETO when the FWD signal is TRUE. The use of NDT (Non-Deterministic Transition) in the transition from the ddMOVETO state to the stEND state models the unknown duration of the motion from one state to another. The NDT event input of the LinearDA function block, which was unassigned in the application, is reserved for non-deterministic transitions in the proposed modelling notation. This approach can provide a more detailed and accurate representation of the system, allowing for more thorough formal verification.

The (multi-) closed-loop model of the two cylinders system using this extension of the IEC 61499 language is shown in Fig. 11. This is nothing else, but a Condition/Event discrete-state model represented by means of IEC 61499.

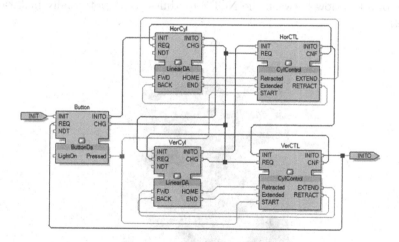

Fig. 11. Complete two cylinders model in the modified FB language.

The fb2smv tool is a model generator that is used to generate SMV (Symbolic Model Verifier) models of function block systems in IEC 61499. It is part of a formal verification tool-chain that includes the model checker NuSMV and a tool for analyzing counterexamples in terms of the original FB system. The tool uses the Abstract State Machine (ASM) as an intermediate model to convert IEC 61499 function blocks expressed in XML format into a formal model. The generated SMV code has a structure that consists of a declaration part and an ASM rules part. The tool can convert both basic and composite function blocks, and also includes additional features such as limiting variable boundaries to reduce the state space, changing the execution order of FBs, and deciding the input event priority by changing its order. Additionally, the tool has been updated to include a proposed non-deterministic transitions notation.

Closed-loop modelling is a powerful approach for the verification of distributed industrial automation systems, as it allows for a comprehensive evaluation of the system's behaviour. However, it requires the creation of a model of the plant, which can be a complex and resource-intensive task, typically done manually. In these papers [36–39], authors show how to generate the plant and controller models automatically using a data-driven approach. The above-mentioned toolchain has been effectively used in these experiments to verify, simulate and analyse counterexamples.

7   Summary and Open Problems

Systems with dynamically created and terminated modules or dynamic connections between modules cannot be efficiently and naturally modelled within the C/ES paradigm and require complicated workarounds.

The idea of modular analysis of NCES has not been developed although the absence of token flow between the NCES modules could potentially facilitate it (Fig. 12).

Fig. 12. Hans-Michael Hanisch (1957–2022).

Acknowledgments. This paper attempts to be a tribute to Professor Hans-Michael Hanisch who has been a co-inventor and a great enthusiast and proponent of NCES as a part of the closed-loop modelling concept.

References

1. Function Blocks for Industrial Process Measurement and Control Systems, IEC 61499 Standard. International Electrotechnical Commission, Tech. Comm. **65**, Working group 6, Geneva (2005)
2. Visual verifier (2008). http://www.fb61499.com/license.html
3. Čengić, G., Åkesson, K.: A control software development method using IEC 61499 function blocks, simulation and formal verification. IFAC Proc. Volumes **41**(2), 22–27 (2008)
4. Cengic, G., Akesson, K.: Definition of the execution model used in the fuber IEC 61499 runtime environment. In: 2008 6th IEEE International Conference on Industrial Informatics, pp. 301–306. IEEE (2008)
5. Cengic, G., Ljungkrantz, O., Akesson, K.: Formal modeling of function block applications running in IEC 61499 execution runtime. In: 2006 IEEE Conference on Emerging Technologies and Factory Automation, pp. 1269–1276. IEEE (2006)
6. Davidrajuh, R.: A new modular petri net for modeling large discrete-event systems: a proposal based on the literature study. Computers **8**(4), 83 (2019)
7. Dehnert, C., Junges, S., Katoen, J.-P., Volk, M.: A storm is coming: a modern probabilistic model checker. In: Majumdar, R., Kunčak, V. (eds.) CAV 2017. LNCS, vol. 10427, pp. 592–600. Springer, Cham (2017). https://doi.org/10.1007/978-3-319-63390-9_31
8. Drozdov, D., Patil, S., Dubinin, V., Vyatkin, V.: Formal verification of cyber-physical automation systems modelled with timed block diagrams. In: 2016 IEEE 25th International Symposium on Industrial Electronics (ISIE), pp. 316–321. IEEE (2016)
9. Dubinin, V., Vyatkin, V., Hanisch, H.M.: Synthesis of safety controllers for distributed automation systems on the basis of reverse safe net condition/event systems. In: 2015 IEEE Trustcom/BigDataSE/ISPA, vol. 3, pp. 287–292. IEEE (2015)
10. Fkaier, S., Khalgui, M., Frey, G.: Modeling methodology for reconfigurable distributed systems using transformations from GR-UML to GR-TNCES and IEC 61499. In: ENASE, pp. 221–230 (2021)
11. Guellouz, S., Benzina, A., Khalgui, M., Frey, G.: Reconfigurable function blocks: extension to the standard IEC 61499. In: 2016 IEEE/ACS 13th International Conference of Computer Systems and Applications (AICCSA), pp. 1–8. IEEE (2016)
12. Guellouz, S., Benzina, A., Khalgui, M., Frey, G., Li, Z., Vyatkin, V.: Designing efficient reconfigurable control systems using IEC61499 and symbolic model checking. IEEE Trans. Autom. Sci. Eng. **16**(3), 1110–1124 (2018)
13. Hadjidj, R., Boucheneb, H.: Rt-studio: a tool for modular design and analysis of realtime systems using interpreted time petri nets. In: PNSE+ ModPE, pp. 247–254. Citeseer (2013)
14. Hanisch, H.M., Hirsch, M., Missal, D., Preuße, S., Gerber, C.: One decade of IEC 61499 modeling and verification-results and open issues. IFAC Proc. Volumes **42**(4), 211–216 (2009)
15. Hensel, C., Junges, S., Katoen, J.P., Quatmann, T., Volk, M.: The probabilistic model checker storm. Int. J. Softw. Tools Technol. Transfer **24**(4), 589–610 (2022)

16. Meng, X.: Modeling of reconfigurable manufacturing systems based on colored timed object-oriented petri nets. J. Manuf. Syst. **29**(2), 81–90 (2010). https://doi.org/10.1016/j.jmsy.2010.11.002, https://www.sciencedirect.com/science/article/pii/S0278612510000518
17. Missal, D., Hanisch, H.M.: A modular synthesis approach for distributed safety controllers, part a: modelling and specification. IFAC Proc. Volumes **41**(2), 14473–14478 (2008)
18. Missal, D., Hanisch, H.M.: A modular synthesis approach for distributed safety controllers, part b: modular control synthesis. IFAC Proc. Volumes **41**(2), 14479–14484 (2008)
19. Ovsiannikova, P., Buzhinsky, I., Pakonen, A., Vyatkin, V.: Oeritte: user-friendly counterexample explanation for model checking. IEEE Access **9**, 61383–61397 (2021)
20. Ovsiannikova, P., Vyatkin, V.: Towards user-friendly model checking of IEC 61499 systems with counterexample explanation. In: 2021 26th IEEE International Conference on Emerging Technologies and Factory Automation (ETFA), pp. 01–04. IEEE (2021)
21. Pang, C., Vyatkin, V.: Towards formal verification of IEC 61499: modelling of data and algorithms in NCES. In: 2007 5th IEEE International Conference on Industrial Informatics, vol. 2, pp. 879–884. IEEE (2007)
22. Pang, C., Vyatkin, V.: Automatic model generation of IEC 61499 function block using net condition/event systems. In: 2008 6th IEEE International Conference on Industrial Informatics, pp. 1133–1138. IEEE (2008)
23. Patil, S., Bhadra, S., Vyatkin, V.: Closed-loop formal verification framework with non-determinism, configurable by meta-modelling. In: IECON 2011–37th Annual Conference of the IEEE Industrial Electronics Society, pp. 3770–3775. IEEE (2011)
24. Patil, S., Dubinin, V., Pang, C., Vyatkin, V.: Neutralizing semantic ambiguities of function block architecture by modeling with ASM. In: Voronkov, A., Virbitskaite, I. (eds.) PSI 2014. LNCS, vol. 8974, pp. 76–91. Springer, Heidelberg (2015). https://doi.org/10.1007/978-3-662-46823-4_7
25. Patil, S., Dubinin, V., Vyatkin, V.: Formal verification of IEC61499 function blocks with abstract state machines and SMV-modelling. In: 2015 IEEE Trustcom/BigDataSE/ISPA, vol. 3, pp. 313–320. IEEE (2015)
26. Patil, S., Vyatkin, V., Pang, C.: Counterexample-guided simulation framework for formal verification of flexible automation systems. In: 2015 IEEE 13th International Conference on Industrial Informatics (INDIN), pp. 1192–1197. IEEE (2015)
27. Petri, C.A.: Kommunikation mit Automaten. Schriften des IIM Nr. 2, Institut fur Instrumentelle Mathematik, Bonn (1962)
28. Rausch, M., Hanisch., H.M.: Net condition/event systems with multiple condition outputs. In: Symposium on Emerging Technologies and Factory Automation, vol. 1, pp. 592–600. INRIA/IEEE, Paris, France, October 1995
29. Rausch, M., Hanisch, H.M.: Net condition/event systems with multiple condition outputs. In: Proceedings 1995 INRIA/IEEE Symposium on Emerging Technologies and Factory Automation. ETFA'95, vol. 1, pp. 592–600. IEEE (1995)
30. Schnakenbourg, C., Faure, J.M., Lesage, J.J.: Towards IEC 61499 function blocks diagrams verification. In: IEEE International Conference on Systems, Man and Cybernetics, vol. 3, 6-p. IEEE (2002)
31. Sreenivas, R.S., Krogh, B.H.: On condition/event systems with discrete state realizations. Discret. Event Dyn. Syst. **1**(2), 209–236 (1991)

32. Starke, P.H., Hanisch, H.M.: Analysis of signal/event nets. In: 1997 IEEE 6th International Conference on Emerging Technologies and Factory Automation Proceedings, EFTA'97, pp. 253–257. IEEE (1997)

33. Vyatkin, V., Hanisch, H.M.: Formal modeling and verification in the software engineering framework of IEC 61499: a way to self-verifying systems. In: ETFA 2001. 8th International Conference on Emerging Technologies and Factory Automation. Proceedings (Cat. No. 01TH8597), vol. 2, pp. 113–118. IEEE (2001)

34. Vyatkin, V., Hanisch, H.M., Pang, C., Yang, C.H.: Closed-loop modeling in future automation system engineering and validation. IEEE Trans. Syst. Man Cybern. Part C (Appl. Rev.) **39**(1), 17–28 (2008)

35. Wu, N., Zhou, M.: Intelligent token petri nets for modelling and control of reconfigurable automated manufacturing systems with dynamical changes. Trans. Inst. Meas. Control. **33**(1), 9–29 (2011)

36. Xavier, M., Dubinin, V., Patil, S., Vyatkin, V.: An interactive learning approach on digital twin for deriving the controller logic in IEC 61499 standard. In: 27th International Conference on Emerging Technologies and Factory Automation (ETFA 2022), Stuttgart, Germany, 6–9 September 2022. IEEE (2022)

37. Xavier, M., Dubinin, V., Patil, S., Vyatkin, V.: Plant model generation from event log using prom for formal verification of cps. arXiv preprint arXiv:2211.03681 (2022)

38. Xavier, M., Dubinin, V., Patil, S., Vyatkin, V.: Process mining in industrial control systems. In: 2022 IEEE 20th International Conference on Industrial Informatics (INDIN), pp. 1–6. IEEE (2022)

39. Xavier, M., Håkansson, J., Patil, S., Vyatkin, V.: Plant model generator from digital twin for purpose of formal verification. In: 2021 26th IEEE International Conference on Emerging Technologies and Factory Automation (ETFA), pp. 1–4. IEEE (2021)

40. Xavier, M., Patil, S., Vyatkin, V.: Cyber-physical automation systems modelling with IEC 61499 for their formal verification. In: 2021 IEEE 19th International Conference on Industrial Informatics (INDIN), pp. 1–6. IEEE (2021)

41. Yoong, L.H., Roop, P.S.: Verifying IEC 61499 function blocks using Esterel. IEEE Embed. Syst. Lett. **2**(1), 1–4 (2010)

42. Zhang, J., et al.: Modeling and verification of reconfigurable and energy-efficient manufacturing systems. Discret. Dyn. Nat. Soc. **2015** (2015)

43. Zhang, J., Khalgui, M., Li, Z., Mosbahi, O., Al-Ahmari, A.M.: R-TNCES: a novel formalism for reconfigurable discrete event control systems. IEEE Trans. Syst. Man Cybern. Syst. **43**(4), 757–772 (2013)

44. Zhang, J., Li, H., Frey, G., Li, Z.: Shortest legal firing sequence of net condition/event systems using integer linear programming. In: 2018 IEEE 14th International Conference on Automation Science and Engineering (CASE), pp. 1556–1561. IEEE (2018)

45. Zhang, J., Li, Z., Frey, G.: Simulation and analysis of reconfigurable assembly systems based on R-TNCES. J. Chin. Inst. Eng. **41**(6), 494–502 (2018)

Process Mining

There and Back Again

On the Reconstructability and Rediscoverability of Typed Jackson Nets

Daniël Barenholz[1]($\boxtimes$), Marco Montali[2], Artem Polyvyanyy[3], Hajo A. Reijers[1], Andrey Rivkin[2,4], and Jan Martijn E. M. van der Werf[1]

[1] Department of Information and Computing Sciences, Utrecht University, Princetonplein 5, 3584 CC Utrecht, The Netherlands
{d.barenholz,h.a.reijers,j.m.e.m.vanderwerf}@uu.nl
[2] Faculty of Computer Science, Free University of Bozen-Bolzano, piazza Domenicani 3, 39100 Bolzano, Italy
montali@inf.unibz.it
[3] The University of Melbourne, Melbourne, VIC 3010, Australia
artem.polyvyanyy@unimelb.edu.au
[4] Department of Applied Mathematics and Computer Science, Technical University of Denmark, Richard Petersens Plads 321, 2800 Kgs. Lyngby, Denmark
ariv@dtu.dk

Abstract. A process discovery algorithm aims to construct a model from data generated by historical system executions such that the model describes the system well. Consequently, one desired property of a process discovery algorithm is *rediscoverability*, which ensures that the algorithm can construct a model that is behaviorally equivalent to the original system. A system often simultaneously executes multiple processes that interact through object manipulations. This paper presents a framework for developing process discovery algorithms for constructing models that describe interacting processes based on typed Jackson Nets that use identifiers to refer to the objects they manipulate. Typed Jackson Nets enjoy the *reconstructability* property which states that the composition of the processes and the interactions of a decomposed typed Jackson Net yields a model that is bisimilar to the original system. We exploit this property to demonstrate that if a process discovery algorithm ensures rediscoverability, the system of interacting processes is rediscoverable.

1 Introduction

Business processes are fundamental to a wide range of systems. A business process is a collection of activities that, when performed, aims to achieve a business objective at an organization. Examples of business processes are an order-to-cash process at a retailer, a medical assessment process at a hospital, or a credit check process at a bank. Business processes are modeled using process modeling languages, such as Petri nets, and used for communication and analysis purposes [1]. Petri nets provide a graphical representation of the flow of activities within a

L. Gomes and R. Lorenz (Eds.): PETRI NETS 2023, LNCS 13929, pp. 37–58, 2023.
https://doi.org/10.1007/978-3-031-33620-1_3

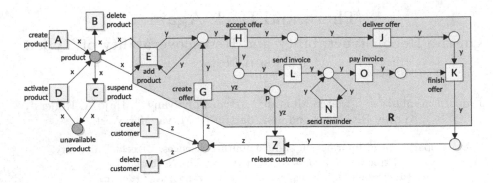

Fig. 1. A retailer system of three interacting processes.

process and can be used to model various types of concurrent and sequential behavior [18].

A process discovery algorithm aims to automatically construct a model from data generated by historical process executions captured in an event log of the system, such that the model describes the system well. A desired property of a discovery algorithm is *rediscoverability*. This property states that if a system S, expressed as a model M, generates an event log L, then a discovery algorithm with the rediscoverability property should construct M from L. In other words, the algorithm can reverse engineer the model of the system from the data the model has generated. Only a few existing algorithms guarantee this property. For example, if the model is a block-structured workflow net, and the event log is directly-follows complete, then the α-Miner algorithm [22] can rediscover the net that generated the event log. Similarly, again under the assumption that the event log is directly-follows complete, Inductive Miner [16] can rediscover process trees without duplicate transitions, self-loops, or silent transitions.

Most existing process discovery algorithms assume that a system executes a single process [4]. Consequently, an event log is defined as a collection of sequences where a sequence describes the execution of a single process instance. However, many information systems, such as enterprise resource planning systems, do not satisfy this assumption. A system often executes multiple interacting processes [11,23]. For example, consider a retailer system that executes three processes: an order, product, and customer management process, as depicted in Fig. 1. These processes are intertwined. Specifically, only available products may be ordered, and customers can only have one order at a time. Consequently, events do not belong to a single process but relate to several processes. For instance, consider an event e in some event log that occurred as transition G was executed for some customer c and created a new order o in the system. Event e relates to the customer process instance c and the order process instance o. Traditional process discovery techniques require event e to be stored in multiple event logs and generate multiple models, one for each process [7].

A different approach is taken in artifact or object-centric process discovery [5,17] and agent system discovery [20,21]. In object-centric process

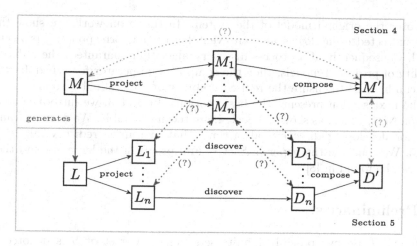

Fig. 2. The framework for rediscoverability of systems of interacting processes.

discovery, instead of linking each event to a single object, events can be linked to multiple objects stored in object-centric event logs [9]. Existing object-centric discovery algorithms project the input event log on each object type to create a set of "flattened" event logs. For each event log, a model is discovered, after which these models are combined into a single model [5]. In general, flattening is lossy [7], as in this step, events can disappear [5], be duplicated (convergence) [3], or lead to wrong event orders (divergence) [3]. In agent system discovery, instead of interacting objects, a system is viewed as composed of multiple autonomous agents, each driving its processes that interact to achieve an overall objective of the system [20]. An agent system discovery algorithm proceeds by decomposing the input event log into multiple event logs, each composed of events performed by one agent (type) and an event log of interactions, and then discovering agent and interaction models and composing them into the resulting system [21].

In this paper, we study under what conditions projections in event logs can guarantee rediscoverability for interacting processes, represented as typed Jackson Nets, a subclass of typed Petri nets with identifiers [19,23]. The class of typed Jackson Nets is inspired by Box Algebra [10] and Jackson Nets [14], which are (representations of) block-structured workflow nets that are *sound* [2] by construction [16]. As we demonstrate, typed Jackson Nets exhibit a special property: they are *reconstructable*. Composing the projections of each type is insufficient for reconstructing a typed Jackson Net. Instead, if the subset-closed set of all type combinations is considered, the composition returns the original model of the system. We show how the reconstructability property can be used to develop a framework for rediscoverability of typed Jackson Nets using traditional process discovery algorithms. The framework builds upon a divide and conquer strategy, as depicted in Fig. 2. The principle idea of this strategy is to project an event log L generated by some model M of the system onto logs $L_1, \ldots, L_n$. Then, if these projected event logs satisfy the conditions of a process discovery algorithm, composition of the resulting models $D_1, \ldots, D_n$ into model D' should rediscover the original model of the system. In this framework, we show that every projected event log is also an event log of the corresponding projected model. Consequently, if a process discovery algorithm guarantees the rediscoverability of projected models, then the composition operator for typed Jackson Nets can be used to ensure the rediscoverability of the original system.

The next section presents the basic notions. In Sect. 3, we introduce typed Jackson Nets, which, as shown in Sect. 4, are reconstructable. We define a framework for developing discovery algorithms that guarantee rediscoverability in Sect. 5. We conclude the paper in Sect. 6. Full proofs of the lemmata and theorems can be found in [8].

2 Preliminaries

Let S and T be two possibly infinite sets. The powerset of S is denoted by $\mathcal{P}(S) = \{S' \mid S' \subseteq S\}$ and $|S|$ denotes the cardinality of S. Two sets S and T are *disjoint* if $S \cap T = \emptyset$, with $\emptyset$ denoting the empty set. The cartesian product of two sets S and T, is defined by $S \times T = \{(a, b) \mid a \in S, b \in T\}$. The generalized cartesian product for some set S and and sets T_s for $s \in S$ is defined as $\Pi_{s \in S} T_s = \{f : S \to \bigcup_{s \in S} T_s \mid \forall s \in S : f(s) \in T_s\}$. Given a relation $R \subseteq S \times T$, its range is defined by $\text{RNG}(R) = \{y \in T \mid \exists x \in S : (x, y) \in R\}$. Similarly, the domain of R is defined by $\text{DOM}(R) = \{x \in S \mid \exists y \in T : (x, y) \in R\}$. Restricting the domain of a relation to a set U is defined by $R_{|U} = \{(a, b) \in R \mid a \in U\}$.

A *multiset* m over S is a mapping of the form $m : S \to \mathbb{N}$, where $\mathbb{N} = \{0, 1, 2, \ldots\}$ denotes the set of natural numbers. For $s \in S$, $m(s) \in \mathbb{N}$ denotes the number of times s appears in multiset m. We write s^n if $m(s) = n$. For $x \notin S$, $m(x) = 0$. We use $S^\oplus$ to denote the set of all finite multisets over S and overload $\emptyset$ to also denote the empty multiset. The size of a multiset is defined by $|m| = \sum_{s \in S} m(s)$. The support of $m \in S^\oplus$ is the set of elements that appear in m at least once: $supp(m) = \{s \in S \mid m(s) > 0\}$. Given two multisets m_1 and m_2 over S: (i) $m_1 \subseteq m_2$ iff $m_1(s) \leq m_2(s)$ for each $s \in S$; (ii) $(m_1 + m_2)(s) = m_1(s) + m_2(s)$ for each $s \in S$; and (iii) if $m_1 \subseteq m_2$, $(m_2 - m_1)(s) = m_2(s) - m_1(s)$ for each $s \in S$.

A *sequence* over S of length $n \in \mathbb{N}$ is a function $\sigma : \{1, \ldots, n\} \to S$. If $n > 0$ and $\sigma(i) = a_i$, for $1 \leq i \leq n$, we write $\sigma = \langle a_1, \ldots, a_n \rangle$. The length of a sequence σ is denoted by $|\sigma|$. The sequence of length 0 is called the *empty sequence*, and is denoted by ϵ. The set of all finite sequences over S is denoted by S^*. We write $a \in \sigma$ if there is $1 \leq i \leq |\sigma|$ such that $\sigma(i) = a$ and $supp(\sigma) = \{a \in S \mid \exists 1 \leq i \leq |\sigma| : \sigma(i) = a\}$. *Concatenation* of two sequences $\nu, \gamma \in S^*$, denoted by $\sigma = \nu \cdot \gamma$, is a sequence defined by $\sigma : \{1, \ldots, |\nu| + |\gamma|\} \to S$, such that $\sigma(i) = \nu(i)$ for $1 \leq i \leq |\nu|$, and $\sigma(i) = \gamma(i - |\nu|)$ for $|\nu| + 1 \leq i \leq |\nu| + |\gamma|$. Projection of sequences on a set T is defined inductively by $\epsilon_{|T} = \epsilon$, $(\langle a \rangle \cdot \sigma)_{|T} = \langle a \rangle \cdot \sigma_{|T}$ if $a \in T$ and $(\langle a \rangle \cdot \sigma)_{|T} = \sigma_{|T}$ otherwise. Renaming a sequence with an injective function $r : S \to T$ is defined inductively by $\rho_r(\epsilon) = \epsilon$, and $\rho_r(\langle a \rangle \cdot \sigma) = \langle r(a) \rangle \cdot \rho_r(\sigma)$. Renaming is extended to multisets of sequences as follows: given a multiset $m \in (S^*)^\oplus$, we define $\rho_r(m) = \sum_{\sigma \in supp(m)} \sigma(m) \cdot \rho_r(\sigma)$. For example, $\rho_{\{x \mapsto a, y \mapsto b\}}(\langle x, y \rangle^3) = \langle a, b \rangle^3$.

A *directed graph* is a pair (V, A) where V is the set of vertices, and $A \subseteq V \times V$ the set of arcs. Two graphs $G_1 = (V_1, A_1)$ and $G_2 = (V_2, A_2)$ are *isomorphic*, denoted by $G_1 \leftrightsquigarrow G_2$, if a bijection $b : V_1 \to V_2$ exists, such that $(v_1, v_2) \in A_1$ iff $(b(v_1), b(v_2)) \in A_2$.

Given a finite set A of (action) labels, a *(labeled) transition system* (LTS) over A is a tuple $\Gamma_A = (S, A, s_0, \to)$, where S is the (possibly infinite) set of *states*, s_0 is the *initial state* and $\to \subset (S \times (A \cup \{\tau\}) \times S)$ is the *transition relation*, where $\tau \notin A$ denotes the silent action [13]. In what follows, we write $s \xrightarrow{a} s'$ for $(s, a, s') \in \to$. Let $r : A \to (A' \cup \{\tau\})$ be an injective, total function. Renaming Γ with r is defined as $\rho_r(\Gamma) = (S, A \setminus A', s_0, \to')$ with $(s, r(a), s') \in \to'$ iff $(s, a, s') \in \to$. Given a set T, hiding is defined as $\hat{\mathsf{H}}_T(\Gamma) = \rho_h(\Gamma)$ with $h : A \to A \cup \{\tau\}$ such that $h(t) = \tau$ if $t \in T$ and $h(t) = t$ otherwise. Given $a \in A$, $p \text{ -}\!\overset{a}{\text{ }}\!\!\triangleright q$ denotes a *weak transition relation* that is defined as follows: *(i)* $p \text{ -}\!\overset{a}{\text{ }}\!\!\triangleright q$ iff $p(\xrightarrow{\tau})^* q_1 \xrightarrow{a} q_2 (\xrightarrow{\tau})^* q$; *(ii)* $p \text{ -}\!\overset{\tau}{\text{ }}\!\!\triangleright q$ iff $p(\xrightarrow{\tau})^* q$. Here, $(\xrightarrow{\tau})^*$ denotes the reflexive and transitive closure of $\xrightarrow{\tau}$.

Let $\Gamma_1 = (S_1, A, s_{01}, \to_1)$ and $\Gamma_2 = (S_2, A, s_{02}, \to_2)$ be two LTSs. A relation $R \subseteq (S_1 \times S_2)$ is called a *strong simulation*, denoted as $\Gamma_1 \prec_R \Gamma_2$, if for every pair $(p, q) \in R$ and $a \in A \cup \{\tau\}$, it holds that if $p \xrightarrow{a}_1 p'$, then there exists $q' \in S_2$ such that $q \xrightarrow{a}_2 q'$ and $(p', q') \in R$. Relation R is a *weak simulation*, denoted by $\Gamma_1 \preccurlyeq_R \Gamma_2$, iff for every pair $(p, q) \in R$ and $a \in A \cup \{\tau\}$ it holds that if $p \xrightarrow{a}_1 p'$, then $a = \tau$ and $(p', q) \in R$, or there exists $q' \in S_2$ such that $q \text{ -}\!\overset{a}{\text{ }}\!\!\triangleright_2 q'$ and $(p', q') \in R$. Relation R is called a strong (weak) *bisimulation*, denoted by $\Gamma_1 \sim_R \Gamma_2$ ($\Gamma_1 \approx_R \Gamma_2$) if both $\Gamma_1 \prec \Gamma_2$ ($\Gamma_1 \preccurlyeq_R \Gamma_2$) and $\Gamma_2 \prec_{R^{-1}} \Gamma_1$ ($\Gamma_2 \preccurlyeq_{R^{-1}} \Gamma_1$). Given a strong (weak) (bi)simulation R, we say that a state $p \in S_1$ is strongly (weakly) rooted (bi)similar to $q \in S_2$, written $p \sim_R^r q$ (correspondingly, $p \approx_R^r q$), if $(p, q) \in R$. The relation is called *rooted* iff $(s_{01}, s_{02}) \in R$. A rooted relation is indicated with a superscript ^{r}.

A *weighted Petri net* is a 4-tuple (P, T, F, W) where P and T are two disjoint sets of *places* and *transitions*, respectively, $F \subseteq ((P \times T) \cup (T \times P))$ is the *flow relation*, and $W : F \to \mathbb{N}^+$ is a *weight function*. For $x \in P \cup T$, we write ${}^\bullet x = \{y \mid (y, x) \in F\}$ to denote the *preset* of x and $x^\bullet = \{y \mid (x, y) \in F\}$ to denote the *postset* of x. We lift the notation of preset and postset to sets element-wise. If for a Petri net no weight function is defined, we assume $W(f) = 1$ for all $f \in F$. A *marking* of N is a multiset $m \in P^\oplus$, where $m(p)$ denotes the number of *tokens* in place $p \in P$. If $m(p) > 0$, place p is called *marked* in marking m. A *marked Petri net* is a tuple (N, m) with N a weighted Petri net with marking m. A transition $t \in T$ is enabled in (N, m), denoted by $(N, m)[t\rangle$ iff $W((p, t)) \leq m(p)$ for all $p \in {}^\bullet t$. An enabled transition can *fire*, resulting in marking m' iff $m'(p) + W((p, t)) = m(p) + W((t, p))$, for all $p \in P$, and is denoted by $(N, m)[t\rangle (N, m')$. We lift the notation of firings to sequences. A sequence $\sigma \in T^*$ is a *firing sequence* iff $\sigma = \epsilon$, or markings $m_0, \ldots, m_n$ exist such that $(N, m_{i-1})[\sigma(i)\rangle(N, m_i)$ for $1 \leq i \leq |\sigma| = n$, and is denoted by $(N, m_0)[\sigma\rangle(N, m_n)$. If the context is clear, we omit the weighted Petri net

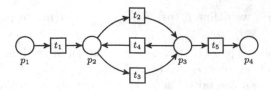

Fig. 3. An example block-structured WF-net. Each block corresponds to a node in the Jackson type $(p_1;(t_1;(((p_2;((t_2+t_3);p_3))\#t_4);(t_5;p_4))))$. As example, the choice between transitions t_2 and t_3 corresponds to the node $(p_2;((t_2+t_3);p_3))$.

N. The set of reachable markings of (N,m) is defined by $\mathcal{R}(N,m) = \{m' \mid \exists\sigma \in T^* : m[\sigma\rangle m'\}$. The set of all possible finite firing sequences of (N,m) is denoted by $\mathcal{L}(N,m_0) = \{\sigma \in T^* \mid m[\sigma\rangle m'\}$. The semantics of a marked Petri net (N,m) with $N = (P,T,F,W)$ is defined by the LTS $\Gamma_{N,m} = (P^\oplus, T, m_0, \rightarrow)$ with $(m,t,m') \in \rightarrow$ iff $m[t\rangle m'$. A Petri net $N = (P,T,F,W)$ has underlying graph $(P\cup T, F)$. Two Petri nets N and N' are isomorphic, denoted using $N \leftrightsquigarrow N'$, if their underlying graphs are.

A *workflow net* (WF-net for short) is a tuple $N = (P,T,F,W,in,out)$ such that: *(i)* (P,T,F,W) is a weighted Petri net; *(ii)* $in, out \in P$ are the source and sink place, respectively, with ${}^\bullet in = out^\bullet = \emptyset$; *(iii)* every node in $P \cup T$ is on a directed path from in to out. N is called k-*sound* for some $k \in \mathbb{N}$ iff *(i)* it is proper completing, i.e., for all reachable markings $m \in \mathcal{R}(N,[in^k])$, if $[out^k] \subseteq m$, then $m = [out^k]$; *(ii)* it is weakly terminating, i.e., for any reachable marking $m \in \mathcal{R}(N,[in^k])$, the final marking is reachable, i.e., $[out^k] \in \mathcal{R}(N,m)$; and *(iii)* it is quasi-live, i.e., for all transitions $t \in T$, there is a marking $m \in \mathcal{R}(N,[in])$ such that $m[t\rangle$. The net is called *sound* if it is 1-sound.

3 Typed Jackson Nets to Model Interacting Processes

In this section, we introduce typed Jackson Nets as subclass of typed Petri nets with identifiers. We show that this class is a natural extension to Jackson Nets, which are representations of block-structured workflow nets. Typed Jackson Nets are identifier sound and live by construction.

3.1 Jackson Nets

Whereas WF-nets do not put any restriction on the control flow of activities, block-structured WF-nets divide the control flow in logical blocks [15]. Each "block" represents a single unit of work that can be performed, where this unit of work is either atomic (single transition), or one involving multiple steps (multiple transitions). An example block-structured WF-net is shown in Fig. 3. The main advantage of block-structured WF-nets, is that the block-structure ensures that the WF-net is sound by definition [14–16]. In this paper, we consider Jackson Types and Jackson Nets [14]. A Jackson Type is a data structure used to capture all information involved in a single execution of a WF-net.

Definition 1 (Jackson Type [14]). *The set of Jackson Types $\mathcal{J}$ is recursively defined by the following grammar:*

$$\mathcal{J} ::= \mathscr{A}^p \mid (\mathscr{A}^p; (\mathcal{J}^t; \mathscr{A}^p))$$
$$\mathcal{J}^t ::= \mathscr{A}^t \mid (\mathcal{J}^t; (\mathcal{J}^p; \mathcal{J}^t)) \mid (\mathcal{J}^t + \mathcal{J}^t)$$
$$\mathcal{J}^p ::= \mathscr{A}^p \mid (\mathcal{J}^p; (\mathcal{J}^t; \mathcal{J}^p)) \mid (\mathcal{J}^p \parallel \mathcal{J}^p) \mid (\mathcal{J}^p \# \mathcal{J}^t)$$

where $\mathscr{A} = \mathscr{A}^p \cup \mathscr{A}^t = \{a, b, c, \ldots\}$ denotes two disjoint sets of atomic types for places and transitions, resp., and symbols $;, \parallel, +, \#$ stand for sequence, parallelism, choices, and loops. ◁

A Jackson Net is a Petri net where each place has an assigned Jackson Type. The class of Jackson Nets is obtained by recursively applying *generation rules*, starting from a singleton net with only one place. These generation rules are similar to those defined by Murata [18] and preserve soundness [14]. Thus, any Jackson Net is sound by construction.

Definition 2 (Jackson Net [14]). *A WF-net $N = (P, T, F, in, out)$ is called a Jackson Net if it can be generated from a single place p by applying the following five generation rules recursively:*

$$J1: p \leftrightarrow (p_1; (t; p_2)) \quad J4: p \leftrightarrow (p_1 \parallel p_2)$$
$$J2: t \leftrightarrow (t_1; (p_1; t_2)) \quad J5: t \leftrightarrow (t_1 + t_2)$$
$$J3: p \leftrightarrow (p \# t)$$

We say that N is generated by p. ◁

As shown in [14], Jackson Nets are completely determined by Jackson Types, and vice versa.

3.2 Petri Nets with Identifiers

Whereas WF-nets describe all possible executions for a single case, systems typically consist of many interacting processes. The latter can be modeled using typed Petri nets with identifiers (t-PNIDs for short) [23]. In this formalism, each object is typed and has a unique identifier to be able to refer to it. Tokens carry vectors of identifiers, which are used to relate objects. Variables on the arcs are used to manipulate the identifiers.

Definition 3 (Identifiers, Types and Variables). *Let $\mathcal{I}$, Λ, and $\mathcal{V}$ denote countably infinite sets of identifiers, type labels, and variables, respectively. We define:*

- *the domain assignment function $I : \Lambda \to \mathcal{P}(\mathcal{I})$, such that $I(\lambda_1)$ is an infinite set, and $I(\lambda_1) \cap I(\lambda_2) \neq \emptyset$ implies $\lambda_1 = \lambda_2$ for all $\lambda_1, \lambda_2 \in \Lambda$;*
- *the id typing function $\text{type}_{\mathcal{I}} : \mathcal{I} \to \Lambda$ s.t. if $\text{type}_{\mathcal{I}}(\text{id}) = \lambda$, then $\text{id} \in I(\lambda)$;*
- *a variable typing function $\text{type}_{\mathcal{V}} : \mathcal{V} \to \Lambda$, prescribing that $x \in \mathcal{V}$ can be substituted only by values from $I(\text{type}_{\mathcal{V}}(x))$.*

When clear from the context, we omit the subscripts of type. *We lift the* type *functions to sets, vectors, and sequences by applying the function on each of its constituents.*   ◁

In a t-PNID, each place is annotated with a label, called the *place type*. A place type is a vector of types, indicating types of identifier tokens the place can carry. Similar to Jackson Types, we use $[p, \lambda]$ to denote that place p has type $\alpha(p) = \lambda$. Each arc is inscribed with a multiset of vectors of identifiers, such that the type of each variable coincides with the place types. If the inscription is empty or contains a single element, we omit the brackets.

Definition 4 (Typed Petri net with identifiers). *A typed Petri net with identifiers (t-PNID) N is a tuple (P, T, F, α, β), where:*

- *(P, T, F) is a classical Petri net;*
- *$\alpha : P \to \Lambda^*$ is the* place typing function*;*
- *$\beta : F \to (\mathcal{V}^*)^\oplus$ defines for each arc a multiset of variable vectors s.t. $\alpha(p) = \text{type}(x)$ for any $x \in supp\,(\beta((p, t)))$ and $\text{type}(y) = \alpha(p')$ for any $y \in supp\,(\beta((t, p')))$ where $t \in T$, $p \in {}^\bullet t$, $p' \in t^\bullet$.*   ◁

A marking of a t-PNID is the configuration of tokens over the set of places. Each token in a place should be of the correct type, i.e., the vector of identifiers carried by a token in a place should match the corresponding place type. The set $C(p)$ defines all possible vectors of identifiers a place p may carry.

Definition 5 (Marking). *Given a t-PNID $N = (P, T, F, \alpha, \beta)$, and place $p \in P$, its id set is $C(p) = \prod_{1 \le i \le |\alpha(p)|} I(\alpha(p)(i))$. A marking is a function $m \in \mathbb{M}\,(N)$, with $\mathbb{M}\,(N) = P \to (\Lambda^*)^\oplus$, such that $m(p) \in C(p)^\oplus$, for each place $p \in P$. The set of identifiers used in m is denoted by $Id(m) = \bigcup_{p \in P} \text{RNG}(supp\,(m(p)))$ The pair (N, m) is called a marked t-PNID.*   ◁

To define the semantics of a t-PNID, the variables need to be valuated with identifiers.

Definition 6 (Variable sets [23]). *Given a t-PNID $N = (P, T, F, \alpha, \beta)$, $t \in T$ and $\lambda \in \Lambda$, we define the following sets of variables:*

- *input variables as $In(t) = \bigcup_{x \in \beta((p,t)), p \in {}^\bullet t} \text{RNG}(supp\,(x))$;*
- *output variables as $Out(t) = \bigcup_{x \in \beta((t,p)), p \in t^\bullet} \text{RNG}(supp\,(x))$;*
- *variables as $Var(t) = In(t) \cup Out(t)$;*
- *emitting variables as $Emit(t) = Out(t) \setminus In(t)$;*
- *collecting variables as $Collect(t) = In(t) \setminus Out(t)$;*
- *emitting transitions as $E_N(\lambda) = \{t \mid \exists x \in Emit(t) \wedge \text{type}(x) = \lambda\}$;*
- *collecting transitions as $C_N(\lambda) = \{t \mid \exists x \in Collect(t) \wedge \text{type}(x) = \lambda\}$;*
- *types in N as $\text{type}(N) = \{\vec{\lambda} \mid \exists p \in P : \vec{\lambda} \in \alpha(p)\}$.*   ◁

A valuation of variables to identifiers is called a *binding*. Bindings are used to inject new fresh data into the net via variables that emit identifiers, i.e., via variables that appear only on the output arcs of that transition. Note that in this definition, freshness of identifiers is local to the marking, i.e., disappeared identifiers (those fully removed from the net through collecting transitions) may be reused, as it does not hamper the semantics of the t-PNID.

Definition 7 (Firing rule for t-PNIDs). *Given a marked t-PNID* (N, m) *with* $N = (P, T, F, \alpha, \beta)$, *a binding for transition* $t \in T$ *is an injective function* $\psi : \mathcal{V} \to \mathcal{I}$ *such that* $\mathrm{type}(v) = \mathrm{type}(\psi(v))$ *and* $\psi(v) \notin Id(m)$ *iff* $v \in Emit(t)$. *Transition* t *is enabled in* (N, m) *under binding* ψ, *denoted by* $(N, m)[t, \psi\rangle$ *iff* $\rho_\psi(\beta(p, t)) \le m(p)$ *for all* $p \in {}^\bullet t$. *Its firing results in marking* m', *denoted by* $(N, m)[t, \psi\rangle(N, m')$, *such that* $m'(p) + \rho_\psi(\beta(p, t)) = m(p) + \rho_\psi(\beta(t, p))$. ◁

The firing rule is inductively extended to sequences. A marking m' is *reachable* from m if there exists $\eta \in (T \times (\mathcal{V} \to \mathcal{I}))^*$ such that $(N, m)[\eta\rangle(N, m')$. We denote with $\mathcal{R}(N, m)$ the set of all markings reachable from m for (N, m). We use $\mathcal{L}(N, m)$ to denote all possible firing sequences of (N, m), i.e., $\mathcal{L}(N, m) = \{\eta \mid (N, m)[\eta\rangle\}$ and $Id(\eta) = \bigcup_{(t,\psi) \in \eta} \mathrm{RNG}(\psi)$ for the set of identifiers used in η. The execution semantics of a t-PNID is defined as an LTS that accounts for all possible executions starting from a given initial marking. We say two t-PNIDs are bisimilar if their induced transition systems are.

Definition 8. *Given a marked t-PNID* (N, m_0) *with* $N = (P, T, F, \alpha, \beta)$, *its induced transition system is* $\Gamma_{N,m_0} = (\mathbb{M}(N), (T \times (\mathcal{V} \to \mathcal{I})), m_0, \to)$ *with* $m \xrightarrow{(t,\psi)} m'$ *iff* $(N, m)[t, \psi\rangle(N, m')$. ◁

Soundness properties for WF-nets typically consist of proper completion, weak termination, and quasi-liveness [6]. Extending soundness to t-PNIDs gives *identifier soundness* [23]. In t-PNIDs, each object of a given type "enters" the system through an emitting transition, binding it to a unique identifier. Identifier soundness intuitively states that it should always be possible to remove objects (weak type termination), and that once a collecting transition fires for an object, there should be no remaining tokens referring to the removed object (proper type completion).

Definition 9 (Identifier Soundness [23]). *Let* (N, m_0) *a marked t-PNID and* $\lambda \in \Lambda$ *some type.* (N, m_0) *is* λ-*sound iff it is*

- *Proper* λ-*completing, i.e., for all* $t \in C_N(\lambda)$, *bindings* $\psi : \mathcal{V} \to \mathcal{I}$ *and markings* $m, m' \in \mathcal{R}(N, m_0)$, *if* $m[t, \psi\rangle m'$, *then for all identifiers* $\mathrm{id} \in \mathrm{RNG}(\psi|_{Collect(t)}) \cap Id(m)$ *and* $\mathrm{type}(\mathrm{id}) = \lambda$, *it holds that* $\mathrm{id} \notin Id(m')$[1];
- *Weakly* λ-*terminating, i.e., for every* $m \in \mathcal{R}(N, m_0)$ *and identifier* $\mathrm{id} \in I(\lambda)$ *such that* $\mathrm{id} \in Id(m)$, *there exists a marking* $m' \in \mathcal{R}(N, m)$ *with* $\mathrm{id} \notin Id(m')$.

If it is λ-*sound for all* $\lambda \in \mathrm{type}(N)$, *then it is* identifier sound. ◁

3.3 Typed Jackson Nets

In general, identifier soundness is undecidable for t-PNIDs [23]. Similar as Jackson Nets restrict WF-nets to blocks, *typed Jackson Nets* (t-JNs) restrict t-PNIDs

---

[1] Here, we constrain ψ only to objects of type λ that are only consumed.

to blocks, while guaranteeing identifier soundness and liveness. For t-JNs, we disallow multiplicity on arcs and variables, i.e., $\beta(f)(v) \leq 1$ for all $f \in F$ and $v \in \mathcal{V}$, and imply a bijection on variables and identifier types. This prevents place types like $\lambda = \langle x, x \rangle$. Assuming a Gödel-like number on types (cf. [14]), place types and arc inscriptions can be represented as sets. Similar as Jackson Types describe Jackson Nets, we apply a notation based on Jackson Types to denote typed Jackson Nets.

Definition 10 (Typed Jackson Net). *A t-PNID N is a typed Jackson Net if it can be generated from a set of transitions T' by applying any of the following six generation rules recursively. If N is generated from a singleton set of transitions (i.e., $|T'| = 1$), N is called* atomic.

R1 Place Expansion: $[p, \lambda] \leftrightarrow ([p_1, \lambda]\,;(t_1\,; [p_2, \lambda]))$

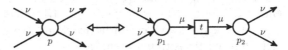

R2 Transition Expansion: $t \leftrightarrow (t_1\,;([p, \lambda]\,; t_2))$, *with* $Var(t) \subseteq \lambda$

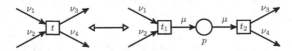

R3 Place Duplication: $(t_1\,;([p, \lambda]\,; t_2)) \leftrightarrow (t_1\,;(([p, \lambda] \parallel [p', \lambda'])\,; t_2))$,
   with $\lambda' \cap Emit(p^\bullet) = \emptyset$

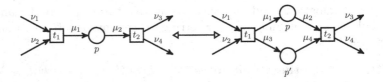

R4 Transition Duplication: $t \leftrightarrow (t + t')$

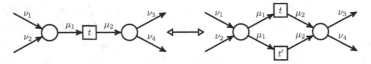

R5 Self Loop Addition: $[p, \lambda] \leftrightarrow ([p, \lambda]\,\#t)$

R6 Identifier Introduction: $t \leftrightarrow (t \lhd (N_1, [p, \lambda], N_2))$, *with* $(N_1; ([p, \lambda]; N_2))$ *a t-JN and* $\lambda \cap Var(t) = \emptyset$

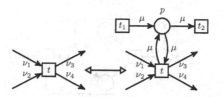

$\lhd$

An example t-JN is given in Fig. 1. Starting with the product process, transitions C and D can be reduced using rule $R2$. The resulting transition is a self-loop transition, and can be reduced using $R5$, resulting in the block $(E \lhd (A, product, B))$. This block can be reduced using $R6$, leaving transition E. Transition E is again a self-loop, and can be reduced using $R5$. The block containing transitions H, J, L O, N and K can be reduced to a single place by applying rules $R1$, $R2$ and $R5$ repeatedly. The remaining place is a duplicate place with respect to place p, and can be reduced using $R3$. Applying $R2$ on G and Z results in the block $(G \lhd (T, customer, V))$, which can be reduced to the transition G. Hence, the net in Fig. 1 is an atomic t-JN.

Theorem 1 (Identifier Soundness of typed Jackson Nets [23]). *Given a t-JN N, then N is* identifier sound *and* live. $\lhd$

4 Decomposability of t-JNs

t-PNIDs specify a class of nets with explicitly defined interactions between objects of different types within one system. However, sometimes one may want to focus only on some behaviors exhibited by a given set of object types, by extracting a corresponding net from the original t-PNID model. We formalize this idea below.

Definition 11 (Type projection). *Let $N = (P_N, T_N, F_N, \alpha, \beta)$ be a t-PNID and $\Upsilon \subseteq \Lambda$ be a set of identifier types. The* type projection *of Υ on N is a t-PNID $\pi_\Upsilon (N) = (P_\Upsilon, T_\Upsilon, F_\Upsilon, \alpha_\Upsilon, \beta_\Upsilon)$, where:*

- $P_\Upsilon = \{p \in P \mid \Upsilon \subseteq \alpha(p)\}$;
- $T_\Upsilon = \{t \in T \mid ({}^{\bullet}t \cup t^{\bullet}) \cap P_\Upsilon \neq \emptyset\}$;
- $F_\Upsilon = F \cap ((P_\Upsilon \times T_\Upsilon) \cup (T_\Upsilon \times P_\Upsilon))$;
- $\alpha_\Upsilon(p) = \Upsilon$, *for each* $p \in P_\Upsilon$;
- $\beta_\Upsilon(f) = \beta(f)|_{\text{type}_{\mathcal{V}}^{-1}(\Upsilon)}$, *for each* $f \in ((P_\Upsilon \times T_\Upsilon) \cup (T_\Upsilon \times P_\Upsilon))$. $\lhd$

With the next lemma we explore a property of typed Jackson nets that, in a nutshell, shows that t-JNs are closed under the type projection. This also indirectly witnesses that t-JNs provide a suitable formalism for specifying and manipulating systems with multiple communicating components.

Lemma 1. *If $N = (P_N, T_N, F_N, \alpha, \beta)$ is a t-JN, then $\pi_\Upsilon (N)$ is a t-JN as well, for any $\Upsilon \subseteq \text{type}_\Lambda(N)$.* $\lhd$

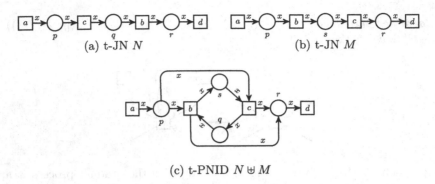

(a) t-JN N (b) t-JN M

(c) t-PNID $N \uplus M$

Fig. 4. Although both N and M are t-JNs, their composition is not

Proof. (sketch) Let us assume for simplicity that N is atomic. Then, using rules from Definition 10, N can be reduced to a single transition. Starting from this transition, one can construct a t-JN following the net graph construction from Definition 11 using the same rules (but the identifier introduction one), proviso that arc inscriptions are always of type Υ. Then, it is easy to check that the constructed net is indeed the type projection of Υ on N.   ∎

We define next how t-PNIDs can be composed and show that t-JNs are not closed under the composition.

Definition 12 (Composition). *Let* $N = (P_N, T_N, F_N, \alpha_N, \beta_N)$ *and* $M = (P_M, T_M, F_M, \alpha_M, \beta_M)$ *be two t-PNIDs. Their* composition *is defined by:*

$$N \uplus M = (P_N \cup P_M, T_N \cup T_M, F_N \cup F_M, \alpha_N \cup \alpha_M, \beta_N \cup \beta_M)$$

It is easy to see that the composition of two t-JNs does not automatically result in a t-JN. Consider nets in Fig. 4. It is easy to see that both N and M can be obtained by applying R2 from Definition 10. However, their composition cannot be reduced to a single transition by consecutively applying rules from Definition 10.

A more surprising observation is that composing type projections of a t-JN may not result in a t-JN. Take for example the net from Fig. 5. Both its projections on $\{\lambda_1\}$ and $\{\lambda_2\}$ are t-JNs. However, bringing them together using the composition operator results in a t-PNID that is not t-JN: indeed, since the "copies" of place p appear in three places, and all such copies have same pre- and post-sets (and only differ by their respective types), it is impossible to apply identifier elimination rule *R6* from Definition 10.

As one may observe from the above example, the only difference between $[p_{xy}, \langle \lambda_1, \lambda_2 \rangle]$ and its copies p_x and p_y is in their respective types, whereas the identifiers carried by p_x and p_y are always contained in p_{xy}, and thus both p_x and p_y can be seen as subsidiary with respect to p_{xy}. We formalize this observation using the notion of *minor places*: a place p is minor to some place q if both p and q have identical pre- and post-sets, and the type of q subsumes the one of p.

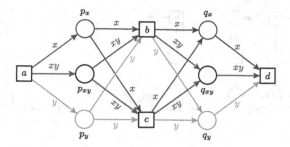

Fig. 5. Composition of the projections on $\{\lambda_1\}$, $\{\lambda_2\}$ and $\{\lambda_1, \lambda_2\}$ on the t-JN $(a; [p, \langle x, y\rangle]; (b||c); [q, \langle x, y\rangle]; d)$. Here, type assignments are as follows: $\alpha(p_x) = \alpha(q_x) = \lambda_1$, $\alpha(p_y) = \alpha(q_y) = \lambda_2$ and $\alpha(p) = \alpha(q) = \lambda_1\lambda_2$.

Definition 13 (Minor places). *Let $N = (P_N, T_N, F_N, \alpha, \beta)$ be a t-PNID. A place $p \in P$ is* minor *to a place $q \in P$ iff the following holds:*

- ${}^\bullet p = {}^\bullet q$, $p^\bullet = q^\bullet$ *and* $\alpha(p) \subset \alpha(q)$;
- $\beta((t,p)) = \beta((t,q))|_{\text{type}^{-1}(\alpha(p))}$, *for each $t \in {}^\bullet p$;*
- $\beta((p,t)) = \beta((q,t))|_{\text{type}^{-1}(\alpha(p))}$, *for each $t \in p^\bullet$.*

$\lhd$

We show next that minor places can be added or removed without altering the overall behavior of the net.

Lemma 2. *Let $N = (P, T, F, \alpha, \beta)$ be a t-PNID with initial marking m_0 s.t. $m_0(p) = m_0(q) = \emptyset$, for $p, q \in P$, where p is minor to q. Let $N' = (P \setminus \{p\}, T, F \setminus (\{(p,t)|t \in p^\bullet\} \cup \{(t,p)|t \in {}^\bullet p\}), \alpha, \beta)$ be a t-PNID obtained by eliminating from N place p. Then $\Gamma_{N,m_0} \sim^r \Gamma_{N',m_0}$.*

$\lhd$

Proof. (sketch) It is enough to define a relation $Q \subseteq \mathcal{R}(N, m_0) \times \mathcal{R}(N', m_0)$ s.t. $(m, m') \in Q$ iff $m(r) = m'(r)$, for $r \in P \setminus \{p\}$, and $m(p)(\text{id}) = m'(q)(\text{id})$, for all $\text{id} \in C(p)$, and $|m(p)| = |m'(q)|$. Then the lemma statement directly follows from the firing rule of t-PNIDs and that pre- and post-sets of p and q coincide. ∎

Let us now address the reconstructability property. In a nutshell, a net is reconstructable if composing all of its type projections returns the same net. This property is not that trivial to obtain. For example, let us consider singleton projections (that is, projections $\pi_{\{\lambda\}}(N)$ obtained for each $\lambda \in \text{type}_\Lambda(N)$) of the net in Fig. 6. It is easy to see that such projections "ignore" interactions between objects (or system components). Thus, the composition of the singleton projections $\pi_{\{\lambda_1\}}(N)$ and $\pi_{\{\lambda_2\}}(N)$ from Fig. 6 does not result in a model that merges p_x and p_y in one place as the composition operator cannot recognize component interactions between such projections. This is reflected in Fig. 6d.

To be able to reconstruct the original model from its projections (or at least do it approximately well), one needs to consider a projection reflecting component interactions. In the case of the net from Fig. 6a, its non-singleton projection

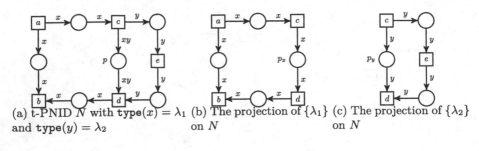

(a) t-PNID N with $\mathbf{type}(x) = \lambda_1$ and $\mathbf{type}(y) = \lambda_2$ (b) The projection of $\{\lambda_1\}$ on N (c) The projection of $\{\lambda_2\}$ on N

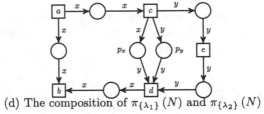

(d) The composition of $\pi_{\{\lambda_1\}}(N)$ and $\pi_{\{\lambda_2\}}(N)$

Fig. 6. t-PNID N (6a), its singleton projections and their composition

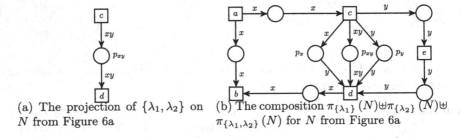

(a) The projection of $\{\lambda_1, \lambda_2\}$ on N from Figure 6a (b) The composition $\pi_{\{\lambda_1\}}(N) \uplus \pi_{\{\lambda_2\}}(N) \uplus \pi_{\{\lambda_1,\lambda_2\}}(N)$ for N from Figure 6a

Fig. 7. Adding the projection $\pi_{\{\lambda_1,\lambda_2\}}(N)$ reflecting interactions to the composition results in the original net N modulo places minor to p (such as p_x and p_y).

$\pi_{\{\lambda_1,\lambda_2\}}(N)$ is depicted in Fig. 7a. Now, using this projection we can obtain a composition (see Fig. 7b) that closely resembles N. Notice that, in this composition, copies of the interaction place p appear three times as places p_x, p_y and p_{xy}, respectively. It is also easy to see that places p_x and p_y are minor to p_{xy}, and $\alpha(p) = \alpha(p_{xy})$ witnesses that $\pi_{\{\lambda_1,\lambda_2\}}(N)$ is the maximal projection defined over types of N s.t. the correct type of p is "reconstructed". This leads us to the following result stipulating the reconstructability property of typed Jackson nets.

Theorem 2. *Let $N = (P, T, F, \alpha, \beta)$ be a t-JN. Then $\Gamma_{N,\emptyset} \sim^r \Gamma_{N',\emptyset}$, where*
$$N' = \biguplus_{\emptyset \subset \Upsilon \subseteq \mathbf{type}_\Lambda(N)} \pi_\Upsilon(N).$$
◁

Proof. (sketch) The proof immediately follows from the next observation. Among all possible projections, for each place $p \in P$ there exists a projection $\pi_\Upsilon(N)$ such that $\alpha(p) = \Upsilon$. This also means that $\pi_\Upsilon(N)$ contains p and that all other projections $\pi_{\Upsilon'}(N)$ with $\Upsilon' \subset \Upsilon$ will at most include the minors of p. Following Definition 12, it is easy to see that the composition of all the projections yields a t-JN identical to N modulo additional place minors introduced by some of the projections. Showing that the obtained net is bisimilar to N can be done by analogy with Lemma 2. ∎

Notice that the above result can be made stronger if all the additional minors (i.e., minors that were not present originally in N) are removed using reduction rules from Definition 10. For simplicity, given a t-PNID N with the set of places P, we denote by $\lfloor P \rfloor$ the set of its minor places.

Corollary 1. *Let N be a t-JN and N' is as in Theorem 2. Then $(N, \emptyset) \rightsquigarrow (N', \emptyset)$, if $\lfloor P \rfloor = \lfloor P' \rfloor$, where P and P' are respectively the sets of places of N and N'.*
                                                                                    ◁

The above result can be obtained by complementing the proof of Theorem 2 with a step that applies finitely many t-JN reduction rules to all the minor places that are in N' and not in N.

5   A Framework for Rediscoverability

In the previous section, we showed that t-JNs enjoy the reconstructability property: given a t-JN, a composition of *all* its (proper) type projections yields a t-JN that is strongly bisimilar to the original one.[2]

In this section, we propose a framework to rediscover systems of interacting processes that rely on this property. The framework builds upon a divide and conquer strategy [21]. The first step of the approach is to divide the event logs over all possible projections. For this, we translate the notion of event logs to event logs of interacting systems, and show that if these event logs are generated by a t-JN, projections on these event logs have a special property: the projected event log can be replayed by the projected net. In other words, there is no distinction between the projection on the event log, or that the projected net generated the event log. This observation forms the basis of the proposed framework for rediscoverability. In the second step, we conquer the discoverability problem of the system of interacting processes by first discovering a model for each of the projections, and then composing these projections into the original system. If the event log and discovery algorithm guarantee the defined properties, composition yields rediscoverability.

5.1   Event Logs and Execution Traces

In process discovery, an event log is represented as a (multi)set of sequences of events (called traces), where each sequence represents an execution history of a

---

[2] Such nets are also isomorphic if minor places of the composition are removed by consecutively applying the reduction rules from Definition 10.

Table 1. Firing sequence for the t-PNID in Fig. 1

transition	x	y	z	transition	x	y	z	transition	x	y	z
A	$p1$			T			$c2$	D	$p1$		
A	$p2$			H		$o1$		V			$c2$
T			$c1$	L		$o1$		K		$o1$	
G		$o1$	$c1$	J		$o1$		Z		$o1$	$c1$
C	$p1$			B	$p2$			V			$c1$
E	$p2$	$o1$		O		$o1$		B	$p1$		

process instance. Traditional process discovery assumes the process to be a WF-net. Consequently, each trace in an event log should correspond to a sequence of transition firings of the workflow net. If this is the case, the event log is said to be generated by the WF-net. We generalize this notion to marked Petri nets.

Definition 14 (Event Log). *Given a set of transitions T, a set of traces $L \subseteq T^*$ is called an* event log. *An event log L is generated by a marked Petri net (N, m) if $(N, m)[\sigma\rangle$ for all $\sigma \in L$, i.e., $L \subseteq \mathcal{L}(N, m_0)$.* ◁

Each sequence in a single process event log assumes to start from the initial marking of the WF-net. A marked t-PNID, instead, represents a continuously executing system, for which, given a concrete identifier, exists a single observable execution that can be recorder in an event log. Thus, event logs are partial observations of a larger execution within the system: an event log for a certain type captures only the relevant events that contain identifiers of that type, and stores these in order of their execution. Since each transition firing consists of a transition and a binding, a t-PNID firing sequence induces an event log for each set of types Υ. Intuitively, this induced event log is constructed by a filtering process. For each possible identifier vector for Υ we keep a firing sequence. Each transition firing is inspected, and if its binding satisfies an identifier vector of Υ, it is added to the corresponding sequence.

Definition 15 (Induced Event Log). *Let (N, m_0) be a marked t-PNID. Given a non-empty set of types $\Upsilon \subseteq \text{type}_\Lambda(N)$, the Υ-induced event log of a firing sequence $\eta \in \mathcal{L}(N, m_0)$ is defined by: $Log_\Upsilon(\eta) = \{\eta_{|i} \mid i \in (Id(\eta) \cap I(\Upsilon))^{|\Upsilon|}\}$, where $\eta_{|i}$ is inductively defined by (1) $\epsilon_{|i} = \epsilon$, (2) $(\langle\langle(t, \psi)\rangle \cdot \eta)_{|i} = \langle\langle(t, \psi)\rangle \cdot \eta_{|i}$ if $supp(i) \subseteq \text{RNG}(\psi)$, and (3) $(\langle\langle(t, \psi)\rangle \cdot \eta)_{|i} = \eta_{|i}$ otherwise.* ◁

Different event logs can be induced from a firing sequence. Consider, for example, the firing sequence of the net from Fig. 1 represented as table in Table 1. As we cannot deduce the types for each of the variables from the firing sequences in Table 1, we assume that there is a bijection between variables and types, i.e., that each variable is uniquely identified by its type, and vice-versa. Like that, we can create an induced log for each variable, as the type and variable name are interchangeable. For example, the x-induced event log is $Log_{\{x\}} = \{\langle A, E, B\rangle, \langle A, C, D, B\rangle\}$, and the z-induced event log is

$Log_{\{z\}} = \{\langle T, G, Z, V \rangle, \langle T, V \rangle\}$. Similarly, event logs can be also induced for combinations of types. In this example, the only non-empty induced event logs on combined types are $Log_{\{y,z\}} = \{\langle G, Z \rangle\}$ and $Log_{\{x,y\}} = \{\langle E \rangle\}$.

As the firing sequence in Table 1 shows, transition firings (and thus also events) only show bindings of variables to identifiers. For example, for firing G with binding $y \mapsto o1$ and $z \mapsto c1$, it is not possible to derive the token types of the consumed and produced tokens directly from the table. Therefore, we make the following assumptions for process discovery on t-PNIDs:

1. There are no "black" tokens: all places carry tokens with at least one type, and all types occur at most once in a place type, i.e., all places refer to at least one process instance.
2. There is a bijection between variables and types, i.e., for each type exactly one variable is used.
3. A Gödel-like number $\mathscr{G}$ is used to order the types in place types, i.e., for any place p, we have $\mathscr{G}(\alpha(p)(i)) < \mathscr{G}(\alpha(p)(j))$ for $1 \le i < j \le |\alpha(p)|$ and $p \in P$.

5.2 Rediscoverability of Typed Jackson Nets

Whereas traditional process discovery approaches relate events in an event log to a single object: the process instance, object-centric approaches can relate events to many objects [12]. Most object-centric process discovery algorithms (e.g., [5,17]) use a divide and conquer approach, where "flattening" is the default implementation to divide the event data in smaller event logs. The flattening operation creates a trace for each object in the data set, and combines the traces of objects of the same type in an event log. As we have shown in Sect. 4, singleton projections, i.e., those just considering types in isolation, are insufficient to reconstruct the t-JN that induced the object-centric event log. A similar observation is made for object-centric process discovery (cf. [3,5,7]): flattening the event data into event logs generates inaccurate models. Instead, reconstructability can only be achieved if all possible combinations of types are considered. Hence, for a divide and conquer strategy, the divide step should involve all possible combinations of types, i.e., each interaction between processes requires their own event log. In the remainder of this section, we show that if all combinations of types are considered, flattening is possible, and traditional process discovery algorithms can be used to rediscover a system of interacting processes.

For a system of interacting processes, we consider execution traces, i.e., a firing sequence from the initial marking. Like that, event logs for specific types or combinations of types are induced from the firing sequence. The projection of the system on a type or combinations of types, results again in a t-JN. Similarly, if we project a firing sequence of a t-JN N on a set of types Υ, then this projection is a firing sequence of the Υ-projection on N. The property follows directly from the result that t-JN N is weakly simulated by its Υ-projection.

Lemma 3. *Let N be a t-JN, and let $\Upsilon \subseteq \mathsf{type}_\Lambda(N)$. Then $\hat{\mathsf{H}}_U(\Gamma_{N,\emptyset}) \preccurlyeq^r \Gamma_{\pi_\Upsilon(N),\emptyset}$, with $U = T_N \setminus T_\Upsilon$.*   ◁

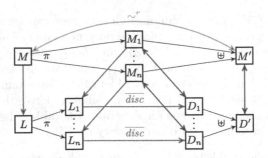

Fig. 8. Framework for rediscoverability of typed Jackson Nets. Model M generates an event log L. Log projections $L_1 \ldots L_n$ are generated from projected nets $M_1 \ldots M_n$. Discovery algorithm $disc$ results in nets $D_1 \ldots D_n$, isomorphic to $M_1 \ldots M_n$, which can be composed in D'. D' is isomorphic to M' and thus to M.

Proof. (sketch) Let $N_\Upsilon = \Upsilon_{|N} = (P_\Upsilon, T_\Upsilon, F_\Upsilon, \alpha_\Upsilon, \beta_\Upsilon)$. We can define a relation $Q \subseteq \mathbb{M}(N) \times \mathbb{M}(\pi_\Upsilon(N))$ s.t. $Q(m)(p)(a_{|I(\Upsilon)}) = m(p)(a)$ if $p \in P_\Upsilon$ and $Q(m)(p) = m(p)$ otherwise. The rooted weak bisimulation of Q follows directly from the firing rule of t-PNIDs. ∎

As the lemma shows, projecting a firing sequence yields a firing sequence for the projected net. A direct consequence of the simulation relation is that, no matter whether we induce an event log from a firing sequence on the original net, or induce it from the projected firing sequence, the resulting event logs are the same.

Corollary 2. *Let (N, m_0) be a marked t-PNID. Given a set of types $\Upsilon \subseteq$ $\mathrm{type}_\Lambda(N)$. Then $Log_\Upsilon(\eta) = Log_\Upsilon(\pi_\Upsilon(\eta))$.* ◁

Hence, it is not possible to observe whether an induced event log stems from the original model, or from its projection. Note that the projection may exhibit more behavior, so the reverse does not hold. In general, not any induced event log from the projection can be induced from the original model.

In general, a projection does not need to be an atomic t-JN (that is, a t-JN that can be reduced by applying rules from Definition 10 to a single transition). However, if the projection is atomic, then its structure is a transition-bordered WF-net: a WF-net that, instead of having source and sink places, has a set of start and finish transitions, such that pre-sets (resp., post-sets) of start (resp., finish) transitions are empty. The closure of a transition-bordered WF-net is constructed by adding a new source place i so that each start transition consumes from i, and a new sink place f so that each finish transition produces in f.

Lemma 4. *Let N be a t-JN and $\pi_\Upsilon(N) = (P_\Upsilon, T_\Upsilon, F_\Upsilon, \alpha_\Upsilon, \beta_\Upsilon)$ for some $\Upsilon \subseteq$ $\mathrm{type}_\Lambda(N)$ such that $\pi_\Upsilon(N)$ is atomic. Let $\eta \in \mathcal{L}(N, \emptyset)$ be a firing sequence. Then $Log_\Upsilon(\eta)$ is generated by $(N_\Upsilon, \emptyset)$ with $N_\Upsilon = (P_\Upsilon \cup \{i, f\}, T_\Upsilon, F_\Upsilon \{(i, t) \mid {}^\bullet t = \emptyset\} \cup \{(t, f) \mid t^\bullet = \emptyset\})$.* ◁

Proof. (sketch) Let $\sigma \in Log_\Upsilon(\eta)$. By construction, each firing sequence in $Log_\Upsilon(\eta)$ has some corresponding identifier vector that generated the sequence. Assume $\vec{v} \in \mathcal{I}^{|\Upsilon|}$ is such a vector for σ.

Observe that for any transition $t \in T$ if ${}^\bullet t = \emptyset$, $Emit(t) \cap \Upsilon \neq \emptyset$, and similarly, if $t^\bullet = \emptyset$, $Collect(t) \cap \Upsilon \neq \emptyset$. As N is identifier sound, only ${}^\bullet\sigma(1) = \emptyset$ and $\sigma(|\sigma|)^\bullet = \emptyset$. Define relation $R = \{(M, m) \mid \forall p \in P : M(p)(v) = m(p)\}$ and $U = \{(t, \psi) \mid v \not\subseteq \text{RNG}(\psi)\}$, i.e., U contains all transitions that do not belong to σ. Then R is a weak simulation, i.e., $\hat{H}_U(\Gamma_{N,\emptyset}) \preccurlyeq^r_R \Gamma_{N_\Upsilon,\emptyset}$ and thus $(N_\Upsilon, \emptyset)[\sigma\rangle$. ∎

Given a set of types Υ, if its projection is atomic, the projection can be transformed into a workflow net, and for any firing sequence of the original net, this WF-net can generate the Υ-induced event log. Suppose we have a discovery algorithm *disc* that can rediscover models, i.e., given an event log L that was generated by some model M, then *disc* returns the original model. Rediscoverability of an algorithm requires some property $P_{disc}(M)$ on the generating model M, and some property $Q_{disc}(L, M)$ on the quality of event log L with respect to the generating model M. In other words, $P(M)$ and $Q(L, M)$ are premises to conclude rediscoverability for discovery algorithm *disc*. For example, α-miner [22] requires for $P(M)$ that model M is well-structured, and for $Q(L, M)$ that event log L is directly-follows complete with respect to model M. Similarly, Inductive Miner [16] requires the generating model M to be a process tree without silent actions or self-loops ($P(M)$), and that event log L is directly-follows complete with respect to the original model M ($Q(L, M)$).

Definition 16 (Rediscovery). *An algorithm disc can* rediscover *WF-net $W = (P, T, F, in, out)$ from event log $L \subseteq T^*$ if $P_{disc}(W)$ and $Q_{disc}(L, W)$ imply $disc(L) \rightsquigarrow W$.* ◁

Thus, suppose there exists a discovery algorithm *disc* that is – under conditions P and Q – able to reconstruct a workflow model given an event log. In other words, given an event log L generated by some model M, *disc* returns a model that is isomorphic to the generating model. Now, suppose we have a firing sequence η of some t-JN N, and some projection Υ. Then, if $P(\pi_\Upsilon(N))$, and $Q(Log_\Upsilon(\eta), \pi_\Upsilon(N))$, then *disc* returns a model that is isomorphic to the closure of $\pi_\Upsilon(N)$, as *disc* only returns WF-nets. With $\overline{disc}$ we denote the model where the source and sink places are removed, i.e., $\overline{disc} \rightsquigarrow \pi_\Upsilon(N)$. Then, as shown in Fig. 8, if we discover for every possible combination of types, i.e., the subset-closed set of all type combinations, a model that is isomorphic to the type-projected model, then the composition results in a model that is bisimilar to the original model.

Theorem 3 (Rediscoverability of typed Jackson Nets). *Let N be a t-JN, and let $\eta \in \mathcal{L}(N, \emptyset)$ without minor places. Let disc be a discovery algorithm with properties P and Q that satisfy Definition 16. If for all $\emptyset \subset \Upsilon \subseteq \text{type}_\Lambda(N)$ the Υ-projection is atomic and satisfies conditions $P(\pi_\Upsilon(N))$ and $Q(Log_\Upsilon(\eta), \pi_\Upsilon(N))$, then $\Gamma_{N,\emptyset} \rightsquigarrow \Gamma_{N',\emptyset}$ with $N' = \biguplus_{\emptyset \subset \Upsilon \subseteq \text{type}_\Lambda(N)} \overline{disc}(Log_\Upsilon(\eta))$.*

Proof. (sketch) Let $\emptyset \subset \Upsilon \subseteq \mathsf{type}_\Lambda(N)$ be a set of types in N. Since $P(\pi_\Upsilon(N))$ and $Q(Log_\Upsilon(\eta)), \pi_\Upsilon(N))$the closure of $\pi_\Upsilon(N)$ and $disc(Log_\Upsilon(\eta))$ are isomorphic. From the closure, places *in* and *out* exist with ${}^\bullet in = \emptyset = out^\bullet$. As the nets are isomorphic, we have $\Upsilon_{|N} \rightsquigarrow \overline{disc}(Log_\Upsilon(\eta))$. Combining the results gives $\biguplus_{\emptyset \subset \Upsilon \subseteq \mathsf{type}_\Lambda(N)} \overline{disc}(Log_\Upsilon(\eta)) \rightsquigarrow \biguplus_{\emptyset \subset \Upsilon \subseteq \mathsf{type}_\Lambda(N)} \pi_\Upsilon(N)$. The statement then follows directly from Corollary 1.   ∎

6   Conclusion

In this paper, we studied typed Jackson Nets to model systems of interacting processes, a class of well-structured process models describing manipulations of object identifiers. As we show, this class of nets has an important property of reconstructability. In other words, the composition of the projections on all possible type combinations returns the model of the original system. Ignoring the interactions between processes results in less accurate, or even wrong, models. Similar problems occur in the discovery of systems of interacting processes, such as object-centric process discovery, where event logs are flattened for each object.

This paper provides a formal foundation for the composition of block-structured nets, and uses this to develop a framework for the discovery of systems of interacting processes. We link the notion of event logs used for process discovery to system executions, and show that it is not possible to observe whether an event log is generated by a system of interacting processes, or by a projection of the system. These properties form the key ingredients of the framework. We show under what conditions a process discovery algorithm (that guarantees rediscoverability) can be used to discover the individual processes and their interactions, and how these can be combined to rediscover a model of interacting processes that is bisimilar to the original system that generated the event logs.

Although typed Jackson Nets have less expressive power than formalisms like Object-centric Petri nets [5], proclets [11] or interacting artifacts [17], this paper shows the limitations and potential pitfalls of discovering interacting processes. This work aims to lay formal foundations for object-centric process discovery. As a next step, we plan to implement the framework and tune our algorithms to discover useful models from industrial datasets.

Acknowledgements. Artem Polyvyanyy was in part supported by the Australian Research Council project DP220101516.

References

1. Aalst, W.M.P.: Workflow verification: finding control-flow errors using petri-net-based techniques. In: van der Aalst, W., Desel, J., Oberweis, A. (eds.) Business Process Management. LNCS, vol. 1806, pp. 161–183. Springer, Heidelberg (2000). https://doi.org/10.1007/3-540-45594-9_11
2. Aalst, W.M.P.: Verification of workflow nets. In: Azéma, P., Balbo, G. (eds.) ICATPN 1997. LNCS, vol. 1248, pp. 407–426. Springer, Heidelberg (1997). https://doi.org/10.1007/3-540-63139-9_48

3. Aalst, W.M.P.: Object-centric process mining: dealing with divergence and convergence in event data. In: Ölveczky, P.C., Salaün, G. (eds.) SEFM 2019. LNCS, vol. 11724, pp. 3–25. Springer, Cham (2019). https://doi.org/10.1007/978-3-030-30446-1_1

4. van der Aalst, W.M.P.: Foundations of process discovery. In: van der Aalst, W.M.P., Carmona, J. (eds.) Process Mining Handbook. LNBIP, vol. 448, pp. 37–75. Springer, Cham (2022). https://doi.org/10.1007/978-3-031-08848-3_2

5. van der Aalst, W.M.P., Berti, A.: Discovering object-centric petri nets. Fund. Inform. 1–4(175), 1–40 (2020)

6. van der Aalst, W.M.P., et al.: Soundness of workflow nets: classification, decidability, and analysis. Formal Asp. Comput. 23(3), 333–363 (2011)

7. Adams, J.N., Park, G., Levich, S., Schuster, D., van der Aalst, W.M.P.: A framework for extracting and encoding features from object-centric event data. In: Troya, J., Medjahed, B., Piattini, M., Yao, L., Fernáindez, P., Ruiz-Cortés, A. (eds.) ICSOC 2022. LNCS, vol. 13740, pp. 36–53. Springer, Cham (2022). https://doi.org/10.1007/978-3-031-20984-0_3

8. Barenholz, D., Montali, M., Polyvyanyy, A., Reijers, H.A., Rivkin, A., van der Werf, J.M.E.M.: On the reconstructability and rediscoverability of typed Jackson nets (extended version) (2023). https://doi.org/10.48550/ARXIV.2303.10039, https://arxiv.org/abs/2303.10039

9. Berti, A., van der Aalst, W.M.P.: OC-PM: analyzing object-centric event logs and process models. Int. J. Softw. Tools Technol. Transf. (2022). https://doi.org/10.1007/s10009-022-00668-w

10. Best, E., Devillers, R., Koutny, M.: The box algebra=petri nets+process expressions. Inf. Comput. 178(1), 44–100 (2002). https://doi.org/10.1006/inco.2002.3117

11. Fahland, D.: Describing behavior of processes with many-to-many interactions. In: Donatelli, S., Haar, S. (eds.) PETRI NETS 2019. LNCS, vol. 11522, pp. 3–24. Springer, Cham (2019). https://doi.org/10.1007/978-3-030-21571-2_1

12. Ghahfarokhi, A.F., Park, G., Berti, A., van der Aalst, W.M.P.: OCEL: a standard for object-centric event logs. In: Bellatreche, L., et al. (eds.) ADBIS 2021. CCIS, vol. 1450, pp. 169–175. Springer, Cham (2021). https://doi.org/10.1007/978-3-030-85082-1_16

13. Glabbeek, R.J.: The linear time—branching time spectrum II. In: Best, E. (ed.) CONCUR 1993. LNCS, vol. 715, pp. 66–81. Springer, Heidelberg (1993). https://doi.org/10.1007/3-540-57208-2_6

14. van Hee, K.M., Hidders, J., Houben, G.J., Paredaens, J., Thiran, P.: On the relationship between workflow models and document types. Inf. Syst. 34(1), 178–208 (2009). https://doi.org/10.1016/j.is.2008.06.003

15. Kopp, O., Martin, D., Wutke, D., Leyman, F.: The difference between graph-based and block-structured business process modelling languages. EMISAJ 4(1), 3–13 (2009)

16. Leemans, S.J.J., Fahland, D., van der Aalst, W.M.P.: Discovering block-structured process models from event logs - a constructive approach. In: Colom, J.-M., Desel, J. (eds.) PETRI NETS 2013. LNCS, vol. 7927, pp. 311–329. Springer, Heidelberg (2013). https://doi.org/10.1007/978-3-642-38697-8_17

17. Lu, X., Nagelkerke, M., van de Wiel, D., Fahland, D.: Discovering interacting artifacts from ERP systems. IEEE Trans. Serv. Comput. 8(6), 861–873 (2015)

18. Murata, T.: Petri nets: properties, analysis and applications. Proc. IEEE 77(4), 541–580 (1989). https://doi.org/10.1109/5.24143

19. Polyvyanyy, A., van der Werf, J.M.E.M., Overbeek, S., Brouwers, R.: Information systems modeling: language, verification, and tool support. In: Giorgini, P., Weber, B. (eds.) CAiSE 2019. LNCS, vol. 11483, pp. 194–212. Springer, Cham (2019). https://doi.org/10.1007/978-3-030-21290-2_13
20. Tour, A., Polyvyanyy, A., Kalenkova, A.A.: Agent system mining: vision, benefits, and challenges. IEEE Access **9**, 99480–99494 (2021)
21. Tour, A., Polyvyanyy, A., Kalenkova, A.A., Senderovich, A.: Agent miner: an algorithm for discovering agent systems from event data. CoRR abs/2212.01454 (2022)
22. van der Aalst, W., Weijters, T., Maruster, L.: Workflow mining: discovering process models from event logs. Knowl. Data Eng. **16**(9), 1128–1142 (2004)
23. van der Werf, J.M.E.M., Rivkin, A., Polyvyanyy, A., Montali, M.: Data and process resonance. In: Bernardinello, L., Petrucci, L. (eds.) PETRI NETS 2022. LNCS, vol. 13288, pp. 369–392. Springer, Cham (2022). https://doi.org/10.1007/978-3-031-06653-5_19

ILP² Miner –
Process Discovery for Partially Ordered Event Logs Using Integer Linear Programming

Sabine Folz-Weinstein[1]([⊠]), Robin Bergenthum[2], Jörg Desel[1], and Jakub Kovář[3]

[1] FernUniversität in Hagen, Lehrgebiet Softwaretechnik und Theorie der
Programmierung, Hagen, Germany
{sabine.folz-weinstein, joerg.desel}@fernuni-hagen.de
[2] FernUniversität in Hagen, Fakultät für Mathematik und Informatik,
Hagen, Germany
robin.bergenthum@fernuni-hagen.de
[3] FernUniversität in Hagen, Lehrgebiet Programmiersysteme, Hagen, Germany
jakub.kovar@fernuni-hagen.de

Abstract. Process mining is based on event logs. Traditionally, an event log is
a sequence of events. Yet, there is a growing amount of work in the literature
that suggests we should extend the notion of an event log and use partially
ordered logs as a basis for process mining. Thus, the need for algorithms able to
handle these partially ordered logs will grow in the upcoming years. In this
paper, we adapt an existing, classical process discovery algorithm to be able to
handle partially ordered logs. We use the ILP Miner [1] as a basis and replace its
region theory part by compact tokenflow regions [2] to introduce the ILP²
Miner. This ILP² Miner handles sequential event logs just like the ILP Miner
but, in addition, is able to directly process partially ordered logs. We prove that
the ILP² Miner provides the same guarantees regarding structural and behavioral
properties of the discovered process models as the ILP Miner. We implement the
ILP² Miner and show experimental results of its runtime using three well-known
example log files from the process mining community literature.

Keywords: Process mining · Process discovery · Synthesis · ILP Miner ·
Partially ordered event log · Compact tokenflow · Integer linear programming

1 Introduction

Process mining aims to identify business processes and to gain insight into their performance and conformance by analyzing recorded behavior [3, 4]. Over time, a wide
variety of process mining algorithms and methods have been introduced as well as many
tools, contests, and case studies. In this paper, we focus on process discovery, which is
often said to be the most interesting, but at the same time the most challenging part of

process mining [5]. The goal of process discovery is to automatically create a process model that adequately describes the underlying process, based on recorded behavior.

Process discovery algorithms are based on two formalisms: event logs as the basis and workflow models as the result. A workflow model is an executable, often Petri net-like, model of a business process. An event log is a sequence of events, where every event is an observed execution of an activity of a business process. Traditionally, an event log is a total order on the activity instances. More recently, event logs are also represented as partially ordered sequences of events. At the moment, it is not yet common to directly record partially ordered event logs, but partially ordered representations are derived from sequential logs in a pre-processing step using attributes stored in the event data like timestamps, activity life-cycle information, resources or other domain knowledge [6, 7].

The advantage is that partially ordered sets of events can directly express concurrency and are able to model specific properties of the underlying activities like non-zero duration, start and end point in time, inherent uncertainty in process data logging etc., which are not supported by a total order assumption. Therefore, Marlon Dumas and Luciano García-Bañuelos suggest recasting all process mining operations based on partially ordered event structures [6]. Furthermore, Leemans, van Zelst, and Lu advocate partial order-based process mining, at least using partial orders as an intermediary data representation [7]. Altogether, we expect the need for algorithms designed to handle partially ordered logs to grow in the upcoming years.

Paper [7] presents a survey and outlook concerning partial order-based process mining. Although several new types of approaches have evolved in this area recently, the number of new publications which in one way or another work with partial orders is still limited compared to traditional, total order-based approaches. Concerning the field of process discovery, most partial order-based work refers to synthesis. One example is the process discovery approach called Prime Miner [8], which can handle partially ordered logs, but has a slightly different goal and does not yet offer the same guarantees and results as the established classical discovery algorithms.

In this paper, instead of developing a new and even more fancy discovery algorithm able to handle partially ordered logs, we extend an existing, classical process discovery algorithm, the ILP Miner [1]. The ILP Miner is well-established and part of every process discovery tutorial and textbook. It works best in applications where the log is of moderate size and of good quality. The main disadvantages are a high runtime complexity and a tendency to produce over-fitting models. This is widely discussed in the literature. By extending the ILP Miner, we obviously inherit all benefits and shortcomings. The goal of this paper, however, is to adapt the classical ILP Miner so that it is able to handle partially ordered input and keep all other characteristics unchanged.

The ILP Miner algorithm has been implemented and is available in the HybridILP-Miner package in the ProM (http://promtools.org) and RapidProM (http://rapidprom. org) toolkits. As a typical process discovery algorithm, the ILP Miner expects sequential event logs as input. The ILP Miner guarantees to discover relaxed sound workflow nets. It uses an integer linear programming (ILP) formulation, i.e., an objective function over a system of inequalities representing the constraints for a region, which is then solved for every causal relation in the event log to find the places of the resulting net.

In the new ILP^2 Miner, we use the algorithm structure and framework of the classical ILP Miner but replace the region part of the integer linear program by compact tokenflow (CTF) regions [2]. The CTF synthesis algorithm, which introduces this type of regions, is the approach which currently offers the best runtime for partial order-based synthesis. It uses labeled Hasse diagrams to represent the partially ordered input. Just like the ILP Miner, the CTF synthesis algorithm uses a region theory-based system of inequalities which is solved for every wrong continuation of the input to find places of the resulting net. Thus, this is a very good fit.

We prove that if we integrate and use the compact region inequality system of the CTF synthesis algorithm within slightly adapted formulations of the ILP Miner, we get a new miner that: (1) generates the same results as the classical ILP Miner if we apply the miner to a sequentially ordered event log, and (2) can perfectly handle partially ordered event logs. We conduct experiments to show that if we have two kinds of event logs, one totally ordered and the other partially ordered, both recording the same business process, the ILP^2 Miner, using the partially ordered event log, outperforms the ILP Miner, using the totally ordered event log.

The remainder of this paper is organized as follows: Since the ILP Miner serves as the basis for the new ILP^2 Miner, we recall the main characteristics of the classical ILP Miner in Sect. 2. There, we also present the core functional aspects of the CTF synthesis algorithm and its compact region formulation. In Sect. 3, we introduce the extended ILP^2 Miner algorithm and prove that the systems of (in)equalities used for finding the regions in the ILP Miner and the ILP^2 Miner produce equivalent sets of places when processing sequential orders. Therefore, the ILP^2 Miner extends but does not alter the classical ILP formulation. Finally, in Sect. 4 we present the implementation and runtime analysis for the extended ILP^2 Miner compared to the classical ILP Miner. We use three example event logs to illustrate benefits of the new approach. Section 5 concludes the paper.

2 Preliminaries

Let $\mathbb{N}$ be the set of non-negative integers and $\mathbb{R}$ the set of real numbers. Let f be a function and B a subset of the domain of f. We write $f|_B$ to denote the restriction of f to B. We denote the transitive closure of an acyclic and finite relation $<$ by $<^+$. We denote the skeleton of $<$ by $<^\diamond$. The skeleton of $<$ is the smallest relation $\lhd$ so that $\lhd^+ = <^+$ holds.

$X = \{x_1, x_2, \ldots, x_n\}$ denotes a set. $\mathcal{B}(X)$ denotes the powerset of X.

A sequence w of length k relates positions to elements $x \in X$, i.e., $w: 1, 2, \ldots, k \to X$. An empty sequence is denoted as ε. We denote a non-empty sequence as $\langle x_1, x_2, \ldots, x_k \rangle$. The set of all possible sequences over a set X is denoted as X^*.

We write a concatenation of sequences w_1 and w_1 as $w_1 w_2$, e.g., $\langle a, b \rangle \langle c, d \rangle = \langle a, b, c, d \rangle$. Let $Y \in X^*$ be a set of sequences. The prefix-closure of Y is defined as: $\overline{Y} = \{w_1 \in X^* | \exists w_2 \in X^* (w_1 w_2 \in Y)\}$.

We assume the reader is familiar with the use of vectors, and all vectors to be column vectors. We write $\mathbf{1}$ to denote the 1-vector and $\mathbf{0}$ to denote the 0-vector.

A Parikh vector $\boldsymbol{p}$ represents the number of occurrences of an element within a sequence, i.e. $\boldsymbol{p}: X^* \to \mathbb{N}^{|X|}$ with $\boldsymbol{p}(w) = (\#x_1(w), \#x_2(w), ..., \#x_n(w))$ and $\#x_i(w) = |\{i' \in \{1, 2, ..., |w|\} \mid w(i') = x_i\}|$.

A multiset m over a set A is a function $m : A \to \mathbb{N}$. We denote the empty multiset as $\emptyset$. Let m be a multiset, we write $m = \sum_{a \in A} m(a) \cdot a$ to denote all multiplicities of m. We extend all set operations to multisets.

Let T be a set of activities. An event records the execution of an activity. A sequence of events over T is a case. A sequence over T. is a trace. An event log is a multiset of traces.

In this paper, we model distributed systems by Petri nets also allowing for arc weights [9–11].

Definition 1 *(Petri net)*: A Petri net is a tuple (P, T, W) where P is a finite set of places, T is a finite set of transitions such that $P \cap T = \emptyset$ holds, and $W : (P \times T) \cup (T \times P) \to \mathbb{N}$ is a multiset of arcs. A marking of (P, T, W) is a multiset $m : P \to \mathbb{N}$. Let m_0 be a marking. We call the tuple $N = (P, T, W, m_0)$ a marked Petri net and m_0 the initial marking of N.

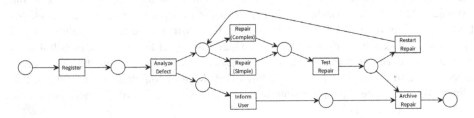

Fig. 1. Petri net of the repair example.

Figure 1 depicts an example Petri net modeling the business process of the so-called repair example, well-known from the ProM Tools tutorial (http://promtools.org). We show transitions as rectangles, places as circles, the multiset of arcs as weighted arcs, and the initial marking as black dots called tokens. This Petri net serves as a running example in this paper and will be discussed in more detail in the experimental results section.

2.1 ILP Miner

The ILP Miner [1, 12] is one of the corner stones of discovery algorithms and thus, part of every textbook on process discovery. The following definitions and descriptions are based on [1]. The algorithm uses integer linear programming to generate places of a Petri net. Every solution of the program, also called a region, is a set of values of binary variables encoding incoming and outgoing arcs of a valid place, as well as its initial marking.

Definition 2 *(ILP Miner region)*: Let L be an event log over a set of activities T, then a region is a triple $r = (m, x, y)$ with $x, y \in \{0, 1\}^{|T|}, m \in \{0, 1\}$ that satisfies:

$$\forall t \in T, \ w = w't \in \bar{L}: \ m + p(w')^\mathsf{T} \cdot x - p(w)^\mathsf{T} \cdot y \geq 0.$$

Paper [1] uses these regions to define the integer linear program of the ILP Miner.

Definition 3 *(ILP Miner process discovery program)*: Let L be an event log over a set of activities T, let $\bar{L}$ be the prefix-closure of the set of traces of L, let $m \in \{0, 1\}$ and let $x, y \in \{0, 1\}^{|T|}$. Let M and M' be two $|\bar{L} \backslash \{\varepsilon\}| \times |T|$ matrices with $M(w, t) = p(w)(t)$ and $M'(w)(t) = p(w')(t)$ where $w = w't \in \bar{L}$. Let M_L be a $|L| \times |T|$ matrix with $M_L(w, t) = p(w)(t)$ for $w \in L$, i.e., the equivalent of M for all complete traces in the event log. Let $c_m \in \mathbb{R}$ and $c_x, c_y \in \mathbb{R}^{|T|}$. The ILP Miner process discovery program is:

(1) *minimize* $z = c_m m + c_x^\mathsf{T} \cdot x + c_y^\mathsf{T} \cdot y$ *objective function*

(2) *such that* $m \cdot \mathbf{1} + M' \cdot x - M \cdot y \geq \mathbf{0}$ *theory of regions*

and

(3.1) $m \cdot \mathbf{1} + M_L \cdot (x - y) = \mathbf{0}$ *place is empty after each trace*
(3.2) $\mathbf{1}^\mathsf{T} \cdot x + \mathbf{1}^\mathsf{T} \cdot y \geq 1$ *at least one arc connected*
(3.3) $\mathbf{0} \leq x \leq \mathbf{1}$ *arc weight restricted to {0,1}*
(3.4) $\mathbf{0} \leq y \leq \mathbf{1}$ *arc weight restricted to {0,1}*
(3.5) $0 \leq m \leq 1$ *initial marking restricted to {0,1}*

The inequalities (2), marked above as "theory of regions", are the inequalities defined in Definition 2. Roughly speaking, they guarantee that every solution relates to a place which can execute all traces of the input event log. The objective function z selects the most expressive region considering the existing constraints. For example, paper [12] proposes an objective function minimizing x values and maximizing y values. Thus, the ILP Miner program minimizes the number of incoming arcs and maximizes the number of outgoing arcs of all the generated places. Other objective functions may be used. The other inequalities guarantee additional properties of the resulting net so that the resulting net is a so-called workflow net.

Finally, to find a finite number of places for the resulting workflow net, the ILP Miner uses the so-called causal relations heuristics [1]. Roughly speaking, it solves the integer linear program for each causal pair of activities of the event log. Two activities a and b are a causal pair if and only if ab is a subsequence of a trace of the event log and there is no trace so that ba is a subsequence. Paper [12] proves that for complete logs, causal dependencies directly relate to places and hence provide a good guide for finding a finite number of solutions.

To construct a net, we add a transition for every activity. A region r translates to a Petri net place p as follows: we add an arc leading from transition t to p if $x(t) = 1$, and an arc from p to transition t if $y(t) = 1$.

2.2 Compact Tokenflow Synthesis

In this section, we present the algorithm for synthesizing Petri nets from Hasse diagrams [2]. We refer to this algorithm as compact tokenflow (CTF) synthesis algorithm. This algorithm applies region theory to partially ordered sets of events. We will use this definition to replace the region theory part of the ILP Miner later. The approach requires a complete specification of the desired behavior as a set of Hasse diagrams.

Definition 4 *(labeled Hasse diagram)*: Let T be a set of labels. A labeled partial order (lpo) is a triple $lpo = (V, <, l)$ where V is a finite set of events, $< \subseteq V \times V$ is a transitive and irreflexive relation, and the labeling function $l : V \to T$ assigns a label to every event. A triple $run = (V, <, l)$ is a labeled Hasse diagram if $(V, <^+, l)$ is an lpo and $<^{\diamond} = <$ holds. We denote the set of all possible labeled Hasse diagrams over a set of labels T as $T^<$.

The CTF synthesis algorithm uses compact regions based on compact tokenflows to build a Petri net from a specification. A compact tokenflow is a distribution of tokens on the arcs of a Hasse diagram. We only distribute tokens over arcs because an event can only consume tokens from its preset to ensure that these are available. If an event produces tokens, it can pass these tokens to its postset. Tokens of the initial marking are free for all, i.e., any event can consume tokens from the initial marking. Such a distribution of tokens is valid if and only if (4) every event receives enough tokens, (5) an event must not pass too many tokens, and (6) the initial marking is not exceeded. A compact region is an abstract representation of a place together with a valid tokenflow. Thus, every region defines a valid place.

Definition 5 *(compact region)*: Let $S = (V_1, <_1, l_1), (V_2, <_2, l_2), \ldots, (V_n, <_n, l_n) \subseteq T^<$ be a set of labeled Hasse diagrams and p be a place. A function $r : (\bigcup_i (V_i \cup <_i) \cup (T \times \{p\}) \cup (\{p\} \times T) \cup \{p\}) \to \mathbb{N}$ is a compact region for S if and only if

(4) $\forall i : \forall v \in V_i : \quad r(v) + \sum_{v' < v} r(v', v) \geq r(p, l_i(v))$,
(5) $\forall i : \forall v \in V_i : \quad \sum_{v < v'} r(v, v') \leq r(v) + \sum_{v' < v} r(v', v) - r(p, l_i(v)) + r(l_i(v), p)$,
(6) $\forall i : \sum_{v \in V_i} r(v) \leq r(p)$ holds.

In Definition 5, a region r is a place together with a valid compact tokenflow for every Hasse diagram. The region has one value for every event and every arc of the Hasse diagrams, as well as values for incoming and outgoing arcs for every transition, and one value for the initial marking. The region satisfies the defined set of inequalities. Here, we have one inequality of type (4) per event to ensure that every transition receives enough tokens, one of type (5) per event to ensure that no event has to pass too many tokens, and additionally one inequality of type (6) per Hasse diagram to ensure that the initial marking is not exceeded. Altogether, the place defined by the region is able to execute all input Hasse diagrams.

3 ILP^2 Miner

In this section, we introduce our new ILP^2 Miner. This miner adapts and extends the classical ILP Miner presented in Sect. 2.1. Like the ILP Miner, we use integer linear programming to generate places of a workflow net. We use the compact region formulation of Definition 5 instead of the sequence based ILP regions of Definition 2 to get a mining algorithm that also supports partially ordered event logs as input.

To replace the region part of the ILP Miner, we need to adapt the compact region inequality system because using (5) and (6) directly, we would not be able to define whether a final marking is empty or not. In contrast to Definition 2, tokens in Definition 5 can disappear from the inequality system. This is because (5) and (6) are formulated as inequalities. The reason for this is that compact tokenflows only produce tokens if they are needed by later events. This has a positive effect on the runtime of a related compact tokenflow verification algorithm [2]. To guarantee the same properties as the ILP Miner, we extend every Hasse diagram by an initial event s and a final event f. Furthermore, we transform inequality (5) into an equality so that superfluous tokens must be passed to the final node. Thus, the resulting equality (8) enforces that no tokens are lost and the tokenflow at the final event f is the final marking. Similarly, we transform inequality (6) into equality (9) so that all the initial tokens are produced by the initial event s, and every initial token is counted.

Definition 6 *(ILP^2 Miner compact region)*: Let $S = \{(V_1, <_1, l_1), (V_2, <_2, l_2), \ldots, (V_n, <_n, l_n)\}$ be a specification, T be its set of labels, p be a place, and s, f two events. A function $r : (\bigcup_i (<_i \cup (\{s\} \times V_i) \cup (V_i \times \{f\}))) \cup (T \times \{p\}) \cup (\{p\} \times T) \cup \{p\}) \to \mathbb{N}$ is a compact region with a final marking for S if and only if

(7) $\forall i : \forall v \in V_i : r(s, v) + \sum_{v' <_v} r(v', v) \geq r(p, l_i(v))$,

(8) $\forall i : \forall v \in V_i : \sum_{v < v'} r(v, v') + r(v, f) = $

$$r(s, v) + \sum_{v' <_v} r(v', v) - r(p, l_i(v)) + r(l_i(v), p),$$

(9) $\forall i : \sum_{v \in V_i} r(s, v) = r(p)$ holds.

A totally ordered event log is a multiset of traces. Now, we define a partially ordered event log as a multiset of Hasse diagrams and apply Definition 6 to define the integer linear program for the ILP^2 Miner.

Definition 7 *(ILP^2 Miner process discovery program)*: Let L be a partially ordered event log over a set of activities T, let $\{(V_1, <_1, l_1), (V_2, <_2, l_2), \ldots, (V_n, <_n, l_n)\}$ be the set of labeled Hasse diagrams of L, p be a place, s and f two events, and r a function $r : (\bigcup_i <_i \cup (\{s\} \times V_i)) \cup (T \times \{p\}) \cup (\{p\} \times T) \cup \{p\}) \to \mathbb{N}$. The ILP^2 Miner process discovery program is:

(10) $minimize \ z = \sum_i \sum_{e \in \bigcup_i (<_i \cup (\{s\} \times V_i))} r(e)$ *objective function*

     such that   *theory of regions*

(11) $\forall i : \forall v \in V_i : r(s,v) + \sum_{v' <_v} r(v',v) \geq r(p, l_i(v))$,

(12) $\forall i : \forall v \in V_i : \sum_{v <_{v'}} r(v,v') = r(s,v) + \sum_{v' <_v} r(v',v) - r(p, l_i(v)) + r(l_i(v), p)$,

(13) $\forall i : \sum_{v \in V_i} r(s,v) = r(p)$,

   and

(14.1)   (-)   *place is empty after each trace*

(14.2) $\sum_i \sum_{e \in \bigcup_i <_i} r(e) \geq 1$,   *at least one token consumed / arc connected*

(14.3) $\forall t \in T : 0 \leq r(t,p) \leq 1$,   *arc weight restricted to {0,1}*

(14.4) $\forall t \in T : 0 \leq r(p,t) \leq 1$,   *arc weight restricted to {0,1}*

(14.5) $0 \leq r(p) \leq 1$,   *initial marking restricted to {0,1}*

In Definition 7, condition (14.1) is empty. Just like the ILP Miner, we want to guarantee that the place is empty after each trace. Therefore, we must ensure that there is no tokenflow from any event v to the final event f. For the ILP^2 Miner, this translates to the restriction $\forall i : \sum_{v \in V_i} r(v,f) = 0$. Thus, we delete all variables $r(v,f)$ from equation (8) of Definition 6 to get to equation (12) of Definition 7, so that there can be no tokenflow to the final event, and the place is empty after each trace.

Every solution of the ILP^2 Miner integer linear program is a region defining a place. $r(p)$ is the initial marking and $r(p, l(v))$ and $r(l(v), p)$ are ingoing and outgoing arcs. To generate a finite set of places, we use the same heuristics as the ILP Miner, the so-called causal relations, and solve the integer linear program for every causal pair of the partially ordered log.

Altogether, both the ILP Miner and the ILP^2 Miner use region theory-based systems of (in)equalities to generate places of the workflow net. In our ILP^2 Miner, we still use the same algorithm framework as the classical ILP Miner and guarantee the same structural properties of the resulting net, but we replace the region theory part, i.e., the region-matrix-form ILP-constraints, with the compact region inequality system based on CTF synthesis to be able to process partial orders.

In the remainder of this section, we prove that the ILP^2 Miner is in fact an extension of the ILP Miner. Introducing the ILP^2 Miner, we do not ruin already established features. Roughly speaking, we still satisfy the same formal guarantees as the original.

We show that if we apply both miners to a totally ordered event log, the region theory parts (2) of Definition 3 and (11)–(13) of Definition 7 produce the same set of feasible places.

As a first step, we look at lines (11), (12), and (13). Inequality (11) ensures every event receives enough tokenflow and equation (12) ensures every event passes the correct number of tokens. In a partially ordered event log one event can obviously have multiple predecessors and multiple successors, thus, the number of ingoing and outgoing tokens is a sum of tokenflows. In a totally ordered event log, every event has at most one predecessor and at most one successor, so that there is no need to sum up.

Figure 2 depicts an example of a totally ordered trace with at most one incoming and outgoing arc, and Figure 3 an example of a partially ordered trace with multiple incoming and outgoing arcs.

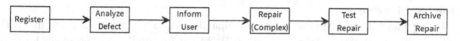

Fig. 2. One trace of the repair example.

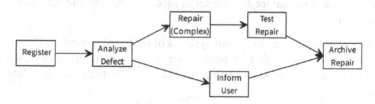

Fig. 3. The most frequent partially ordered trace of the repair example.

Let e be an event of a totally ordered event log, we denote $•e$ the predecessor and $e•$ the successor of e. Thus, we simplify (11) and (12) for totally ordered event logs as follows:

(11′) $\forall i : \forall e \in V_i : r(s,e) + r(•e,e) \geq r(p,l_i(e))$,
(12′) $\forall i : \forall e \in V_i : r(e,e•) = r(s,e) + r(•e,e) - r(p,l_i(e)) + r(l_i(e),p)$.

Equation (13) ensures that all tokens from the initial marking are distributed over all events of the Hasse diagram. In a partially ordered event log, one Hasse diagram can have multiple initial events as well as several alternative, concurrent paths (see Figure 3), thus, we need to be able to distribute initial tokens. In a totally ordered event log, every trace has one initial event and tokens distributed to this event can reach every other event in a straight line (see Figure 2). There is no need to distribute initial tokens anymore.

Let e be an event of a totally ordered event log. If $^\bullet e$ is empty, i.e., e has no predecessor, we can simplify the inequality system so that this event is the only event receiving all the tokens from the initial marking. Obviously, this event does not get tokens from any other event so that in this case we also write $r(^\bullet e, e)$ to denote the variable $r(s, e)$. Thus, we simplify the compact region ILP^2 Miner program for totally ordered event logs even further:

(11*) $\forall i : \forall e \in V_i : r(^\bullet e, e) \geq r(p, l_i(e))$,
(12*) $\forall i : \forall e \in V_i : r(e, e^\bullet) = r(^\bullet e, e) - r(p, l_i(e)) + r(l_i(e), p)$, and
(13*) $\forall i : \forall e \in V_i, {}^\bullet e = \emptyset : r(^\bullet e, e) = r(p)$.

Please note that these simplifications alter the inequalities as well as the variables of the ILP^2 Miner (in)equalities. Thus, they also alter the set of ILP^2 Miner regions concerning tokenflow variables but, obviously, they do not alter the set of places related to all regions.

Using these simplifications, we prove that the set of places defined by the ILP Miner is the set of places defined by the ILP^2 Miner for totally ordered event logs.

Theorem 1: Let L be a totally ordered event log. The set of places defined by the ILP Miner integer linear program for L is the set of places defined by the ILP^2 Miner integer linear program for L.

Proof. In a first step, we assume both systems of (in)equalities define the same set of places for L. We now prove that by adding one event to some trace of L, we add the same restrictions to both systems of (in)equalities so that both sets of related places remain equal.

Choose an arbitrary trace $w = \langle t_1, t_2, \ldots, t_n \rangle \in L$, let t be an action, and let $L' := L - w + wt$. Using ILP Miner notations, wt is a trace in L'. Obviously, all inequalities for w in L are still in the ILP Miner system for wt in L' because Definition 2 adds inequalities for every prefix of a trace.

Using the ILP^2 Miner notations, let there be a trace of sequentially ordered events $e_1 e_2 \ldots e_n$ in L so that $l(e_i) = t_i$ holds. We append a new event e with $l(e) = t$ to e_n by adding an arc (e_n, e) to construct the new trace $wt = e_1 e_2 \ldots e_n e$ of L'. Obviously, all (in)equalities for w in L are still in the ILP^2 Miner system for wt in L' because Definition 7 adds inequalities for every event of a trace.

Adding wt to L, the ILP Miner system of inequalities changes as stated in Definition 2. If $w = \varepsilon$ holds, the system for L' is obtained from the system for L by adding the following inequality:

$$m + \boldsymbol{p}(\langle \varepsilon \rangle)^\mathsf{T} \cdot x - \boldsymbol{p}(\langle t \rangle)^\mathsf{T} \cdot y \geq 0 \Leftrightarrow$$
$$m - y(t) \geq 0 \Leftrightarrow$$
$$m \geq y(t).$$

That is, the initial marking must be greater or equal to the arc-weight of the arc starting at the place to be constructed and leading to transition t.

If $w \neq \varepsilon$ holds, the system for L' is obtained from the system for L by adding the following inequality:

$$m + p(w)^\top \cdot x - p(wt)^\top \cdot y \geq 0 \Leftrightarrow$$
$$m + \Sigma_i x(t_i) - \Sigma_i y(t_i) - y(t) \geq 0.$$

That is, the initial marking of the place to be constructed plus the accumulated changes caused by firing all t_i of w, is greater or equal to the arc-weight of the arc starting at the place to be constructed and leading to transition t.

Adding wt to L, the ILP^2 Miner system of (in)equalities changes as stated in Definition 7. L' is totally ordered, thus, (11), (12), and (13) are equivalent to (11*), (12*), and (13*).

If $w = \varepsilon$ holds, the system for L' is obtained from the system for L by adding the following inequality and equations:

$$r(s, e) \geq r(p, t),$$
$$r(e, f) = r(s, e) - r(p, t) + r(t, p),$$
$$r(s, e) = r(p).$$
$$\Leftrightarrow$$
$$r(p) \geq r(p, t),$$
$$r(e, f) = r(p) - r(p, t) + r(t, p).$$

Like for Definition 2, the initial marking must be greater or equal to the arc-weight of the arc starting at the place to be constructed and leading to transition t. The additional condition which defines $r(e, f)$, i.e., the number of tokens remaining after firing t, is a new, not yet bound variable; this does not restrict the solution space.

If $w \neq \varepsilon$ holds, the system for L' is obtained from the system for L by replacing the equality

$$r(e_n, f) = r(e_{n-1}, e_n) - r(p, l(e_n)) + r(l(e_n), p)$$

by

$$r(e_n, e) = r(e_{n-1}, e_n) - r(p, l(e_n)) + r(l(e_n), p),$$

and adding the following inequality and equation:

$$r(e_n, e) \geq r(p, t),$$

$$r(e, f) = r(e_n, e) - r(p, t) + r(t, p),$$

We replace the equation to detach the final event and append it to the new last event of the sequence. Again, the second equation does not restrict the solution space because $r(e, f)$ is new and unbound.

The prefix w of wt is in L and in L' so that we have the following equations in both systems:

$$r(e_n, e) = r(e_{n-1}, e_n) - r(p, l(e_n)) + r(l(e_n), p)$$
$$r(e_{n-1}, e_n) = r(e_{n-2}, e_{n-1}) - r(p, l(e_{n-1})) + r(l(e_{n-1}), p),$$
$$\dots$$
$$r(e_2, e_3) = r(e_1, e_2) - r(p, l(e_2)) + r(l(e_2), p),$$
$$r(e_1, e_2) = r(s, e_1) - r(p, l(e_1)) + r(l(e_1), p),$$
$$r(s, e_1) = r(p).$$

Thus, if we recursively replace the variables $r(e_{i-1}, e_i)$ in the new inequality above by the right sides of their equations, we get:

$$r(e_n, e) \geq r(p, t) \Leftrightarrow$$

$$\sum_{i=1}^{n} (r(l(e_i), p) - r(p, l(e_i))) + r(p) \geq r(p, t).$$

Like for Definition 2, that is, the initial marking of the place to be constructed, plus the accumulated changes caused by firing all t_i of w, is greater or equal to the arc-weight of the arc starting at the place to be constructed and leading to transition t.

Altogether, whenever we add an event to a trace of an event log, we add the same restrictions to the set of places related to the solution space of the ILP Miner and to the set of places related to the solution space of the ILP^2 Miner. Finally, we add the recursive argument that we can build every event log by adding events to the empty log. ∎

The ILP Miner and ILP^2 Miner define the same set of places for totally ordered logs. We add inequalities (3.1)–(3.5) of Definition 3 and inequalities (14.2)–(14.5) of Definition 7, so that both algorithms guarantee the same additional properties of the resulting net. Furthermore, both miners apply heuristics based on causal relations to find a finite set of places. Thus, the ILP^2 Miner generates the same results as the ILP Miner on regular event logs. Obviously, in contrast to the ILP Miner, the ILP^2 Miner can process partially ordered logs.

Another important advantage of the ILP^2 Miner is its objective function. The ILP Miner selects optimal regions by maximizing the expressiveness of places by, roughly speaking, counting and weighing connected arcs and the initial marking. The ILP^2 Miner can now count tokens of the tokenflow directly because the variables are available. Thus, we minimize tokens present in the sum of every reachable marking.

4 Experimental Results

The goal of this paper is to extend the well-known ILP Miner, working on totally ordered inputs, to a new miner, able to process partially ordered inputs. Consequently, we inherit all its benefits and shortcomings. These benefits and shortcomings, as well as various comparisons of region-based miners to other discovery algorithms, are extensively discussed in the literature. Thus, since the ILP Miner and the ILP^2 Miner produce similar results, there is no point in comparing the quality of these results to other approaches again. Therefore, in this section, we only compare the runtime of our new ILP^2 Miner to the runtime of the ILP Miner.

As stated in the introduction, partially ordered logs, in contrast to totally ordered logs, are increasingly recommended as an expressive data representation and their benefits are undisputed.

If we have a totally ordered log, we can apply a so-called concurrency oracle to construct a partially ordered log. A concurrency oracle uses information like timestamps, life-cycle information, localities, resources or even user input to derive causal and concurrency information. For more information on different approaches for partial order extraction, we refer the reader to [7].

A partially ordered log is a much more compact representation of the observed behavior if the oracle faithfully mines the underlying concurrency relation.

Assuming a partially ordered log as the basis for discovery, the existing classical process discovery algorithms must construct a totally ordered log in a pre-processing step. Here, every single partially ordered trace can induce a high number of totally ordered traces. The number of traces is exponential in the length and factorial in the breadth of the partially ordered trace. Thus, directly working on partial orders has a significant positive effect on the runtime of a discovery procedure.

Altogether, we can go from totally ordered logs to partially ordered logs and the other way around. But for the purpose of runtime experiments, starting with one or the other would be unfair to one of the mining algorithms. Thus, in the remainder of this section, we assume there are two versions of every event log: one totally ordered event log and one partially ordered event log, both faithfully modeling the underlying workflow process.

To illustrate the benefits of using partially ordered logs, we have a look at a well-known example: the *repair example* included in the ProM framework tutorial [13, 14]. In the totally ordered version of the repair example, the event log records 11855 events, 1104 cases, and 39 different totally ordered traces. In the partially ordered version of the repair example, the event log records the same set of events and cases, but only 9 partially ordered traces (Hasse diagrams) [8]. Figure 3 depicts the Hasse diagram of the most frequent trace of the partially ordered log of the repair example. This trace alone represents 524 of the 1104 cases, which means half of the observed behavior.

For the following experiments, we have implemented the ILP Miner and the ILP^2 Miner as modules of the I ♥ Petri Nets website. The website is available at www. fernuni-hagen.de/ilovepetrinets/. The 🐾 module implements the ILP Miner, the 🐾 module implements the ILP^2 Miner.

In this section, we compare the runtime of both algorithms using three example log files quite famous in the process mining community. We consider the *reviewing example* [15, 16], the already mentioned *repair example* [13, 14], and the *teleclaims example* [15, 16]. We construct the partially ordered versions of these three totally ordered event logs by an implementation of the so-called alpha oracle. This oracle exploits the directly-follows-relation of activities to determine a concurrency relation [8, 17]. All .xes-files of the logs used in the experiments are available on our website.

Figure 4 depicts the ILP² Miner module. We start the ILP² Miner by dragging an. xes-file to the ★ symbol. We download the synthesized model by clicking the ♥ button. We can visualize the result using the "show a Petri net" module of the website. Clicking the 🐬 button downloads a report of the mining procedure.

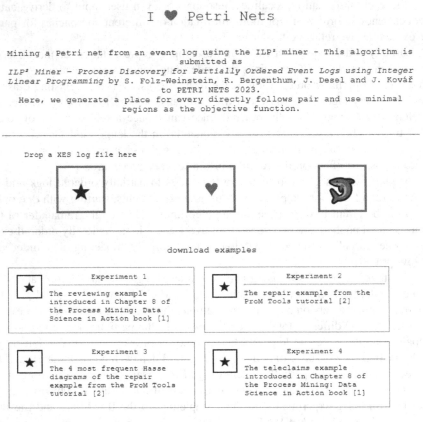

Fig. 4. The ILP² Miner module in the I ♥ Petri Nets toolkit.

We performed all the experiments on an Intel Core i5-8350U 1.70 GHz (4 CPUs) machine with 16 GB RAM running a Linux Mint 20.2 operating system. The source code of all modules is available on GitHub [18, 19].

In the ILP Miner, we use a simple objective function. We minimize 30 times the initial marking plus 10 times the sum of ingoing arc-weights minus the sum of outgoing arc-weights of possible solutions. Please note that the ProM tool (http://promtools.org) offers different implementations of the ILP Miner using a large selection of objective functions and additional heuristics that may provide better looking results in more complex examples. However, for the sake of runtime comparison, we choose a simple objective function and compute one region for every causal pair of the event log.

In the ILP² Miner, we could use the same objective function as in the ILP Miner to obtain identical results. But every region produced by the ILP² Miner also contains variables describing the related compact tokenflow. Thus, we minimize the sum of these variables to directly get minimal regions, and therefore very good-looking results. These results are equivalent to the results obtained by the ILP Miner implementations in ProM which use more sophisticated functions and heuristics. Again, we compute one region for every causal relation.

Experiment 1: The *reviewing example* contains 100 cases. These relate to 96 totally ordered traces or to 93 partially ordered traces. Thus, there is very limited concurrency, and we consider this to be the worst-case scenario for the ILP² Miner. The ILP Miner constructs an integer linear program containing 863 equations and inequalities and uses 27 variables. The average runtime to construct and solve the integer linear program for all causal pairs is round about 1100 ms. The ILP² Miner constructs an integer linear program containing 2378 equations and inequalities and introduces 1236 variables. It is important to note that although the system is about triple the size, the individual equations and inequalities are much simpler. The average runtime of the ILP² Miner is round about 750 ms. This is an improvement of 30%.

Figure 5 depicts the Petri net generated by the ILP² Miner using the minimal regions objective function.

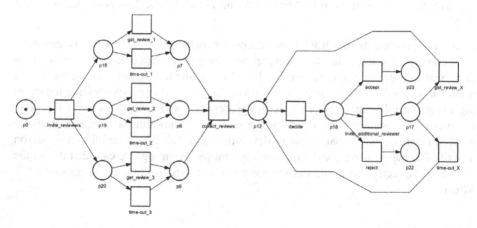

Fig. 5. Petri net for the reviewing example (ILP² Miner).

Experiment 2: The *repair example* contains 1104 cases. These relate to 39 different totally ordered traces, or 9 partially ordered traces. The ILP Miner constructs an integer linear program containing 57 equations and inequalities and uses 16 variables. The average runtime of the ILP Miner is round about 110 ms. The ILP² Miner constructs an integer linear program containing 90 equations and inequalities and introduces 66 variables. The average runtime of the ILP² Miner is round about 70 ms. This is an improvement of 36%.

Experiment 3: In the *repair example*, the four most frequent Hasse diagrams are sufficient to mine a model representing the behavior of the complete log. These four diagrams represent 913 cases of the original log file and 26 different totally ordered traces. Here, we use this reduced repair example. The ILP Miner constructs an integer linear program containing 33 equations and inequalities and uses 16 variables. The average runtime of the ILP Miner is round about 47 ms. The ILP^2 Miner constructs an integer linear program containing 44 equations and inequalities and introduces 43 variables. Filtering based on frequencies of Hasse diagrams is easy. Please note that in comparison to Experiment 2, both miners benefit from a reduction of the number of inequalities but the ILP^2 Miner also benefits from a reduction of the number of variables. The average runtime of the ILP^2 Miner is round about 26 ms. This is an improvement of 44%.

Figure 1 depicts the workflow net discovered by the ProM implementation of the ILP Miner using the totally ordered repair example. The same net is discovered by the ILP^2 Miner using the partially ordered repair example in Experiments 2 and 3.

Experiment 4: The *teleclaims example* contains 3512 cases. These relate to 12 totally ordered traces, or 8 partially ordered traces. The ILP Miner constructs an integer linear program containing 31 equations and inequalities and uses 22 variables. The average runtime of the ILP Miner is round about 93 ms. The ILP^2 Miner constructs an integer linear program containing 62 equations and inequalities and introduces 58 variables. The average runtime of the ILP^2 Miner is round about 45 ms. This is an improvement of 51%.

In our experiments, the ILP^2 Miner outperforms the ILP Miner in every example. The more concurrency in the example, the bigger the speed-up. However, it is important to note that the input for the ILP Miner is the totally ordered event log, and the input for the ILP^2 Miner is the set of Hasse diagrams after pre-processing the event log using a concurrency oracle. This pre-processing obviously takes extra time and resources. But assuming the input for the discovery algorithm is already a partially ordered event log is the main point of developing the ILP^2 Miner and the main reason to write this paper. If we consider a partially ordered event log, we can feed this to the ILP^2 Miner directly and would have to pre-process it to feed all interleavings to the ILP Miner.

5   Conclusion

The use of partially ordered event logs is increasingly recommended within process mining and thus, in process discovery. The need for algorithms which can directly process partial orders is expected to grow. However, the amount of such tools and algorithms currently available is still limited, and the existing new approaches are not yet comparable to the classical, well-known approaches based on regular, totally ordered event logs. Classical approaches can only process sequential decompositions of partially ordered data representations which contradicts the intentions and benefits of these representations.

Instead of developing an entirely new algorithm for process discovery, the goal of this paper was to adapt and extend an established classical process discovery algorithm so that it can directly process partially ordered inputs. In this paper, we focused on the ILP Miner and showed that it is indeed possible to adapt and extend the algorithm to partially ordered inputs by replacing the region theory core of its integer linear program. We proved that the resulting ILP² Miner can provide the same guarantees concerning structural and behavioral properties of the discovered process models as the classical ILP Miner, and that they find equivalent sets of places when applied to an ordinary, totally ordered event log. On top of that, as we now have the tokenflow variables available, it is possible to use an improved objective function in the ILP² Miner which finds minimal regions without employing additional heuristics or post-processing of the resulting net.

We implemented the ILP Miner and the ILP² Miner and conducted four experiments with well-established standard logs. The experiments show that there is a considerable improvement concerning runtime. The more concurrency in the event log, the bigger the speed-up using the ILP² Miner.

In future work, we plan to further fold identical prefixes and suffixes of the Hasse diagrams so that identical subgraphs are represented by fewer variables. Obviously, just like the ILP Miner, the ILP² Miner tends to over-fitting and has runtime issues for very large event logs. This is another aspect we would like to address in future work. As suggested in [7], we also hope that the process mining community will publish more partially ordered event logs and will extend their algorithms, contests, and case studies to partially ordered logs. Using these, we would further optimize and compare the ILP² Miner to other partial order-based approaches.

References

1. van Zelst, S.J., van Dongen, B.F., van der Aalst, W.M.P., Verbeek, H.M.W: Discovering Workflow Nets Using Integer Linear Programming. In: Computing 100, pp. 529–556. Springer (2018). https://doi.org/10.1007/s00607-017-0582-5
2. Bergenthum, R.: Synthesizing Petri Nets from Hasse Diagrams. In: Carmona, J., Engels, G., Kumar, A. (eds.) BPM 2017. LNCS, vol. 10445, pp. 22–39. Springer, Cham (2017). https://doi.org/10.1007/978-3-319-65000-5_2
3. van der Aalst, W.M.P.: Process Mining: Discovery, Conformance and Enhancement of Business Processes. Springer (2011)
4. van der Aalst, W., et al.: Process Mining Manifesto. In: Daniel, F., Barkaoui, K., Dustdar, S. (eds.) BPM 2011. LNBIP, vol. 99, pp. 169–194. Springer, Heidelberg (2012). https://doi.org/10.1007/978-3-642-28108-2_19
5. van der Aalst, W. M. P., Carmona, J.: Process Mining Handbook. Springer (2022)
6. Dumas, M., García-Bañuelos, L.: Process Mining Reloaded: Event Structures as a Unified Representation of Process Models and Event Logs. In: Devillers, R., Valmari, A. (eds.) PETRI NETS 2015. LNCS, vol. 9115, pp. 33–48. Springer, Cham (2015). https://doi.org/10.1007/978-3-319-19488-2_2
7. Leemans, S. J. J., van Zelst, S. J., Lu, X.: Partial-order-based process mining: a survey and outlook. Knowledge and Information Systems, vol. 65, pp. 1–29. Springer (2023).

8. Bergenthum, R.: Prime Miner - Process Discovery using Prime Event Structures. ICPM 2019, pp. 41–48 (2019)

9. Reisig, W.: Understanding Petri Nets - Modeling Techniques, Analysis Methods. Springer, Case Studies (2013)

10. Desel, J., Juhás, G.: What is a Petri Net? In: Ehrig, H., Juhás, G., Padberg, J., Rozenberg, G. (eds.) Unifying Petri Nets, Advances in Petri Nets, LNCS 2128, pp. 1–25. Springer (2001). https://doi.org/10.1007/978-0-387-09766-4_134

11. Peterson, J.L.: Petri Net Theory and the Modeling of Systems. Prentice-Hall, Englewood Cliffs (1981)

12. van der Werf, J. M. E. M., van Dongen, B. F., Hurkens, C. A. J., Serebrenik, A.: Process discovery using integer linear programming. Fundamenta Informaticae, vol. 94 no. 3–4, pp. 387–412. IOS Press (2009)

13. ProM Tools Documentation. https://www.promtools.org/doku/docs/index.html. Accessed 22 Dec 2022

14. ProM Tools example log files. https://www.promtools.org/doku/prom6/downloads/example-logs.zip. Accessed 22 Dec 2022

15. van der Aalst, W.M.P.: Process Mining: Data Science in Action. Springer (2016). https://doi.org/10.1007/978-3-662-49851-4

16. Event logs and models used in Process Mining book. https://processmining.org/old-version/event-book.html. Accessed 22 Dec 2022

17. Armas-Cervantes, A., Dumas, M., La Rosa, M., Maaradji, A.: Local concurrency detection in business process event logs. In: ACM Transactions on Internet Technology, vol. 19, no. 1, pp. 1–23 (2019)

18. ILP Miner module repository. https://github.com/ILPN/ILPN-Module-ILP-miner. Accessed 22 Dec 2022

19. ILP^2 Miner module repository. https://github.com/ILPN/ILPN-Module-ILP2-miner. Accessed 22 Dec 2022

Modelling Data-Aware Stochastic Processes - Discovery and Conformance Checking

Felix Mannhardt[1]([✉]) [iD], Sander J.J. Leemans[2] [iD], Christopher T. Schwanen[2] [iD], and Massimiliano de Leoni[3] [iD]

[1] Eindhoven University of Technology, Eindhoven, Netherlands
f.mannhardt@tue.nl
[2] RWTH Aachen University, Aachen, Germany
[3] University of Padova, Padua, Italy

Abstract. Process mining aims to analyse business process behaviour by discovering process models such as Petri nets from process executions recorded as sequential traces in event logs. Such discovered Petri nets capture the process behaviour observed in a log but do not provide insights on the likelihood of behaviour: the stochastic perspective. A stochastic Petri net extends a Petri net to explicitly encode the occurrence probabilities of transitions. However, in a real-life processes, the probability of a trace may depend on data variables: e.g., a higher requested loan amount will trigger additional checks. Such dependencies are not described by current stochastic Petri nets and corresponding stochastic process mining techniques. We extend stochastic Petri nets with data-dependent transition weights and provide a technique for learning them from event logs. We discuss how to evaluate the quality of these discovered models by deriving a stochastic data-aware conformance checking technique. The implementations are available in ProM, and we show on real-life event logs that the discovery technique is competitive with existing stochastic process discovery approaches, and that new types of stochastic data-based insights can be derived.

Keywords: Stochastic labelled data Petri nets · Process mining · stochastic data-aware process discovery · stochastic data-aware conformance checking

1 Introduction

The largest portion of research in Process Mining has focused on the discovery, conformance checking and enhancement of processes that do not consider the likelihood of the behavior allowed by the process model. In other words, when multiple activities are enabled according to the current state of the process model, they are assumed to have the same probability to occur. This is often unrealistic: even if multiple steps are possible as next, some are more common than others. As an example, in a loan application, when the model allows the

L. Gomes and R. Lorenz (Eds.): PETRI NETS 2023, LNCS 13929, pp. 77–98, 2023.
https://doi.org/10.1007/978-3-031-33620-1_5

notification of the application's acceptance or rejection as next activities, they cannot be associated with the same probability to occur.

These considerations motivate the importance of stochastic process mining, which little research has been carried on. Existing works on stochastic process discovery [3,9,21] and stochastic conformance checking [16] take the frequencies of the event log, which are a sample of the full process behaviour, into account and enable several analysis tasks, e.g., computing the occurrence probability for a trace or obtaining the probability that a marking can be reached [18].

These and other works on stochastic process mining have only focused on mining the activity occurrence probabilities on the basis of the sequence of activities that have happened beforehand. This is certainly valuable. However, in reality, the computation of the probability of an activity to occur as next within a set of enabled ones depends on the current state of the process data variables, as well. For instance, the probability to execute the activity of acceptable notification of a loan applicant will likely depend on the amount requested by the applicants, and on the his/her wealthiness.

We address this shortcoming and enable the discovery of models that, stochastically, fit better to the underlying distribution of the actual process. In particular, the methods rely on process models that are implemented as stochastic Data Petri nets, which are a variation on Data Petri nets [23] to encode the occurrence probabilities of transitions. This requires new methods for both process discovery and conformance checking. Our proposed discovery method learns data-dependant weight functions by building a set of regression problems that are fitted on the observed transition occurrences and the observed data values. To determine the quality of the resulting discovered Stochastic Labelled Data Petri nets (SLDPN), we design a new conformance checking technique that allows to compare the learned process behavior expressed by an SLDPN with that observed in an event log.

In contrast to existing work our methods leverage the information encoded in data attributes from the event log. In particular, our conformance checking technique overcomes the problem of stochastically comparing potentially infinite behaviour defined by an SLDPN with finite and sparse behaviour observed in an event log. The technique has been implemented as plug-ins of ProM, the largest open-source process mining framework. The evaluation has been carried out via a large set of publicly available event logs. For conformance checking, we illustrate that the technique follows the intuition of stochastic conformance and is a proper generalization of existing measures. For discovery of SLDPNs, the inclusion of the data variables for the computation of the activity occurrence probability is shown to improve the stochastic fitness for event logs. Of course, this holds for event logs that include data variables.

Section 2 discusses related work on stochastic process mining. Section 3 reports on the notation and concepts used in the paper. Section 4 introduces SLDPN. Sections 5 and 6 illustrates the techniques proposed for discovery and conformance checking methods of SLDPN. Section 7 reports on the evaluation with many real-life process event logs, while Sect. 8 concludes this paper, summarizing the paper's contributions and delineating potential future work.

2   Related Work

A large body of work exists on the discovery of data-dependent guards for activities of business process models, including transitions. This research field is often referred to as Decision Mining, starting from the seminal work by Rozinat et al. [31]. Batoulis et al. [5] focus on extracting guards for the outgoing arcs of XOR splits of BPMN models, while Bazhenova et al. [6] aims to discover Decision Model and Notation tables. The discovery of guards for causal nets is discussed in [24]. All of these approaches focus on ensuring that exactly one transition is enabled when a decision point (i.e., an XOR split) is reached. Mannhardt et al. [23] is the only approach that attempts to discover overlapping guards for Petri Nets, namely such that multiple transitions may be enabled in certain data states. However, this work does not provide a probability for transitions, such that the most reasonable assumption is that every enabled transition has the same probability to occur, whereas this paper aims to discover probabilities of transitions to fire when being given a data state. Thus, this paper does not consider guards, but generalisations of guards.

Within the realm of conformance checking, a few research works aim to check the conformance of process executions with respect to a process model represented as a Data Petri Net [13,22], but the conformance of each event-log trace is computed in isolation. This contrasts the notion of stochastic conformance checking that this paper tackles: the determination of the suitability of the overall stochastic behaviour requires the consideration of all traces together.

Stochastic process discovery aims to find a stochastic model such as a stochastic Petri net from an event log. Approaches include those that take a Petri net and estimate their weights, using alignments or frequencies [8], or based on time [29]. Our discovery technique falls into this category, but adds data awareness. Another approach starts from a model with the stochastic behaviour of the log, and reduces this into a smaller model repeatedly [9].

Examples of stochastic conformance checking techniques include the Earth Movers' Stochastic Conformance [20], Entropic Relevance [28] and Probabilistic Trace Alignments [7]. It would be challenging to adapt these to data-aware settings, as our models do not exhibit a stochastic language without data sequences as input. Stochastic models that are declarative have been proposed in [3]; these models express families of stochastic languages.

Key differentiators between stochastic process models and existing Markov-based stochastic models are concurrency, silent transitions and arc-based labels, the combination of which is not the focus of the latter [4,30]. Even though stochastic model checkers such as [15] do not typically consider these three aspects, they could still be applicable after appropriate translations.

Stochastic process discovery also relates to building a model that can compute the firing probability of each enabled transition, as a function of the sequence of fired transitions and data variables. This falls into the realm of predictive process monitoring (cf. [12,25,27]), and several techniques can be leveraged to compute the transition weights. However, the predictive monitoring techniques rely on the typical evaluation of machine-learning techniques, which looks at each transition in isolation and cannot be used for conformance checking against stochastic process models, which is conversely a global property that looks at traces as whole.

3   Preliminaries

In this section, we introduce required existing concepts.

A multiset is a function mapping its elements to the natural numbers. For a set A, $\mathcal{M}(A)$ denotes the set of all multisets over A. For instance, $[a^2, b]$ is a multiset containing two as and one b. Let X and Y be multisets, then $X \subseteq Y$ if and only if $\forall_a X(a) \leq Y(a)$. The multiset union is $\forall_a (X \uplus Y)(a) = X(a) + Y(a)$. The multiset difference is $\forall_a (X \setminus Y)(a) = \max(0, X(a) - Y(a))$. The set view $\widetilde{X} = \{a \mid X(a) > 0\}$.

Let Σ be an alphabet of activities, i.e. process tasks, such that $\tau \notin \Sigma$. A data state is an assignment to numeric[1] variables; let Δ be the set of all data states.

An event denotes the occurrence of an activity in a process, and a trace denotes the sequence of events that were executed for a particular case. A stochastic language is a weighted set of traces, such that their weights sum up to 1.

A data event is an event annotated with a data state, which indicates the data state after the event happened. A data trace denotes all data events belonging to a particular case. Formally, let $a_1, \ldots, a_n \in \Sigma$ and $d_0, \ldots, d_n \in \Delta$, then a data trace is a pair of lists $(\langle a_1, \ldots, a_n \rangle, \langle d_0, \ldots, d_n \rangle)$, in which each a_i indicates that event i involved activity a_i, and in which d_0 indicates the data state at the start of the trace, while subsequent $d_{i>0}$ indicate data states after occurrence of event i. Given a data trace $\sigma = (\langle a_1, \ldots, a_n \rangle, \langle d_0, \ldots, d_n \rangle)$, we refer to the sequence $\langle a_1, \ldots, a_n \rangle$ as the activity sequence (σ_Σ) and to the sequence $\langle d_0, \ldots, d_n \rangle$ as the data sequence (σ_Δ). We refer to the multisets of activity sequences and data sequences of a log L as L_Σ and L_Δ.

For instance, $(\langle a, b, c \rangle, \langle x = 10, x = 15, x = 20, x = 0 \rangle)$ indicates a data trace with three activities (a, b and c), where the variable x is 10 before a, 15 after a, 20 after b and 0 after c.

A labelled Petri net (LPN) is a tuple (P, T, F, λ, S_0), in which P is a set of places, T is a set of transitions such that $P \cap T = \emptyset$, $F \in \mathcal{M}(P \times T \cup T \times P)$ is a flow relation, $\lambda : T \to \Sigma \cup \{\tau\}$ is a labelling function, and $S_0 \in \mathcal{M}(P)$ is an initial marking. For a node $n \in P \cup T$, we denote ${}^\bullet n = [n' \mid (n', n) \in F]$ and $n^\bullet = [n' \mid (n, n') \in F]$. We assume the standard semantics of Petri nets here: a marking consisting of tokens on places indicates the state of the net. A transition $t \in T$ is enabled in a marking S if ${}^\bullet t \subseteq S$. Let $E(S)$ be the set of all enabled transitions in a marking S. An enabled transition t can fire in a marking S, which changes the marking to $S' = S \uplus t^\bullet \setminus {}^\bullet t$. The firing of a transition such that $\lambda(t) \neq \tau$ indicates the execution of the mapped activity. A path of the net

---

[1] Note that our technique only considers numeric variables. Other types of variables can be mapped using a suitable encoding, such as one-hot-encoding.

is a sequence of transitions that brings the marking from S_0 to a marking in which no transition is enabled. The corresponding activity sequence is obtained by mapping the path using λ, while removing all transitions mapped to τ:

$$\langle t_1, \ldots, t_n \rangle \downarrow_\lambda = \begin{cases} \langle \rangle & \text{if } n < 1 \\ \lambda(t_1) \cdot \langle t_2, \ldots, t_n \rangle \downarrow_\lambda & \text{if } \lambda(t_1) \neq \tau \\ \langle t_2, \ldots, t_n \rangle \downarrow_\lambda & \text{otherwise} \end{cases}$$

A stochastic labelled Petri net (SLPN) is a tuple $(P, T, F, \lambda, S_0, w)$ such that (P, T, F, λ, S_0) is an LPN and $w : T \to \mathbb{R}^+$ is a weight function. In a marking S, the probability to fire $t \in E(S)$ is $\frac{w(t)}{\sum_{t' \in E(S)} w(t')}$. Note that this probability depends on all other enabled transitions, and as such also expresses likelihoods on the order of transitions, even when they are concurrent. The probability of a path $\langle t_1, \ldots, t_n \rangle$ is, due to the independence of subsequent transitions, $\prod_{i=1}^{n} \frac{w(t_i)}{\sum_{t' \in E} w(t')}$. Note that the silent transitions make this a little-studied class of models [18].

In order to validate the quality of a stochastic model, a useful measure is the overlap in probability mass between the stochastic language of an event log and the stochastic language of the model. For stochastic process models, such a measure has been defined as the Unit Earth Movers' Stochastic Conformance (uEMSC) measure [20]. uEMSC measures the overlap in probability mass between a log and a stochastic language, by, for each trace σ of the log L, taking the positive difference between the probability of that trace in the log and the probability of that trace in the SLPN M [20]:

$$\text{uEMSC}(L, M) = 1 - \sum_{\sigma \in L} \max(L(\sigma) - M(\sigma), 0) \tag{1}$$

This rather simple formula uses the probability of a trace σ in a stochastic process model ($M(\sigma)$), which is not trivial to compute. $M(\sigma)$ indicates the sum of all paths through the model that yield the trace σ, however in case of silent transitions labelled τ there may be infinitely many such paths. A solution proposed in [18] – for bounded SLPNs – is to explicitly construct a state space of paths, and compute the trace probability using standard Markov reduction techniques.

The Earth Movers' Distance (EMD) is also known as the Wasserstein distance (W_1) of order 1. For the present special case where we consider unit distances, the EMD is also equivalent to the total variation distance (TV). A proof of the coupling between EMD and TV is for example shown in [14]. Thus, $\text{uEMSC}(L, M) = 1 - \text{TV}(L, M)$.

4 SLDPN

In this section, we extend SLPNs with data-based weight functions to Stochastic Labelled Data Petri nets (SLDPN). Syntactically, SLDPNs are similar to SLPNs, but utilise a weight function that is dependent on a data state.

Definition 1 (Stochastic Labelled Data Petri Net - syntax). *A stochastic labelled data Petri net (SLDPN) is a tuple* $(P, T, F, \lambda, S_0, \mathfrak{w})$, *such that* (P, T, F, λ, S_0) *is a Petri net and* $\mathfrak{w} : T \times \Delta \to \mathbb{R}^+$ *is a weight function.*

The state of an SLDPN is the combination of a marking and a data state $(d \in \Delta)$. The marking determines which transitions are enabled, while the data state influences the probabilities of transitions.

Definition 2 (Stochastic Labelled Data Petri Net - semantics). *Let* $(P, T, F, \lambda, S_0, \mathfrak{w})$ *be an SLDPN, and let* $\langle d_0, d_1, \ldots \rangle$ *be a data sequence. The SLDPN starts in state* (S_0, d_0). *Suppose the SLDPN is in state* (S_i, d_i). *The probability to fire* $t \in E(S_i)$ *is:*

$$\frac{\mathfrak{w}(t, d_i)}{\sum_{t' \in E(S_i)} \mathfrak{w}(t', d_i)}.$$

When a transition t fires, then the new state is (S_{i+1}, d_{i+1}) *with* $S_{i+1} = S_i \uplus t^\bullet \setminus {}^\bullet t$.

An SLDPN is not executable without further data modelling: the data state influences the likelihood of decisions, but the model does neither describe how the data state is initialised, nor how it changes with the execution of transitions. Thus, an SLDPN potentially has infinitely many stochastic languages.

Furthermore, these definitions do not specify *when* the data state is considered. In a real-life process, the data state may change in between the executions of visible transitions; for instance based on temperature, blood pressure or weather events, time, etc. Our semantics abstracts from the timing of such a decision point, however assumes that a stochastic decision between transitions is made given a data state that does not change at the moment of choice. In future work, this could be extended to choices at arbitrary moments.

Example. Figure 1 shows an example of an SLDPN. The control flow of this SLDPN consists of a choice between a and b, followed by a choice between c and d. The transitions are annotated with weight functions: the weight of a and b depend on the continuous variable X, while c and d depend on the categorical variable Y.

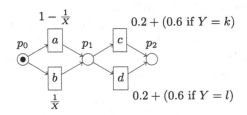

Fig. 1. Example of an SLDPN.

4.1   Trace-Based Execution Semantics & XES Logs

In order to use SLDPNs in a process mining setting, we need to further operationalise the execution semantics. To this end, in this section, we draw links with event logs of the XES standard [2] explicitly. Notice that we assume that the log fits the LPN underlying the SLDPN.

An XES log (XLog) consists of XES traces (XTraces), which are sequences of XES events (XEvents). All XLogs, XTraces and XEvents are annotated with key-value pairs of data attributes. One of the attributes of an XEvent – typically concept:name – is designated as the activity. There are also other attributes indicating the time of occurrence and the identifier of the process case.

The activity sequences A and data sequences D of a trace σ can be directly obtained from XES traces. The initial data state d_0 is obtained from the attributes of the XTrace. Note that in the context of our work typically a selection of considered attributes will need to be made. Only attributes that can be assumed to be available at the start of the process case should be considered; however, XTraces of real-life logs may also contain attributes that are the result of the process case executing (e.g., a decision or outcome of the case).

Subsequent data states $d_{i>0}$ are obtained by updating the previous data state with the values from the numeric attributes of that each of the XEvents provides. The activities $a_{i>0}$ are obtained from the designed activity attribute, which is not used for the data state. In our operationalisation, we assume that this data state represents the data *directly after* the event happened. This is not limiting as the mapping could be adapted for other interpretations. Finally, silent transitions are not observed in event logs; thus, there is no information about the data state at the moment of their execution. Therefore, in our operationalisation, silent transitions do not change the data state.

Example. Table 1 shows an example of an event log. In this log, the attribute X is continuously uniform distributed between 1 and 10, and Y is a categorical attribute of $\{k, l\}$ with equal likelihood. Their distribution is shown in Fig. 3a. The complete log has 10 000 traces.

Table 1. Running example of an event log with two attributes.

Trace attributes	$\langle$event #1,	Event #2$\rangle$
$X = 5.381523$	$\langle a^{X=5.381523,Y=l}$,	$d^{X=5.381523,Y=l}\rangle$
$X = 8.214670$	$\langle a^{X=8.214670,Y=l}$,	$d^{X=8.214670,Y=l}\rangle$
$X = 2.463189$	$\langle b^{X=2.463189,Y=l}$,	$d^{X=2.463189,Y=l}\rangle$
$X = 6.361540$	$\langle a^{X=6.361540,Y=k}$,	$c^{X=6.361540,Y=k}\rangle$
$X = 3.125406$	$\langle a^{X=3.125406,Y=l}$,	$d^{X=3.125406,Y=l}\rangle$
$X = 4.099525$	$\langle b^{X=4.099525,Y=k}$,	$c^{X=4.099525,Y=k}\rangle$
...		

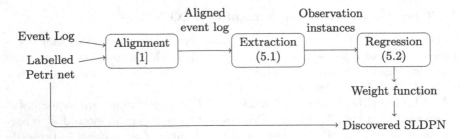

Fig. 2. The proposed method uses an alignment between a Petri net and an event log to extract observation instances for inferring a weight function through regression. This weight function extend the input Petri net to an SLDPN.

5   Data-Based Stochastic Discovery

In this section, we define a method to discover an SLDPN: Data-Based Stochastic Discovery (DSD). DSD takes as input an LPN $N = (P, T, F, \lambda, S_0)$ as well as an event log L, as indicated in Fig. 2. Our discovery method learns the weight function $\mathfrak{w}$ from the activity and data traces observed in the log and yields an SLDPN $= (P, T, F, \lambda, S_0, \mathfrak{w})$.

The weight function needs to be learned based on the data values and transition occurrences observed in the log, i.e., the data sequences σ_Δ and their corresponding activity sequences σ_Σ for each trace $\sigma \in L$. For a transition t, the learned function $\mathfrak{w}(t)$ should return a higher weight for those data states $d \in \Delta$ for which t is more likely to occur compared to other transitions that may be enabled in the same marking.

As shown in Fig. 2, we transform this problem to a regression problem. The first step is to build a set of observation instances (a training set) for each transition t, where each instance is an observation in the log of t being enabled in the LPN, with the corresponding data state. The second step is to fit a regression model to each of the sets observations, and to combine the learned regression models to the weight function of the SLDPN. Both steps are detailed in the remainder of this section.

5.1   Extracting Observation Instances

To extract observation instances for the data traces in a log L and the transitions of an LPN N we firstly relate the observed activity sequences L_Σ to paths of N. Secondly, we relate the observed data states in the data traces L_Δ to sequences of transition firings.

An activity sequence $A \in L_\Sigma$ has no direct correspondence to a path of the LPN: there may be steps required in N that are not present in A, N may contain silent transitions, or there may be activities in A that cannot be mapped to a transition in N. Therefore, we use alignments [1] to establish a mapping between L_Σ and N. That is, each activity $a \in A$ is either mapped to a transition $t \in N$

Table 2. Example of an alignment computed for a data trace and our example LPN (Fig. 1)

(a) Alignment notation.

model (transitions)	a	c
log (activity)	a	$\gg$
log (data)	$X = 10, Y = k \gg$	

(b) Matrix notation.

$$\gamma = \begin{bmatrix} a & c \\ a & \gg \\ X = 10, Y = k \gg \end{bmatrix}$$

such that $a = \lambda(t)$, or to a log move $\gg$. The thus-mapped transitions must form a path of N, and may need intermediate transitions that are not represented in A (model moves $\gg$). An alignment is such a mapping, such that the number of log and model moves is minimised. We provide an example in Table 2, but do not further detail the computation of the alignments; please refer to [1] for more details. Please note that we index the matrix notation starting from 1.

Without loss of generality, we may assume that the alignment γ does not contain column vectors in which only the log has an activity, without the model having a corresponding transition $(\forall_i \gamma(i, 1) = \gg \Rightarrow \gamma(i, 2) \neq \gg)$. That is, that the alignment contains no log moves. From such an alignment γ, we construct a data sequence that corresponds to the followed path, by taking a previous data state if none is present:

$$D(\gamma, 0) = \sigma_{\Delta 0}$$

$$D(\gamma, i \geq 1) = \begin{cases} D(\gamma, i-1) & \text{if } \gamma(i,1) = \gg \vee \lambda(\gamma(i,1)) = \tau \\ \gamma(i,3) & \text{otherwise} \end{cases}$$

Then, we build observation instances for each transition. For a transition $t \in T$, we collect all observations (d, t') of transition t' firing while t was enabled, with the corresponding data state d. That is, here $d \in \Delta$ is the observed data state before transition t' fired. Note that t' may be the same as t. To collect observations, we define an observation instance builder $O_\Gamma(t)$ that provides a multiset of instances from a collection of alignments Γ.

$$O(\Gamma, t) = \biguplus_{\gamma \in \Gamma \wedge \gamma(i,1) = t \wedge t' \in E(S_i)} [(D(\gamma, i-1), t')]$$

with

$$S_{i \geq 1} = \begin{cases} S_{i-1} & \text{if } \gamma(i, 0) = \gg \\ S_{i-1} \uplus \gamma(i,1)^\bullet \setminus {}^\bullet\gamma(i,1) & \text{otherwise} \end{cases} \tag{2}$$

This gives us a multiset of data states with positive and negative samples concerning transition t – that is, t was enabled and fired (positive) or t was enabled but another transition fired (negative). The multiset frequencies also inform on the occurrences of transitions.

Example. From our running example (Fig. 1 and Table 1), consider the data trace $\sigma_e = (\langle a, d \rangle, \langle \{X = 5.381523\}, \{X = 5.381523, Y = l\}, \{X = 5.381523, Y = l\} \rangle)$.

The observation points derived from this data trace are $(\{X = 5.381523\}, a)$ and $(\{X = 5.381523\}, b)$ for a; and $(\{X = 5.381523, Y = l\}, c)$ and $(\{X = 5.381523, Y = l\}, d)$ for d.

5.2 Learning Weight Functions

In this section, we use the multisets of observations to discover weight functions for transitions. This involves two steps for each transition: 1) choosing a weight function $\mathfrak{w}$, and 2) estimating the parameters of the weight function. In principle, any machine learning approach could be used, including regression and classification, that eventually provides a numeric value. The positive or negative cases with their attached data states can be used to learn the chosen weight function. The choice for a weight function $\mathfrak{w}$ also sets the types of variables in the data states that can be supported: in principle, any data type up to images, sound and even video can be supported, as long as there is a weight function available that transforms a datum into a numeric weight.

We do not aim to cover a broad range of possible weight functions, however, in order to illustrate SLDPNs, we consider numeric, categorical and boolean variables, as such variables are typically found as attributes in event logs. As weight function, we choose the simple logistic model with parameters β_0 (the intercept) and $\beta_1, \ldots, \beta_n$ (coefficients). Let $x_1, \ldots, x_n$ be the variables of the data state, then

$$\mathfrak{w}(t) = \frac{1}{1 + e^{-(\beta_0 + \beta_1 x_1 + \ldots + \beta_n x_n)}} \tag{3}$$

As this weight function only supports numerical variables, categorical and boolean variables are included using one-hot encoding. As such, in the remainder of this paper, we only consider numerical variables. Variables that have not been assigned a value, e.g., because they are only observed later in the process, are handled in the learning procedure through mean imputation; to distinguish these cases, an additional variable is recorded that indicates whether the variable has been assigned in the data state.

The use of the simple logistic model also implies that there is no need to consider all transitions together: global approaches could learn the entire weight function for all transitions together. Instead, a local approach learns the weight function for each transition in isolation, thereby limiting the search space considerably.

To estimate the parameters of the simple logistic weight function – one for each transition –, we leverage the observation instances. For each observation instance $(d, t') \in O(\Gamma, t)$ we obtain a data point in our training set as (d, c) with the to-be predicted independent variable c encoded as:

$$c = \begin{cases} 0 & \text{if } t \neq t' \\ 1 & \text{if } t = t' \end{cases}.$$

Using simple logistic regression, the intercept β_0 and a set of coefficients $\beta_1, \ldots, \beta_n$ are fitted.

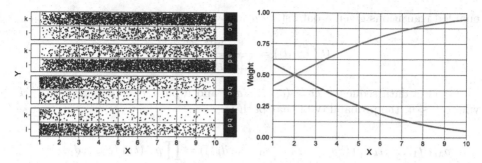

(a) Scatter plot of the example log.　　(b) Plot of the weight function of transitions a and b in the discovered SLDPN.

Fig. 3. Distributions of our running example log and SLDPN.

There may be cases in which we cannot collect observation instances for either the positive or the negative case, such as when no other transition is enabled when t is enabled, when t was never observed, or when none of the variables have been assigned (yet). In these case no sensible logistic weight function can be learned from the data states and we resolve to setting $\mathfrak{w}(t)$ to the support of transition t, i.e., the relative frequency of occurrences of t when it was enabled.

Example. For our running example of (Fig. 1 and Table 1), the regressed parameters for transition a are as follows: The intercept β_0 is -0.716, while the coefficient on X β_1 is 0.359. For b, this is 0.716 and -0.359, respectively. Figure 3b shows that the weight of a and b depend on X, e.g., the weight of b reduces with increasing X. Note that we started the example with a function $1 - \frac{1}{X}$ for transition a, and the fitted logistic function $\frac{1}{1+e^{-(-0.716+0.356X)}}$ on it; this is the best-fitting logistic function, however it may be possible to fit other functions as well.

6 Conformance Checking

In this section, we introduce a technique to check the conformance of an SLDPN and an event log. If the SLDPN was discovered from an event log, preferably, a test log that has not been used in the discovery of the SLDPN should be used for conformance checking. To evaluate the agreement between an SLDPN and a log, we need to compare their respective probability distributions: whereas a trace has a certain probability in a log, an SLDPN expresses a trace having a probability *for a particular data sequence*. In this section, we first derive conditional probabilities for SLDPNs, then for logs, and we finish with a conformance measure.

6.1 Conditional Probabilities in SLDPNs

Given an SLDPN $M = (P, T, F, \lambda, S_0, \mathfrak{w})$ in a marking S, the probability of an enabled transition $t \in E(S)$ to fire can be determined from the weights of all enabled transitions given a data state d following Definition 2, i.e.,

$$p_M(t \mid (S, d)) = \frac{\mathfrak{w}(t, d)}{\sum_{t' \in E(S)} \mathfrak{w}(t', d)}.$$

Given a data sequence $D = \langle d_0, \ldots, d_n \rangle$ and a path $P = \langle t_1, \ldots, t_k \rangle$ of M where $k \leq n$, the probability of that path is

$$p_M(\langle t_1, \ldots, t_k \rangle \mid (\langle S_0, \ldots, S_k \rangle, \langle d_0, \ldots, d_k \rangle)) = \prod_{i=1}^{k} p_M(t_i \mid (S_{i-1}, d_{i-1}))$$

Given a path and the initial marking S_0, the sequence of markings is deterministic (see Eq. (2)). Thus, we may omit the sequence of markings.

However, in conformance checking we need to compare activity sequences rather than paths of transitions. Given an activity sequence A, the conditional probability $p_M(A \mid D)$ of the activity sequence given the data sequence D equals the sum of the probabilities of all paths P such that $P{\downarrow}_\lambda = A$. However, there may be infinitely many corresponding paths for a given activity sequence A, due to duplicate labels, silent transitions and loops. We use the same technique as in [20] to compute the conditional trace probability $p_M(A \mid D)$, which – for bounded SLDPNs – explicitly constructs a state space of the cross product of A and M under assumption of D, and then computes the probability of reaching a deadlock state using standard Markov techniques. Note that the computation requires the data sequence to be at least as long as the longest path taken into consideration, which is easily guaranteed by replicating the last data state in D a sufficient number of times.

Example. From our running example (Fig. 1 and Table 1), consider again the data trace $\sigma_e = (\langle a, d \rangle, \langle \{X = 5.381523\}, \{X = 5.381523, Y = l\}, \{X = 5.381523, Y = l\} \rangle)$. As $\langle a, d \rangle$ is the only path in our SLDPN that corresponds to σ_e, we could directly compute $p_M(\sigma_{e\Sigma} \mid \sigma_{e\Delta})$:

$$p_M(\langle a, d \rangle \mid (\langle [p_0], [p_1], [p_2] \rangle, \sigma_{e\Delta})) = p_M(a \mid ([p_0], \{X = 5.381523\}))$$
$$\cdot\, p_M(d \mid ([p_1], \{X = 5.381523, Y = l\}))$$
$$= \frac{1 - \frac{1}{X}}{\frac{1}{X} + 1 - \frac{1}{X}}$$
$$\cdot\, \frac{0.2 + (0.6 \text{ if } Y = l)}{0.2 + (0.6 \text{ if } Y = k) + 0.2 + (0.6 \text{ if } Y = l)}$$
$$= 0.651$$

To compute this probability when multiple paths would be present, we compute the cross product of the SLDPN and σ_e, which is shown in Fig. 4. The probability of reaching the end state $[p_3]$ from the initial state $[p_0]$ is $0.814 \cdot 0.8 = 0.651$. Thus, the conditional probability $p_M(\sigma_{e\Sigma} \mid \sigma_{e\Delta})$ is 0.651.

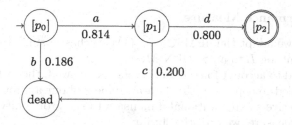

Fig. 4. Cross product of the likelihood of σ_e in our running example.

6.2   Conditional Probabilities in Logs

A log can be seen as a multiset of pairs of an activity sequence A and a data sequence D:

$$L = [(A_0, D_0)^{x_0}, \ldots, (A_n, D_n)^{x_n}].$$

From such a multiset, the probabilities we derive directly are conjunctive. That is, each pair (A, D) is observed a number of times, and the corresponding joint probability concerns both A and D:

$$p_L(A \wedge D) = \frac{L((A, D))}{|L|}$$

The probability of a data sequence is therefore:

$$p_L(D) = \sum_{A \in L_\Sigma} p_L(A \wedge D) = \frac{|[D \mid (A, D) \in L]|}{|L|}$$

Their ratio is the conditional probability of a trace σ given a data sequence D:

$$p_L(A \mid D) = \frac{p_L(A \wedge D)}{p_L(D)}$$

Notice, however, that if D is unique in L, then $p_L(A \mid D) = 1$ for any A, which makes direct comparisons with an SLDPN challenging.

Example. From our running example (Fig. 1 and Table 1), consider again the data trace $\sigma_e = (\langle a, d \rangle, \langle \{X = 5.381523\}, \{X = 5.381523, Y = l\}, \{X = 5.381523, Y = l\} \rangle)$. As X is continuous, the data sequence σ_Σ is unique in our example log. Then:

$$p_L(\sigma_{e\Sigma} \wedge \sigma_{e\Delta}) = {}^1/_{10\,000}$$
$$p_L(\sigma_{e\Delta}) = {}^1/_{10\,000}$$
$$p_L(\sigma_{e\Sigma} \mid \sigma_{e\Delta}) = 1$$

6.3   A Conformance Measure

In this section, we adapt the uEMSC (Eq. (1)) stochastic similarity measure to compare an event log L to an SLDPN M.

Since we need to account for the data sequences as well, the uEMSC measure has to be extended to cope with the data-awareness of our approach. By adding the data perspective as an additional dimension to the probability distributions in the uEMSC measure, we directly obtain:

$$duEMSC(L, M) = 1 - \sum_{D \in L_\Delta} \sum_{A \in \widetilde{L_\Sigma}} \max(p_L(A \wedge D) - p_M(A \wedge D), 0)$$

We can rewrite the joint probabilities using conditional probabilities:

$$p_M(A \wedge D) = p_M(A \mid D)p_M(D)$$

In absence of a data distribution in M, $p_M(D)$ is not defined. However, intuitively, we compare the likelihood of the activity sequences in L ($\widetilde{L_\Sigma}$) with the likelihoods of those activity sequences in M, under the *same* data distribution. Henceforth, we can assume the data distribution of L (L_Δ) for M, and thus $p_M(D) = p_L(D)$. Then, the duEMSC measure results to

$$duEMSC(L, M) = 1 - \sum_{D \in L_\Delta} \sum_{A \in \widetilde{L_\Sigma}} \max(p_L(A \wedge D) - p_M(A \mid D)p_L(D), 0) \quad (4)$$

Notice that if all data sequences in the log are equal, then $duEMSC$ is equal to $uEMSC$, and as such, $duEMSC$ is a proper generalisation of $uEMSC$, and can be used interchangeably.

Example. For our running example (Fig. 1 and Table 1), the overall value of duEMSC is 0.997. This value is not precisely 1, which, given the large sample size of the log (10 000), indicates that a logistic formula is not able to capture the distributions in the log perfectly.

7   Evaluation

In this section, we validate our approach threefold: we show its feasibility using an implementation, we compare the discovered models with existing stochastic process discovery techniques, and we illustrate the new types of insights that can be obtained using SLDPNs.

7.1   Implementation

We implemented discovery and conformance checking methods for SLDPNs as plug-ins of the ProM framework[2], in the StochasticLabelledDataPetriNet

---

[2] Available in the nightly builds at https://promtools.org/.

package. Further functionality for SLDPNs provided by the package are plug-ins to import, export, and visualise and interact with SLDPNs (see Sect. 7.2). The source code is available at http://svn.win.tue.nl/repos/prom/Packages/ StochasticLabelledDataPetriNet/Trunk.

The discovery plug-in first uses the alignments provided by ProM [1] to obtain the observation instances, after which the logistic regression implementation provided by Weka 3.8 [32] based on ridge regression [11] is leveraged for inferring the weight function. A parameter adjusts the one-hot encoding for categorical event log attributes: it sets a maximum on the number of categories that are considered for one-hot encoding for a single variable. Another parameter avoids using one-hot encoding altogether and only considers numerical variables. This is useful to avoid attempting to create a model with a very large number of variables which poses the risk of over-fitting and excessive run times.

The conformance checking plug-in implements $duEMSC$, by extending the EarthMoversStochasticConformance [20] implementation.

7.2 Insights

We illustrate the kind of insights provided by the data-dependant stochastic perspective by presenting an example of a discovered SLDPN on a real-life event log indicating a road fines handling process that is known to contain process relevant data attributes [23]. Using the Directly Follows Model Miner (DFM), an SLDPN was discovered using our ProM plug-in using only numeric attributes.

In the interactive visualisation of our ProM Package the discovered SLDPN can explore influence of data variables on the likelihood of transitions. Figure 5a shows the stochastic perspective for the variable points being 0, while Fig. 5b shows the stochastic perspective for the variable points being 2 with all other variables unchanged. This variable indicates the number of penalty points deducted from the driving license. In total a driver has 20 points and a new driving exam needs to be taken if all points are lost.

One can observe the difference in probability in the highlighted choice between Payment and Send Fine. Here the occurrence of Send Fine indicates that the fine was not directly paid [23]. In the SLDPN, we can observe that if a fine corresponds to 2 penalty points deducted from the license, then it is much less likely that the fine is paid on the spot without being sent out (1%) vs. if the fine does not correspond to any points (36%). These types of insights can be obtained with neither common process mining techniques, nor stochastic process mining techniques, nor data-aware process mining techniques.

7.3 Quantitative

In this experiment, we compare the models of our technique with existing stochastic discovery techniques. Figure 6 shows the set-up of this experiment: from several of real-life logs, we first discover control-flow models. Second, on a random 50% trace-based sample, we apply stochastic discovery techniques, including ours. These stochastic process models are then measured with respect

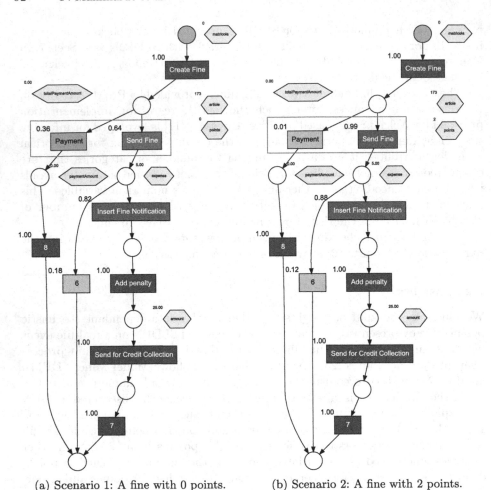

(a) Scenario 1: A fine with 0 points.   (b) Scenario 2: A fine with 2 points.

Fig. 5. An SLDPN discovered by DFM from the road fines event log visualised interactively in ProM. Variables are shown as yellow hexagon shaped nodes with their assignment next to them. The assignment can be changed to investigate the impact of a data state on the transition weights. Transitions are coloured according to their weights. Note that to give a quick overview the marking is not considered. Nodes have been repositioned for better legibility. (Color figure online)

to the remaining 50% of the log. The entire procedure is repeated 10 times to nullify random effects. Table 3 shows the details of the set-up.

To study the impact of using more variables we not only use our technique (DSD), but also include a variant (DSDwe) that does not use one-hot-encoding. The stochastic discovery was bounded by a timeout of 6 h, which was never reached. The experiments were conducted on an AMD EPIC 2 GHz CPU with 100 GB RAM available; the logs were taken from https://data.4tu.nl/search?q=:

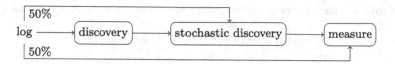

Fig. 6. Set-up of a single quantitative experiment.

Table 3. Details of the quantitative experiment's set-up.

(a) Logs.

Log	traces	events	activities
bpic12-a	6562	30 541	10
bpic13-incidents	3786	32 825	13
bpic13-open problems	412	1 179	5
bpic13-closed problems	748	3 361	7
bpic17-offer log	21 455	96 752	8
bpic20-domestic declarations	5 228	28 108	16
bpic20-international declarations	3 204	35 815	34
bpic20-prepaid travel cost	1 052	9 164	29
bpic20-request for payment	3 447	18 458	18
sepsis	526	7 7615	16
road fines	75 167	280 779	11

(b) Discovery techniques.

Directly Follows Model Miner [19]	DFM
Inductive Miner - infrequent [17] (0.8)	IMf
Flower model: a model that allows for any be-haviour of the observed alphabet	FM

(c) Stochastic discovery techniques.

Baseline: uniform choices	BUC
Alignment-based estimator	ABE
Frequency-based estimator	FBE
Data-based stochastic discovery without one-hot encoding (Section 5)	DSDwe
Data-based stochastic discovery (Section 5)	DSD

(d) Measures.

Number of transitions	transitions
Number of transitions with non-1 weights	weights
Number of transitions with data-dependent weights	data weights
unit Eerth Movers' Stochastic Conformance [20]	uEMSC
Data-aware uEMSC (Section 6)	duEMSC

keyword:%20%22real%20life%20event%20logs%22. We archived the code and the full results at Zenodo[3].

Results. Table 4 summarises the full results that are available in the Zenodo archive. The values obtained by uEMSC for BUC, ABE and FBE were equivalent to the values obtained by duEMSC for these stochastic discovery techniques, as shown in Sect. 6. Therefore, uEMSC is not shown or further discussed.

From these summarised results, it is clear that the data-aware stochastic process discovery techniques can compete with existing stochastic discovery techniques on model quality. In particular, they are – in most cases – able to better represent the behaviour in real-life event logs than existing stochastic discovery techniques. Out of 36 experiments DSD achieves most often the highest duEMSC with 19 runs. Comparing to DSDwe it seems that considering categorical attributes is useful in two cases but has, overall, a limited impact. This motivates future research on using categorical attributes. A potential pitfall is that by adding more variables, or using other regression functions, the likelihood of over-fitting increases, which would lead to lower scores in this experiment. Unsurprisingly the state-of-the-art on non-data-aware discovery ABE is

---

[3] https://dx.doi.org/10.5281/zenodo.7578655.

Table 4. Summary of our quantitative results for 36 experimental runs.

Stochastic algorithm	Fastest	Highest duEMSC
BUC	36	6
FBE	0	9
ABE	0	12
DSDwe	0	17
DSD	0	19

the second best algorithm with several times achieving the same score. Note that in contrast to typical application scenarios we did not investigate or manually select particular attributes for their relevance. Neither did we select event logs for the suitability to data-aware techniques. Thus, it is expected that DSD cannot always achieve better results.

We discuss the Sepsis log in a bit more detail. For IMf, the model contains quite some concurrency, which involves many potential traces, especially with local loops within concurrent blocks. As alignment-based stochastic discovery techniques are not sensitive to concurrent behaviour – they only consider how *often* transitions are executed, not *when* –, all tested stochastic discovery techniques obtain low duEMSC scores. For the DFM miner, the poor performance may be explained by the repeated blood, leucocytes, lactic acid and CRP measurements are taken regularly throughout the process, which makes control-flow without concurrency challenging. Furthermore, they are performed regularly, that is, they are not dependent on data. For the flower model – in theory – any activity that is executed based on data rather than other activities (control flow), should contribute to the stochastic perspective. Hence, the low duEMSC score for all stochastic models shows that the sepsis log describes a structured process.

Figure 7 shows the distribution of stochastic discovery run times in the experiment. We observe that it takes more time to discover an SLDPN compared to the non-data-aware approaches BUC, FBE and ABE. BUC does not consider the log at all and simply assigns a weight of 1 to each transition, which takes very little time. FBE traverses the log, and ABE creates an alignment. Thus, DSD is expected to take at least as long as ABE. Still, all the SLDPNs could be discovered within a maximum of 14 s, which is highly feasible. Please note that for some logs, such as bpic11 and bpic15, alignments are hard to compute, which keeps these logs out of reach for ABE and DSD.

Figure 8 shows the distribution of the stochastic conformance checking run times in our experiment. In the worst case the conformance checking took 573 798 milliseconds for the bpic20-international declarations event log and discovered SLDPN, which took into account 24 variables. Overall, conformance checking of the models discovered by FM takes consistently much longer than their respective IMf and DFM counterparts. However, this difference can also be observed in the non-data-aware approaches. With the exception of bpic20-international declarations, the run times stay in most cases within a limit of 1 to 2 min. Notably, up to 80 GB of RAM was required for these computations.

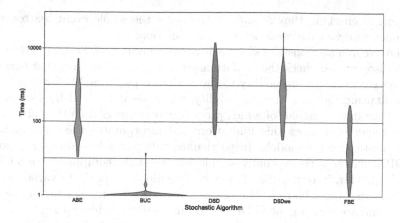

Fig. 7. Run times of the stochastic process discovery.

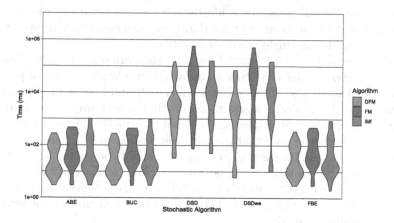

Fig. 8. Run times of stochastic conformance checking for different algorithms.

8   Conclusion

Process models that are typically used in business process management and mining do not incorporate stochasticity: when multiple activities are enabled, no information is incorporated into the model that defines the likelihood of each activity to fire. As a consequence, each activity has the same probability to fire. This is oftentimes not realistic: some activities are more probable than others.

This paper is centered around stochastic process mining, and provides a twofold contribution. On the one hand, it puts forward a technique to discover stochastic models that incorporate a characterization of the probability of each enabled activity to fire. On the other hand, it defines stochastic conformance checking, which do not only aim to verify the compliance of each execution with respect to a model, but also considers whether the distribution of traces in the event log is consistent with the probability distribution of model executions.

Conformance checking thus requires to consider the whole event log together, and cannot analyse each event-log trace in isolation.

Some research also exists in stochastic process mining (cf. Sects. 1 and 2), and aims to discover and check the conformance of stochastic models that correlate the activity occurrence probability to the activities performed beforehand. This is often limiting, because this probability might be influenced by the current values of the data variables of which process executions change the values.

This paper overcomes this limitation and incorporates the data variables into stochastic process models. In particular, this paper introduces the notion of SLDPN, which is conceptually simple but yet fully equipped to model the process' behavior, in terms of activities and manipulation of data variables, and transition firing probabilities. The paper contributes techniques for discovering and conformance checking of SLDPNs. About discovery, the experiments shows that by including relevant data variables into the computation of the firing probability of SLDPN's transitions can yield a more accurate characterization of transition firing probabilities. In conformance checking, the technique follows the intuition of stochastic conformance that computes metrics at event-log level, rather than considering single traces in isolation.

SLDPNs are very suited to model business simulation models [26]. Business Process Simulation enables to generate an arbitrarily large number of potential process executions. It also allows process analysts to implement various process' modifications with the aim to assess their correlation with process performance. By trying several process modifications without putting them in real production, analysts can determine those that improve the process' performance with little or no consequences. As future work, we plan to exploit the technique to discover transition firing probabilities to mine more accurate and realistic simulation models, compared with the state of the art (cf., e.g., [10]). Indeed, more accurate firing probabilities allow analysts to better model the run-time characterisation of business simulation models.

The discovery of the transition firing probabilities builds on logistic regression as an oracle to find the transition's weights and consequently the transition's firing probabilities. This has shown to be beneficial to better compute weights. Logistic regression also has the advantage to naturally explain how weights are computed in each and every case. However, generally it is not the best regression technique in several settings, especially when the variables are correlated. Here, we intend to evaluate alternative regression techniques, including those based on neural networks, with the goal to improve the weight accuracy.

References

1. van der Aalst, W.M.P., Adriansyah, A., van Dongen, B.F.: Replaying history on process models for conformance checking and performance analysis. WIREs Data Mining Knowl. Discov. **2**(2), 182–192 (2012)
2. Acampora, G., Vitiello, A., Stefano, B.N.D., van der Aalst, W.M.P., Günther, C.W., Verbeek, E.: IEEE 1849: the XES standard: the second IEEE standard sponsored by IEEE computational intelligence society [society briefs]. IEEE Comput. Intell. Mag. **12**(2), 4–8 (2017). https://doi.org/10.1109/MCI.2017.2670420

3. Alman, A., Maggi, F.M., Montali, M., Peñaloza, R.: Probabilistic declarative process mining. Inf. Syst. **109**, 102033 (2022)
4. Baier, C., Katoen, J.: Principles of Model Checking. MIT Press, Cambridge (2008)
5. Batoulis, K., Meyer, A., Bazhenova, E., Decker, G., Weske, M.: Extracting decision logic from process models. In: Zdravkovic, J., Kirikova, M., Johannesson, P. (eds.) CAiSE 2015. LNCS, vol. 9097, pp. 349–366. Springer, Cham (2015). https://doi.org/10.1007/978-3-319-19069-3_22
6. Bazhenova, E., Buelow, S., Weske, M.: Discovering decision models from event logs. In: Abramowicz, W., Alt, R., Franczyk, B. (eds.) BIS 2016. LNBIP, vol. 255, pp. 237–251. Springer, Cham (2016). https://doi.org/10.1007/978-3-319-39426-8_19
7. Bergami, G., Maggi, F.M., Montali, M., Peñaloza, R.: Probabilistic trace alignment. In: ICPM, pp. 9–16. IEEE (2021)
8. Burke, A., Leemans, S.J.J., Wynn, M.T.: Stochastic process discovery by weight estimation. In: Leemans, S., Leopold, H. (eds.) ICPM 2020. LNBIP, vol. 406, pp. 260–272. Springer, Cham (2021). https://doi.org/10.1007/978-3-030-72693-5_20
9. Burke, A., Leemans, S.J.J., Wynn, M.T.: Discovering stochastic process models by reduction and abstraction. In: Buchs, D., Carmona, J. (eds.) PETRI NETS 2021. LNCS, vol. 12734, pp. 312–336. Springer, Cham (2021). https://doi.org/10.1007/978-3-030-76983-3_16
10. Camargo, M., Dumas, M., González-Rojas, O.: Automated discovery of business process simulation models from event logs. Decis. Support Syst. **134**, 113284 (2020). https://www.sciencedirect.com/science/article/pii/S0167923620300397
11. le Cessie, S., van Houwelingen, J.: Ridge estimators in logistic regression. Appl. Stat. **41**(1), 191–201 (1992)
12. Di Francescomarino, C., Ghidini, C.: Predictive process monitoring. In: van der Aalst, W.M.P., Carmona, J. (eds.) Process Mining Handbook. LNBIP, vol. 448, pp. 320–346. Springer, Cham (2022). https://doi.org/10.1007/978-3-031-08848-3_10
13. Felli, P., Gianola, A., Montali, M., Rivkin, A., Winkler, S.: CoCoMoT: conformance checking of multi-perspective processes via SMT. In: Polyvyanyy, A., Wynn, M.T., Van Looy, A., Reichert, M. (eds.) BPM 2021. LNCS, vol. 12875, pp. 217–234. Springer, Cham (2021). https://doi.org/10.1007/978-3-030-85469-0_15
14. Gibbs, A.L., Su, F.E.: On choosing and bounding probability metrics. Int. Stat. Rev. **70**(3), 419–435 (2002). https://onlinelibrary.wiley.com/doi/abs/10.1111/j.1751-5823.2002.tb00178.x
15. Hensel, C., Junges, S., Katoen, J., Quatmann, T., Volk, M.: The probabilistic model checker storm. Int. J. Softw. Tools Technol. Transf. **24**(4), 589–610 (2022). https://doi.org/10.1007/s10009-021-00633-z
16. Leemans, S.J.J., van der Aalst, W.M.P., Brockhoff, T., Polyvyanyy, A.: Stochastic process mining: earth movers' stochastic conformance. Inf. Syst. **102**, 101724 (2021). https://doi.org/10.1016/j.is.2021.101724
17. Leemans, S.J.J., Fahland, D., van der Aalst, W.M.P.: Discovering Block-Structured Process Models from Event Logs Containing Infrequent Behaviour. In: Lohmann, N., Song, M., Wohed, P. (eds.) BPM 2013. LNBIP, vol. 171, pp. 66–78. Springer, Cham (2014). https://doi.org/10.1007/978-3-319-06257-0_6
18. Leemans, S.J.J., Maggi, F.M., Montali, M.: Reasoning on labelled petri nets and their dynamics in a stochastic setting. In: Di Ciccio, C., Dijkman, R., del Rio Ortega, A., Rinderle-Ma, S. (eds.) BPM 2022. LNCS, vol. 13420, pp. 324–342. Springer, Cham (2022). https://doi.org/10.1007/978-3-031-16103-2_22

19. Leemans, S.J.J., Poppe, E., Wynn, M.T.: Directly follows-based process mining: exploration & a case study. In: International Conference on Process Mining, ICPM 2019, Aachen, Germany, 24–26 June 2019, pp. 25–32. IEEE (2019). https://doi.org/10.1109/ICPM.2019.00015

20. Leemans, S.J.J., Syring, A.F., van der Aalst, W.M.P.: Earth movers' stochastic conformance checking. In: Hildebrandt, T., van Dongen, B.F., Röglinger, M., Mendling, J. (eds.) BPM 2019. LNBIP, vol. 360, pp. 127–143. Springer, Cham (2019). https://doi.org/10.1007/978-3-030-26643-1_8

21. Leemans, S.J.J., Tax, N.: Causal reasoning over control-flow decisions in process models. In: Franch, X., Poels, G., Gailly, F., Snoeck, M. (eds.) CAiSE 2022. LNCS, vol. 13295, pp. 183–200. Springer, Cham (2022). https://doi.org/10.1007/978-3-031-07472-1_11

22. Mannhardt, F., de Leoni, M., Reijers, H.A., van der Aalst, W.M.P.: Balanced multi-perspective checking of process conformance. Computing **98**(4), 407–437 (2016). https://doi.org/10.1007/s00607-015-0441-1

23. Mannhardt, F., de Leoni, M., Reijers, H.A., van der Aalst, W.M.P.: Decision mining revisited - discovering overlapping rules. In: Nurcan, S., Soffer, P., Bajec, M., Eder, J. (eds.) CAiSE 2016. LNCS, vol. 9694, pp. 377–392. Springer, Cham (2016). https://doi.org/10.1007/978-3-319-39696-5_23

24. Mannhardt, F., de Leoni, M., Reijers, H.A., van der Aalst, W.M.P.: Data-driven process discovery - revealing conditional infrequent behavior from event logs. In: Dubois, E., Pohl, K. (eds.) CAiSE 2017. LNCS, vol. 10253, pp. 545–560. Springer, Cham (2017). https://doi.org/10.1007/978-3-319-59536-8_34

25. Márquez-Chamorro, A.E., Resinas, M., Ruiz-Cortés, A.: Predictive monitoring of business processes: a survey. IEEE Trans. Serv. Comput. **11**(6), 962–977 (2018)

26. Melão, N., Pidd, M.: Use of business process simulation: a survey of practitioners. J. Oper. Res. Soc. **54**(1), 2–10 (2003)

27. Park, G., Song, M.: Prediction-based resource allocation using LSTM and minimum cost and maximum flow algorithm. In: International Conference on Process Mining (ICPM), pp. 121–128 (2019)

28. Polyvyanyy, A., Moffat, A., García-Bañuelos, L.: An entropic relevance measure for stochastic conformance checking in process mining. In: ICPM, pp. 97–104. IEEE (2020)

29. Rogge-Solti, A., van der Aalst, W.M.P., Weske, M.: Discovering stochastic petri nets with arbitrary delay distributions from event logs. In: Lohmann, N., Song, M., Wohed, P. (eds.) BPM 2013. LNBIP, vol. 171, pp. 15–27. Springer, Cham (2014). https://doi.org/10.1007/978-3-319-06257-0_2

30. Rogge-Solti, A., Weske, M.: Prediction of business process durations using non-Markovian stochastic petri nets. Inf. Syst. **54**, 1–14 (2015). https://doi.org/10.1016/j.is.2015.04.004

31. Rozinat, A., van der Aalst, W.M.P.: Decision mining in ProM. In: Dustdar, S., Fiadeiro, J.L., Sheth, A.P. (eds.) BPM 2006. LNCS, vol. 4102, pp. 420–425. Springer, Heidelberg (2006). https://doi.org/10.1007/11841760_33

32. Witten, I.H., Frank, E., Hall, M.A.: Data Mining: Practical Machine Learning Tools and Techniques, 3rd edn. Morgan Kaufmann, Elsevier, Amsterdam (2011)

Exact and Approximated Log Alignments
for Processes with Inter-case Dependencies

Dominique Sommers[✉], Natalia Sidorova, and Boudewijn van Dongen

Department of Mathematics and Computer Science,
Eindhoven University of Technology, Eindhoven, The Netherlands
{d.sommers,n.sidorova,b.f.v.dongen}@tue.nl

Abstract. The execution of different cases of a process is often restricted by inter-case dependencies through e.g., queueing or shared resources. Various high-level Petri net formalisms have been proposed that are able to model and analyze coevolving cases. In this paper, we focus on a formalism tailored to conformance checking through alignments, which introduces challenges related to constraints the model should put on interacting process instances and on resource instances and their roles. We formulate requirements for modeling and analyzing resource-constrained processes, compare several Petri net extensions that allow for incorporating inter-case constraints. We argue that the Resource Constrained ν-net is an appropriate formalism to be used the context of conformance checking, which traditionally aligns cases individually failing to expose deviations on inter-case dependencies. We provide formal mathematical foundations of the globally aligned event log based on theory of partially ordered sets and propose an approximation technique based on the composition of individually aligned cases that resolves inter-case violations locally.

Keywords: Petri nets · Conformance checking · Inter-case dependencies · Shared resources

1 Introduction

Event logs record which activity is executed at which moment of time, and additionally they often include indications which resources were involved in which activity, mentioning the exact person(s) or machine(s). The availability of such event logs enables the use of conformance checking for resource-constrained processes, analyzing not only the single instance control-flow perspective, but also checking whether and where the actual process behavior recorded in an event log deviates from the resource constraints prescribed by a process model.

Process models, and specifically Petri nets with their precise semantics, are often used to describe and reason about the execution of a process. In many approaches, a process model considers a process instance (a case) in isolation from other cases [1]. In practice, however, a process instance is usually subject to interaction with other cases and/or resources, whose availability puts additional constraints on the process execution. In order to expose workflow deviations caused by inter-case dependencies, it is crucial to use models considering multiple cases simultaneously.

L. Gomes and R. Lorenz (Eds.): PETRI NETS 2023, LNCS 13929, pp. 99–119, 2023.
https://doi.org/10.1007/978-3-031-33620-1_6

There are several approaches to modeling and analysis of processes with inter-case dependencies. In [7] and [12], Petri nets are extended with resources to model availability of durable resources, with multiple cases competing by claiming and releasing these *shared* resources. To distinguish the cases, ν-Petri nets [22] incorporate name creation and management as a minimal extension to classical Petri nets, with the advantage that coverability and termination are still decidable, opposed to more advanced Petri net extensions. The functionality of ν-Petri nets is inherited in other extensions such as Catalog Petri nets [11], synchronizing proclet models [10], resource and instance-aware workflow nets (RIAW-nets) [18], DB-nets [19] and resource constrained ν-Petri nets [24], all with the ability to handle multiple cases simultaneously. For the latter, the cases are assumed to follow the same process, interacting via (abstract) shared resources in a one-to-many relation, i.e., a resource instance can be claimed by one case at a time. More sophisticated extensions allow for cases from various perspectives with many-to-many interactions, via e.g., concepts from databases, shared resources and proclet channels. This may impose, however, problems of undecidability during conformance checking, which we discuss in this work.

Many conformance checking techniques use *alignments* to expose where the behavior recorded in a log and the model agree, which activities prescribed by the model are missing in the log and which log activities should not be performed according to the model [3,8]. The usual focus is on the control flow of the process. In more advanced techniques [6,15–17], data and/or resource information is additionally incorporated in the alignments by considering these perspective only after the control flow [15], by balancing the different perspectives in a customizable manner [16] or by considering all perspectives at once [17]. These three types of techniques operate on a case-by-case basis, which can lead to misleading results in case of shared resources, e.g., when multiple cases claim the same resource simultaneously.

In our previous work we considered the execution of all process instances by aligning the complete event log to a resource constrained ν-Petri nets [24]. In this paper, we present our further steps: (1) We compare how the existing Petri net extensions support modeling and analysis of processes with inter-case dependencies by formulating the requirements to such models, and we argue that ν-nets are an appropriate formalism. (2) We employ the poset theory to provide mathematical foundations for aligning the complete event log and exposing deviations of inter-case dependencies; (3) We propose an approximation method for computing optimal alignments in practice, which tackles the limitation of the computational efficiency when computing the complete event log alignment. The approximation method is based on composing alignments for isolated cases first and then resolving inter-case conflicts and deviations in the log locally.

The paper is organized as follows. In Sect. 2 we introduce basic concepts of the poset theory, Petri nets and event logs. In Sect. 3 we compare different Petri net extensions. We provide the mathematical foundations of the complete event log alignment in Sect. 4. Section 5 presents the approximation method for computing alignments. We discuss implications of our work in Sect. 6.

2 Preliminaries

In this section, we introduce basic concepts related to Petri nets and event logs and present the notations that we will use throughout the paper.

2.1 Multisets and Posets

We start with definitions and notation regarding multisets and partially ordered sets.

Definition 1 *(Multiset). A multiset m over a set X is $m : X \to \mathbb{N}$. $X^{\oplus}$ denotes the set of all multisets over X. We define the support $supp(m)$ of a multiset m as the set $\{x \in X \mid m(x) > 0\}$. We list elements of the multiset as $[m(x) \cdot x \mid x \in X]$, and write $|x|$ for $m(x)$, when it is clear from context which multiset it concerns.*

For two multisets m_1, m_2 over X, we write $m_1 \leq m_2$ if $\forall_{x \in X} m_1(x) \leq m_2(x)$, and $m_1 < m_2$ if $m_1 \leq m_2 \wedge m_1 \neq m_2$. We define $m_1 + m_2 = [(m_1(x) + m_2(x)) \cdot x \mid x \in X]$, and $m_1 - m_2 = [\max(0, m_1(x) - m_2(x)) \cdot x \mid x \in X]$ for $m_1 \geq m_2$.

Furthermore, $m_1 \sqcup m_2 = [\max(m_1(x), m_2(x)) \cdot x \mid x \in X]$, $m_1 \sqcap m_2 = [\min(m_1(x), m_2(x)) \cdot x \mid x \in X]$.

In some cases, we consider multisets over a set X as vectors of length $|X|$, assuming an arbitrary but fixed ordering of elements of X.

Definition 2 *(Partial order, Partially ordered set, Antichains). A partially ordered set (poset) $X = (\bar{X}, \prec_X)$ is a pair of a set $\bar{X}$ and a partial order $\prec_X \subseteq X \times X$. We overload the notation and write $x \in X$ if $x \in \bar{X}$. For $x, y \in X$, we write $x \|_X y$ if $x \not\prec y \wedge y \not\prec x$ and $x \preceq y$ if $x \prec y \vee x = y$.*

Given $\prec_X$, we define $\prec_X^+$ to be the smallest transitively closed relation containing $\prec_X$. Thus $\prec_X^+$ is a partial order with $\prec_X \subseteq \prec_X^+$.

We extend the standard set operations of union, intersection, difference and subsets to posets: for any two posets X and Y, $X \circ Y = (\bar{X} \circ \bar{Y}, (\prec_X \circ \prec_Y)^+)$, with $\circ \in \{\cup, \cap, \setminus\}$ and $Y \subseteq X$ iff $\bar{Y} \subseteq \bar{X}$ and $\prec_Y = \prec_X \cap (\bar{Y} \times \bar{Y})$.

A poset A is an antichain if no elements of A are comparable, i.e., $\forall_{x,y \in A} x \|y$. For poset X, $\mathcal{A}(X)$ denotes the set of all antichains $A \subseteq X$, and $\mathcal{A}^+(X)$ is the set of all maximal antichains: $\mathcal{A}^+(X) = \{A \mid A \in \mathcal{A}(X), \forall_{B \in \mathcal{A}(X)} B \subseteq A \implies B = A\}$.

Two special maximal antichains are the minimum and maximum elements of X, defined by $\min(X) = \{x \mid x \in X, \forall_{y \in X} y \not\prec x\} \in \mathcal{A}^+(X)$ and $\max(X) = \{x \mid x \in X, \forall_{y \in X} x \not\prec y\} \in \mathcal{A}^+(X)$.

We define $X^< = \{(\bar{Y}, \prec_Y) \mid \bar{Y} = \bar{X}, \prec_X \subseteq \prec_Y, \forall_{a,b \in Y, a \neq b} a \not\|_Y b\}$ to be the set of totally ordered permutations of X that respect the partial order.

Definition 3 *(Interval, prefix and postfix in a poset). With a poset X and two antichains $A, B \in \mathcal{A}(X)$, the closed jhkcbvinterval from A to B is the subposet defined as follows: $[A, B] = (\overline{AB}, \prec_X \cap (\overline{AB} \times \overline{AB}))$ with $\overline{AB} = \{x \mid x \in X, A \preceq x \preceq B\}$, and the half open and open intervals: $(A, B] = [A, B] \setminus A$, $[A, B) = [A, B] \setminus B$ and $(A, B) = [A, B) \setminus A$.*

Artificial minimal and maximal elements are denoted as $\bot$ and $\top$ respectively, i.e., $\forall_{x \in X} \bot \prec x \prec \top$. $(\bot, A]$, $(\bot, A)$, $[A, \top)$ $(A, \top)$ denote the corresponding prefixes and postfixes of an antichain $A \in \mathcal{A}(X)$ in X.

2.2 Petri Nets

Petri nets can be used as a tool for the representation, validation and verification of workflow processes to provide insights in how a process behaves [21].

Definition 4 *(Labeled Petri nets, Pre-set, Post-set). A labeled Petri net [20] is a tuple $N = (P, T, \mathcal{F}, \ell)$, with sets of places and transitions P and T, respectively, such that $P \cap T = \emptyset$, and a multiset of arcs $\mathcal{F} : (P \times T) \cup (T \times P) \to \mathbb{N}$ defining the flow of the net. $\ell : T \to \Sigma^\tau = \Sigma \cup \{\tau\}$ is a labeling function, assigning each transition t a label $\ell(t)$ from alphabet Σ or $\ell(u) = \tau$ for silent transitions.*

We assume that the intersection, union and subsets are only defined for two labeled Petri nets N_1, N_2 where $\forall_{t \in T_1 \cap T_2} \ell_1(t) = \ell_2(t)$.

Given an element $x \in P \cup T$, its pre- and post-set $^\bullet x$ ($x^\bullet$) are multisets defined by $^\bullet x = [\mathcal{F}(y, x) \cdot y \mid y \in P \cup T]$ and $x^\bullet = [\mathcal{F}(x, y) \cdot y \mid y \in P \cup T]$ resp.

Definition 5 *(Marking, Enabling and firing of transitions, Reachable markings). A marking $m \in P^\oplus$ of a (labeled) Petri net $N = (P, T, \mathcal{F}, \ell)$ assigns how many tokens each place contains and defines the state of N.*

With m and N, a transition $t \in T$ is enabled for firing iff $m \geq {}^\bullet t$. We denote the firing of t by $m \xrightarrow{t} m'$, where m' is the resulting marking after firing t and is defined by $m' = m - {}^\bullet t + t^\bullet$. For a transition sequence $\sigma = \langle t_1, \ldots, t_n \rangle$ we write $m \xrightarrow{\sigma} m'$ to denote the consecutive firing of t_1 to t_n. We say that m' is reachable from m and write $m \xrightarrow{*} m'$ if there is some $\sigma \in T^*$ such that $m \xrightarrow{\sigma} m'$.

$\mathcal{M}(N) = P^\oplus$ and it denotes the set of all markings in net N and $\mathcal{R}(N, m)$ the set of markings reachable in net N from marking m.

Definition 6 *(Place invariant). Let $N = (P, T, \mathcal{F}, \ell)$ be a Petri net. A place invariant [14] is a row vector $I : \mathbf{P} \to \mathbb{Q}$ such that $I \cdot \mathbf{F} = \mathbf{0}$, with $\mathbf{P}$ and $\mathbf{F}$ vector representations of P and $\mathcal{F}$. We denote the set of all place invariants as $\mathcal{I}_N$, which is a linear subspace of $\mathbb{Q}^P$.*

The main property of a place invariant I in a net N with initial marking m_i is that $\forall_{m_1, m_2 \in \mathcal{R}(N, m_i)} I \cdot m_1 = I \cdot m_2$.

Definition 7 *(Net system, Execution poset and sequence, Language). A net system is a tuple $SN = (N, m_i, m_f)$, where N is a (labeled) Petri net, and m_i and m_f are respectively the initial and final marking. An execution sequence in a net system $SN = (N, m_i, m_f)$ is a firing sequence from m_i to m_f. Additionally, an execution poset is a poset of transition firings, where each totally ordered permutation is a firing sequence. The language of a net system SN is the set of all execution sequences in SN.*

2.3 Event Logs

An event log records activity executions as events including at least the occurred activity, the time of occurrence and the case identifier of the corresponding case. Often resources are also recorded as event attributes, e.g., the actors executing the action. It is generally known beforehand in which activities specific *resource roles R* are involved and which resource instances Id_r are involved in the process for each role $r \in R$. We assume that each resource has only one role (function) allowing to execute a predefined number of tasks, and therefore define the set Id_R of resource instances of all roles as the disjoint union of resource instance sets of roles: $\mathrm{Id}_R = \biguplus_{r \in R} \mathrm{Id}_r$. A *resource instance* $\rho \in \mathrm{Id}_R$ with role $r \in R$ is equipped with capacity, making Id_r and Id_R both multisets.

Definition 8 *(Event, Event log, Trace). An* event *e is a tuple* $(a, t, c, \mathrm{Id}'_R)$*, with an* activity $a = \mathrm{activity}(e) \in \Sigma$*, a* timestamp $t = time(e) \in \mathbb{R}$*, a* case identifier $c = \mathrm{case}(e) \in \mathrm{Id}_c$ *and a multiset of resource instances* $\mathrm{Id}'_R = \mathrm{Res}(e) \leq \mathrm{Id}_R$*. Such an event represents that activity a occurred at timestamp t for case c and is executed by resource instances from* Id'_R *belonging to possibly different resource roles.*

An event log *L is a set of events with partial order* $\prec_L$ *that respects the chronological order of the events, i.e.,* $\forall_{e_1, e_2 \in L} time(e_1) < time(e_2) \implies e_2 \nprec_L e_1$*. An event log can be partitioned into* traces*, defined as projections e.g., on the case identifiers or on the resources names. For every* $c \in \mathrm{Id}_c$*,* L_c *denotes a trace projected on the case identifier c defined by* $L_c = (\{e \mid e \in L, \mathrm{case}(e) = c\}, \prec_{L_c})$ *with* $\prec_{L_c} = \{(e, e') \mid (e, e') \in \prec_L, \mathrm{case}(e) = \mathrm{case}(e') = c\}$*.*

Alternatively, we write $\langle e_1, e_2, \cdots \rangle$ for an event log which is totally ordered, and $a^{\mathrm{Id}'_R}$ and $a^{\mathrm{Id}'_R}$ for events where the case is identified by the activity color (and bar position) and the time of occurrence is abstracted away from.

For a (labeled) Petri net modeling a process, the transitions' names or labels correspond to the activity names found in the recorded event log.

3   Modeling, Analysis and Simulation of Case Handling Systems with Inter-case Dependencies

A classical Petri net models a process execution using transition firings and the corresponding changes of markings without making distinctions between different cases on which the modeled system works simultaneously. To create a case view, Workflow nets [2] model processes from the perspective of a single case. Systems in which cases interact with each other, e.g., by queueing or sharing resources, need to be modeled in a different way. We show from a modeling point how this boils down to multiple cases competing over shared tokens representing resources in a Petri net, which requires an extension on the formalism of the classical Petri nets. In Sect. 3.1, we motivate the requirements by providing examples, after which, in Sect. 3.2, we discuss whether existing Petri net extensions satisfy these requirements. We end, in Sect. 3.3 by proposing a *minimal extension* based on ν-Petri nets [22] that meets each requirement for simulation and analysis of resource-constrained processes.

3.1   Requirements Imposed by Inter-case Dependencies

When modeling systems with inter-case dependencies, i.e., shared resources, simultaneous cases can interfere in each other's processing via the resources, causing inter-case dependencies. To model, simulate and analyze such behavior, the cases and resources,

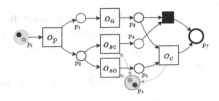

Fig. 1. Example Petri net N_1 to argue the requirements, with token colors denoting different instances.

represented as tokens in a Petri net, should be handled together and simultaneously in the process model. This introduces the need for case (R1) and resource isolation (R2) as well as durable resources (R3) and case-resource correlations (R4), which regular Petri nets are not capable of. For analysis, like computing alignments (see Sect. 4), non-invertible functions can cause state-space explosions (R5). We show for each requirement, when not satisfied, how simulation and/or analysis concerning multiple simultaneous cases fails:

R1 *Distinguishable cases* are required when dealing with multiple cases. Tokens involved in a firing of a transition should not belong to different cases, unless case batching is used. Mixing tokens from different cases, possible in classical Petri nets, can potentially cause model behavior that is not possible in the modeled system: Suppose we have a simple operation process modeled by Petri net N_1, shown in Fig. 1, where a patient undergoes an operation involving the activities of preparation (o_p), assistance (o_a), closed surgery (o_{sc}) and open surgery (o_{so}) which is followed by closeup (o_c). We assume case tokens to be indistinguishable. The language of $(N_1, [p_i, 2p_s], [p_f, 2p_s])$ is $\{\langle o_p, o_a, o_{sc}\rangle, \langle o_p, o_{sc}, o_a\rangle, \langle o_p, o_a, o_{so}, o_c\rangle, \langle o_p, o_{so}, o_a, o_c\rangle\}$ and the language of the same net processing two cases with sufficient resources has to consist of all possible interleaving of two traces belonging to single cases. However, $\{\langle o_p, o_a, o_{sc}, \underline{o_p}, \underline{o_a}, \underline{o_{so}}, \mathbf{o_c}\rangle\}$ is included in the language of $(N_1, [2p_i, 2p_s], [2p_f, 2p_s])$, which is impossible to obtain by an interleaving of two single cases, as o_c is never enabled after o_{sc} fires. Here and later we use underlined symbols when referring to the second case in examples. From now on, we assume case tokens are distinguishable and we have $m_i(p_i) = (c, \underline{c})$;

R2 *Distinguishable resources* are required when resource instances are uniquely identifiable. If the tokens in p_s are indistinguishable, $\langle \ldots, o_{so}^{\{x\}}, \underline{o_{sc}}^{\{x\}}, o_c^{\{x\}}\rangle$ belongs to the language of $(N_1, [2p_i, 2p_s], [2p_f, 2p_s])$. However, resource instance x can only be claimed by the second case after it has been released by the first case (by firing transition o_c), hence it should not be included in the language. From now on, we assume resource tokens are distinguishable and we have $m_i(p_s) = (x, y)$;

R3 Resources are required to be *durable* when having a variable number of cases in the system simultaneously. In N_1, the resource instances in p_s are modeled to be durable, since these instances are always released after being claimed. However, were arc (o_c, p_s) to be removed, problems arise when observed behavior concerns more than two cases, since after transition o_{so} fired twice, it is never enabled again, causing a deadlock;

R4 Capturing *case-resource correlation* is required when dealing with multiple distinguishable cases and resources in order to keep track of which resource handles
which case. Without it, the language of $(N_1, [2p_i, 2p_s], [2p_f, 2p_s])$ includes e.g.,
$\langle \ldots, o_{so}^{\{x\}}, \underline{o_{so}}^{\{y\}}, \underline{o_c}^{\{x\}}, o_c^{\{y\}} \rangle$, which is undesirable as resources x and y have
switched cases after transition o_{so} is fired twice. Case-resource correlation should
ensure, in this case, that transition o_c can only be fired using the same resource as
was claimed by firing transition o_{so};
R5 Operations on token values (e.g., guards, arc inscriptions) should be *invertible* and
computable when aligning observed and modeled behavior in order to keep the
problem decidable. Consider e.g., that patients enter the process by their name
and birthdate v, which is transformed to an identifier c in the first transition by an
operation $f(v)$ on (o_p). When activity o_p is missing for a patient, it is undecidable
which value v should be inserted for the firing of o_p when f is not invertible.

3.2 Existing Petri Net Extensions

Several extensions on Petri nets have been proposed focusing on multi-case and/or
multi-resource processes able to handle (some) inter-case dependencies. We go over
each extension, describing how they satisfy (and violate) requirements listed in
Sect. 3.1. We propose an extension, which combines concepts of the described extensions and satisfies all requirements.

Resource constrained workflow nets (RCWF-nets) [12] are Petri nets extended with
resource constraints, where resources are durable units: they are claimed and then
released again (R3). They define structural criteria for its correctness.

Definition 9 *(Resource-constrained workflow net [12]). Let R be a set of* resource
roles. *A net system* $N = (P_p \uplus P_r, T, \mathcal{F}_p \uplus \mathcal{F}_r, m_i, m_f)$ *is a* resource-constrained
workflow net *(RCWF-net)* with the set P_p of production places *and the set* $P_r = \{p_r \mid$
$r \in R\}$ of resource places *iff*

- $\mathcal{F}_p : (P_p \times T) \cup (T \times P_p) \to \mathbb{N}$ *and* $\mathcal{F}_r : (P_r \times T) \cup (T \times P_r) \to \mathbb{N}$;
- $N_p = (P_p, T, \mathcal{F}_p, [m_i(p) \cdot p \mid p \in P_p], [m_f(p) \cdot p \mid p \in P_p])$ *is a net system, called
the production net of N.*

The semantics of Petri nets is extended by having colored tokens on production places
(R1) and as resources are shared across all cases, tokens on resource places are colorless
($\neg$R2, $\neg$R4). A transition is enabled if and only if there are sufficient tokens on its
incoming places using tokens of the same color on production places.

ν-Petri nets [22] are an extension of Petri nets with pure name creation and name
management, strictly surpassing the expressive power of regular Petri nets and they
essentially correspond to the minimal object-oriented Petri nets of [13]. In a ν-Petri
net, names can be created, communicated and matched which can be used to deal with
authentication issues [23], correlation or instance isolation [9]. Name management is
formalized by replacing ordinary tokens by distinguishable ones, thus adding color the
Petri net.

Definition 10 *(ν-Petri net [22]). Let Var be a fixed set of variables. A ν-Petri net is a tuple ν-N = ⟨P, T, $\mathcal{F}$⟩, with a set of places P, a set of transitions T with P∩T = ∅, and a flow function $\mathcal{F}$: (P×T)∪(T×P) → Var$^{\oplus}$ such that $\forall_{t \in T}$, Υ∩•t = ∅ ∧ t•\Υ ⊆ •t, where* •t = $\bigcup_{p \in P}$ supp($\mathcal{F}$(p,t)) *and* t• = $\bigcup_{p \in P}$ supp($\mathcal{F}$(t,p)). Υ ⊂ Var *denotes a set of special variables ranged by ν, ν$_1$, . . . to instantiate fresh names.*

A marking of ν-N is a function m : P → Id$^{\oplus}$. Id(m) *denotes the set of names in* m, *i.e.* Id(m) = $\bigcup_{p \in P}$ supp(m(p)).

A mode μ of a transition t is an injection μ : Var(t) → Id, *that instantiates each variable to an identifier.*

For a firing of transition t with mode μ, we write m $\xrightarrow{t_\mu}$ m'. t *is enabled with mode* μ *if* μ($\mathcal{F}$(p,t)) ⊆ m(P) *for all* p ∈ P *and* μ(ν) ∉ Id(m) *for all* ν ∈ Υ ∩ Var(t) = supp(∪$_{p \in P}\mathcal{F}$(p,t)). *The reached state after the firing of t with mode μ is the marking* m', *given by:*

$$m'(p) = m(p) - \mu(\mathcal{F}(p,t)) + \mu(\mathcal{F}(t,p)) \text{ for all } p \in P \tag{1}$$

We denote T$_\mu$ to be the set of all possible transition firings.

ν-Petri nets support instance isolation for cases and resources requiring the tokens involved in a transition firing to have matching colors (R1, R2). Due to the tokens having singular identifiers, correlation between cases and resources can not be captured (¬R4).

Resource and instance-aware workflow nets (RIAW-nets) [18], are Petri nets combining the notions from above by defining similar structural criteria for handling resource constraints on top of ν-Petri nets. However, the resource places are assumed to only carry black tokens, not allowing for resource isolation and properly capturing the case-resource correlation.

Synchronizing proclets [10] are a type of Petri net that describe the behavior of processes with many-to-many interactions: unbounded dynamic synchronization of transitions, cardinality constraints limiting the size of the synchronization, and history-based correlation of token identities (R1,R2). This correlation is captured by message-based interaction, specifying attributes of a message as correlation attributes (R4). The correlation constraints are C_{init}, $C_{match}^{\subseteq}$ and $C_{match}^{=}$, for initializing the attributes, partially and fully matching them. ν-Petri nets are at the basis of proclets handling multiple objects by separating their respective subnets. While the proclet formalism is sufficient for satisfying all requirements listed above, they extend to many-to-many relations, which lifts the restriction that a resource can only be claimed by a single case.

Object-centric Petri nets [4], similarly to synchronizing proclets, describe the behavior of processes with multiple perspectives and one-to-many and many-to-many relations between the different object types. These nets are a restricted variant of colored Petri nets where places are typed, tokens are identifiable referring to objects (R1,R2), and transitions can consume and produce a variable number of tokens. Correlation can be achieved with additional places of combined types (R4). Again, due to many-to-many relations, our one-to-many restriction on resources is lifted.

Database Petri nets (DB-nets) [19] are extensions of ν-Petri nets with multi-colored tokens that allows for multiple types of objects and their correlation (R1,R2,R4). Additionally, they support underlying read-write persistent storage consisting of a relational database with full-fledged constraints. Special "view" places in the net are used to inspect the content of the underlying data, while transitions are equipped with database update operations. These are in the general sense not invertible causing undecidability (¬R5).

Catalog Petri nets (CLog-nets) [11] are similar to DB-nets, but without the "write" operations (R1,R2,R4). The queries from view places in DB-nets have been relocated to transition guards, relying solely on the "read-only" modality for a persistent storage, however suffering from the same undecidability problem as these guards are not invertible in the general sense (¬R5).

3.3 Resource Constrained ν-Petri Net with Fixed Color Types

We combine conceptual ideas from the extensions described above, by extending RIAW-nets, which inherit the modeling restrictions from RCWF-nets and name management from ν-Petri nets, using concepts from DB-nets and CLog-nets.

The resource places from RCWF-nets model the availability of resource instances by tokens, which is insufficient to capture correlation of cases by which they are claimed and released. We propose a minimal extension *resource constrained ν-Petri nets* (RC ν-net) which additionally contain busy places $\bar{P}_r = \{\bar{p}_r \mid r \in R\}$ for each resource role. Token moves from p_r to $\bar{p}_r$ show that the resource gets occupied, and moves from $\bar{p}_r$ to p_r show that the resource becomes available. Also tests whether there are free/occupied resources can be modeled. A structural condition is imposed on the net to guarantee that resources are *durable*, meaning that resources can neither be created nor destroyed. This also implies that in the corresponding net system with initial and final marking m_i and m_f, $m_i(p_r) = m_f(p_r)$ and $m_i(\bar{p}_r) = m_f(\bar{p}_r)$, for any resource role $r \in R$.

Furthermore, similar to DB-nets and CLog-nets, we extend the tokens from carrying single data values to multiple. Where DB-nets and CLog-nets allow for a variable number of predefined color types, we restrict ourselves to two which are strictly typed, to distinguish between both cases and resources.

Definition 11 *(Resource-constrained ν-Petri net). Let C^ε be the set of case ids Id_c extended with ordinary tokens, i.e., $\varepsilon \in \mathrm{Id}_c$, and $\mathrm{Id}_R^\varepsilon$ be the set of resource ids extended with ordinary tokens. A resource-constrained ν-Petri net $N = (P, T, \mathcal{F}, m_i, m_f)$ is a Petri net system with $\mathcal{F} : (P \times T) \cup (T \times P) \to (Var_c^\varepsilon \times Var_r^\varepsilon)^\oplus$, where Var_c denote case variables and Var_r denote resource variables, allowing for two colored tokens. $P = (P_p \uplus P_r \uplus \bar{P}_r)$, with production places P_p and resource availability and busy places $P_r = \{p_r \mid r \in R\}$ and $\bar{P}_r = \{\bar{p}_r \mid r \in R\}$. The following modeling restrictions are imposed on N for each $r \in R$:*

1. *${}^\bullet p_r + {}^\bullet \bar{p}_r = p_r^\bullet + \bar{p}_r^\bullet$, i.e., $\forall_{t \in T}\ \mathcal{F}(p_r, t) + \mathcal{F}(\bar{p}_r, t) = \mathcal{F}(t, p_r) + \mathcal{F}(t, \bar{p}_r)$;*
2. *$m_i(p_r) = m_f(p_r)$ and $m_i(\bar{p}_r) = m_f(\bar{p}_r) = 0$;*

A marking of N is a function $m : P \to (C^\varepsilon \times R^\varepsilon)^\oplus$ with case ids C and resources R, which is a mapping from places to multisets of colored tokens.

A mode of a transition t is an injection $\mu : (Var_c^\varepsilon \times Var_r^\varepsilon)(t) \to (C^\varepsilon \times R^\varepsilon)$, that instantiates each variable to an identifier.

Proposition 1. *The resource-constrained ν-Petri nets as defined in Definition 11 satisfy requirements R1-R5, i.e., they allow to* distinguish cases *and* resource instances *which are* durable, *and capture* case-resource correlation *while restricting to operations that are* invertible.

Proof. The two-colored strictly typed tokens distinguish both the cases (R1) and resource instances (R2) in the system. The modeling restrictions imposed on the RC ν-net enforce that for each resource role $r \in R$, tokens can only move between p_r and $\bar{p}_r$, i.e., we have the place invariant $(1, 1)$ on p_r and $\bar{p}$, implying that $m(p_r) + m(\bar{p}_r) = m_i(p_r)$ for any reachable marking m, and that all resource tokens are returned to p_r when the net reaches its final marking, ensuring that resources are durable (R3). The two colors on tokens residing in $\bar{p}$ capture correlation between cases and resources instances (R4), denoting by which case a resource instance is claimed throughout their interaction. As the transition firing's modes are bijective functions, each operation on N is invertible (R5).   □

Note that the RC ν-net formalism is a restricted version of DB-nets, CLog-net and synchronizing proclets, as all three can capture the behavior that can be modeled by RC ν-nets. DB-nets and CLog-nets additionally have database operations which we deem not relevant for our purposes. Synchronizing proclets allow for many-to-many interactions, while we assume that a resource instance cannot be shared by several cases at the same time.

4   Complete Event Logs Alignments

Several state-of-the-art techniques in conformance checking use alignments to relate the recorded executions of a process with a model of this process [5]. An alignment shows how a log or trace can be replayed in a process model, which can expose deviations explaining either how the process model does not fit reality or how the reality differs from what should have happened.

Traditionally, this is computed for individual traces, however, as we show in previous work [24], this fails to expose deviations on a multi-case and -resource level in processes with inter-case dependencies as described in Sect. 3.3. In this section, we go over the foundations of alignments in Sect. 4.1 and show how we extend this to compute alignments of complete event logs in Sect. 4.2.

4.1   Foundations of Alignments

At the core of alignments are three types of moves: log, model, and synchronous moves (cf. Definition 12), indicating, respectively, that an activity from the log can not be mimicked in the process model, that the model requires the execution of some activity not observed in the log, and that observed and modeled behavior of an activity agree.

Definition 12 *(Log, model and synchronous moves). Let L be an event log and $N = (P, T, \mathcal{F}, \ell, m_i, m_f)$ be a labeled ν-Petri net with T_μ the set of all possible firings in N. We define the set of log moves $\Gamma_l = \{(e, \gg) \mid e \in L\}$, the set of model moves $\Gamma_m =$*

$\{(\gg, t_\mu) \mid t_\mu \in T_\mu\}$ and the set of synchronous moves $\Gamma_s = \{(e, t_\mu) \mid e \in L, t_\mu \in T_\mu, \text{activity}(e) = \ell(t)\}$. As abbreviations, we write $\Gamma_{ls} = \Gamma_l \cup \Gamma_s$, $\Gamma_{lm} = \Gamma_l \cup \Gamma_m$, $\Gamma_{ms} = \Gamma_m \cup \Gamma_s$, and $\Gamma_{lms} = \Gamma_l \cup \Gamma_m \cup \Gamma_s$.

Log moves and model moves can expose deviations of the real behavior from the model, by an *alignment* (cf. Definition 13) on a net (N, m_i, m_f) and event log L (possibly a single trace) which is a poset of moves from Definition 12 incorporating the event log and execution sequences in N from m_i to m_f:

Definition 13 *(Alignment). An alignment* $\gamma = \text{align}(N, L)$ *of an event log* $L = (\bar{L}, \prec_L)$ *) and a labeled Petri net* $N = (P, T, \mathcal{F}, \ell, m_i, m_f)$ *is a poset* $\gamma = (\bar{\gamma}, \prec_\gamma)$*, where* $\bar{\gamma} \subseteq (\Gamma_l \cup \Gamma_s \cup \Gamma_m^\oplus)$*, having the following properties:*

1. $\overline{\gamma\restriction_L} = \bar{L}$ *and* $\prec_L \subseteq \prec_{\gamma\restriction_L}$
2. $m_i \xrightarrow{\gamma\restriction_T} m_f$*, i.e.,* $\forall_{\sigma \in (\gamma\restriction_T)^<}, m_i \xrightarrow{\sigma} m_f$

with alignment projections on the log events $\gamma\restriction_L$ *and on the transition firings* $\gamma\restriction_{T_\mu}$*:*

$$\gamma\restriction_L = (\{e \mid (e, t_\mu) \in \gamma \cap \Gamma_{ls}\}, \{(e, e') \mid ((e, t_\mu), (e', t'_\mu)) \in \prec_\gamma \cap (\Gamma_{ls} \times \Gamma_{ls})\}) \quad (2)$$

$$\gamma\restriction_T = (\{t_\mu \mid (e, t_\mu) \in \gamma \cap \Gamma_{ms}\}, \{(t_\mu, t'_\mu) \mid ((e, t_\mu), (e', t'_\mu)) \in \prec_\gamma \cap (\Gamma_{ms} \times \Gamma_{ms})\}) \quad (3)$$

Note the slight difference in the definition of an alignment as opposed to our previous work in [24], where the alignment is simplified from a distributed run to a poset of moves. The process's history of states (markings) as it has supposedly happened in reality can be extracted from the alignment. For the general case, we introduce the *pseudo-firing* of transitions from corresponding alignment's non-log moves in the process model, to obtain a pseudo-marking, which can be unreachable or contain a negative number of tokens:

Definition 14 *(Pseudo-markings). A pseudo-marking* m *of a Petri net* $N = (P, T, \mathcal{F})$ *is a multiset* $P \to \mathbb{Z}$*, i.e., the assigned number of tokens a place contains can be negative.* $\widetilde{\mathcal{M}}(N)$ *denotes the set of all pseudo-markings in* N*.*

Definition 15 *(Pseudo-firing of posets). Let* $N = (P, T, \mathcal{F}, m_i, m_f)$ *be a RC ν-net and* γ *an alignment on* N*. We define a function* $\widetilde{m} : \mathcal{P}(\gamma) \to \widetilde{\mathcal{M}}(N)$*, with powerset* $\mathcal{P}$*, to obtain the model pseudo-marking of every subposet of* γ*. For every subposet* $\gamma' \subseteq \gamma$*, we have for every* $p \in P$*:*

$$\widetilde{m}(\gamma')(p) = m_i(p) + \sum_{(e, t_\mu) \in \gamma': t_\mu \neq \epsilon} (\mu(\mathcal{F}(t, p)) - \mu(\mathcal{F}(p, t))) \quad (4)$$

i.e., the pseudo-marking is obtained by firing all the transitions of γ' *with corresponding modes. Note that it is not necessarily reachable.*

An antichain in an alignment denotes a possible point in time, and therefore a state of the process. By pseudo-firing the respective (open) prefix of the antichain, we obtain the corresponding *pre- (or post-)antichain marking*:

Definition 16 *(Pre- and post-antichain marking). Let* γ *be an alignment and* $G \in \mathcal{A}(\gamma)$ *an antichain in* γ*. The pre- (post-)antichain marking defines the marking reached after the pseudo-firing of* $(\bot, G)$ $((\bot, G])$*, i.e.,* $\widetilde{m}((\bot, G))$ $(\widetilde{m}((\bot, G]))$*.*

4.2 Alignments Extended to Include Inter-case Dependencies

The foundational work on constructing alignments is presented in [5] and it relies on the synchronous product of the Petri net $N = (P, T, \mathcal{F}, \ell, m_i, m_f)$ modeling a process and a trace Petri net $N_\sigma = (P^{(\sigma)}, T^{(\sigma)}, \mathcal{F}^{(\sigma)}, \ell^{(\sigma)}, m_i^{(\sigma)}, m_f^{(\sigma)})$ (a Petri net representation of a trace in the event log). The synchronous product consists of the union of N and N_σ, and a transition t_s for each pair of transitions $(t_m, t_l) \in T \times T^{(\sigma)}$ with ${}^\bullet t_s = {}^\bullet t_m + {}^\bullet t_l$ and $t_s^\bullet = t_m^\bullet + t_l^\bullet$, iff t_m and t_l share the same label and variables on the incoming arcs, i.e., $\ell(t_m) = \ell^{(\sigma)}(t_l)$ and $Var(t_m) = Var(t_l)$. The alignment is then computed by a depth-first search on the synchronous product net from $m_i + m_i^{(\sigma)}$ to $m_f + m_f^{(\sigma)}$ using the A^* algorithm, with the firings of transition from $T^{(\sigma)}, T$ and $T^{(s)}$ corresponding to the log, model and synchronous moves from Definition 12 [5].

With $c : \Gamma_{lms} \to \mathbb{R}^+$ a cost function, usually defined for each $(e, t_\mu) \in \Gamma_{lms}$ as follows:

$$c((e, t_\mu)) = \begin{cases} 0 & (e, t_\mu) \in \Gamma_s \\ 1 & (e, t_\mu) \in \Gamma_{lm} \land \ell(t) \neq \tau \\ \epsilon & \ell(t) = \tau \end{cases} \tag{5}$$

The *optimal alignment* is an alignment γ such that $\sum_{g \in \gamma} c(g) \leq \sum_{g \in \gamma'} c(g)$ holds for any alignment γ', which prefers synchronous moves over model and log moves. In terms of conformance checking and exposing realistic deviations, the optimal alignment provides the "best" explanation for the relation between observed and modeled behavior.

In Sect. 3.3, we have shown how a RC ν-net is a Petri net formalism with capability of modeling inter-case dependencies and suitability for conformance checking. We extend the alignment problem in order to expose inter-case deviations by adapting the synchronous product net to ν-nets: an RC ν-net and the log ν-net:

Definition 17 (*Log ν-Petri net*). *Given an event log L, a log ν-Petri net $N^{(L)} = (P^{(L)}, T^{(L)}, \mathcal{F}^{(L)}, \ell^{(L)}, m_i^{(L)}, m_f^{(L)})$ is a labeled ν-net constructed as follows. For every $e \in L$, we make a transition $t_e \in T^{(L)}$ with $\ell^{(L)}(t) = $ activity(e), and for each resource instance $\rho_r \in supp(Res(e))$ we make a place $p \in P^{(L)}$ with ${}^\bullet p = \emptyset$, ${}^\bullet p = [|\rho_r| \cdot t]$, $\mathcal{F}^{(L)}(p, t) = [|\rho_r| \cdot (\varepsilon, r)]$ and $m_i^{(L)}(p)((\varepsilon, \rho)) = |\rho|$. Further, for every pair $(e_1, e_2) \in \prec_L$, we make a place $p \in P^{(L)}$ with ${}^\bullet p = [t_{e_1}], p^\bullet = [t_{e_2}]$ and*

$$\mathcal{F}^{(L)}(t_{e_1}, p) = \mathcal{F}^{(L)}(p, t_{e_2}) = \begin{cases} [(c, \varepsilon)] & case(e_1) = case(e_2) \\ [(\varepsilon, \varepsilon)] & otherwise \end{cases} \tag{6}$$

For every $e^- \in \min(L)$, we make a place $p^- \in P^{(L)}$ with ${}^\bullet p^- = \emptyset, p^{-\bullet} = [t_{e^-}]$ and $m_i^{(L)}(p^-)((case(e^-), \varepsilon)) = 1$. Similarly, for every $e^+ \in \max(L)$, we make a place $p^+ \in P^{(L)}$ with ${}^\bullet p^+ = [t_{e^+}], p^{+\bullet} = \emptyset$ and $m_f^{(L)}(p^+)((case(e^+), \varepsilon)) = 1$.

Computing the *complete event log alignment* is again a matter of finding a path from the initial to the final marking in the synchronous product net, i.e., from $m_i + m_i^{(L)}$ to $m_f + m_f^{(L)}$, for which we can use *any* of the existing methods as described before. The optimal alignment is again the one with lowest cost. In terms of complexity, the alignment problem with an empty event log and an all-zero cost function can be reduced to the reachability problem for bounded Petri nets from m_i to m_f, which has exponential worst-case complexity [20]. Adding event to the log ν-Petri net and a non-zero costs on moves makes the problem strictly more complex.

Note that while ν-Petri nets are inherently unbounded in general due to the generation of fresh tokens, we can retain boundedness in the context of alignments, since the bound is predicated by the event log and we can get this information by preprocessing it.

For our running example, modeled in Fig. 2, we extend the small operation process from Fig. 1 with an assistant resource during the operation, an intake subprocess (i_s, i_p) involving a general practitioner (GP), and a prescription subprocess with a FIFO waiting room (p_w, w_e, w_l, p_r), where the prescription can only be written by the GP involved in the intake, if appropriate. Both the intake and operation subprocesses can be skipped via silent transitions τ_1 and τ_3 respectively in N. Figure 3 shows the recorded event log L of this process which concerns two patients. An optimal complete event log alignment on N and L, computed by the method above is presented in Fig. 4.

5   Approximation by Composition and Local Realignments

Since multiple cases are executed in parallel, computing the alignment on the complete event log L, as described in Sect. 4, is a computationally expensive task. At the same time, one can see that the multi-case and -resource alignment only deviates from the classical individual alignments when violations occur on the inter-case dependencies, e.g., when a resource is claimed while it is already at maximal capacity.

We can approximate the alignment of a complete event log L and a Petri net N by using a composition of individually aligned cases. An overview of this method is illustrated in Fig. 7, which we subdivide into two parts, described respectively in Sects. 5.1 and 5.2.

1. L is decomposed into the individual cases (L_c, L_c), which are aligned to N (γ_c, γ_c) and composed using the event log's partial order $\prec_L$ $(\widetilde{\gamma})$. The result is not necessarily an alignment as inter-case deviations may be left unresolved;
2. We transform this composed alignment into a valid alignment by taking a permutation $(\widetilde{\gamma}')$ and realigning parts $([A_1, B_1], [A_2, B_2], [A_3, B_3])$ of the event log locally to resolve the violations. The approximated alignment (γ^*) is obtained by substituting the realignments $(\gamma_{AB_1}, \gamma_{AB_2}, \gamma_{AB_3})$.

The implementation of both the original method from [24] and the approximation method for computing complete event log alignments is available at gitlab.com/dominiquesommers/mira, including the examples used in this paper and some additional examples.

5.1   Composing Individual Alignments

For every case $c \in \mathrm{Id}_c$, we have the trace L_c (cf. Definition 8) projected on the case identifier c. As described in Sect. 4, the optimal complete event log alignment γ_L consists of individual alignments γ_c, on N and L_c for every $c \in \mathrm{Id}_c$, composed together

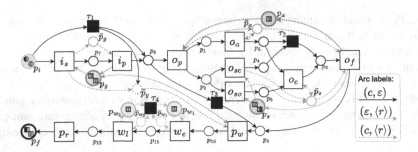

Fig. 2. Process model RC ν-net N, with initial and final marking, annotated with circular and square tokens respectively.

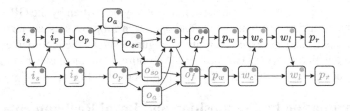

Fig. 3. Event log L.

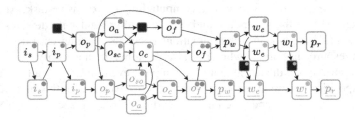

Fig. 4. Complete event log alignment γ, with the colors depicting the move types; green, purple, and yellow for synchronous, model, and log moves respectively. (Color figure online)

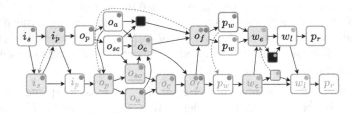

Fig. 5. Composed alignment $\tilde{\gamma}$ with annotated permutation and realignment intervals.

respecting the event log's partial order $\prec_L$, where each γ_c is not necessarily optimal with regard to L_c.

It is computationally less expensive to compute the optimal alignments $\gamma_c = \text{align}(N, L_c)$ for each $c \in \text{Id}_c$ and then approximate γ_L. We create a composed

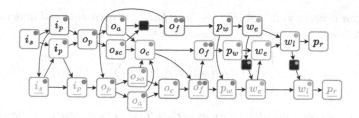

Fig. 6. Approximated alignment γ^*.

$$(N, m_i, m_f) \overbrace{\qquad\qquad\qquad}$$

$$L \left\{ \begin{array}{l} L_c \\ L_c \end{array} \right. \begin{array}{l} \gamma_c \\ \gamma_c \end{array} \right\} \tilde{\gamma} \to \tilde{\gamma}' \left\{ \begin{array}{l} [A_1, B_1] \to \gamma_{AB_1} \\ [A_2, B_2] \to \gamma_{AB_2} \\ [A_3, B_3] \to \gamma_{AB_3} \end{array} \right\} \gamma^*$$

Fig. 7. Overview of our approximation method.

alignment $\tilde{\gamma}$ with the optimal individual alignments and the event log's partial order, as defined in Definition 18. Figure 5 shows the composed alignment for the running example with additional annotations (in red) which we cover later.

Definition 18 (*Composed alignment*). *Given a Petri net N and an event log L with traces L_c for $c \in \mathrm{Id}_c$, let $\gamma_c = \mathrm{align}(N, L_c)$ be the corresponding optimal individual alignments. The* composed alignment $\tilde{\gamma} = \uplus_{c\in\mathrm{Id}_c}\gamma_c$ *is the union of individual alignments with the extended partial order on the synchronous moves, defined as the transitive closure of the union of partial orders from the individual alignments and the partial order on moves imposed by the partial order $\prec_L$ of the event log:*

$$\prec_{\tilde{\gamma}} = \left(\bigcup_{c\in\mathrm{Id}_c} \prec_{\gamma_c} \cup \prec_{\gamma_L} \right)^+ \tag{7}$$

with $\prec_{\gamma_L} = \{((e, t_\mu), (e', t'_\mu)) \mid e \prec_L e', (e, t_\mu), (e', t'_\mu) \in (\gamma \cap \Gamma_{ls})\}.$

Recall that for every sequence $\sigma \in \tilde{\gamma}|_T$ of an alignment $\tilde{\gamma}$, we have $m_i \xrightarrow{\sigma} m_f$, i.e., σ is a firing sequence in N. This property is not guaranteed for a composed alignment, even in the absence of inter-case deviations. In the presence thereof, we say that a composed alignment is *violating* as there exists no such sequence.

Definition 19 (*Violating composed alignment*). *Let $\rho_r \in supp(\mathrm{Id}_R)$ be a resource instance and $\tilde{\gamma} = \uplus_{c\in\mathrm{Id}_c}\gamma_c$ a composed alignment. We define*

$$\mathcal{S}(\tilde{\gamma}) = \{(\bar{\tilde{\gamma}}', \prec_{\tilde{\gamma}'}) \mid \tilde{\gamma} = \bar{\tilde{\gamma}}', \prec_{\tilde{\gamma}} \subseteq \prec_{\tilde{\gamma}'}, \prec_{\tilde{\gamma}'} = (\prec_{\tilde{\gamma}'})^+, \forall_{g\in\tilde{\gamma}} \not\prec_{\tilde{\gamma}'} g\} \tag{8}$$

as the set of transitively closed and acyclic antichain permutations of $\tilde{\gamma}$ that respect the partial order $\prec_{\tilde{\gamma}}$.

   $\tilde{\gamma}$ *is in violation with any of the resource instances if and only if:*

$$\forall_{\tilde{\gamma}'\in\mathcal{S}(\tilde{\gamma})} \exists_{G\in\mathcal{A}^+(\tilde{\gamma}')} \mathrm{viol}(G) \tag{9}$$

with violation criteria viol : $\mathcal{A}^+(\widetilde{\gamma}) \to \mathbb{B}$ *defined for each maximal antichain* $G \in \mathcal{A}^+(\widetilde{\gamma})$ *as follows:*

$$\text{viol}(G) = \exists_{\rho_r \in supp(\text{Id}_R)} \left[\widetilde{m}((\bot, G))(p_r)((\varepsilon, \rho_r)) < \sum_{(e, t_\mu) \in G} \mathcal{F}(p_r, t)(\mu^{-1}((\varepsilon, \rho_r))) \right]$$

$$(10)$$

i.e., there is no way of firing all transitions in the alignment such that at all times enough capacity is available.

In Fig. 5, antichains meeting the violation criteria are the single moves with an incoming red arc. In Theorem 1 we show that for every sequence of transitions $\sigma \in \widetilde{\gamma}|_T$ in violating composed alignment $\widetilde{\gamma}$, we have $m_i \not\xrightarrow{} m_f$, i.e., $\widetilde{\gamma}$ is not firable.

Theorem 1. *(A violating composed alignment is not firable) Let $\widetilde{\gamma} = \mathbb{U}_{c \in \text{Id}_c} \gamma_c$ be a composed alignment on RC ν-net $N = (P, T, \mathcal{F}, m_i, m_f)$ and event log L, such that $\widetilde{\gamma}$ is violating. Then there exists no firing sequence σ in $\widetilde{\gamma}$ such that $m_i \xrightarrow{\sigma} m_f$.*

Proof. $\widetilde{\gamma}$ is violating, therefore, for every $\widetilde{\gamma}' \in \mathcal{S}(\widetilde{\gamma})$, there is a maximal antichain $G \in \mathcal{A}^+(\widetilde{\gamma}')$ and resource instance $\rho_r \in supp(\text{Id}_R)$, such that

$$\widetilde{m}((\bot, G))(p_r)((\varepsilon, \rho_r)) < \sum_{(e, t_\mu) \in G} \mathcal{F}(p_r, t)(\mu^{-1}((\varepsilon, \rho_r)))$$

$$(11)$$

$$\widetilde{m}((\bot, G))(p_r)((\varepsilon, \rho_r)) - \sum_{(e, t_\mu) \in G} \mathcal{F}(p_r, t)(\mu^{-1}((\varepsilon, \rho_r))) < 0$$

$$(12)$$

hence firing the transitions in G leads to a negative marking for (ε, ρ_r) in place p_r, which is invalid.   □

With an antichain $G \subseteq \mathcal{A}(\widetilde{\gamma})$, we show in Lemma 1 that $\widetilde{m}((\bot, G))$ (and $\widetilde{m}(\bot, G]$) is reachable if an only if the prefix $(\bot, G)$ $((\bot, G])$ is not violating.

Lemma 1. *(A pre- (and post-)antichain marking in a composed alignment is reachable iff the corresponding prefix is not violating). Let $\widetilde{\gamma} = \mathbb{U}_{c \in \text{Id}_c} \gamma_c$ be a composed alignment on RC ν-net $N = (P, T, \mathcal{F}, m_i, m_f)$ and event log L and let $G \in \mathcal{A}(\widetilde{\gamma})$ be an antichain. Then the pre- (and post-)antichain marking $\widetilde{m}((\bot, G))$ ($\widetilde{m}((\bot, G])$) is reachable if and only if $(\bot, G)$ $((\bot, G])$ is not violating.*

Proof. We prove the lemma by proving both sides of the bi-implication:

($\implies$) $m_G = \widetilde{m}((\bot, G))$ is reachable, hence there exists a sequence $\sigma \in (\bot, G)^*$ with $\prec_{(\bot, G)} \subseteq \prec_\sigma$ such that $m_i \xrightarrow{\sigma} m_G$. Let $\widetilde{\gamma}' \in \mathcal{S}(\widetilde{\gamma})$ be an antichain permutation with $\prec_\sigma \subseteq \prec_{\widetilde{\gamma}'}$. Then by definition of reachable marking, for every maximal antichain $G \in \mathcal{A}^+(\widetilde{\gamma}')$ and every resource instance $\rho_r \in supp(\text{Id}_R)$, we have $\widetilde{m}((\bot, G))(p_r)((\varepsilon, \rho_r)) \geq \sum_{(e, t_\mu) \in G} \mathcal{F}(p_r, t)(\mu^{-1}((\varepsilon, \rho_r)))$. Thus $(\bot, G)$ is not violating.

($\impliedby$) $(\bot, G]$ is not violating, hence there exists a $\widetilde{\gamma}' \in \mathcal{S}((\bot, G])$, such that for all $G' \in \mathcal{A}^+(\widetilde{\gamma}')$ and all $\rho \in supp(\text{Id}_R)$ we have:

$$\widetilde{m}((\bot, G))(p_r)((\varepsilon, \rho_r)) \geq \sum_{(e, t_\mu) \in G} \mathcal{F}(p_r, t)(\mu^{-1}((\varepsilon, \rho_r)))$$

$$(13)$$

$m_i \xrightarrow{\sigma} \widetilde{m}((\bot, G])$ with σ respecting the partial order $\prec_{\widetilde{\gamma}'}$.   □

5.2 Resolving Violations in the Composed Alignment

Let $\tilde{\gamma}' \in \mathcal{S}(\gamma)$ be an antichain permutation of $\tilde{\gamma}$. Then, by Definition 19, we have a set of violating maximal antichains (which is empty when $\tilde{\gamma}$ is not violating) where the corresponding transitions are not enabled. Instead of needing to align the complete event log, we show that we can resolve violations locally around such antichain. For each violating antichain G, there exists an interval $[A, B] \subseteq \tilde{\gamma}'$ with $A \preceq G \preceq B$ such that $[A, B]$ is alignable, formally defined in Definition 20.

Definition 20 (*Alignable interval*). *Let $\gamma = \uplus_{c \in \mathrm{Id}_c} \gamma_c$ be a composed alignment on RC ν-net $N = (P, T, \mathcal{F}, m_i, m_f)$ and event log L, and let $A, B \in \mathcal{A}(\gamma)$ be two antichains. We say that the interval $[A, B]$ is alignable if and only if $m_B = \widetilde{m}((\bot, B])$ is reachable from $m_A = \widetilde{m}((\bot, A))$, i.e., $m_A \xrightarrow{*} m_B$, assuming m_A is reachable.*

Note that $[\min(\gamma'), \max(\gamma')]$ is always an alignable interval. We use our running example to show that it can be taken locally around G instead, e.g., $[\{i_s\}, \{i_p\}]$ with $G = \{i_s\}$ (cf. Fig. 5). Note how the violation can be resolved by substituting $[A, B]$ by a subalignment from $m_A = \widetilde{m}((\bot, A))$ to $m_B = \widetilde{m}((\bot, B])$.

In order to prove statements that do not depend on a chosen realignment mechanism, we now assume that there exists a function $f_{\tilde{\gamma}} : \mathcal{A}^+(\tilde{\gamma}) \rightarrow \mathcal{P}(\tilde{\gamma})$ that produces an alignable interval $[A, B]$ for an arbitrary $G \in \mathcal{A}^+(\tilde{\gamma})$.

$$W(\tilde{\gamma}'_V) = \{[\min(\gamma_v), \max(\gamma_v)] \mid \gamma_v \subseteq \tilde{\gamma}'_V, \forall_{g \in \gamma_v, g' \in \tilde{\gamma}'_V \setminus \gamma_v} g \|_{\tilde{\gamma}'_V} g', \quad (14)$$

$$\forall_{g \in \gamma_v} \exists_{g' \in \gamma_v} g \not\|_{\gamma_v} g'\}$$

with $\tilde{\gamma}'_V = \bigcup_{G \in \mathcal{A}^+(\tilde{\gamma}')} f_{\tilde{\gamma}'}(G)$, denotes the set of alignable intervals covering every violating antichain in $\tilde{\gamma}'$, and it is annotated in red for the running example in Fig. 5, with the three intervals $[\{i_s\}, \{i_p\}]$, $[\{o_p\}, \{o_f\}]$, and $[\{w_e\}, \{\tau\}]$ covering the violating antichains $\{i_s\}$, $\{o_p\}$, $\{o_{so}\}$, and $\{w_e\}$.

We resolve the violations in $\tilde{\gamma}'$ by substituting every interval $[A, B] \in W(\tilde{\gamma}'_V)$ by an alignment γ_{AB} on N and $[A, B]\upharpoonright_L$ from $m_A = \widetilde{m}((\bot, A))$ to $m_B = \widetilde{m}((\bot, B])$.

Since, for now, we assume that every interval $f(G)$ is alignable, a subalignment γ_{AB} exists. The approximated alignment $\gamma^* = (\tilde{\gamma}^*, \prec_{\gamma^*})$ is then defined as follows:

$$\tilde{\gamma}^* = \bigcup_{[A,B] \in W(\tilde{\gamma}'_V)} \tilde{\gamma}_{AB} \cup (\tilde{\gamma} \setminus \tilde{\gamma}'_V) \quad (15)$$

$$\prec_{\gamma^*} = \left(\bigcup_{[A,B] \in W(\tilde{\gamma}'_V)} \prec_{\gamma_{AB}} \cup \{(g_1, g_2) \mid g_1, g_2 \in \gamma \setminus \tilde{\gamma}'_V, g_1 \prec_{\tilde{\gamma}'} g_2\} \right)^+ \quad (16)$$

γ^* for the running example is shown in Fig. 6 with substituted realignments for the intervals annotated in red from Fig. 5. Note that γ^* is an approximation of the optimal alignment γ from Fig. 4 as $c(\gamma^*) \geq c(\gamma)$, due to the local realignments. In Theorem 2 we show that γ^* is a valid alignment.

Theorem 2 (γ^* *is an alignment*). *Let $\tilde{\gamma} = \uplus_{c \in \mathrm{Id}_c} \gamma_c$ be a composed alignment on RC ν-net $N = (P, T, \mathcal{F}, m_i, m_f)$ and event log L and let $\tilde{\gamma}' \in \mathcal{S}(\tilde{\gamma})$ be an antichain* *permutation of* $\widetilde{\gamma}$, *with* $W(\widetilde{\gamma}'_V)$ *the set of alignable intervals covering every violating antichain in* $\widetilde{\gamma}'$.

$\gamma^* = (\bar{\gamma}^*, \prec_{\gamma^*})$, *following Eqs. 15 and 16, is a valid alignment, i.e., it has properties (1), (2) and (3) from Definition 13.*

Proof. We prove that γ^* is an alignment by induction on the size of $W(\widetilde{\gamma}'_V)$. For the base case with $|W(\widetilde{\gamma}'_V)| = 0$, we have $\bar{\gamma}^* = \widetilde{\bar{\gamma}}$ and $\prec_{\gamma^*} = \prec_{\widetilde{\gamma}'}$. By definition, $\widetilde{\gamma}\!\restriction_L = \bar{L}$ and $\prec_L \subseteq \prec_{\widetilde{\gamma}\restriction_L}$. Furthermore, since $|W(\widetilde{\gamma}'_V)| = 0$, we know that for all $G \in \mathcal{A}^+(\widetilde{\gamma}')$, we have $\neg \operatorname{viol}(G)$, implying that $m_i \xrightarrow{\widetilde{\gamma}'} m_f$.

Let us assume that γ^* is an alignment for $|W(\widetilde{\gamma}'_V)| = w$. We prove the statement for $W'(\widetilde{\gamma}'_V) = W(\widetilde{\gamma}'_V) \cup \{[A, B]\}$ with $|W'(\widetilde{\gamma}'_V)| = w+1$ and $[A, B] \in \min(W'(\widetilde{\gamma}'_V))$. For every maximal antichain $G \in \mathcal{A}^+((\bot, A))$ before A, i.e., $G \prec A$, we have $\neg \operatorname{viol}(G)$, which we prove by contradiction. Assume $\operatorname{viol}(G)$, then by our assumption of the existence of $f_{\widetilde{\gamma}'}$, there is an alignable interval $[A', B'] \subseteq \widetilde{\gamma}'$ with $A' \preceq G \preceq B'$, thus, by $G \prec A$, we have $[A', B'] \prec [A, B]$, implying that $[A, B] \notin \min(W'(\widetilde{\gamma}'))$ which is a contradiction. By Lemma 1 and the assumption that $f_{\widetilde{\gamma}'}(G)$ is an alignable interval, $m_i \xrightarrow{*} m_A \xrightarrow{*} m_B$ and $[A, B]$ can be substituted by γ_{AB} without violations in $(\bot, B]$, completing the proof. $\qquad\square$

5.3 Obtaining Minimal Local Alignable Intervals

We propose a method to find an antichain permutation of a composed alignment $\widetilde{\gamma}$ together with the intervals $W(\widetilde{\gamma}_V)$ such that all violations can be resolved by realigning these intervals as described in Sect. 5.2. For computational efficiency, we choose to minimize the number of moves in the intervals that need to be realigned.

We formulate this as an Integer Linear Programming (ILP) problem. The objective of the ILP problem is to adjust the partial order of $\widetilde{\gamma}$, such that alignable intervals can be identified around violating antichains, preferring intervals with fewer moves.

Let there be a (possibly arbitrary) fixed order in $\widetilde{\gamma}$ and Id_R such that each element has a unique index, i.e., for every $1 \le i \le n_{\widetilde{\gamma}}$, $\widetilde{\gamma}(i)$ and $(e(i), t_\mu(i))$ both denote the ith move in $\widetilde{\gamma}$, with $n_{\widetilde{\gamma}} = |\widetilde{\gamma}|$. Furthermore, for every $1 \le j \le n_r$, $\operatorname{Id}_R(j)$ denotes the jth resource instance, with $n_r = |supp(\operatorname{Id}_R)|$.

Let $\mathbf{R}$ be a $n_{\widetilde{\gamma}} \times n_{\widetilde{\gamma}}$ matrix, with $\mathbf{R}$ defined for every two indices $1 \le i, j \le n_{\widetilde{\gamma}}$ such that $\mathbf{R}_{ij}$ is a binary value denoting $(\widetilde{\gamma}(i), \widetilde{\gamma}(j)) \in \prec_{\widetilde{\gamma}}$. For each $c \in \operatorname{Id}_c$, we introduce the set I_c of indices corresponding to moves in $\widetilde{\gamma}\!\restriction_{\gamma_c}$. Furthermore, we use $[1..n] = \{1, \ldots, n\}$ as an abbreviation for the set of all indices from 1 to n.

The set of minimal alignable intervals containing all violations, denoted by $W(\widetilde{\gamma}'_V)$, with $\widetilde{\gamma}'_V$ given by

$$\widetilde{\gamma}'_V = \bigcup_{i,j \in [1..n_{\widetilde{\gamma}}]:\mathbf{X}_{ij} - \mathbf{R}_{ij} = 1} [\widetilde{\gamma}(j), \widetilde{\gamma}(i)] \tag{17}$$

where $\mathbf{X}$ denotes the new partial order relation between alignment moves which respects the resources capacities and provides the solution to

$$\text{Minimize} \sum_{i,j \in [1..n_{\widetilde{\gamma}}]} (1 - \mathbf{R}_{ij})\mathbf{R}_{ji}\mathbf{X}_{ij} + \epsilon \cdot (1 - \mathbf{R}_{ij})(1 - \mathbf{R}_{ji})\mathbf{X}_{ij} \tag{18}$$

subject to

$$\forall_{i,j\in[1..n_{\widetilde{\gamma}}]} \qquad\qquad\qquad\qquad \mathbf{X}_{ij} \in \{0,1\} \qquad (19)$$

$$\forall_{c\in\mathrm{Id}_c}\forall_{i,j\in I_c} \qquad\qquad\qquad\qquad \mathbf{X}_{ij} = \mathbf{R}_{ij} \qquad (20)$$

$$\forall_{i,j\in[1..n_{\widetilde{\gamma}}]} \qquad \mathbf{R}_{ij} + (1 - \mathbf{X}_{ij}) - \mathbf{X}_{ji} \leq 1 \qquad (21)$$

$$\forall_{i,j,k\in[1..n_{\widetilde{\gamma}}]} \qquad \mathbf{X}_{ij} + \mathbf{X}_{jk} - \mathbf{X}_{ik} \leq 1 \qquad (22)$$

$$\forall_{i\in[1..n_{\widetilde{\gamma}}]} \qquad (1 - \mathbf{X}_{i\bullet})\mathbf{C}^{\downarrow} - \mathbf{X}_{\bullet i}^{T}\mathbf{C}^{\uparrow} \leq \mathbf{k} \qquad (23)$$

with $\mathbf{C}^{\downarrow}$ and $\mathbf{C}^{\uparrow}$ both $n_{\widetilde{\gamma}} \times n_r$ matrices counting how many resource instances are claimed and released respectively for every alignment move. Both are defined for every $i \in [1..n_{\widetilde{\gamma}}]$ and $k \in [1..n_r]$ with $(e, t_\mu) = \widetilde{\gamma}(i)$ and $\rho_r = \mathrm{Id}_R(k)$:

$$\mathbf{C}_{ik}^{\downarrow} = \mathcal{F}(p_r, t)((\varepsilon, \mu^{-1}(\rho_r))) \text{ and } \mathbf{C}_{ik}^{\uparrow} = \mathcal{F}(t, p_r)((\varepsilon, \mu^{-1}(\rho_r))) \qquad (24)$$

and capacity vector $\mathbf{k}$ of length n_r, defined as $\mathbf{k}_k = |\mathrm{Id}_R(k)|$ for every $k \in [1..n_r]$.

$\mathbf{X}$ provides the solution of a new partial order of moves in $\widetilde{\gamma}$ such that all violations are resolved and the least number of partial order relations is removed. For the running example, the additional arcs from the solution $\mathbf{X}$ are shown in red in Fig. 5.

We refer to App. A in [25] for the correctness proof of the ILP problem, where we show (1) the effectiveness of each constraint, (2) that there always exists a solution, (3) that the optimal solution has zero cost if and only if the composed alignment is not violating, and (4) that each interval obtained in $W(\widetilde{\gamma}_V')$ is alignable.

6 Conclusion

We have formulated the requirements for modeling and analyzing processes with inter-case dependencies and argued that our previously proposed Petri net extension named Resource Constrained ν-Petri nets meets them. This paper continues on work presented in [24], where we showed that the traditional methods of aligning observed behavior with the modeled one fall short when dealing with coevolving cases, as they consider isolated cases only. The technique we present here aligns multiple cases simultaneously, exposing violations on inter-case dependencies. We developed and implemented an approximation technique based on a composition of individual alignments and local resolution of violations, which is an important advancement for the use of the technique in practice.

There can be ambiguity in the interpretation of the exposed violations, e.g., was the activity executed but not recorded, executed by an "incorrect" resource instance, or not executed at all? In [24], we briefly touched upon relaxations of the synchronous product model as a means to improve the deviations' interpretability. One such relaxation helps to detect situations when a step required by the model was skipped in a process execution, and the resources needed for the step were not available at the time when it should have been executed. Adding "resource-free" model moves for transitions allows to capture such deviations. Such special moves, when present in the alignment, reduce the ambiguity and provide a better explanation, e.g., that the activity was not executed at all, rather than it might also have been executed but not recorded. For future work, we plan to extend and formalize the relaxations, and evaluate the insights obtained with the alignments based on a real-life case study.

Acknowledgments. This work is done within the project "Certification of production process quality through Artificial Intelligence (CERTIF-AI)", funded by NWO (project number: 17998).

References

1. van der Aalst, W.M.P.: The application of Petri nets to workflow management. J. Circuits Syst. Comput. **8**(01), 21–66 (1998)
2. van der Aalst, W.M.P.: Data science in action. In: van der Aalst, W.M.P. (ed.) Process Mining, pp. 3–23. Springer, Heidelberg (2016). https://doi.org/10.1007/978-3-662-49851-4_1
3. van der Aalst, W.M.P., Adriansyah, A., van Dongen, B.F.: Replaying history on process models for conformance checking and performance analysis. Wiley Interdisc. Rev. Data Min. Knowl. Discov. **2**(2), 182–192 (2012)
4. van der Wil, M.P.: Aalst and Alessandro Berti. Discovering object-centric Petri nets. Fundamenta informaticae **175**(1–4), 1–40 (2020)
5. Adriansyah, A.: Aligning observed and modeled behavior. Ph.D. thesis, Mathematics and Computer Science (2014)
6. Alizadeh, M., Lu, X., Fahland, D., Zannone, N., van der Aalst, W.M.P.: Linking data and process perspectives for conformance analysis. Comput. Secur. **73**, 172–193 (2018)
7. Barkaoui, K., Petrucci, L.: Structural analysis of workflow nets with shared resources (1998)
8. Carmona, J., van Dongen, B.F., Solti, A., Weidlich, M.: Conformance Checking. Springer, Cham (2018). https://doi.org/10.1007/978-3-319-99414-7
9. Decker, G., Weske, M.: Instance isolation analysis for service-oriented architectures. In: 2008 IEEE International Conference on Services Computing, vol. 1, pp. 249–256. IEEE (2008)
10. Fahland, D.: Describing behavior of processes with many-to-many interactions. In: Donatelli, S., Haar, S. (eds.) PETRI NETS 2019. LNCS, vol. 11522, pp. 3–24. Springer, Cham (2019). https://doi.org/10.1007/978-3-030-21571-2_1
11. Ghilardi, S., Gianola, A., Montali, M., Rivkin, A.: Petri nets with parameterised data: modelling and verification (extended version). arXiv preprint arXiv:2006.06630 (2020)
12. van Hee, K., Sidorova, N., Voorhoeve, M.: Resource-constrained workflow nets. Fundamenta Informaticae **71**(2, 3), 243–257 (2006)
13. Kummer, O.: Undecidability in object-oriented Petri nets. In: Petri Net Newsletter. Citeseer (2000)
14. Lautenbach, K.: Liveness in Petri Nets. Bonn Interner Bericht ISF. Selbstverl, GMD (1975)
15. de Leoni, M., van der Aalst, W.M.P.: Aligning event logs and process models for multiperspective conformance checking: an approach based on integer linear programming. In: Daniel, F., Wang, J., Weber, B. (eds.) BPM 2013. LNCS, vol. 8094, pp. 113–129. Springer, Heidelberg (2013). https://doi.org/10.1007/978-3-642-40176-3_10
16. Mannhardt, F., de Leoni, M., Reijers, H.A., van der Aalst, W.M.P.: Balanced multiperspective checking of process conformance. Computing **98**(4), 407–437 (2016)
17. Mozafari Mehr, A.S., de Carvalho, R.M., van Dongen, B.: Detecting privacy, data and control-flow deviations in business processes. In: Nurcan, S., Korthaus, A. (eds.) CAiSE 2021. LNBIP, vol. 424, pp. 82–91. Springer, Cham (2021). https://doi.org/10.1007/978-3-030-79108-7_10

18. Montali, M., Rivkin, A.: Model checking Petri nets with names using data-centric dynamic systems. Form. Asp. Comput. **28**(4), 615–641 (2016)
19. Montali, M., Rivkin, A.: DB-nets: on the marriage of colored petri nets and relational databases. In: Koutny, M., Kleijn, J., Penczek, W. (eds.) Transactions on Petri Nets and Other Models of Concurrency XII. LNCS, vol. 10470, pp. 91–118. Springer, Heidelberg (2017). https://doi.org/10.1007/978-3-662-55862-1_5
20. Murata, T.: Petri nets: properties, analysis and applications. Proc. IEEE **77**(4), 541–580 (1989)
21. Peterson, J.L.: Petri Net Theory and the Modeling of Systems. Prentice Hall PTR, Hoboken (1981)
22. Rosa-Velardo, F., de Frutos-Escrig, D.: Decision problems for Petri nets with names. arXiv preprint arXiv:1011.3964 (2010)
23. Rosa-Velardo, F., de Frutos-Escrig, D., Marroquín-Alonso, O.: On the expressiveness of mobile synchronizing Petri nets. Electron. Notes Theor. Comput. Sci. **180**(1), 77–94 (2007)
24. Sommers, D., Sidorova, N., van Dongen, B.F.: Aligning event logs to resource-constrained ν-Petri nets. In: PETRI NETS 2022. LNCS, vol. 13288, pp. 325–345. Springer, Cham (2022). https://doi.org/10.1007/978-3-031-06653-5_17
25. Sommers, D., Sidorova, N., van Dongen, B.F.: Exact and approximated log alignments for processes with inter-case dependencies, arXiv (2023)

Semantics

Taking Complete Finite Prefixes to High Level, Symbolically

Nick Würdemann[1]($\boxtimes$) (ID), Thomas Chatain[2](ID), and Stefan Haar[2]

[1] Department of Computing Science, University of Oldenburg, Oldenburg, Germany
wuerdemann@informatik.uni-oldenburg.de
[2] Université Paris-Saclay, INRIA and LMF, CNRS and ENS Paris-Saclay,
Gif-sur-Yvette, France
{thomas.chatain,stefan.haar}@inria.fr

Abstract. Unfoldings are a well known partial-order semantics of P/T
Petri nets that can be applied to various model checking or verification
problems. For *high-level* Petri nets, the so-called *symbolic* unfolding generalizes this notion. A complete finite prefix of the unfolding of a P/T
Petri net contains all information to verify, e.g., reachability of markings.
We unite these two concepts and define complete finite prefixes of the
symbolic unfolding of high-level Petri nets. For a class of safe high-level
Petri nets, we generalize the well-known algorithm by Esparza et al. for
constructing small such prefixes. Additionally, we identify a more general class of nets with infinitely many reachable markings, for which an
approach with an adapted cut-off criterion extends the complete prefix
methodology, in the sense that the original algorithm cannot be applied
to the P/T net represented by a high-level net.

1 Introduction

Petri nets [17], also called P/T (for Place/Transition) Petri nets or low-level Petri
nets, are a well-established formalism for describing distributed systems. *High-
level Petri nets* [12] (also called *colored Petri nets*) are a concise representation
of P/T Petri nets, allowing the places to carry tokens of different colors. Every
high-level Petri net represents a P/T Petri net, here called its *expansion*[1], where
the process of constructing this P/T net is called *expanding* the high-level net.

    Unfoldings of P/T Petri nets are introduced by Nielsen et al. in [15]. Engel-
friet generalizes this concept in [9] by introducing the notion of *branching processes*, and shows that the unfolding of a net is its maximal branching process.
In [14], McMillan gives an algorithm to compute a complete finite prefix of the
unfolding of a given Petri net. In a well-known paper [10], Esparza, Römer, and
Vogler improve this algorithm by defining and exploiting a total order on the set
of configurations in the unfolding. We call the improved algorithm the "ERV-
algorithm". It leads to a comparably small complete finite prefix of the unfolding.

---

[1] In the literature, the represented Petri net is often called the *unfolding* of the high-
level Petri net. To avoid a clash of notions, we use the term expansion as, e.g., in [4].

© The Author(s), under exclusive license to Springer Nature Switzerland AG 2023
L. Gomes and R. Lorenz (Eds.): PETRI NETS 2023, LNCS 13929, pp. 123–144, 2023.
https://doi.org/10.1007/978-3-031-33620-1_7

In [13], Khomenko and Koutny describe how to construct the unfolding of the expansion of a high-level Petri net without first expanding it.

High-level representations on the one hand and processes (resp. unfoldings) of P/T Petri nets on the other, at first glance seem to be conflicting concepts; one being a more concise, the other a more detailed description of the net('s behavior). However, in [8], Ehrig et al. define processes of high-level Petri nets, and in [5], Chatain and Jard define *symbolic branching processes* and *unfoldings* of high-level Petri nets. The work on the latter is built upon in [4] by Chatain and Fabre, where they consider so-called "puzzle nets". Based on the construction of a symbolic unfolding, in [6], complete finite prefixes of safe time Petri nets are constructed, using time constraints associated with timed processes. In [3], using a simple example, Chatain argues that in general there exists no complete finite prefix of the symbolic unfolding of a high-level Petri net. However, this is only true for high-level Petri nets with infinitely many reachable markings such that the number of steps needed to reach them is unbounded, in which case the same arguments work for P/T Petri nets.

In this paper, we lift the concepts of complete prefixes and adequate orders to the level of symbolic unfoldings of high-level Petri nets. We consider the class of *safe* high-level Petri nets (i.e., in all reachable markings, every place carries at most one token) that have decidable guards and finitely many reachable markings. This class generalizes safe P/T Petri nets, and we obtain a generalized version of the ERV-algorithm creating a complete finite prefix of the symbolic unfolding of such a given high-level Petri net. Our results are a generalization of [10] in the sense that if a P/T Petri net is viewed as a high-level Petri net, the new definitions of adequate orders and completeness of prefixes on the symbolic level, as well as the algorithm producing them, all coincide with their P/T counterparts.

We proceed to identify an even more general class of so-called *symbolically compact* high-level Petri nets; we drop the assumption of finitely many reachable markings, and instead assume the existence of a bound on the number of steps needed to reach all reachable markings. In such a case, the expansion is possibly not finite, and the original ERV-algorithm from [10] therefore not applicable. We adapt the generalized ERV-algorithm by weakening the cut-off criterion to ensure finiteness of the resulting prefix. Still, in this cut-off criterion we have to compare infinite sets of markings. We overcome this obstacle by symbolically representing these sets, using the decidability of the guards to decide cut-offs.

Due to spatial limitations, we move some proofs to the full version [18] of this paper.

2 High-Level Petri Nets and Symbolic Unfoldings

In [5], symbolic unfoldings for high-level Petri nets are introduced. We recall definitions and formalism for high-level Petri nets and symbolic unfoldings.

Multi-sets. For a set X, we call a functions $A : X \to \mathbb{N}$ a *multi-set over X*. We denote $x \in A$ if $A(x) \geq 1$. For two multi-sets A, A' over the same set X, we write $A \leq A'$ iff $\forall x \in X : A(x) \leq A'(x)$, and denote by $A + A'$ and $A - A'$

the multi-sets over X given by $(A + A')(x) = A(x) + A'(x)$ and $(A - A')(x) = \min(A(x) - A'(x), 0)$. We use the notation $\{\!|\ldots|\!\}$ as introduced in [13]: elements in a multi-set can be listed explicitly as in $\{\!| x_1, x_1, x_2 |\!\}$, which describes the multi-set A with $A(x_1) = 2$, $A(x_2) = 1$, and $A(x) = 0$ for all $x \in X \setminus \{x_1, x_2\}$. A multi-set A is finite if there are finitely many $x \in X$ such that $A(x) \geq 0$. In such a case, $\{\!| f(x) \mid x \in A |\!\}$, with $f(x)$ being an object constructed from $x \in X$, denotes the multi-set A' such that $A' = \sum_{x \in X} A(x) \cdot f(x)$, where the $A(x) \cdot y$ is the multi-set containing exactly $A(x)$ copies of y.

2.1 High-Level Petri Nets

We assume two given sets Col (colors) and Var (variables). A *high-level net struc-ture* is a tuple $\mathcal{N} = \langle P, T, F, \iota \rangle$, with disjoint sets of places P and transitions T, a flow function $F : (P \times Var \times T) \cup (T \times Var \times P) \to \mathbb{N}$, and a function ι mapping each $t \in T$ to a predicate $\iota(t)$ on $Var(t) := \{v \in Var \mid \langle p, v, t \rangle \in F \vee \langle t, v, p \rangle \in F\}$, called the *guard* of t. A *marking* in $\mathcal{N}$ is a multi-set M over $P \times Col$, describing how often each color $c \in Col$ currently lies on each place $p \in P$. A *high-level Petri net* $N = \langle \mathcal{N}, \mathcal{M}_0 \rangle$ is a net structure $\mathcal{N}$ together with a set $\mathcal{M}_0$ of *initial markings*, where we assume $\forall M_0, M_0' \in \mathcal{M}_0 : \{\!| p \mid \langle p, c \rangle \in M_0 |\!\} = \{\!| p \mid \langle p, c \rangle \in M_0' |\!\}$, i.e., in all initial markings, the same places are marked with *the same number of col-ors*.

For two nodes $x, y \in P \cup T$, we write $x \to y$, if there exists a variable v such that $\langle x, v, y \rangle \in F$. The reflexive and irreflexive transitive closures of $\to$ are denoted respectively by $\leq$ and $<$. For a transition $t \in T$, we denote by $pre(t) := \{\!| \langle p, v \rangle \mid \langle p, v, t \rangle \in F |\!\}$ and $post(t) := \{\!| \langle p, v \rangle \mid \langle t, v, p \rangle \in F |\!\}$ the *preset* and *postset* of t. A *firing mode* of t is a mapping $\sigma : Var(t) \to Col$ such that $\iota(t)$ evaluates to *true* under the substitution given by σ, denoted by $\iota(t)[\sigma] \equiv true$. We then denote $pre(t, \sigma) := \{\!| \langle p, \sigma(v) \rangle \mid \langle p, v \rangle \in pre(t) |\!\}$ and $post(t, \sigma) := \{\!| \langle p, \sigma(v) \rangle \mid \langle p, v \rangle \in post(t) |\!\}$. The set of modes of t is denoted by $\Sigma(t)$. t can *fire* in such a mode σ from a marking M if $M \geq pre(t, \sigma)$, denoted by $M[t, \sigma\rangle$. This firing leads to a new marking $M' = (M - pre(t, \sigma)) + post(t, \sigma)$, which is denoted by $M[t, \sigma\rangle M'$. We collect in the set $\mathcal{R}(\mathcal{N}, \mathcal{M})$ the markings reachable by firing a sequence of transitions in $\mathcal{N}$ from any marking in a set of markings $\mathcal{M}$. $\mathcal{N}$ resp. N is called *finite* if P, T and F are finite.

Let $\mathcal{N} = \langle P, T, F, \iota \rangle$ and $\mathcal{N}' = \langle P', T', F', \iota' \rangle$ be two net structures. A func-tion $h : P \cup T \to P' \cup T'$ is called a *high-level net homomorphism*, if:

i) it maps places and transitions in $\mathcal{N}$ into the corresponding sets in $\mathcal{N}'$, i.e., $h(P) \subseteq P' \wedge h(T) \subseteq T'$;

ii) it is "compatible" with the preset, postset, and guard of transitions, i.e., for all $t \in T$ we have $pre(h(t)) = \{\!| \langle h(p), v \rangle \mid \langle p, v \rangle \in pre(t) |\!\}$, $post(h(t)) = \{\!| \langle h(p), v \rangle \mid \langle p, v \rangle \in post(t) |\!\}$, and $\iota(t) = \iota'(h(t))$.

For $N = \langle \mathcal{N}, \mathcal{M}_0 \rangle$ and $N' = \langle \mathcal{N}', \mathcal{M}_0' \rangle$, the homomorphisms between N and N' are the homomorphisms between $\mathcal{N}$ and $\mathcal{N}'$. Such a homomorphism h is called *initial* if additionally $\{\{\!| \langle h(p), c \rangle \mid \langle p, c \rangle \in M_0 |\!\} \mid M_0 \in \mathcal{M}_0\} = \mathcal{M}_0'$ holds.

We define *P/T Petri nets* as high-level Petri nets with singletons $Col = \{\bullet\}$ and $Var = \{v_\bullet\}$ for colors and variables, i.e., in a marking, every place holds a number of tokens •, which is the only value ever assigned to the variable $v_\bullet$ on every arc. The guard of every transition in a P/T Petri net is *true*.

2.2 Symbolic Branching Processes and Unfoldings

We recall the definition of symbolic branching processes and unfoldings from [5]. It is a generalization of branching processes and unfoldings for P/T Petri nets.

A net structure $\mathcal{N} = \langle P, T, F, \iota \rangle$ is called *ordinary* if there is at most one arc connecting any two nodes in $\mathcal{N}$, i.e., $\forall x, y \in P \cup T : \sum_{v \in Var} F(x, v, y) \leq 1$. For such an ordinary net structure, analogously to the low-level case described, e.g., in [10], two nodes $x, y \in P \cup T$ are in *structural conflict*, denoted by $x \sharp y$, if $\exists p \in P \, \exists t, t' \in T : t \neq t' \wedge p \to t \wedge p \to t' \wedge t \leq x \wedge t' \leq y$.

A *high-level occurrence net* is a high-level Petri net $O = \langle B, E, G, \iota, \mathcal{K}_0 \rangle$ with an ordinary net structure $\langle B, E, G, \iota \rangle$, where B is a set of *conditions* (places), E is a set of *events* (transitions), G is a flow relation, and $\mathcal{K}_0$ is the set of initial *cuts* (markings), having the following properties:

i) No event is in structural self-conflict, i.e., $\forall e \in E : \neg(e \sharp e)$.
ii) No node is its own causal predecessor, i.e., $\forall x \in B \cup E : \neg(x < x)$;
iii) The flow relation is well-founded, i.e., $\forall x \in B \cup E : |\{y \mid y \leq x\}| < \infty$;
iv) For every $b \in B$, exactly one of the following holds:
   a) $\forall K_0 \in \mathcal{K}_0 : \sum_{c \in Col} K_0(b, c) = 0$ and there exists a unique pair $\langle e, v \rangle$ called $pre(b)$ s.t. $\langle e, v, b \rangle \in G$, and for this pair we have $G(e, v, b) = 1$.
   b) $\forall K_0 \in \mathcal{K}_0 : \sum_{c \in Col} K_0(b, c) = 1$ and $\{e \mid e \to b\} = \emptyset$.
   In this case we denote $pre(b) := \langle \bot, v^b \rangle$.

The properties *i)* – *iii)* are exactly as in the low-level case and concern solely the net structure. Property *iv)* generalizes the requirement of low-level occurrence nets that every condition has at most one event in its preset, and that the conditions with empty preset constitute the initial cut.

In a crucial notation for what follows in later sections, we identify in case *iv.a)* the event e by $\mathbf{e}(b)$ and the variable v by $\mathbf{v}(b)$, and in case *iv.b)* we define $\mathbf{e}(b) := \bot$, and $\mathbf{v}(b) := v^b$. We abbreviate $\mathbf{v_e}(b) := \mathbf{v}(b)_{\mathbf{e}(b)}$. We denote by $B_0 := \{b \in B \mid \exists K_0 \in \mathcal{K}_0, c \in Col : \langle b, c \rangle \in K_0\}$ the conditions from *iv.b)* occupied in all initial cuts. $\bot$ can be seen as a "special event" that fires only once to initialize the net, and produces the initial cuts $K_0 \in \mathcal{K}_0$ by assigning values to the variables v^b on "special arcs" $\langle \bot, v^b, b \rangle$ towards the conditions $b \in B_0$.

For a high-level occurrence net, we define the mappings *loc-pred* and *pred* equipping events with predicates. For any $e \in E$, $pred(e)$ is satisfiable iff e is not dead, i.e., there are cuts $K_0, \ldots, K_n$ with $K_0 \in \mathcal{K}_0$ and events $e_1, \ldots, e_n$, s.t. $K_0[e_1\rangle \ldots [e_n\rangle K_n[e\rangle$. This predicate is obtained by building a conjunction over all *local predicates* of events e' with $e' \leq e$ (including the special event $\bot$). The local predicate of e is, in its turn, a conjunction of two predicates expressing that (i) the guard of the event e is satisfied, and (ii) that for any $\langle b, v \rangle \in pre(e)$, the value of the variable v coincides with the color that the event $\mathbf{e}(b)$ placed b.

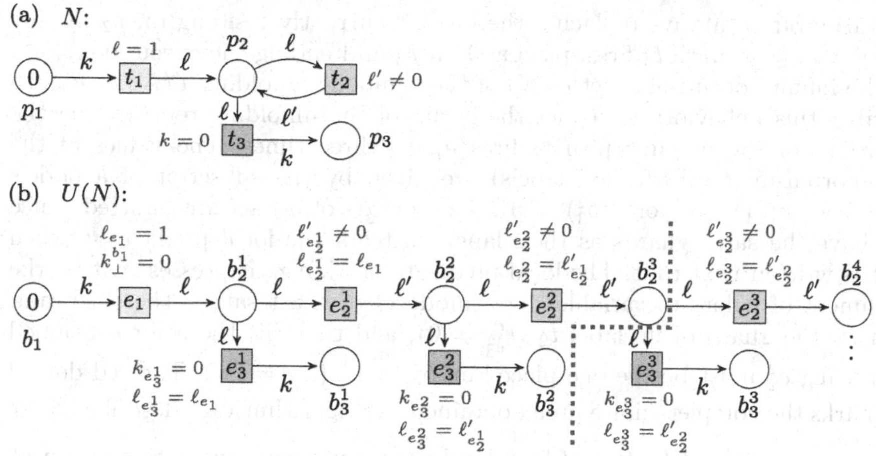

(a) N:

(b) $U(N)$:

Fig. 1. A safe high-level Petri net N in (a), and (a prefix of) the infinite symbolic unfolding $U(N)$ in (b). We have $Col = \{0, \ldots, m\}$ and $Var = \{k, \ell, \ell'\}$.

To realize this, the variables $v \in Var(e)$ are instantiated by the index e, so that v_e describes the value assigned to v by a mode of e. Formally, we have

$$loc\text{-}pred(e) \quad := \quad \iota(e)[v \leftarrow v_e]_{v \in Var(e)} \quad \wedge \bigwedge_{\langle b,v \rangle \in pre(e)} v_e = \mathbf{v_e}(b)$$

$$pred(e) \quad := \quad pred(\bot) \wedge \bigwedge_{e' \leq e} loc\text{-}pred(e'),$$

where $pred(\bot) := \bigvee_{K_0 \in \mathcal{K}_0} \bigwedge_{\langle b,c \rangle \in K_0} (v_\bot^b = c)$ describes the set of initial cuts.

A *symbolic branching process* of a high-level Petri net N is a pair $\beta = \langle O, h \rangle$ with an occurrence net $O = \langle B, E, G, \iota, \mathcal{K}_0 \rangle$ in which $pred(e)$ is satisfiable for all $e \in E$, and an initial homomorphism $h : O \to N$ that is injective on events with the same preset, i.e., $\forall e, e' \in E : (pre(e) = pre(e') \wedge h(e) = h(e')) \Rightarrow e = e'$.

For two symbolic branching processes $\beta = \langle O, h \rangle$ and $\beta' = \langle O', h' \rangle$ of a high-level Petri net, β is a *prefix* of β' if there exists an injective initial homomorphism ϕ from O into O', such that $h' \circ \phi = h$. In [5] it is argued that for any given high-level Petri net N there exists a unique maximal branching process (maximal w.r.t. the prefix relation and unique up to isomorphism). This branching process is called the *symbolic unfolding*, and denoted by $\Upsilon(N) = \langle U(N), \pi \rangle$.

Example 1. Let $Col = \{0, \ldots, m\}$ for a fixed m, and $Var = \{k, \ell, \ell'\}$ be the given sets of colors and variables. In Fig. 1a, the running example N of a high-level Petri net[2] is depicted. Places are drawn as circles, and transitions as squares. The flow is described by labeled arrows, and the guards are written next to the respective transition. N has just one initial marking $M_0 = \{\!| \langle p_1, 0 \rangle |\!\}$, which is depicted in the net. From M_0 only t_1 can fire, and only in the mode $[k \leftarrow 0, \ell \leftarrow 1]$, taking 0 from p_1 and placing a token of color 1 on p_2. From there, t_2 can fire

---

[2] The structure of this example is taken from Figure D.4.5 in [2].

arbitrarily often, always replacing the color ℓ currently residing on p_2 by any color $0 < \ell' \leq m$, until t_3 fires, placing 0 on p_3 and ending the execution.

The infinite occurrence net $U(N)$ of the symbolic unfolding $\Upsilon(N)$ in Fig. 1b describes this behavior: we depict the prefix of the unfolding representing the executions of the net in which t_2 fires up to three times. The values of the homomorphism π (also called labels) are given by the subscript of a node's name, e.g., $\pi(e_1) = t_1$ or $\pi(b_3^2) = p_3$. The guards of events are omitted, since they have the same guards as their label. Instead, the local predicate of each event is written next to it. The local predicate of e_2^2, e.g., expresses that (i) the assignment of colors to variables by a mode of e_2^2 must satisfy the constraint given by the guard of its label t_2 ($\ell'_{e_2^2} \neq 0$), and that (ii) the color consumed when firing e_2^2 must be the one placed on b_2^2 by e_2^1 ($\ell_{e_2^2} = \ell'_{e_2^1}$). The red dotted line marks the complete finite prefix obtained by Algorithm 1, as described later.

As we see in the definition of high-level occurrence nets, the notion of causality and structural conflict are the same as in the low-level case. However, a set of events in an occurrence net can also be in what we call *color conflict*, meaning that the conjunction of their predicates is not satisfiable. In a symbolic branching process, this means that the constraints on the values of the firing modes, coming from the guards of the transitions, prevent joint occurrence of all events from such a set in any *one* run of the net:

The nodes in a set $X \subseteq E \cup B$ and are in *color conflict* if $\bigwedge_{e \in X \cap E} pred(e) \wedge \bigwedge_{b \in X \cap B} pred(\mathbf{e}(b))$ is not satisfiable. The nodes of X are *concurrent* if they are *not* in color conflict, and for each $x, x' \in X'$, neither $x < x'$, nor $x' < x$, nor $x \sharp x'$ holds. A set of concurrent conditions is called a *co-set*.

Note that while a set of nodes is defined to be in structural conflict if and only if two nodes in it are in structural conflict, the same does not hold for color conflict: it is possible to have a set $\{x_1, x_2, x_3\}$ of nodes that are in color conflict, but for which every subset of cardinality 2 is *not* in color conflict.

Definition 1 (Configuration [5]). *A (symbolic) configuration is a set of high-level events that is free of structural conflict and color conflict, and causally closed. The configurations in a symbolic branching process β are collected in the set $\mathcal{C}(\beta)$.*

For a configuration C, we define by $\mathrm{cut}(C) := (B_0 \cup (C \rightarrow)) \setminus (\rightarrow C)$ the high-level conditions that are occupied after any concurrent execution of C. Note that $\mathrm{cut}(C)$ is a co-set, and that $\emptyset$ is a configuration with $\mathrm{cut}(\emptyset) = B_0$.

Let $e \in E$ be a high-level event. We define the so-called *cone configuration* $[e] := \{e' \in E \mid e' \leq e\}$. Additionally, we define the sets $Var_e := \{v_e \mid v \in Var(e)\}$ and $Var_\perp := \{v_\perp^b \mid b \in B_0\}$ of indexed variables, and for a set $E' \subseteq E \cup \{\perp\}$ we denote $Var_{E'} := \bigcup_{e \in E'} Var_e$. Note that, for every event e, $pred(e)$ is a predicate over the variables $Var_{[e] \cup \{\perp\}}$.

2.3 Properties of the Symbolic Unfolding

Having recalled the definitions and formal language from [5], we now delve into the novel aspects of this paper. We state three analogues of well-known properties of the Unfolding of P/T Petri nets for the symbolic unfolding of high-level nets. These properties are: (i) the cuts in the unfolding represent precisely the reachable markings in the net, (ii) for every transition that can occur in the net, there is an event in the unfolding with corresponding label (and vice versa), and (iii) the unfolding is complete in the sense that for any configuration, the part of the unfolding that "lies after" that configuration is the unfolding of the original net with the initial markings being the ones represented by the configurations cut. The properties are stated in Proposition 1, 2, and 3, respectively. Their proofs are moved to [18].

To express these properties, we introduce the notion of *instantiations* of configurations C, choosing a mode for every event in C without creating color conflicts. This is realized by assigning to each variable $v_e \in Var_{C \cup \{\perp\}}$ a value in Col, such that the above defined predicates evaluate to *true*. For each $e \in C$, the assignment of values to the indexed variables in Var_e corresponds to a mode of e.

Definition 2 (Instantiation). *For a given configuration C, an* instantiation *of C is a function $\theta : Var_{C \cup \{\perp\}} \to Col$, such that $\forall e \in C \cup \{\perp\} : pred(e)[\theta] \equiv true$, i.e., it satisfies all predicates in the configuration. The set of instantiations of a given configurations C is denoted by $\Theta(C)$.*

Note that, by definition, every configuration C has an instantiation θ. We denote by $\mathrm{cut}(C, \theta) := \{\langle b, c \rangle \mid b \in \mathrm{cut}(C) \wedge \theta(\mathbf{v_e}(b)) = c\} \subseteq B \times Col$ the *cut* of an "instantiated configuration", and by $\mathrm{mark}(C, \theta) := \{\!| \langle h(b), c \rangle \mid \langle b, c \rangle \in \mathrm{cut}(C.\theta) |\!\}$ its *marking*. We collect both of these in $\mathcal{K}(C) := \{\mathrm{cut}(C, \theta) \mid \theta \in \Theta(C)\}$ and $\mathcal{M}(C) := \{\mathrm{mark}(C, \theta) \mid \theta \in \Theta(C)\}$. Note that in this notation, for the empty configuration we have $\mathcal{K}(\emptyset) = \mathcal{K}_0$ and $\mathcal{M}(\emptyset) = \mathcal{M}_0$.

Proposition 1. *Let N be a high-level Petri net and Υ its symbolic unfolding. Then $\mathcal{R}(N) = \{\mathrm{mark}(C.\theta) \mid C \in \mathcal{C}(\Upsilon), \theta \in \Theta(C)\}$.*

Proposition 2. *The symbolic unfolding $\Upsilon = \langle U, \pi \rangle$ with events E of a high-level Petri net $N = \langle P, T, F, \iota, \mathcal{M}_0 \rangle$ satisfies $\forall C \in \mathcal{C}(\Upsilon) \, \forall \theta \in \Theta(C) \, \forall t \in T \, \forall \sigma \in \Sigma(t) :$*

$$\mathrm{mark}(C, \theta)[t, \sigma\rangle \; \Leftrightarrow \; \exists e \in E : \pi(e) = t \wedge \mathrm{cut}(C, \theta)[e, \sigma\rangle.$$

Given a configuration C of a symbolic branching process $\beta = \langle O, h \rangle$, we define $\Uparrow C$ as the pair $\langle O', h' \rangle$, where O' is the unique subnet of O whose set of nodes is $\{x \in B \cup E \mid x \notin (C \cup \to C) \wedge \forall y \in C : \neg(y \sharp x) \wedge (C \cup \{x\}$ is not in color conflict)$\}$ with the set $\mathcal{K}(C)$ of initial cuts, and h' is the restriction of h to the nodes of O'. The branching process $\Uparrow C$ is referred to as the *future* of C.

Proposition 3. *If β is a symbolic branching process of $\langle \mathcal{N}, \mathcal{M}_0 \rangle$ and C is a configuration of β, then $\Uparrow C$ is a branching process of $\langle \mathcal{N}, \mathcal{M}(C) \rangle$. Moreover, if β is the unfolding of $\langle \mathcal{N}, \mathcal{M}_0 \rangle$, then $\Uparrow C$ is the unfolding of $\langle \mathcal{N}, \mathcal{M}(C) \rangle$.*

3 Finite and Complete Prefixes of Symbolic Unfoldings

We combine ideas from [10] (computing small finite and complete prefixes of unfoldings) with results from [5] (symbolic unfoldings of high-level Petri nets) to define and construct complete finite prefixes of symbolic unfoldings of high-level Petri nets. We generalize the concepts and the ERV-algorithm from [10] for safe P/T Petri nets to a class of safe high-level Petri nets, and compare this generalization to the original. We will see that for P/T nets interpreted as high-level nets, all generalized concepts (i.e., complete prefixes, adequate orders, cut-off events), and, as a consequence, the result of the generalized ERV-algorithm all coincide with their P/T counterparts.

We start by lifting the definition of completeness to the level of symbolic unfoldings. Together with Propositions 1 and 2, this can be seen as a direct translation from the low-level case described, e.g., in [10].

Definition 3 (Complete Prefix). *Let $\beta = \langle O, h \rangle$ be a prefix of the symbolic unfolding of a high-level Petri net N, with events E'. Then β is called* complete *if for every reachable marking M in N there exists $C \in \mathcal{C}(\beta)$ and $\theta \in \Theta(C)$ s.t.*

i) $M = \mathrm{mark}(C, \theta)$,   and
ii) $\forall t \in T \, \forall \sigma \in \Sigma(t): \ M[t, \sigma\rangle \ \Rightarrow \ \exists e \in E': h(e) = t \ \wedge \ \mathrm{cut}(C.\theta)[e, \sigma\rangle.$

We now define the class $\mathbf{N_f}$ of high-level Petri nets for which we generalize the construction of finite and complete prefixes of the unfolding of *safe* P/T Petri nets from [10]. We discuss the properties defining this class, and describe how it generalizes safe P/T nets.

Definition 4 (Class $\mathbf{N_f}$). *The class $\mathbf{N_f}$ contains all finite high-level Petri nets $N = \langle P, T, F, \iota, \mathcal{M}_0 \rangle$ satisfying the following three properties:*

(1) The net is safe*, i.e., in every reachable marking there lies at most 1 color on every place (formally; $\forall M \in \mathcal{R}(N) \, \forall p \in P: \sum_{c \in Col} M(p, c) \leq 1$).*
(2) Guards are written in a decidable first-order theory with the set Col as its domain of discourse.
(3) The net has finitely many reachable markings (formally; $|\mathcal{R}(N)| < \infty$).

We require the safety property *(1)* for two reasons; on the one hand, to avoid adding to the already heavy notation. On the other hand, while we think that a generalization to bounded high-level Petri nets is possible, it comes with all the troubles known from going from safe to k-bounded in the P/T case in [10], plus the problems arising from the expressive power of the high-level formalism. We therefore postpone this generalization to future work. Note that, under the safety condition, we can w.l.o.g. assume $\mathcal{N}$ to be ordinary (i.e., $\forall x, y \in P \cup T: \sum_{v \in Var} F(x, v, y) \leq 1$), since transitions violating this property could never fire. The finiteness of $\mathcal{N}$ implies that we can assume Var to be finite.

While property *(2)* seems very strong, the goal is an algorithm that generates a complete finite prefix of the symbolic unfolding of a given high-level Petri net. The definition of symbolic branching processes requires the predicate of every event added to the prefix to be satisfiable, and the predicates are build from the guards in the given net. Thus, satisfiability checks in the generation of the prefix seem for now inevitable. An example for such a theory is Presburger arithmetic [16], which is a first order theory of the natural numbers with addition. The guards in the example from Fig. 1a are written in Presburger arithmetic.

We need Property *(3)* to ensure that the generalized version of the cut-off criterion from [10] yields a finite prefix constructed in the generalized ERV-Algorithm. $|\mathcal{R}(N)| < \infty$ can be ensured by having a finite set *Col* of colors. In Sect. 4, we identify a class of high-level Petri nets with infinitely many reachable markings for which the algorithm works with an adapted cut-off criterion.

Under these three assumptions we generalize the finite safe P/T Petri nets considered in [10]: every such P/T net can be seen as a high-level Petri net with $Col = \{\bullet\}$ and all guards being *true*, and thus satisfying the three properties above. Replacing the safety property *(1)* by a respective "k-bounded property" would result in a generalization of k-bounded P/T nets. In Sect. 3.3, we compare the result of the generalized ERV-algorithm Algorithm 1 applied to a high-level net to the result of the original ERV-algorithm from [10] applied to the nets expansion.

For the rest of the section let $N = \langle P, T, F, \iota, \mathcal{M}_0 \rangle \in \mathbf{N_f}$ with symbolic unfolding $\Upsilon = \langle U, \pi \rangle = \langle B, E, G, \iota, \mathcal{K}_0, \pi \rangle$.

3.1   Generalizing Adequate Orders and Cut-Off Events

We lift the concept of adequate orders on the configurations of an occurrence net to the level of symbolic unfoldings. A main property of adequate orders is the preservation by finite *extensions*, which are defined as for P/T-nets (cp. [10]):

Given a configuration C, we denote by $C \oplus D$ the fact that $C \cup D$ is a configuration such that $C \cap D = \emptyset$. We say that $C \oplus D$ is an *extension* of C, and that D is a *suffix* of C. Obviously, for a configuration C', if $C \subsetneq C'$ then there is a nonempty suffix D of C such that $C \oplus D = C'$. For a configuration $C \oplus D$, denote by $O(C|D) = \langle \mathrm{cut}(C) \cup \rightarrow D \cup D \rightarrow, D, G', \mathcal{K}(C) \rangle$ the occurrence net around D from $\mathrm{cut}(C)$, where G' is the restriction of G to the nodes of $O(C|D)$. Note that for every finite configuration C with an extension $C \oplus D$, we have that D is a configuration of $\Uparrow C$.

For better readability, we abbreviate for a marking M the fact $C[\![M]\!]D :\Leftrightarrow \exists \theta \in \Theta(C \oplus D) : \mathrm{mark}(C, \theta|_{Var_{C \cup \{\bot\}}}) = M$. Thus, $C[\![M]\!]D$ means that the transitions corresponding to the events in D can fire from $M \in \mathcal{M}(C)$.

The now stated Proposition 4 is a weak version of the arguments in [10], where Esparza et al. follow from the low-level version of Proposition 3 that if the cuts of two low-level configurations represent the same marking in the low-level net, then their futures are isomorphic, and the respective (unique) isomorphism maps the suffixes of one configuration to the suffixes of the other.

Proposition 4. *Let C_1 and C_2 be two finite configurations in Υ, and let D be a suffix of C_1. If there is a marking $M \in \mathcal{M}(C_1) \cap \mathcal{M}(C_2)$ s.t. $C_1[\![M]\!]D$, then there is a unique monomorphism $\varphi_{1,D}^2 : O(C_1|D) \to \Uparrow C_2$ that satisfies $\varphi_{1,D}^2(\mathrm{cut}(C_1)) = \mathrm{cut}(C_2)$ and preserves the labeling π.*
For this monomorphism we have that $\varphi_{1,D}^2(D)$ is a suffix of C_2.

The proof is an induction over the size of D (cp. [18]).

Equipped with Proposition 4, we can now lift the concept of adequate order to the level of symbolic branching processes. Compared to [10,14], the monomorphism $\varphi_{1,D}^2$ defined above replaces the isomorphism I_1^2 between $\Uparrow C_1$ and $\Uparrow C_2$ for two low-level configurations C_1, C_2 representing the same marking.

Definition 5 (Adequate order). *A partial order $\prec$ on the finite configurations of the symbolic unfolding of a high-level Petri net is an* adequate order *if:*

- *i)* $\prec$ *is well-founded,*
- *ii)* $C_1 \subset C_2$ *implies* $C_1 \prec C_2$, *and*
- *iii)* $\prec$ *is preserved by finite extensions in the following way: if* C_1, C_2 *are two finite configurations, and* $C_1 \oplus D$ *is a finite extension of* C_1 *such that there is a marking* $M \in \mathcal{M}(C_1) \cap \mathcal{M}(C_2)$ *satisfying* $C_1 \llbracket M \rrbracket D$, *then the monomorphism* $\varphi_{1,D}^2$ *from above satisfies* $C_1 \prec C_2 \Rightarrow C_1 \oplus D \prec C_2 \oplus \varphi_{1,D}^2(D)$.

In the case of a P/T net interpreted as a high-level net, we have $|\mathcal{M}(C)| = 1$ for every configuration C, and therefore, Definition 5 coincides with its P/T version [10]. We could alternatively generalize the P/T case by replacing '$\exists M \in \mathcal{M}(C_1) \cap \mathcal{M}(C_2)$ s.t. $C_1 \llbracket M \rrbracket D$' by '$\mathcal{M}(C_1) = \mathcal{M}(C_2)$', and use the isomorphism I_1^2 between $\Uparrow C_1$ and $\Uparrow C_2$ to define preservation by finite extension. However, in the upcoming generalization of the ERV-algorithm from [10], the generalized cut-off criterion exploits property iii) of adequate orders. Using '$\mathcal{M}(C_1) = \mathcal{M}(C_2)$' would produce an exponential blowup of the generated prefix's size. This is circumvented by using '$\exists M \in \mathcal{M}(C_1) \cap \mathcal{M}(C_2)$ s.t. $C_1 \llbracket M \rrbracket D$', which however leads to obtaining merely a monomorphism that depends on the considered suffix, instead of an isomorphism between the futures. We now show that this monomorphism sufficient:

The upcoming proof that the generalized ERV-algorithm is complete is structurally analogous to the respective proof in [10]. It uses that, under the conditions of Definition 5 *iii)*, we also have $C_2 \prec C_1 \Rightarrow C_2 \oplus \varphi_{1,D}^2(D) \prec C_1 \oplus D$. This result would directly be obtained if $\varphi_{1,D}^2$ was an isomorphism, as I_1^2 is in the low-level case. However, a monomorphism is an isomorphism when its codomain is restricted to its range. This idea is used in the proof (cp. App [18]) of the following proposition, which states that $\varphi_{1,D}^2$ indeed satisfies the above property.

Proposition 5. *Let $\prec$ be an adequate order. Under the conditions of Definition 5 iii) the monomorphism $\varphi_{1,D}^2$ also satisfies $C_2 \prec C_1 \Rightarrow C_2 \oplus \varphi_{1,D}^2(D) \prec C_1 \oplus D$.*

In [10], Esparza et al. discuss three adequate orders on the configurations of the low-level unfolding. In particular, they present a *total* adequate order that uses the *Foata normal form* of configurations. Using such a total order in the algorithm limits the size of the resulting finite and complete prefix; It contains at most $|\mathcal{R}(N)|$ non cut-off events. All three adequate orders presented in [10] can be directly lifted to the configurations of the symbolic unfolding by exchanging every low-level term by its high-level counterpart. The lifted order using the Foata normal form is still a total order. We include these discussions in [18].

We now define cut-off events in a symbolic unfolding. In the low-level case [10], e is a cut-off event if there is another event e' satisfying $[e'] \prec [e]$ and mark($[e]$) = mark($[e']$), which ensures that the future of e needs not be considered further. In the high-level case, we generalize these conditions to high-level events e. However, we do not require the existence of *one* other high-level event e' with $[e'] \prec [e]$ and $\mathcal{M}([e]) = \mathcal{M}([e'])$. While this would still be a valid cut-off criterion and would lead to finite and complete prefixes, the upper bound on the size of such a prefix would be exponential in the number of markings in the original net. Instead, we check whether $\mathcal{M}([e])$ is contained in the union of *all* $\mathcal{M}([e'])$ with $[e'] \prec [e]$. This criterion expresses that we have already seen every marking in $\mathcal{M}([e])$ in the prefix β under construction, and therefore need not consider the future of e any further. By this, we obtain the same upper bounds as in [10], as discussed later.

Definition 6 (Cut-off event). *Let $\prec$ be an adequate order on the configurations of the symbolic unfolding of a high-level Petri net. Let β be a prefix of the symbolic unfolding containing a high-level event e. The high-level event e is a cut-off event in β (w.r.t. $\prec$) if $\mathcal{M}([e]) \subseteq \bigcup_{[e'] \prec [e]} \mathcal{M}([e'])$.*

When interpreting P/T nets as high-level nets, this definition corresponds to the cut-off events defined in [10], since then $|\mathcal{M}([e])| = 1$ for all events e.

3.2   The Generalized ERV-Algorithm

We present the algorithm for constructing a finite and complete prefix of the symbolic unfolding of a given high-level Petri net. It is a generalization of the ERV-algorithm from [10], and is structurally equal (and therefore looks very similar). However, the algorithm is contingent upon the previous section's work of generalizing adequate orders and cut-off events, which ultimately enables us to adopt this structure.

A crucial concept of the ERV-algorithm is the notion of "possible extensions", i.e., the set of individual events that extend a given prefix of the unfolding. In Definition 7, we lift this concept to the level of symbolic unfoldings. We do so by isolating the procedure of adding high-level events in the algorithm from [5] which generates the complete symbolic unfolding of a given high-level Petri net (but does not terminate if the symbolic unfolding is infinite).

We define the data structures similarly to [10]. There, an event is given by a tuple $e = \langle t, B' \rangle$ with $h(e) = t \in T$ and $pre(e) = B' \subseteq B$, and a condition given by a tuple $b = \langle p, e \rangle$ with $h(b) = p \in P$ and $pre(b) = \{e\} \subseteq E$. The finite and complete prefix is a set of such events and transitions.

In the high-level case, we need more information inside the tuples. A high-level event is given by a tuple $e = \langle t, X, pred \rangle$ described by $h(e) = t$, $pre(e) = X \subseteq B \times Var$, and $pred(e) = pred$. Analogously, a high-level condition is given by a tuple $b = \langle p, \langle e, v \rangle, pred \rangle$, where $h(b) = p$, $pre(b) = \langle e, v \rangle \in (E \times Var) \cup (\{\bot\} \times \{v^b \mid b \in B_0\})$, and $pred(e(b)) = pred$.

Definition 7 (Possible Extensions). *Let $\beta = \langle O, h \rangle$ be a branching process of a high-level Petri net N. The possible extensions $PE(\beta)$ are the set of tuples $e = \langle t, X, pred \rangle$ where t is a transition of N, and $X \subseteq B \times Var$ satisfying*

- *$\{b \mid \langle b, v \rangle \in X\}$ is a co-set, and $pre(t) = \{\langle h(b), v \rangle \mid \langle b, v \rangle \in X\}$,*
- *$pred = loc\text{-}pred \wedge \big(\bigwedge_{\langle b,v \rangle \in X} pred(e(b)) \big)$ is satisfiable,*
    where $loc\text{-}pred = \iota(t)[v \leftarrow v_e]_{v \in Var(e)} \wedge \big(\bigwedge_{\langle b,v \rangle \in X} v_e = \mathbf{v_e}(b) \big)$,
- *Fin does not contain $\langle t, X, pred \rangle$.*

Since the notion of co-set in high-level occurrence nets is achieved by the direct translation from low-level occurrence nets plus the "color conflict freedom", possible extensions in a prefix β can be found by searching first for sets of conditions that are not in structural conflict as in the low-level case, and then checking whether these sets are in color conflict.

Algorithm 1 is a generalization of the ERV-Algorithm in [10] for complete finite prefixes of the low-level unfolding. The structure is taken from there, with the only difference being the special initial transition $\perp$. It takes as input a high-level Petri net $N \in \mathbf{N_f}$ and assumes a given adequate order $\prec$.

Algorithm 1: Generalization of the ERV-Algorithm from [10] for complete finite prefixes.

Data: High-level Petri net $N = \langle P, T, F, \iota, \mathcal{M}_0 \rangle \in \mathbf{N_f}$.
Result: A complete finite prefix Fin of the symbolic unfolding of N.
$Fin \leftarrow \{\perp\}$;
$pred(\perp) \leftarrow \bigvee_{M_0 \in \mathcal{M}_0} \bigwedge_{\langle p,c \rangle \in M_0} v_\perp^{b_p} = c$;
foreach $p \in P_0$ **do**
    Create a fresh condition $b_p = \langle p, \langle \perp, v^{b_p} \rangle, pred(\perp) \rangle$;
    $Fin \leftarrow Fin \cup \{b\}$;
$pe \leftarrow PE(Fin)$;
$cut\text{-}off \leftarrow \emptyset$;
while $pe \neq \emptyset$ **do**
    Pick $e = \langle t, X, pred \rangle$ from pe such that $[e]$ is minimal w.r.t. $\prec$;
    if $[e] \cap cut\text{-}off = \emptyset$ **then**
        $Fin \leftarrow Fin \cup \{e\}$;
        foreach $\langle p, v \rangle \in post(t)$ **do**
            Create a fresh condition $b = \langle p, \langle e, v \rangle, pred \rangle$;
            $Fin \leftarrow Fin \cup \{b\}$;
        $pe \leftarrow PE(Fin)$;
        if e *is a cut-off event of* Fin **then**
            $cut\text{-}off \leftarrow cut\text{-}off \cup \{e\}$;
    else
        $pe \leftarrow pe \setminus \{e\}$

We now prove correctness of Algorithm 1 analogously to [10], by stating two propositions – one each to show that the prefix is finite and complete, respectively. The proof structure is also as in [10], but adapted to the setting of high-level Petri nets and symbolic unfoldings.

Proposition 6. *Fin is finite.*

Proof (Sketch). As in [10], we prove the following results (1) – (3):

(1) For every event e of Fin, $d(e) \leq |\mathcal{R}(N)| + 1$, where d is the *depth* of e.
(2) For every event e of Fin, the sets $pre(e)$ and $post(e)$ are finite, and
(3) For every $k \geq 0$, Fin contains only finitely many events e such that $d(e) \leq k$

This works exactly as in [10], as shown in [18], with minor adaptations to the generalization of cut-offs in the symbolic unfolding in (1). □

Proposition 7. *Fin is complete.*

The proof also has the same general structure as the respective proof in [10]. However, since here we use the generalizations of adequate order, possible extensions, and the cut-off criterion to symbolic branching processes, we include the complete proof in the body of the paper.

Notation. For functions $f : X \to Y$ and $f' : X' \to Y$ with $X \cap X' = \emptyset$ we define $f \uplus f' : X \cup X' \to Y$ by mapping x to $f(x)$ if $x \in X$ and to $f'(x)$ if $x \in X'$.

Proof of Proposition. 7 We first show that for every reachable marking in N there exists a configuration in Υ satisfying a) from the definition of complete prefixes, and then show that one of these configurations (a minimal one) also satisfies b).

(1) Let M be an arbitrary reachable marking in N. Then by Proposition 1, we have that there is a $C_1 \in \mathcal{C}(\Upsilon)$ s.t. $M \in \mathcal{M}(C_1)$. Let $\theta_1 \in \Theta(C_1)$ s.t. $M = \mathrm{mark}(C_1.\theta_1)$. If C is not a configuration in *Fin*, then it contains a cut-off event e_1, and so $C_1 = [e_1] \oplus D$ for some set D of events. Let $M_1 = \mathrm{mark}([e_1].\theta_1|_{Var_{[e_1] \cup \{\perp\}}}) \in \mathcal{M}([e_1])$. By the definition of cut-off event, there exists an event e_2 with $[e_2] \prec [e_1]$ and $M_1 \in \mathcal{M}([e_2])$. Since we have $C_1 \llbracket M_1 \rrbracket D$, we get by Proposition 4 that the monomorphism $\varphi_1 := \varphi_{[e_1],D}^{[e_2]} : O([e_1]|D) \to \Uparrow[e_2]$ exists and that $\varphi_1(D)$ is a suffix of $[e_2]$. By Proposition 5 we know

$$C_2 := [e_2] \oplus \varphi_1(D) \prec [e_1] \oplus D = C_1.$$

Let $\theta_2' \in \Theta([e_2])$ s.t. $M_1 = \mathrm{mark}([e_2], \theta_2')$. Define now $\theta_2 \in \Theta(C_2)$ by $\theta_2 = \theta_2' \uplus \theta_2''$, where $\theta_2'' : Var_{\varphi_1(D)} \to Col$ is given by $\theta_2''(v_{\varphi_1(e)}) = \theta_1(v_e)$. By this construction we get $M = \mathrm{mark}(C_2, \theta_2) \in \mathcal{M}(C_2)$.
If C_2 is not a configuration of *Fin*, then we can iterate the procedure and find a configuration C_3 such that $C_3 \prec C_2$ and $M \in \mathcal{M}(C_3)$. The procedure cannot be iterated infinitely often because $\prec$ is well-founded. Therefore, it terminates in a configuration of *Fin*.

(2) Let now C be a minimal configuration w.r.t. $\prec$ s.t. $M \in \mathcal{M}(C)$, and let $t \in T$, $\sigma \in \Sigma(t)$ s.t. $M[t, \sigma\rangle$. If C contains some cut-off event, then we can apply the arguments of a) to conclude that *Fin* contains a configuration $C' \prec C$ such that $M \in \mathcal{M}(C')$. This contradicts the minimality of C. So C contains no cut-off events. Let $\theta \in \Theta(C)$ s.t. $M = \mathrm{mark}(C, \theta)$. Since $pre(t.\sigma) \subseteq M$, we have that there is a co-set $B_{t,\sigma} \subseteq \mathrm{cut}(C)$ s.t. $pre(t, \sigma) = \{\langle h(b), \theta(\mathbf{v_e}(b)) \rangle \mid b \in B_{t,\sigma}\}$. Let now $X := \{\langle b, v \rangle \mid b \in B_{t,\sigma}, \langle h(b), v \rangle \in pre(t)\}$. We then have $\forall \langle b, v \rangle \in X : \sigma(v) = \theta(\mathbf{v_e}(b))$.
We now show that $pred := \iota(t)[v \leftarrow v_e]_{v \in Var(e)} \land \left(\bigwedge_{\langle b,v \rangle \in X} v_e = \mathbf{v_e}(b) \right) \land \bigwedge_{\langle b,v \rangle \in X} pred(\mathbf{e}(b))$ is satisfiable. Let $\theta' := \theta \uplus (\sigma \circ [v_e \mapsto v]_{v \in Var(e)})$. Then
- $\iota(t)[v \leftarrow v_e]_{v \in Var(e)}[\theta'] \equiv \iota(t)[\sigma] \equiv true$, and
- $\left(\bigwedge_{\langle b,v \rangle \in X} v_e = \mathbf{v_e}(b) \right)[\theta'] \equiv \left(\bigwedge_{\langle b,v \rangle \in X} \sigma(v) = \theta(\mathbf{v_e}(b)) \right) \equiv true$, and

- $\bigwedge_{\langle b,v \rangle \in X} pred(\mathbf{e}(b))[\theta'] \equiv \bigwedge_{\langle b,v \rangle \in X} pred(\mathbf{e}(b))[\theta] \equiv true$, since $\theta \in \Theta(C)$.

Thus, $pred[\theta'] \equiv true$. Therefore, $e = \langle t, X, pred \rangle$ is a possible extension and added in the execution of the algorithm. Then we directly have $e \notin C$, $h(e) = t$, and with the same arguments as in a), we get $C \cup \{e\} \in \mathcal{C}(Fin)$ and $\theta \uplus (\sigma \circ [v_e \mapsto v]_{v \in Var(e)}) \in \Theta(C \cup \{e\})$, which means $cut(C, \theta)[e, \sigma]$. Since we chose θ independently of t and σ, this concludes the proof. $\square$

Notice that by this construction, as described in [10], we get that if $\prec$ is a total order, then Fin contains at most $|\mathcal{R}(N)|$ non cut-off events. As already discussed in Sect. 3.1, the total adequate order defined in [10] can be lifted to the configurations in the symbolic unfolding, where it again is total (cp. App ??). Thus, we generalized the possibility to construct such a small complete finite prefix by application of Algorithm 1 with $\prec$ being a total adequate order.

Running Example. For the example N from Fig 1a, the algorithm produces the complete finite prefix marked by the red, dotted line in Fig 1b: starting with the initial condition b_1, the event e_1 is the only possible extension and added to Fin. Since e_1 is obviously not a cut-off event, e_2^1 and e_3^1 are possible extensions and also added. Now we have $\mathcal{M}([e_2^1]) = \{\!|\ \langle p_2, i \rangle\ |\} \mid 0 < i \leq m\}$, and $\mathcal{M}([e_1]) = \{\{\langle p_2, 1 \rangle\}\}$, so e_2^1 is also not a cut-off event, and the possible extensions e_3^2 and e_2^2 are added. Now, however, we have that $\mathcal{M}([e_2^2]) = \{\!|\ \langle p_2, i \rangle\ |\} \mid 0 < i \leq m\} = \mathcal{M}([e_2^1])$, and therefore, e_2^2 is a cut-off event.

3.3 High-Level Versus P/T Expansion

Every high-level Petri net represents a P/T Petri net with the same behavior, in which the places can only carry a number tokens with color •. Markings in a P/T Petri net describe only how many tokens lie on each place. Each transitions only has one possible firing mode that takes and/or lays a fixed number of tokens from resp. onto each connected place.

In this section we state in Lemma 2 that the expansion of a finite complete prefix of the unfolding of a high-level Petri net is a finite and complete prefix of the unfolding of the expanded high-level Petri net. This means the generalization of complete prefixes is "canonical", and compatible with the established low-level concepts. We then shortly compare the results of

- applying the generalized ERV-algorithm Algorithm 1 to obtain a complete finite prefix of the symbolic unfolding of a given high-level Petri net, and
- first expanding a given high-level Petri net and then applying the ERV-algorithm from [10] for a complete finite prefix of the (P/T) unfolding.

The procedure of constructing the represented P/T Petri net $Exp(N)$ (called the *expansion*) of a high-level Petri net N is well established (cp., e.g., Chap. 2.4 in [12]), and we describe it here only briefly; the places of $Exp(N)$ are given by $\mathsf{P} = \{p.c \mid p \in P, c \in Col\}$, and its transitions by $\mathsf{T} = \{t.\sigma \mid t \in T, \sigma \in \Sigma(t)\}$. There is an arc from $p.c$ to $t.\sigma$ iff $\langle p, c \rangle \in pre(t, \sigma)$, and analogously for arcs from transitions to places. Markings in $Exp(N)$ are functions $\mathsf{M} : \mathsf{P} \to \mathbb{N}$, describing how often the only color • lies on each place $p.c$. Every such marking corresponds

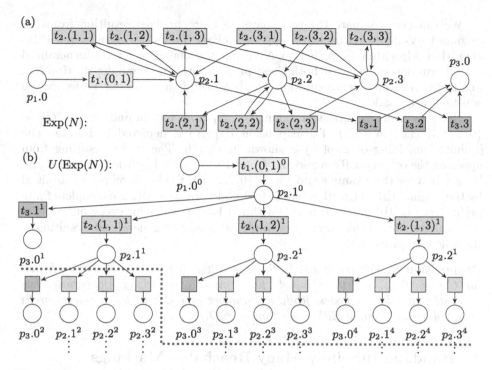

Fig. 2. The expansion $\mathrm{Exp}(N)$ of the running example N from Fig. 1a for $Col = \{0, 1, 2, 3\}$ in (a), and (a prefix of) the respective unfolding $\Upsilon(\mathrm{Exp}(N))$ in (b).

to a marking M in the high-level net N, with $M(p, c) = \mathsf{M}(p.c)$, and a transition t can fire in mode σ from M iff $t.\sigma$ can fire from M. Thus, we say that N and $\mathrm{Exp}(N)$ have the same behavior. For a finite high-level Petri net N, the expansion $\mathrm{Exp}(N)$ is finite iff Col is finite.

For a high-level occurrence net O, we define the P/T net $\mathrm{Exp}_O(O) := U(\mathrm{Exp}(O))$. The operator Exp_O maps high-level occurrence nets to occurrence nets, which is shown in [4]. We denote $\Upsilon(\mathrm{Exp}(O)) = \langle \mathrm{Exp}_O(O), \pi_O \rangle$. Let now $\beta = \langle O, h \rangle$ be a symbolic branching process of N. Then we can define the *expanded symbolic branching process* $\mathrm{Exp}_O(\beta) = \langle \mathrm{Exp}_O(O), h_O \rangle$ of $\mathrm{Exp}(N)$ with the homomorphism $h_O : \mathrm{Exp}_O(O) \rightarrow \mathrm{Exp}(N)$, defined by $h_O(\mathsf{e}) = t.\sigma \Leftrightarrow \pi_O(\mathsf{e}) = e.\sigma \wedge h(e) = t$ and $h_O(\mathsf{b}) = p.c \Leftrightarrow \pi_O(\mathsf{b}) = e.c \wedge h(b) = p$ for events e resp. conditions b in $\mathrm{Exp}_O(O)$. The following result is shown in [4].

Lemma 1 ([4], Sec. 4.1). $\Upsilon(\mathrm{Exp}(N)) \simeq \mathrm{Exp}_O(\Upsilon(N))$

With this result, we state the following:

Lemma 2. *Let N be a high-level Petri net and β be a prefix of $\Upsilon(N)$. Then β is finite and complete iff $\mathrm{Exp}_O(\beta)$ is a finite and complete prefix of $\Upsilon(\mathrm{Exp}(N))$.*

The detailed proof is moved to [18]. It mainly uses the results from Propositions 1 and 2, since the definition of completeness on the symbolic level is a direct translation from its P/T analogue.

We can now compare the two complete finite prefixes resulting from the original ERV-algorithm from [10] applied to $\text{Exp}(N)$ and the generalized ERV-algorithm Algorithm 1 applied to N. From the definition of the generalized cut-off criterion we get that both these prefixes have the same depth. However, due to the high-level representation, the breadth of the symbolic prefix can be substantially smaller.

Running Example. Consider again $N \in \mathbf{N_f}$ from Fig. 1a and assume $Col = \{0, 1, 2, 3\}$ (i.e., $m = 3$). The expansion $\text{Exp}(N)$ is depicted in Fig. 2a. The (infinite) unfolding of $\text{Exp}(N)$ is shown in Fig. 2b. The prefix resulting from applying the original ERV-algorithm from [10] is marked by the red dotted line. We see that for this example and $Col = \{0, \ldots, m\}$, the low-level prefix obtained by the original ERV-algorithm has $O(m^2)$ nodes. In contrast, the complete finite prefix (cp. Fig. 1b) obtained by Algorithm 1 has 11 nodes for every m.

The structure of this running example can easily be generalized, resulting in the following proposition.

Proposition 8. *For every $a \in \mathbb{N}$ there is a high-level net $N \in \mathbf{N_f}$ such that for $Col = \{0, \ldots, m\}$ the complete finite prefix obtained by Algorithm 1 has a constant number of nodes, while the number of nodes in the low-level prefix obtained by the original ERV-algorithm is in $O(m^a)$.*

4   Handling Infinitely Many Reachable Markings

Unfoldings of unbounded P/T Petri nets (i.e., with infinitely many markings) have been investigated in [1,7], and in [11] concurrent well-structured transition systems with infinite state space are unfolded. When applying the generalized ERV-algorithm, Algorithm 1, to high-level Petri nets with infinitely many reachable markings (therefore violating *(3)* from the definition of $\mathbf{N_f}$), the proof for finiteness of the resulting prefix does not hold anymore: the proof of Proposition 6, step (1), is a generalization of the proof of the respective claim in [10] (which uses the pigeonhole principle). It is argued that we cannot have $|\mathcal{R}(N)| + 1$ consecutive events s.t. their cone configurations each generate a marking in the net not seen before, and we thus have a cut-off event. When we deal with infinitely many markings, this argument cannot be made.

In this section, we introduce a class $\mathbf{N_{sc}}$ of safe high-level nets, called symbolically compact, that have possibly infinitely many reachable markings (and therefore an infinite expansion), generalizing the class $\mathbf{N_f}$. We then proceed to make adaptions to Algorithm 1 (i.e., to the used cut-off criterion), so that it generates a finite and complete prefix of the symbolic unfolding for any $N \in \mathbf{N_{sc}}$.

The following Lemma precisely describes the finite high-level Petri nets for which a finite and complete prefix of the symbolic unfolding exists.

Lemma 3. *For a finite high-level Petri net $N = \langle \mathcal{N}, \mathcal{M}_0 \rangle$ there exists a finite and complete prefix of $\Upsilon(N)$ if and only if there exists a bound $n \in \mathbb{N}$ such that every marking in $\mathcal{R}(N)$ is reachable from a marking in $\mathcal{M}_0$ by firing at most n transitions.*

For the proof (cp. App ??), we argue that in the case of such a bound, the symbolic unfolding up to depth $n + 1$ is a finite and complete prefix, and that in the absence of such a bound no depth of a prefix is enough for it to be complete.

4.1   Symbolically Compact High-Level Petri Nets

We use the result of Lemma 3 to define the class $\mathbf{N_{sc}}$ of high-level nets for which we adapt the algorithm for constructing finite and complete prefixes of the symbolic unfolding.

Definition 8 (Class $\mathbf{N_{sc}}$). *A finite high-level Petri net N is called symbolically compact if it satisfies (1) and (2) from Definition 4, and*

(3*) There is a bound $n \in \mathbb{N}$ on the number of transition firings needed to reach all markings in $\mathcal{R}(N)$.

The class $\mathbf{N_{sc}}$ contains all symbolically compact high-level Petri nets.

Note that in the case of a (finite, safe) P/T net, property *(3*)* is equivalent to *(3)* (i.e., $|\mathcal{R}(N)| < \infty$). However, this is *not* true for all high-level nets N: while $|\mathcal{R}(N)| < \infty$ still implies *(3*)* (meaning $\mathbf{N_f} \subseteq \mathbf{N_{sc}}$), the reverse implication does not hold, as our running example from Fig. 1a demonstrates when we change the set of colors to $Col = \mathbb{N}$: it still satisfies *(1)* and *(2)*, with $\mathcal{R}(N) = \{\!|\ \langle p_1, 0 \rangle\ |\!\}, \{\!|\ \langle p_3, 0 \rangle\ |\!\}\} \cup \{\{\!|\ \langle p_2, \ell \rangle\ |\!\}\ |\ \ell \in \mathbb{N}\}$. So we have infinitely many markings that can all be reached by firing at most two transitions, meaning the net satisfies *(3*)* and is therefore symbolically compact.

Lemma 3 implies that the class $\mathbf{N_{sc}}$ of symbolically compact nets contains exactly all high-level Petri nets satisfying *(1)* and *(2)* for which a finite and complete prefix of the symbolic unfolding exists (independently of the number of reachable markings). Since the reachable markings of a high-level Petri net and its expansion correspond to each other, this observation leads to an interesting subclass $\mathbf{N_{sc}} \setminus \mathbf{N_f}$ of symbolically compact high-level Petri nets that have infinitely many reachable markings (such as our running example from Fig. 1a with $Col = \mathbb{N}$). For every net N in this subclass

- there exists a finite and complete prefix of $\Upsilon(N)$,   but
- there does *not* exist a finite and complete prefix of $\Upsilon(\mathrm{Exp}(N))$.

In particular, the original ERV-algorithm cannot be applied to $\mathrm{Exp}(N)$, since the expansion is an infinite net.

For the rest of the paper, let $N = \langle P, T, F, \iota, \mathcal{M}_0 \rangle \in \mathbf{N_{sc}}$ with symbolic unfolding $\Upsilon = \langle U, \pi \rangle = \langle B, E, G, \iota, \mathcal{K}_0, \pi \rangle$.

4.2   The Finite Prefix Algorithm for Symbolically Compact Nets

As previously discussed, the argument that states the existence of one event in a chain of $|\mathcal{R}(N)| + 1$ consecutive events, such that every marking represented by its cone configuration is contained in the union of all markings represented by previous cone configurations, cannot be applied in the case of an infinite number of reachable markings. Consequently, Algorithm 1 may not terminate when applied to a net in $\mathbf{N_{sc}} \setminus \mathbf{N_f}$. However, condition (3^*) guarantees that every marking reached by a cone configuration $[e]$ with depth $> n$ can be reached by a configuration C containing no more than n events.

For the algorithm to terminate, we need to adjust the cut-off criterion since we do not know whether C is also a cone configuration, as demanded in Definition 6. Therefore, we define *cut-off* events*, that generalize cut-off events. They only require that every marking in $\mathcal{M}([e])$ has been observed in a set $\mathcal{M}(C)$ for *any* configuration $C \prec [e]$, rather than just considering cone configurations:

Definition 9 (Cut-off* event). *Under the assumptions of Definition 6, the high-level event e is a cut-off* event (w.r.t. $\prec$) if $\mathcal{M}([e]) \subseteq \bigcup_{C \prec [e]} \mathcal{M}(C)$.*

We additionally assume that the used adequate order satisfies $|C_1| < |C_2| \Rightarrow C_1 \prec C_2$, so that every event with depth $> n$ will be a cut-off event. Since all adequate orders discussed in [10] satisfy this this property (cp. App ??), this is a reasonable requirement. This adaption and assumption now lead to:

Theorem 1. *Assume a given adequate order $\prec$ to satisfy $|C_1| < |C_2| \Rightarrow C_1 \prec C_2$. When replacing in Algorithm 1 the term "cut-off event" by "cut-off* event", it terminates for any input net $N \in \mathbf{N_{sc}}$, and generates a complete finite prefix of $\Upsilon(N)$.*

Proof. The properties of symbolic unfoldings that we stated in Sect. 2.3 are independent on the class of high-level nets. Definition 10 only uses that the considered net is safe, and so do Propositions 4 and 5. We therefore only have to check that the correctness proof for the algorithm still holds. In the proof of Proposition 6 (*Fin* is finite), the steps (2) and (3) are independent of the used cut-off criterion. In step (1), however, it is shown that the depth of events never exceeds $|\mathcal{R}(N)| + 1$. This is not applicable when $|\mathcal{R}(N)| = \infty$, as argued above. Instead we show:

(1*) For every event e of *Fin*, $d(e) \leq n + 1$, where n is the bound on the number of transitions needed to reach all markings in $\mathcal{R}(N)$.

This is done in detail in [18] and proves that *Fin* is finite. In the proof of Proposition 7, the cut-off criterion is used to show (by an infinite descent approach), for any marking $M \in \mathcal{R}(N)$ the existence of a minimal configuration $C \in$ *Fin* with $M \in \mathcal{M}(C)$. Due to the similarity of cut-off and cut-off*, this proof can easily be adapted to work as before.

The only thing remaining to show is termination. In the case of nets in $\mathbf{N_f}$, every object is finite, which, together with Proposition 6, leads to termination of the algorithm. For nets in $\mathbf{N_{sc}} \setminus \mathbf{N_f}$, however, there is at least one event e in *Fin* s.t. $|\mathcal{M}([e])| = \infty$. Thus, we have to show that we can check the cut-off* criterion in finite time. This follows from Corollary 2 in the next section, which is dedicated to symbolically representing markings generated by configurations. $\square$

4.3   Checking Cut-offs Symbolically

We show how to check whether a high-level event e is a cut-off* event in finite time. By definition, this means checking whether $\mathcal{M}([e]) \subseteq \bigcup_{C \prec [e]} \mathcal{M}(C)$. However, since the cut of a configuration can represent infinitely many markings, we cannot simply store the set $\mathcal{M}(C)$ for every $C \in \mathcal{C}(Fin)$. Instead, we now define constraints that symbolically describe the markings represented by a configuration's cut. Checking the inclusion above then reduces to checking an implication of these constraints. Since we consider high-level Petri nets with guards written in a decidable first order theory, such implications can be checked in finite time.

We first define for every condition b a new predicate $pred^{\odot}(b)$ by

$$pred^{\odot}(b) := pred(\mathbf{e}(b)) \wedge (b = \mathbf{v_e}(b)).$$

This predicate now has (in an abuse of notation) an extra variable, called b. The remaining variables in $pred(\mathbf{e}(b))$ are $Var_{[\mathbf{e}(b)] \cup \{\bot\}}$, and $pred(\mathbf{e}(b))$ evaluates to $true$ under an assignment $\theta : Var_{[\mathbf{e}(b)] \cup \{\bot\}} \rightarrow Col$ if and only if a concurrent execution of $[\mathbf{e}(b)]$ with the assigned modes is possible (i.e., under every instantiation of $[\mathbf{e}(b)]$). In such an execution, $\theta(\mathbf{v_e}(b)) \in Col$ is placed on b.

For a co-set $B' \subseteq B$ of high-level conditions, the constraint on B' is an expression over B' describing which color combinations can lie on the high-level conditions. We build the conjunction over all predicates $pred^{\odot}(b)$ for $b \in B'$ and quantify over all appearing variables v_e: the *constraint on* B' is defined by

$$\kappa(B') := \exists_{\bigcup_{b \in B'} Var_{[\mathbf{e}(b)] \cup \{\bot\}}} : \bigwedge_{b \in B'} pred^{\odot}(b),$$

where B' serves as the set of free variables in $\kappa(B')$.

We denote by $\Xi(B')$ the set of variable assignments $\vartheta : B' \rightarrow Col$ that satisfy $\kappa(B')[\vartheta] \equiv true$. Note that for a configuration C, we have $\bigcup_{b \in \text{cut}(C)} Var_{[\mathbf{e}(b)]} = Var_C$, i.e., the bounded variables in $\kappa(\text{cut}(C))$ are exactly the variables appearing in predicates in C. For every instantiation θ of C we define a variable assignment $\vartheta_\theta : \text{cut}(C) \rightarrow Col$ by setting $\forall b \in \text{cut}(C) : \vartheta_\theta(b) = \theta(\mathbf{v_e}(b))$. Instantiations of a configuration and the constraint on its cut are now related as follows.

Lemma 4. *Let* $C \in \mathcal{C}(\Upsilon)$. *Then* $\Xi(\text{cut}(C)) = \{\vartheta_\theta \mid \theta \in \Theta(C)\}$.

The proof is moved to [18], and follows by construction of $pred^{\odot}$ and ϑ_θ. From the definition of $\mathcal{K}(C)$ and $\mathcal{M}(C)$ we get:

Corollary 1. *Let* $C \in \mathcal{C}(\Upsilon)$. *Then* $\mathcal{K}(C) = \{\{\langle b, \vartheta(b)\rangle \mid b \in \text{cut}(C)\} \mid \vartheta \in \Xi(\text{cut}(C))\}$ *and* $\mathcal{M}(C) = \{\!\{\langle \pi(b), \vartheta(b)\rangle \mid b \in \text{cut}(C)\}\!\} \mid \vartheta \in \Xi(\text{cut}(C))\}$.

We now show how to check whether an event is a cut-off* event via the constraints defined above. For that, we first look at general configurations in Theorem 2, and then explicitly apply this result to cone configurations $[e]$ in Corollary 2.

Since we consider safe high-level Petri nets, we can relate two cuts representing the same marking in the following way:

Definition 10. *Let* $C_1, C_2 \in \mathcal{C}(\Upsilon)$ *with* $\pi(\text{cut}(C_1)) = \pi(\text{cut}(C_2))$. *Then there is a unique bijection* $\phi : \text{cut}(C_1) \rightarrow \text{cut}(C_2)$ *preserving* π. *We call this mapping* $\phi_{C_1}^{C_2}$.

Theorem 2. *Let* $C, C_1, \ldots, C_n$ *be finite configurations in the symbolic unfolding of a safe high-level Petri net s.t.* $\forall 1 \leq i \leq n : \pi(\text{cut}(C)) = \pi(\text{cut}(C_i))$. *Then*

$$\mathcal{M}(C) \subseteq \bigcup_{i=1}^{n} \mathcal{M}(C_i) \quad \text{if and only if} \quad \kappa(\text{cut}(C)) \Rightarrow \bigvee_{i=1}^{n} \kappa(\text{cut}(C_i))[\phi_{C_i}^C].$$

Proof. Denote $\phi_i := \phi_{C_i}^C$. Assume $\mathcal{M}(C) \subseteq \bigcup_{i=1}^n \mathcal{M}(C_i)$ and let $\vartheta \in \Xi(\text{cut}(C))$. By Corollary 1 we have that $M_\vartheta := \{\langle \pi(b), \vartheta(b)\rangle \mid b \in \text{cut}(C)\} \in \mathcal{M}(C)$. Thus, $\exists 1 \leq i \leq n : M_\vartheta \in \mathcal{M}(C_i)$. This, again by Corollary 1, means $\exists \vartheta_i \in \Xi(\text{cut}(C_i))$:

$$M_\vartheta = \{\langle \pi(b'), \vartheta_i(b')\rangle \mid b' \in \text{cut}(C_i)\} = \{\langle \pi(\phi_i^{-1}(b)), \vartheta_i(\phi_i^{-1}(b))\rangle \mid b \in \text{cut}(C)\}$$
$$= \{\langle \pi(b), (\vartheta_i \circ \phi_i^{-1})(b)\rangle \mid b \in \text{cut}(C)\}.$$

This shows that $\vartheta|_{\text{cut}(C)} = \vartheta_i \circ \phi_i^{-1}$. Thus, $\kappa(\text{cut}(C_i))[\phi_i][\vartheta] \equiv \kappa(\text{cut}(C_i))[\vartheta \circ \phi_i] \equiv \kappa(\text{cut}(C_i))[\vartheta_i \circ \phi_i^{-1} \circ \phi_i] \equiv \kappa(\text{cut}(C_i))[\vartheta_i] \equiv true$, which proves the implication.

Assume on the other hand $\kappa(\text{cut}(C)) \Rightarrow \bigvee_{i=1}^n \kappa(\text{cut}(C_i))[\phi_i]$. Let $M \in \mathcal{M}(C)$. Then $\exists \vartheta \in \Xi(\text{cut}(C)) : M = \{\langle \pi(b), \vartheta(b)\rangle \mid b \in \text{cut}(C)\}$. Thus, $\exists 1 \leq i \leq n :$ $\kappa(\text{cut}(C_i))[\phi_i][\vartheta] \equiv true$. Let $\vartheta_i = \vartheta \circ \phi_i$. Then $\vartheta_i \in \Xi(\text{cut}(C_i))$, and $M_{\vartheta_i} :=$ $\{\langle \pi(b'), \vartheta_i(b')\rangle \mid b' \in \text{cut}(C_i)\} \in \mathcal{M}(C_i)$. Since

$$M_{\vartheta_i} = \{\langle \pi(\phi_i^{-1}(b)), \vartheta \circ \phi_i(\phi_i^{-1}(b))\rangle \mid b \in \text{cut}(C)\} = \{\langle \pi(b), \vartheta(b)\rangle \mid b \in \text{cut}(C)\},$$

we have $M = M_{\vartheta_i} \in \mathcal{M}(C_i)$, which completes the proof. $\qquad\square$

The following Corollary now gives us a characterization of cut-off* events in a symbolic branching process. It follows from Theorem 2 together with the facts that $\mathcal{M}(C_1) \cap \mathcal{M}(C_2) \neq \emptyset \Rightarrow \pi(\text{cut}(C_1)) = \pi(\text{cut}(C_2))$, and that $\prec[e]$ is finite.

Corollary 2. *Let β be a symbolic branching process and e an event in β. Then e is a cut-off* event in β if and only if*

$$\kappa(\text{cut}([e])) \Rightarrow \bigvee_{\substack{C \prec [e] \\ h(\text{cut}(C))=h(\text{cut}([e]))}} \kappa(\text{cut}(C))[\phi_C^{[e]}].$$

Thus, we showed how to decide for any event e added to a prefix of the unfolding whether it is a cut-off* event, namely, by checking the above implication in Corollary 2. Note that we can also check whether e is a *cut-off* event (w.r.t. Definition 6) by the implication in Corollary 2 when we replace all occurrences of "C" by "$[e']$" .

5   Conclusions and Outlook

We introduced the notion of complete finite prefixes of symbolic unfoldings of high-level Petri nets. We identified a class of 1-safe high-level nets generalizing 1-safe P/T nets, for which we generalized the well-known algorithm by Esparza et al. to compute such a finite and complete prefix. This constitutes a consolidation and generalization of the concepts of [3–5,10]. While the resulting symbolic prefix has the same depth as a finite and complete prefix of the unfolding of the represented P/T net, it can be significantly smaller due to less branching. In the case of infinitely many reachable markings (where the original algorithm is not applicable) we identified the class of so-called *symbolically compact* nets for which an adapted version of the generalized algorithm works. For that, we showed how to check an adapted cut-off criterion by symbolically describing sets of markings.

The next step is an implementation of the generalized algorithm. Future works also include the generalization for k-bounded high-level Petri nets.

References

1. Abdulla, P.A., Iyer, S.P., Nylén, A.: Unfoldings of unbounded Petri nets. In: Emerson, E.A., Sistla, A.P. (eds.) CAV 2000. LNCS, vol. 1855, pp. 495–507. Springer, Heidelberg (2000). https://doi.org/10.1007/10722167_37
2. Best, E., Grahlmann, B.: Programming Environment based on Petri nets - Documentation and User Guide Version 1.4 (1995). https://uol.de/f/2/dept/informatik/ag/parsys/PEP1.4_man.ps.gz?v=1346500853
3. Chatain, T.: Symbolic unfoldings of high-level Petri nets and application to supervision of distributed systems, Ph. D. thesis, Universit é de Rennes (2006). https://www.sudoc.fr/246936924
4. Chatain, T., Fabre, E.: Factorization properties of symbolic unfoldings of colored Petri nets. In: Lilius, J., Penczek, W. (eds.) PETRI NETS 2010. LNCS, vol. 6128, pp. 165–184. Springer, Heidelberg (2010). https://doi.org/10.1007/978-3-642-13675-7_11
5. Chatain, T., Jard, C.: Symbolic diagnosis of partially observable concurrent systems. In: de Frutos-Escrig, D., Núñez, M. (eds.) FORTE 2004. LNCS, vol. 3235, pp. 326–342. Springer, Heidelberg (2004). https://doi.org/10.1007/978-3-540-30232-2_21
6. Chatain, T., Jard, C.: Complete finite prefixes of symbolic unfoldings of safe time Petri nets. In: Donatelli, S., Thiagarajan, P.S. (eds.) ICATPN 2006. LNCS, vol. 4024, pp. 125–145. Springer, Heidelberg (2006). https://doi.org/10.1007/11767589_8
7. Desel, J., Juhás, G., Neumair, C.: Finite unfoldings of unbounded Petri nets. In: Cortadella, J., Reisig, W. (eds.) ICATPN 2004. LNCS, vol. 3099, pp. 157–176. Springer, Heidelberg (2004). https://doi.org/10.1007/978-3-540-27793-4_10
8. Ehrig, H., Hoffmann, K., Padberg, J., Baldan, P., Heckel, R.: High-level net processes. In: Brauer, W., Ehrig, H., Karhumäki, J., Salomaa, A. (eds.) Formal and Natural Computing. LNCS, vol. 2300, pp. 191–219. Springer, Heidelberg (2002). https://doi.org/10.1007/3-540-45711-9_12
9. Engelfriet, J.: Branching processes of Petri nets. Acta Informatica **28**(6), 575–591 (1991). https://doi.org/10.1007/BF01463946
10. Esparza, J., Römer, S., Vogler, W.: An improvement of McMillan's unfolding algorithm. Formal Methods Syst. Des. **20**(3), 285–310 (2002). https://doi.org/10.1023/A:1014746130920
11. Herbreteau, F., Sutre, G., Tran, T.Q.: Unfolding concurrent well-structured transition systems. In: Grumberg, O., Huth, M. (eds.) TACAS 2007. LNCS, vol. 4424, pp. 706–720. Springer, Heidelberg (2007). https://doi.org/10.1007/978-3-540-71209-1_55
12. Jensen, K.: Coloured Petri nets - basic concepts, analysis methods and practical use - volume 1, Second Edition. Monographs in Theoretical Computer Science. An EATCS Series. Springer, Heidelberg (1996). https://doi.org/10.1007/978-3-662-03241-1
13. Khomenko, V., Koutny, M.: Branching processes of high-level Petri nets. In: Garavel, H., Hatcliff, J. (eds.) TACAS 2003. LNCS, vol. 2619, pp. 458–472. Springer, Heidelberg (2003). https://doi.org/10.1007/3-540-36577-X_34
14. McMillan, K.L.: A technique of state space search based on unfolding. Formal Methods Syst. Des. **6**(1), 45–65 (1995). https://doi.org/10.1007/BF01384314
15. Nielsen, M., Plotkin, G.D., Winskel, G.: Petri nets, event structures and domains, part I. Theor. Comput. Sci. **13**, 85–108 (1981). https://doi.org/10.1016/0304-3975(81)90112-2

16. Presburger, M.: über die Vollständigkeit eines gewissen Systems der Arithmetik ganzer Zahlen, in welchem die Addition als einzige Operation hervortritt. In: Proc. Comptes-rendus du I Congrés des Mathématiciens des Pays Slaves, Varsovie 1929, pp. 92–101 (1930)
17. Reisig, W.: Understanding petri nets - modeling techniques, analysis methods, case studies. Springer, Heidelberg (2013). https://doi.org/10.1007/978-3-642-33278-4
18. Würdemann, N., Chatain, T., Haar, S.: Taking complete finite prefixes to high level, symbolically (Full Version) (2023). https://hal.inria.fr/hal-04029490

Interval Traces with Mutex Relation

Ryszard Janicki[1] , Maciej Koutny[2] , and Łukasz Mikulski[3]([⊠])

[1] Department of Computing and Software, McMaster University,
Hamilton, ON L8S 4K1, Canada
janicki@mcmaster.ca
[2] School of Computing, Newcastle University, Newcastle upon Tyne NE4 5TG, UK
maciej.koutny@ncl.ac.uk
[3] Faculty of Mathematics and Computer Science, Nicolaus Copernicus University,
Toruń 12/18, Poland
lukasz.mikulski@mat.umk.pl

Abstract. Interval traces can model sophisticated behaviours of concurrent systems under the assumptions that all observations/system runs are represented by interval orders and simultaneity is not necessarily transitive. What they cannot model is the case when a and b are considered independent, interleavings ab and ba are deemed equivalent, but simultaneous execution of a and b is disallowed. We introduce a new kind of interval traces, incorporating a *mutex relation*, that can model these kind of cases. We discuss the soundness of this concept and show how it can be applied in the domain of Petri nets.

Keywords: interval order · interval sequence · inhibitor net · mutex relation · semantics

1 Introduction

In concurrency theory, *traces* are quotient equational monoids over various types of sequences. The sequences represent observations or system runs and traces themselves represent sets of sequences that are interpreted as equivalent, so only one sequence can represent the entire trace. This approach was pioneered by Mazurkiewicz [27].

Mazurkiewicz traces (or traces) are partially commutative quotient monoids over sequences [3,27,28]. They have been used to model various aspects of concurrency theory and since the late 1970 s s their theory has been substantially developed [5,6]. Traces can be interpreted as partial orders and can model 'true concurrency', i.e., the case where a simultaneous execution of events a and b, and the orders a followed by b, and b followed by a, are all considered equivalent. As a model of concurrent behaviours, traces correspond to *vector firing sequences* [38] that have been used to model concurrent behaviours in the *path expressions* model [17]. The theory of traces has been used to tackle problems from diverse areas including combinatorics, graph theory, algebra and logic [6]. However, not all important aspects of concurrency can be adequately

L. Gomes and R. Lorenz (Eds.): PETRI NETS 2023, LNCS 13929, pp. 145–166, 2023.
https://doi.org/10.1007/978-3-031-33620-1_8

modelled by the traces. For example, they can neither model 'not later than' relation-ships nor the case when system runs are represented by interval orders, i.e., where simultaneity is not transitive [14, 15, 21].

For the standard traces, the basic monoid is just a free monoid of sequences. This means that generators, i.e., elements of a trace alphabet, have no visible internal struc-ture that could be used to define appropriate equations. This is a limitation, as when the generators have some internal structure (for example, if they are sets, or they are divided into two distinct sets with different properties), this internal structure may be used when defining the set of equations generating the quotient monoid.

One natural extension is to just assume that generators are sets, i.e., we have some monoid of step sequences. An underlying assumption behind this approach is that simultaneity is transitive and simultaneous executions are represented *explicitly* by steps.

In trace theory, if the events a and b are independent, i.e., a-followed-by-b and b-followed-by-a are equivalent in some trace, then the a and b are incomparable in the partial order defined by this trace. Hence, simultaneity can be expressed, though implicitly. Following this idea, the standard traces can be extended to step sequences in a quite natural manner ([29, 41] and, implicitly, in [18, 19]). This extension is useful; for example, it allows analyses of greedy maximal concurrency (cf. [18]), but it still keeps the model within the standard 'true concurrency paradigm'.

As long as it has some interpretation in concurrency theory, one might use freely set theory operators in the equations that define equivalent generators. Exploiting this idea has led to the concepts of *comtraces* [15], *g-comtraces* [19], and *step traces* [12, 13]. Here, step traces are the most advanced and general model. Although step traces are quite a new notion, they have already been successfully utilized in computational biology [33], digital graphics [32], and model checking [25]. Still, they can only be used if event simultaneity is transitive, i.e., all observations of a concurrent system can be represented by stratified orders.

It was argued by Wiener in 1914 [43] (and later, more formally, in [14]) that any execution that can be observed by a single observer must be an interval order, and so the most precise observational semantics is based on interval orders, where simultaneity is often non-transitive. Trace generators are sequences and representing interval orders by sequences is a little bit tricky; for example, one may use sequences of maximal antichains, or sequences of *beginnings* and *endings* of events involved [9, 14], and the latter appears to be more suitable [16]. To model this with traces, we assume that gen-erators are divided into two classes of objects, one interpreted as the *beginnings* and the other as the *endings* (of events/actions, etc.). This has led to the concepts of *interval sequences* and *interval traces* discussed in [21, 22].

While each stratified order is interval, the relationship between comtraces [15], g-comtraces [19], step traces [12] is more complex. Each comtrace can be represented by an appropriate interval trace [21], but there are g-comtraces and step traces that *cannot* be represented as any interval trace [20].

Interval traces *cannot* represent the case when a and b are considered independent, interleavings ab and ba are deemed equivalent, but the simultaneous execution of a

and b is not allowed[1], for example, $a : x = x+1$ and $b : x = x+2$ (cf. [14,26]). Both g-comtraces and step traces can easily model such cases. They both have a kind of explicit *mutex* relation that always forbids simultaneity.

In this paper, we add a *mutex relation* to interval traces, based on the concept of this relation for step traces [12,24]. We will prove validity of this concept, analyse its relationship to interval traces and show how it can be used to model the properties of Petri nets with mutex relation (cf. [23]). Adding mutex relation required some modification of interval traces of [21], and extensions of standard theory of interval orders.

2 Preliminaries

Throughout the paper we mainly use the standard notions of sets, relations and formal languages, extended with a very few new notations.

A binary relation $\equiv$ over a set X is an *equivalence relation* if it is reflexive, symmetric and transitive, i.e., for all $a,b,c \in X$, $a \equiv a, a \equiv b \iff b \equiv a$, and $a \equiv b \equiv c \implies a \equiv c$. An equivalence class containing $a \in X$ is denoted as $[a]_\equiv$.

A *(strict) partial order* is a pair $po = (X, \prec)$ such that X is a set and $\prec$ is a binary relation over X which is irreflexive and transitive, i.e., for all $a,b,c \in X$, $a \not\prec a$ and $a \prec b \prec c \implies a \prec c$. We also define a binary *incomparability* relation on the elements of X: $a \frown b$ if $a \not\prec b \not\prec a \neq b$.

Let $po = (X, \prec)$ and $po' = (X, \prec')$ be partial orders. Then (cf. [9]):

- po is *total* if $\frown = \varnothing$, i.e., for all $a \neq b \in X$, $a \prec b$ or $b \prec a$.
- po is *stratified* if, for all $a,b,c \in X$, $a \frown b \frown c \implies a \frown c$ or $a = c$, i.e., $\frown \cup\, id_X$ is an equivalence relation, where $id_X = \{(x,x) \mid x \in X\}$ is the *identity relation* on X.
- po' is an *extension* of po if $\prec$ is a subset of $\prec'$.

Finite total orders are equivalent to finite sequences of elements without repetitions: if $po = (X, \prec)$ is a total order such that $X = \{a_1,\ldots,a_k\}$ and $a_1 \prec \cdots \prec a_k$ then the corresponding sequence is $seq(po) = a_1 \ldots a_k$. Similarly, finite stratified orders are equivalent to finite sequences of mutually disjoint nonempty sets: if $po = (X, \prec)$ is a stratified order then the corresponding step sequence of po is $sseq(po) = A_1 \ldots A_k$, where $A_1,\ldots,A_k$ is a unique partition of X such that $\prec$ is equal to $\bigcup_{i<j} A_i \times A_j$.

The following result provides basis for various *greedy* canonical forms for concurrent behaviours.

Theorem 1 ([15]). *Each partial order $po = (X, \prec)$ has exactly one stratified order extension spo such that $sseq(spo) = A_1 \ldots A_k$ and, for all $i \geq 2$ and $b \in A_i$, there is $a \in A_{i-1}$ satisfying $a \prec b$.*

The unique stratified order spo is often interpreted as a greedy maximally concurrent representation of po, and will be denoted by $gmcr(po)$. Note that po is a stratified order iff $gmcr(po) = po$.

---

[1] Mazurkiewicz traces do not forbid simultaneous executions of a and b for independent a and b, they just do not express it explicitly.

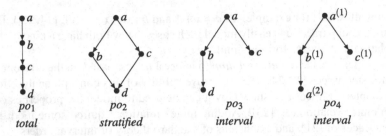

Fig. 1. Partial orders represented as Hasse diagrams (cf. [9]). po_1 is total and uniquely represented by sequence $v = abcd$, i.e., $po_1 = \text{sseq}^{-1}(v)$. Partial order po_2 is stratified and uniquely represented by step sequence $w = \{a\}\{b,c\}\{d\}$, i.e., $po_2 = \text{sseq}^{-1}(w)$. po_3 is interval, but does not have a corresponding sequence or step sequence. po_4 is an interval order on *enumerated* events and has a corresponding (*non*-unique) sequence of begins and ends of events $x = a_{\restriction} a_{\lfloor} c_{\restriction} b_{\restriction} b_{\lfloor} a_{\restriction} a_{\lfloor} c_{\lfloor}$, i.e., $po_4 = \text{intord}(x)$ (see Sect. 6 for details).

3   Interval Orders

A partial order $(X, \prec)$ is *interval* if, for all $a, b, c, d \in X$,

$$a \prec c \wedge b \prec d \implies a \prec d \vee b \prec c.$$

Example 1. Fig. 1 shows different types of partial orders. Note that po_2 is an extension of po_3, and po_1 extends both po_2 and po_3. Also, $gmcr(po_3) = po_2$.   ◇

The adjective 'interval' derives from the following Fishburn's Theorem:

Theorem 2 (Fishburn [8]). *A countable partial order $(X, \prec)$ is interval iff there exists a total order $(Y, \lessdot)$ and two injective mappings $\beta, \varepsilon : X \to Y$ such that $\beta(X) \cap \varepsilon(X) = \varnothing$ and, for all $a, b \in X$, $\beta(a) \lessdot \varepsilon(a)$ and $a \prec b \iff \varepsilon(a) \lessdot \beta(b)$.*

The mappings β and ε are interpreted as the 'beginning of' and 'ending of' actions represented by the elements of X.

The relevance of interval orders in concurrency theory follows from an observation, credited to Wiener [43], that any execution of a physical system that can be observed by a single observer is an interval order. Hence the most precise observational semantics should be defined in terms of interval orders or their suitable representations (cf. [14]).

Example 2. The interval order po_3 from Fig. 1 can be represented by exactly four different ways of totally ordering the values of mappings β and ε, as follows:

$$\beta(a) \lessdot \varepsilon(a) \lessdot \beta(b) \lessdot \beta(c) \lessdot \varepsilon(b) \lessdot \beta(d) \lessdot \varepsilon(c) \lessdot \varepsilon(d)$$
$$\beta(a) \lessdot \varepsilon(a) \lessdot \beta(c) \lessdot \beta(b) \lessdot \varepsilon(b) \lessdot \beta(d) \lessdot \varepsilon(c) \lessdot \varepsilon(d)$$
$$\beta(a) \lessdot \varepsilon(a) \lessdot \beta(b) \lessdot \beta(c) \lessdot \varepsilon(b) \lessdot \beta(d) \lessdot \varepsilon(d) \lessdot \varepsilon(c)$$
$$\beta(a) \lessdot \varepsilon(a) \lessdot \beta(c) \lessdot \beta(b) \lessdot \varepsilon(b) \lessdot \beta(d) \lessdot \varepsilon(d) \lessdot \varepsilon(c).$$

In the case of finite interval orders, the characterisation provided by Theorem 2 can be modified as well as made more concrete so that there is essentially a unique way of ordering the begins and ends of actions, which helps in some proofs.

Theorem 3. *Let* $po = (X, \prec)$ *be a finite partial order.*

1. *po is interval iff there exist two mappings* $\beta, \varepsilon : X \to \{1, 2, 3, \ldots, 2 \cdot n - 1, 2 \cdot n\}$ *($n \geq 0$) such that*

$$\beta^{-1}(1), \beta^{-1}(3), \ldots, \beta^{-1}(2 \cdot n - 1) \text{ and } \varepsilon^{-1}(2), \varepsilon^{-1}(4), \ldots, \varepsilon^{-1}(2 \cdot n)$$

are both partitions of X and, for all $a, b \in X$,
   (a) $\beta(a) < \varepsilon(a)$,
   (b) $a \prec b \iff \varepsilon(a) < \beta(b)$.
2. *If po is interval then* β *and* ε *as above are unique.*

Proof. $(1, \implies)$ Let β and ε be as in Theorem 2. Then, since *po* is finite and $\vartriangleleft$ total as well as the two mappings are injective and have disjoint codomains, we have, for some $f_1, \ldots, f_m \in \{\beta, \varepsilon\}$ ($m = 2 \cdot |X|$), the following:

$$f_1(a_1) \vartriangleleft f_2(a_2) \vartriangleleft \cdots \vartriangleleft f_m(a_m),$$

where $\{f_1(a_1), f_2(a_2), \ldots, f_m(a_m)\} = \beta(X) \cup \varepsilon(X)$. Hence there are indices (indicating places where we switch between β and ε or vice versa)

$$1 = l_1 < l_2 < \cdots < l_k \leq m$$

such that $f_{l_j} = \cdots = f_{l_{j+1}-1} \neq f_{l_{j+1}}$ (for $j < k$) and $f_{l_k} = \cdots = f_m$. We then define $\beta'(a_i) = j$ for odd j and $l_j \leq i < l_{j+1}$, and $\varepsilon'(a_i) = j$ for even j and $l_j \leq i < l_{j+1}$ (assuming that $l_{k+1} = m + 1$). It is then straightforward to check that β' and ε' satisfy the requirements.
$(1, \impliedby)$ It is straightforward to derive from β and ε mappings satisfying the assumptions in Theorem 2. This can be done in

$$|\beta^{-1}(1)|! \cdot |\varepsilon^{-1}(2)|! \cdot \ldots \cdot |\beta^{-1}(2 \cdot n - 1)|! \cdot |\varepsilon^{-1}(2 \cdot n)|!$$

different ways.
   (2) Follows directly from the definitions and the construction in part (1).   □

We stress that the uniqueness of β and ε in Theorem 3(2) was possible since the two mappings are no longer required to be injective as in Theorem 2. We will denote these unique mappings and the integer n in the formulation of Theorem 3 by β_{po}, ε_{po} and n_{po}.

Example 3. For the interval order po_3 from Fig. 1 we have:

$$\beta(a) < \varepsilon(a) < \beta(b) = \beta(c) < \varepsilon(b) < \beta(d) < \varepsilon(c) = \varepsilon(d)$$
$$\ 1 \quad\ < 2 \quad\ < 3 \qquad\qquad < 4 \quad\ < 5 \quad\ < 6$$

The mappings from Theorem 3 are related to the ordering of elements.

Proposition 1. *Let $po = (X, \prec)$ be a finite interval order and $a, b \in X$.*

$$\begin{aligned}
\beta_{po}(a) = \beta_{po}(b) &\iff \{c \mid c \prec a\} = \{c \mid c \prec b\} \\
\varepsilon_{po}(a) = \varepsilon_{po}(b) &\iff \{c \mid a \prec c\} = \{c \mid b \prec c\} \\
\beta_{po}(a) < \beta_{po}(b) &\iff a \prec b \vee \exists c : a \frown c \prec b \\
\varepsilon_{po}(a) < \varepsilon_{po}(b) &\iff a \prec b \vee \exists c : a \prec c \frown b.
\end{aligned}$$

Proof. Follows directly from the definitions.   □

Proposition 2. *Let $po_1 = (X_1, \prec_1)$ and $po_2 = (X_2, \prec_2)$ be finite interval orders. Then:*

$$po_1 = po_2 \iff X_1 = X_2 \wedge \beta_{po_1} = \beta_{po_2} \wedge \varepsilon_{po_1} = \varepsilon_{po_2}.$$

Proof. From Theorem 3(2).   □

Also, being a total or stratified order is directly represented by the two mappings.

Proposition 3. *Let $po = (X, \prec)$ be a finite interval order.*

1. po is total iff for $i = 1, 3, \ldots, 2 \cdot n_{po} - 1$,

$$\beta_{po}^{-1}(i) = \varepsilon_{po}^{-1}(i+1) \text{ are singleton sets}.$$

2. po is stratified iff for $i = 1, 3, \ldots, 2 \cdot n_{po} - 1$

$$\beta_{po}^{-1}(i) = \varepsilon_{po}^{-1}(i+1).$$

Proof. Follows directly from the definitions.   □

Note that finite interval orders do not have simple sequential representations in the same way as the total and stratified orders do (cf. po_3 from Fig. 1).

4   Sequences and Partial Orders

Let X be a nonempty set (of symbols). The set of all finite sequences over X, including the empty sequence λ (i.e., the sequence of length zero), is denoted by X^*. We will use the standard notions of concatenation of two sequences w and u, denoted by wu, as well as the notions a prefix of w. A step sequence over X is a finite sequence of nonempty subsets of X.

Associating sequential representations to finite total and stratified orders was straightforward. The converse is not true for arbitrary finite (step) sequences, due to the possibility of repeated occurrences of symbols. To address this problem, one usually proceeds by introducing individual occurrences of symbols, where $a^{(i)}$ represents the i-th occurrence of a. We also denote $\widehat{X} = \{a^{(i)} \mid a \in X \wedge i \geq 1\}$.

The following are useful notions associated with a sequence $w \in X^*$:

- $\text{len}(w)$ is the length of w.
- $\pi_Y(w)$ is the sequence obtained from w after deleting all the symbols in $X \setminus Y$.
- $\text{alph}(w)$ is the set of the symbols occurring within w.

- $\#_a(w)$ is the *number of occurrences* of $a \in X$ within w.
- $\widehat{\text{alph}}(w) = \{a^{(i)} \mid a \in \text{alph}(w) \wedge 1 \leq i \leq \#_w(a)\}$ is the set of symbol occurrences associated with w.
- $\text{pos}_w(a^{(i)}) = \text{len}(u) + 1$ is the *position* of a symbol occurrence $a^{(i)} \in \widehat{\text{alph}}(w)$, where u is the longest prefix of w such that $\#_u(a) = i - 1$.
- $\widehat{w} = \text{pos}_w^{-1}(1) \ldots \text{pos}_w^{-1}(\text{len}(w))$ is the *enumerated representation* of w.
- $\text{totord}(w) = (\widehat{\text{alph}}(w), \{(a^{(i)}, b^{(j)}) \mid \text{pos}_w^{-1}(a^{(i)}) < \text{pos}_w^{-1}(b^{(j)})\})$ is the *total order induced* by w.

The definition of $\text{totord}(w)$ is sound as we have $\text{seq}(\text{totord}(w)) = \widehat{w}$.

5   Mazurkiewicz Traces

Let Σ be a nonempty alphabet of actions (symbols) fixed throughout the rest of this paper. The finite sequences in Σ^* will be called *words*, and the indexed actions in $\widehat{\Sigma}$ will be called *events*.

A *concurrent alphabet* is a pair $\Psi = (\Sigma, ind)$, where $ind \subseteq \Sigma \times \Sigma$ is a reflexive and symmetric *independence* relation on the actions in Σ. The corresponding *dependence* relation is given by $dep = (\Sigma \times \Sigma) \setminus ind$.

A concurrent alphabet Ψ defines an equivalence relation $\equiv_\Psi$ identifying words which differ only by the ordering of independent actions. Two words, $w, v \in \Sigma^*$, satisfy $w \equiv_\Psi v$ if there exists a finite sequence of commutations of adjacent independent actions transforming w into v. More precisely, $\equiv_\Psi$ is a binary relation over Σ^* which is the reflexive and transitive closure of the relation $\sim_\Psi$ such that $w \sim_\Psi v$ if there are $u, z \in \Sigma^*$ and $(a, b) \in ind$ satisfying $w = uabz$ and $v = ubaz$.

Equivalence classes of $\equiv_\Psi$ are called *(Mazurkiewicz) traces* (see [6,27,28]), and the trace containing a given word w is denoted by $[w]_\Psi$. The set of all traces over Ψ is denoted by $\Sigma^*/_{\equiv_\Psi}$, and the pair $(\Sigma^*/_{\equiv_\Psi}, \circ)$ is a (trace) monoid, where $\tau \circ \tau' = [ww']_\Psi$, for any words $w \in \tau$ and $w' \in \tau'$, is the concatenation operation for traces. Note that trace concatenation is well-defined as $[ww']_\Psi = [vv']_\Psi$, for all $w, v \in \tau$ and $w', v' \in \tau'$. Similarly, for every trace $\tau = [w]_\Psi$ and every action $a \in \Sigma$, we can define $\text{alph}(\tau) = \text{alph}(w)$ and $\#_a(\tau) = \#_a(w)$.

Trace equivalence can be characterised in at least two different ways, given below: (i) by considering projections onto binary dependent subalphabets (i.e., $\{a, b\}$ such that $(a, b) \in dep$); and (ii) by considering positions of the occurrences of dependent actions.

Theorem 4. *The following statements are equivalent for all $u, w \in \Sigma^*$:*

1. $u \equiv_\Psi w$.
2. $\pi_{\{a,b\}}(u) = \pi_{\{a,b\}}(w)$, *for all* $(a, b) \in dep$.
3. $\widehat{\text{alph}}(u) = \widehat{\text{alph}}(w)$ *and, for all* $a^{(i)}, b^{(j)} \in \widehat{\text{alph}}(w)$ *satisfying* $(a, b) \in dep$:

$$\text{pos}_u(a^{(i)}) < \text{pos}_u(b^{(j)}) \iff \text{pos}_w(a^{(i)}) < \text{pos}_w(b^{(j)}).$$

Proof. (1) $\iff$ (2) follows from [39], and (1) $\iff$ (3) from [23].   □

Let $\ll$ be an arbitrary total order on Σ extended lexicographically to Σ^*. A sequence $w \in \Sigma^*$ is in *Foata canonical form* w.r.t. the dependence relation *dep* and a lexicographical order $\ll$ on Σ^*, if $w = w_1 \ldots w_n$ $(n \geq 0)$, where:

- each w_i is a nonempty word without multiple occurrences of actions such that the actions of $\mathrm{alph}(w_i)$ are pairwise independent and w_i minimal w.r.t. lexicographical order $\ll$ among $[w_i]_\Psi$, and
- for each $i > 1$ and a occurring in w_i, there is b occurring in w_{i-1} such that $(a,b) \in dep$.

The intuition behind the Foata canonical form is that it groups actions into greedy maximally concurrent steps.

Theorem 5 ([3]). *Every Mazurkiewicz trace has a unique representation in the Foata canonical form.*

The above result is a simple consequence of Theorem 1; however, it was proven independently and before Theorem 1.

6   Interval Sequences

Interval traces — introduced in [22] and substantially refined in [20,21] — have their roots in Mazurkiewicz traces [6,27,28] and Fishburn's representation of interval orders [8]. The latter allows to represent interval orders by suitable sequences of event beginnings and event endings that we call interval sequences. In principle, interval traces are specialized Mazurkiewicz traces over the domain of interval sequences.

For each $a \in \Sigma$ (or $a \in \widehat{\Sigma}$), we will use $a_\uparrow$ to denote the *beginning* of a, and $a_\downarrow$ to denote the *ending* of a. Moreover, for every set $A \subseteq \Sigma$ (or $A \subseteq \widehat{\Sigma}$), we denote $A_\uparrow = \{a_\uparrow \mid a \in A\}$, $A_\downarrow = \{a_\downarrow \mid a \in A\}$, and $A_{\uparrow\downarrow} = A_\uparrow \cup A_\downarrow$.

We would like to emphasise the difference between the notations $\beta(a), \varepsilon(a)$ and $a_\uparrow, a_\downarrow$. The first notation is used for partial orders, so each $\beta(a), \varepsilon(a)$ are unique, while the second notation is for sequences, so both $a_\uparrow$ and $a_\downarrow$ may occur many times.

Definition 1. *A sequence x over $\Sigma_{\uparrow\downarrow}$ is interval if $\pi_{\{a_\uparrow, a_\downarrow\}}(x) \in (a_\uparrow a_\downarrow)^*$, for every $a \in \Sigma$. We then denote by $\mathrm{ev}(x)$ the subset of $\widehat{\Sigma}$ such that $\mathrm{ev}(x) = \widehat{\mathrm{alph}}(x)$. All interval sequences are denoted by IntSeq.*

Example 4. $w = a_\uparrow b_\uparrow b_\downarrow a_\downarrow c_\uparrow a_\uparrow b_\uparrow c_\downarrow b_\downarrow a_\downarrow a_\uparrow a_\downarrow$ is an interval sequence, but neither $a_\downarrow b_\uparrow b_\downarrow a_\uparrow$ nor $b_\uparrow b_\downarrow a_\uparrow c_\downarrow$ is. Moreover, $\mathrm{ev}(w) = \{a^{(1)}, a^{(2)}, a^{(3)}, b^{(1)}, b^{(2)}, c^{(1)}\}$.   ◇

Interval sequences are closed under concatenation.

Proposition 4 ([22]). *For all $x, y \in IntSeq$, $xy \in IntSeq$.*

Interval sequences provide a simple sequence representation of interval orders via the Fishburn representation. They are conceptually close to ST-traces [40,42] proposed earlier. The difference is that ST-traces were defined for Petri nets, whereas interval sequences do not assume any system model.

Every interval sequence x generates a total order $totord(\widehat{alph}(x), \lhd_x)$ as defined at the end of Sect. 4. For example, $x = a_\uparrow b_\uparrow a_\downarrow b_\downarrow a_\uparrow a_\downarrow$ generates

$$a_\uparrow^{(1)} \lhd_x b_\uparrow^{(1)} \lhd_x a_\downarrow^{(1)} \lhd_x b_\downarrow^{(1)} \lhd_x a_\uparrow^{(2)} \lhd_x a_\downarrow^{(2)} \, .$$

Although an interval sequence generates a total order on the beginnings and ends of events it represents, in general there is no similar representation for the events in $\widehat{\Sigma}$ it represents. To achieve the desired result, one needs to switch to interval orders.

Definition 2 ([21,22]). *The* interval order generated *by an interval sequence* $x \in IntSeq$ *is defined as* $intord(x) = (ev(x), \blacktriangleleft_x)$, *where, for all* $a^{(i)}, b^{(j)} \in ev(x)$,

$$a^{(i)} \blacktriangleleft_x b^{(j)} \iff a_\downarrow^{(i)} \lhd_x b_\uparrow^{(j)} \, .$$

Note that, by Theorem 2, $intord(x)$ is an interval order.

Example 5. In Fig. 1, $po_4 = intord(a_\uparrow a_\downarrow c_\uparrow b_\uparrow b_\downarrow a_\uparrow a_\downarrow c_\downarrow)$. In Fig. 2, we have $po_4 = intord(a_\uparrow a_\downarrow b_\uparrow b_\downarrow c_\uparrow d_\uparrow d_\downarrow c_\downarrow)$. Moreover, po_5 is generated by $a_\uparrow a_\downarrow b_\uparrow c_\uparrow b_\downarrow d_\uparrow c_\downarrow d_\downarrow$ as well as thirteen other interval sequences.   ◇

A characterisation of interval orders generated by interval sequences is provided by the next result.

Theorem 6. $intord(x) = (ev(x), \{(a^{(i)}, b^{(j)}) \mid pos_x(a_\downarrow^{(i)}) < pos_x(b_\uparrow^{(j)})\})$, *for every* $x \in IntSeq$.

Proof. Follows directly from the definitions.   □

There is an alternative way of associating 'position' to an event of an interval sequence $x \in IntSeq$ based on Theorem 3. More precisely, for all $a^{(i)}, b^{(j)} \in ev(x)$:

$$\widehat{pos}_x(a_\uparrow^{(i)}) = \beta_{intord(x)}(a^{(i)}) \quad and \quad \widehat{pos}_x(b_\downarrow^{(j)}) = \varepsilon_{intord(x)}(b^{(j)}) \, .$$

For example, if $x = a_\uparrow b_\uparrow a_\downarrow b_\downarrow a_\uparrow a_\downarrow$ then $\widehat{pos}_x(a_\uparrow^{(2)}) = 3$ whereas $pos_x(a_\uparrow^{(2)}) = 5$.

Interval sequences generating the same interval orders assign the same modified positions to events, and they also form a (Mazurkiewicz) trace. To show the latter, let $\Psi_{iseq} = (\Sigma_{\uparrow\downarrow}, ind_{iseq})$ be a concurrent alphabet such that

$$ind_{iseq} = \{(a_\uparrow, b_\uparrow) \mid a \neq b \in \Sigma\} \cup \{(a_\downarrow, b_\downarrow) \mid a \neq b \in \Sigma\} \, .$$

Theorem 7. *The following statements are equivalent, for all* $x, y \in IntSeq$:

1. $intord(x) = intord(y)$.
2. $x \equiv_{\Psi_{iseq}} y$.
3. $ev(x) = ev(y)$ *and* $\widehat{pos}_x = \widehat{pos}_y$.

Proof. $(2) \implies (1)$ By the definition of ind_{iseq} we have $x \sim_{\Psi_{iseq}} y \implies \blacktriangleleft_x = \blacktriangleleft_y$ and $ev(x) = ev(y)$.
$(1) \implies (2)$ Suppose that $x \not\equiv_{\Psi_{iseq}} y$ and $ev(x) = ev(y)$. From the definition of ind_{iseq} it follows that there are $a^{(i)}, b^{(j)} \in ev(x) = ev(y)$ such that $a_\downarrow^{(i)} \lhd_x b_\downarrow^{(j)}$ or $b_\uparrow^{(j)} \lhd_y a_\uparrow^{(i)}$, so by Definition 2, $\blacktriangleleft_x \neq \blacktriangleleft_y$.
$(1) \iff (3)$ It is a consequence of Proposition 2.   □

That is, $[x]_{\equiv_{\Psi_{iseq}}}$ comprises interval sequences generating the same interval order.

7   Interval Traces

An *interval trace alphabet* is a tuple $\Phi = (\Sigma, wind)$, where $wind \subseteq \Sigma \times \Sigma$ is an irreflexive relation called *weak independence*. Intuitively, if $(a,b) \in wind$ then a and b may occur simultaneously, or a may occur before b, with both executions being equivalent. In general, *wind* is not symmetric.

Example 6. If $(a,b) \in wind$ then the interval sequences $a_\uparrow a_\downarrow b_\uparrow b_\downarrow$ (representing a before b) as well as $a_\uparrow b_\uparrow a_\downarrow b_\downarrow$, $b_\uparrow a_\uparrow a_\downarrow b_\downarrow$, $a_\uparrow b_\uparrow b_\downarrow a_\downarrow$, $b_\uparrow a_\uparrow b_\downarrow a_\downarrow$ (all representing simultaneous execution of a and b), will be considered equivalent.   ◇

The following rendering of *wind* as a relation over $\Sigma_{\uparrow\downarrow}$ leads directly to the concept of interval traces.

Definition 3. *Let* $\Phi = (\Sigma, wind)$ *be an interval trace alphabet. Then* $\phi = (\Sigma_{\uparrow\downarrow}, ind_\Phi)$ *is an* internal interval trace alphabet, *where* ind_Φ *is a relation over* $\Sigma_{\uparrow\downarrow}$ *given by:*

$$ind_\Phi = \{(a_\uparrow, b_\uparrow), (a_\downarrow, b_\downarrow) \mid a \neq b \in \Sigma\} \cup \{(a_\downarrow, b_\uparrow), (b_\uparrow, a_\downarrow) \mid (a,b) \in wind\}.$$

The corresponding interval dependence *relation is*

$$\begin{aligned} dep_\Phi &= (\Sigma_{\uparrow\downarrow} \times \Sigma_{\uparrow\downarrow}) \setminus ind_\Phi \\ &= \{(a_\downarrow, b_\uparrow), (b_\uparrow, a_\downarrow) \mid (a,b) \notin wind\} \cup \{(a_\uparrow, a_\uparrow), (a_\downarrow, a_\downarrow) \mid a \in \Sigma\}. \end{aligned}$$

We will skip 'internal' and just write 'interval trace alphabet' for $\phi = (\Sigma_{\uparrow\downarrow}, ind_\Phi)$ which is a well-defined concurrent alphabet. The first component in the formula for ind_Φ follows from the generalisation of the observation that the interval sequences $a_\uparrow b_\uparrow a_\downarrow b_\downarrow$, $b_\uparrow a_\uparrow a_\downarrow b_\downarrow$, $a_\uparrow b_\uparrow b_\downarrow a_\downarrow$, and $b_\uparrow a_\uparrow b_\downarrow a_\downarrow$ represent the same relationships between events, namely that $a^{(1)}$ and $b^{(1)}$ are simultaneous.

As interval traces are a class of Mazurkiewicz traces, we also adapt the standard trace notation of the latter. Moreover, for the reminder of this section, we assume that $\Phi = (\Sigma, wind)$ and $\phi = (\Sigma_{\uparrow\downarrow}, ind_\Phi)$ are fixed.

Definition 4 ([21,22]). *A Mazurkiewicz trace* $[x]_\phi$ *over* $\phi = (\Sigma_{\uparrow\downarrow}, ind_\Phi)$ *is called an* interval trace *if* $[x]_\phi \subseteq IntSeq$.

The soundness of the above definition is due to the following result.

Proposition 5 ([21]). *Let* $\phi = (\Sigma_{\uparrow\downarrow}, ind_\Phi)$ *be an interval trace alphabet, and let* $x, y \in IntSeq$.

1. $[x]_\phi \subseteq IntSeq$.
2. $[x]_\phi [y]_\phi \subseteq [xy]_\phi \subseteq IntSeq$.
3. $\mathrm{intord}(x) = \mathrm{intord}(y) \implies x \equiv_\phi y$.

A result similar to Proposition 4 also holds for interval traces.

Theorem 8. *The following statements are equivalent for all* $x, y \in IntSeq$:

1. $x \equiv_\phi y$.

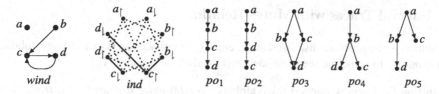

Fig. 2. A weak independence relation *wind*, the interval independence relation *ind* derived from *wind* (the default part given by Definition 3 is represented by dotted lines), and interval orders generated by interval sequences from the interval trace $[a_\uparrow a_\downarrow b_\uparrow b_\downarrow c_\uparrow c_\downarrow d_\uparrow d_\downarrow]_\phi$.

2. $\pi_{\{a_\downarrow, b_\uparrow\}}(x) = \pi_{\{a_\downarrow, b_\uparrow\}}(y)$, *for all* $(a, b) \notin wind$.

3. $\mathrm{ev}(x) = \mathrm{ev}(y)$ *and, for all* $a^{(i)}, b^{(j)} \in \mathrm{ev}(x)$ *such that* $(a, b) \notin wind$:

$$\mathrm{pos}_x(a_\downarrow^{(i)}) < \mathrm{pos}_x(b_\uparrow^{(j)}) \iff \mathrm{pos}_y(a_\downarrow^{(i)}) < \mathrm{pos}_y(b_\uparrow^{(j)}).$$

Proof. (1) $\iff$ (2) follows from [20], and (1) $\iff$ (3) from Proposition 2. □

Since interval traces are a special case of Mazurkiewicz traces, the concept of canonicity applies for them as well. Assuming that $\ll$ is a total ordering of Σ, we extend it to $\Sigma_{\uparrow\downarrow}$ as follows: $a_\uparrow \ll b_\uparrow \ll a_\downarrow \ll b_\downarrow$ whenever $a \ll b$. This new order is called the *natural ordering* of $\Sigma_{\uparrow\downarrow}$, which is then extended to a *lexicographical* order on $\Sigma_{\uparrow\downarrow}^*$ in the standard fashion. Let $y \in [x]_\phi$ be an interval sequence in Foata canonical form with respect to the dependence relation dep_Φ and the natural lexicographical order on $\Sigma_{\uparrow\downarrow}^*$ given above. Since any interval trace is also a Mazurkiewicz trace, there is exactly one such a canonical sequence in $[x]_\phi$.

While interval traces can model a broad range of concurrent behaviours where the observations are represented by interval orders, there are cases that cannot be handled by them. Consider the following example.

Example 7. Let $E = \{a, b, c\}$ where a, b and c are three atomic operations defined as follows (we assume simultaneous reading is allowed):

$$a: x \leftarrow x + 1, \quad b: x \leftarrow x + 2, \quad c: y \leftarrow y + 1.$$

It is reasonable to consider them all as 'concurrent' as any order of their executions, yields exactly the same results. Note that while simultaneous execution of a and c, and b and c are allowed, the simultaneous execution of a and b *is not!*.   ◇

This case cannot be modelled by any interval trace. Had such trace exist, we would have: $a_\uparrow a_\downarrow b_\uparrow b_\downarrow \equiv_\Phi b_\uparrow b_\downarrow a_\uparrow a_\downarrow$, as ab and ba are equivalent executions, but $(a_\uparrow, b_\downarrow) \notin ind_\Phi$ and $(b_\uparrow, a_\downarrow) \notin ind_\Phi$, as the simultaneous execution of a and b is not allowed, resulting in a contradiction.

This case can easily be modelled by *g-comtraces* of [19] and powerful *step traces* of [12], but these two models assume that observations are fully represented by step sequences (i.e., stratified orders), a subclass of interval orders, so they still do not cover the most general case.

8   Interval Traces with Mutex Relation

A solution to the problem discussed at the end of the last section is to add a new relation called *mutex*. In principle the same idea was used for step traces [12].

Definition 5. *A* mutex interval trace alphabet *(or* MI-*trace alphabet) is a triple* $\Phi = (\Sigma, wind, mut)$, *where* $wind \subseteq \Sigma \times \Sigma$ *is an irreflexive relation called* weak independence *and* $mut \subseteq \Sigma \times \Sigma$ *is an irreflexive and symmetric relation called* mutual exclusion.

If $(a,b) \in mut$ then the executions orders a followed by b, and b followed by a are equivalent. E.g., if $(a,b) \in mut$ then the interval sequences $a_{\uparrow} a_{\downarrow} b_{\uparrow} b_{\downarrow}$ - which represents a before b, and $b_{\uparrow} b_{\downarrow} a_{\uparrow} a_{\downarrow}$ - which represents a after b, will be considered equivalent.

For the case from Example 7, we would have $wind = \{(a,c),(c,a),(b,c),(c,b)\}$ and $mut = \{(a,b),(b,a)\}$.

Definition 6. *Let* $\Phi = (\Sigma, wind, mut)$ *be a* MI-*trace alphabet. Then* $\phi = (\Sigma_{\uparrow\downarrow}, ind_{\Phi}, mut_{\Phi})$ *is an* internal mutex interval trace alphabet *(or* MI-trace alphabet*), where* $ind_{\Phi} \subseteq \Sigma_{\uparrow\downarrow} \times \Sigma_{\uparrow\downarrow}$ *and* $mut_{\Phi} \subseteq \Sigma_{\uparrow} \times \Sigma_{\downarrow} \times \Sigma_{\uparrow} \times \Sigma_{\downarrow}$ *are relations given by:*

$$ind_{\Phi} = \{(a_{\uparrow}, b_{\uparrow}),(a_{\downarrow}, b_{\downarrow}) \mid a \neq b \in \Sigma\} \cup \{(a_{\downarrow}, b_{\uparrow}),(b_{\uparrow}, a_{\downarrow}) \mid (a,b) \in wind\}$$
$$mut_{\Phi} = \{(a_{\uparrow}, a_{\downarrow}, b_{\uparrow}, b_{\downarrow}) \mid (a,b) \in mut\}.$$

We then introduce equivalences on sequences following ideas behind the original traces model.

Definition 7. *Let* $\phi = (\Sigma_{\uparrow\downarrow}, ind_{\Phi}, mut_{\Phi})$ *be a* MI-*trace alphabet. We define binary relations* $\approx_{ind}, \approx_{mut}, \approx_{\phi}, \equiv_{\phi}$ *over* $\Sigma_{\uparrow\downarrow}^{*}$, *as follows:*

1. *For all* $x,y \in \Sigma_{\uparrow\downarrow}^{*}$:
   - $x \approx_{ind} y$ *if* $x = zefw$ *and* $y = zfew$, *for some* $z,w \in \Sigma_{\uparrow\downarrow}^{*}$ *and* $(e,f) \in ind_{\Phi}$.
   - $x \approx_{mut} y$ *if* $x = za_{\uparrow} a_{\downarrow} b_{\uparrow} b_{\downarrow} w$ *and* $y = zb_{\uparrow} b_{\downarrow} a_{\uparrow} a_{\downarrow} w$, *for some* $z,w \in \Sigma_{\uparrow\downarrow}^{*}$ *and* $(a_{\uparrow}, a_{\downarrow}, b_{\uparrow}, b_{\downarrow}) \in mut_{\Phi}$.
2. $\approx_{\phi} = \approx_{ind} \cup \approx_{mut}$.
3. $\equiv_{\phi} = (\approx_{\phi})^{*}$.

Clearly the relation $\equiv_{\phi}$ is an equivalence relation and an equivalence class of $\equiv_{\phi}$, $\tau \in \Sigma/_{\equiv_{\phi}}$, will be called a *mutex interval trace* (or MI-*trace*) if $\tau \subseteq IntSeq$.

The trace containing an interval sequence w is denoted by $[w]_{\phi}$, and the trace concatenation $\circ$ is defined as $\tau \circ \tau' = [ww']_{\phi}$, for any interval sequences $w \in \tau$ and $w' \in \tau'$. Clearly, we have $[ww']_{\phi} = [vv']_{\phi}$, for all $w,v \in \tau$ and $w',v' \in \tau'$. However, this is not enough to prove that the definition of MI-trace is sound. For this we need to show a result similar to Proposition 5.

Proposition 6. *Let* $\phi = (\Sigma_{\uparrow\downarrow}, ind_{\Phi}, mut_{\Phi})$ *be a* MI-*trace alphabet, and* $x,y \in IntSeq$.

1. $[x]_{\phi} \subseteq IntSeq$.
2. $[x]_{\phi} [y]_{\phi} \subseteq = [xy]_{\phi} \subseteq IntSeq$.
3. $\text{intord}(x) = \text{intord}(x) \implies x \equiv_{\phi} y$.

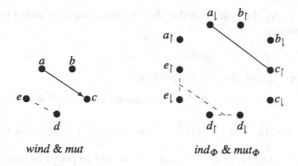

Fig. 3. An example of relations *wind*, *mut*, *ind*$_\Phi$ and *mut*$_\Phi$ for $\Sigma = \{a,b,c,d,e\}$. The relation *wind* is represented by an arrow, *mut* by dashed line, the part of *ind*$_\Phi$ from Definition 6 by solid line, quadruples of *mut*$_\Phi$ are connected by dashed lines, and the part of *ind*$_\Phi$ from Definition 6 is omitted (dotted lines in Fig. 2).

Proof. (1) From Definition 6 we know that $a_\uparrow$ and $a_\downarrow$ cannot commute for any $a \in \Sigma$. Similarly, from Definition 6, we have that *mut* also does not change the orders between $a_\uparrow$ and $a_\downarrow$ for any $a \in \Sigma$. Hence, if $\pi_{\{a_\uparrow,a_\downarrow\}}(x) \in (a_\uparrow a_\downarrow)^*$ then also $\pi_{\{a_\uparrow,a_\downarrow\}}(s) \in (a_\uparrow a_\downarrow)^*$ for each $s \in [x]_\phi$.

(2) A consequence of Proposition 4 and (1) above.

(3) From Proposition 5(3), it follows $\mathrm{intord}(x) = \mathrm{intord}(x) \implies x(\approx_{ind})^* y$ and from Definition 7(3), we have $x(\approx_{ind})^* y \implies x \equiv_\phi y$. $\square$

Note that if *mut* $\neq \varnothing$ then there always exists an interval sequence x such that $[x]_\equiv$ is not a Mazurkiewicz trace. This follows from the fact that the set of interval sequences $\{a_\uparrow a_\downarrow b_\uparrow b_\downarrow, b_\uparrow b_\downarrow a_\uparrow a_\downarrow\}$ is not a Mazurkiewicz trace, for any *ind*. However, if *wind* $= \varnothing$ and *mut* $= \{(a,b),(b,a)\}$, then $\{a_\uparrow a_\downarrow b_\uparrow b_\downarrow, b_\uparrow b_\downarrow a_\uparrow a_\downarrow\} = [a_\uparrow a_\downarrow b_\uparrow b_\downarrow]_\equiv$, so it is a legal MI-trace.

Example 8. Let $\Phi = (\Sigma, wind, mut)$ be a MI-trace alphabet, with $\Sigma = \{a,b,c,d,e\}$ and *wind*, *mut* as defined in Fig. 3. Consider an interval sequence $a_\uparrow c_\uparrow a_\downarrow b_\uparrow c_\downarrow b_\downarrow d_\uparrow d_\downarrow e_\uparrow e_\downarrow$ which generates the interval order po_7 in Fig. 4. Let $\tau = [a_\uparrow c_\uparrow a_\downarrow b_\uparrow c_\downarrow b_\downarrow d_\uparrow d_\downarrow e_\uparrow e_\downarrow]_\equiv$ for the relations from Fig. 3. One can show by inspection that τ comprises the following sequences:

$$
\begin{array}{lll}
\underline{a_\uparrow c_\uparrow a_\downarrow b_\uparrow c_\downarrow b_\downarrow d_\uparrow d_\downarrow e_\uparrow e_\downarrow} & c_\uparrow a_\uparrow a_\downarrow b_\uparrow c_\downarrow b_\downarrow d_\uparrow d_\downarrow e_\uparrow e_\downarrow & a_\uparrow c_\uparrow a_\downarrow b_\uparrow b_\downarrow c_\downarrow d_\uparrow d_\downarrow e_\uparrow e_\downarrow \\
c_\uparrow a_\uparrow a_\downarrow b_\uparrow b_\downarrow c_\downarrow d_\uparrow d_\downarrow e_\uparrow e_\downarrow & a_\uparrow c_\uparrow a_\downarrow b_\uparrow c_\downarrow b_\downarrow e_\uparrow e_\downarrow d_\uparrow d_\downarrow & a_\uparrow c_\uparrow a_\downarrow b_\uparrow b_\downarrow c_\downarrow e_\uparrow e_\downarrow d_\uparrow d_\downarrow \\
a_\uparrow c_\uparrow a_\downarrow b_\uparrow b_\downarrow c_\downarrow e_\uparrow e_\downarrow d_\uparrow d_\downarrow & c_\uparrow a_\uparrow a_\downarrow b_\uparrow b_\downarrow c_\downarrow e_\uparrow e_\downarrow d_\uparrow d_\downarrow & a_\uparrow a_\downarrow b_\uparrow c_\uparrow b_\downarrow c_\downarrow d_\uparrow d_\downarrow e_\uparrow e_\downarrow \\
a_\uparrow a_\downarrow c_\uparrow b_\uparrow b_\downarrow c_\downarrow d_\uparrow d_\downarrow e_\uparrow e_\downarrow & a_\uparrow a_\downarrow b_\uparrow c_\uparrow c_\downarrow b_\downarrow d_\uparrow d_\downarrow e_\uparrow e_\downarrow & a_\uparrow a_\downarrow c_\uparrow b_\uparrow c_\downarrow b_\downarrow d_\uparrow d_\downarrow e_\uparrow e_\downarrow \\
a_\uparrow a_\downarrow b_\uparrow c_\uparrow b_\downarrow c_\downarrow e_\uparrow e_\downarrow d_\uparrow d_\downarrow & a_\uparrow a_\downarrow c_\uparrow b_\uparrow b_\downarrow c_\downarrow e_\uparrow e_\downarrow d_\uparrow d_\downarrow & a_\uparrow a_\downarrow b_\uparrow c_\uparrow c_\downarrow b_\downarrow e_\uparrow e_\downarrow d_\uparrow d_\downarrow \\
a_\uparrow a_\downarrow c_\uparrow b_\uparrow c_\downarrow b_\downarrow e_\uparrow e_\downarrow d_\uparrow d_\downarrow
\end{array}
$$

The interval orders generated by the interval traces from τ are exactly the partial orders $po_7 - po_{10}$ presented on the right hand side of Fig. 4. ◇

We will now prove the result similar (but weaker, one way only) to that of Theorem 8, but for MI-traces.

Proposition 7. *Let $x, y \in IntSeq$ and $x \equiv_\phi y$. Then:*

1. $\pi_{\{a_\downarrow, b_\uparrow\}}(x) = \pi_{\{a_\downarrow, b_\uparrow\}}(y)$, for all $(a, b) \notin wind \cup mut$.

2. $ev(x) = ev(y)$ and, for all $a_\downarrow^{(i)}, b_\uparrow^{(j)} \in ev(x)$ such that $(a, b) \notin wind \cup mut$:

$$\mathrm{pos}_x(a_\downarrow^{(i)}) < \mathrm{pos}_x(b_\uparrow^{(j)}) \iff \mathrm{pos}_y(a_\downarrow^{(i)}) < \mathrm{pos}_y(b_\uparrow^{(j)}). \tag{*}$$

Proof. (1) Since $\equiv_\phi$ equals $(\approx_{ind} \cup \approx_{mut})^*$, it suffices to prove the result for $\approx_{ind}$ and $\approx_{mut}$. For $x \approx_{ind} y$ it follows from Theorem 8. For $x \approx_{mut} y$, we have:

$$x = z a_\uparrow a_\downarrow b_\uparrow b_\downarrow w \wedge y = z b_\uparrow b_\downarrow a_\uparrow a_\downarrow w \wedge (a_\uparrow, a_\downarrow, b_\uparrow, b_\downarrow) \in mut_\phi.$$

For all c, d such that $\{c, d\} \neq \{a, b\}$ we have $\pi_{\{c_\downarrow, d_\uparrow\}}(x) = \pi_{\{c_\downarrow, d_\uparrow\}}(z) \cdot \pi_{\{c_\downarrow, d_\uparrow\}}(w) = \pi_{\{c_\downarrow, d_\uparrow\}}(y)$, so we are done.

(2) Again, it suffices to prove for $\approx_{ind} \cup \approx_{mut}$. For $x \approx_{ind} y$ it follows from Theorem 8. Consider $x \approx_{mut} y$, i.e. $x = z a_\uparrow a_\downarrow b_\uparrow b_\downarrow w \wedge y = z b_\uparrow b_\downarrow a_\uparrow a_\downarrow w \wedge (a_\uparrow, a_\downarrow, b_\uparrow, b_\downarrow) \in mut_\phi$. For each $\alpha \in ev(z)$ we have $\mathrm{pos}_x(\alpha) = \mathrm{pos}_y(\alpha) = \mathrm{pos}_z(\alpha)$, while for each $\alpha \in ev(w)$ we have $\mathrm{pos}_x(\alpha) = \mathrm{pos}_y(\alpha) = len(z) + 4 + \mathrm{pos}_w(\alpha)$. This means the formula (*) holds for all $a_\downarrow^{(i)}, b_\uparrow^{(j)}, a_\uparrow^{(i)}, b_\downarrow^{(j)} \in ev(z) \cup ev(w)$. Assume $u = z a_\uparrow a_\downarrow b_\uparrow b_\downarrow$ and $\widehat{u} = \widehat{z} a_\uparrow^{(k)} a_\downarrow^{(k)} b_\uparrow^{(l)} b_\downarrow^{(l)}$, for some k, l. Clearly for every $c_\downarrow^{(i)} \in ev(z)$, we have $\mathrm{pos}_x(c_\downarrow^{(i)}) < \mathrm{pos}_x(a_\uparrow^{(k)})$ and $\mathrm{pos}_y(c_\downarrow^{(i)}) < \mathrm{pos}_x(a_\uparrow^{(k)})$. Similarly for $b_\downarrow^{(l)}$. Now consider $c_\uparrow^{(j)} \in ev(w)$, we have $\mathrm{pos}_x(a_\downarrow^{(k)}) < \mathrm{pos}_x(c_\uparrow^{(j)})$ and $\mathrm{pos}_y(a_\downarrow^{(k)}) < \mathrm{pos}_x(c_\uparrow^{(j)})$, and similarly for $b_\downarrow^{(l)}$. $\square$

Although MI-traces are no longer Mazurkiewicz traces, a version of Foata canonical form can still be introduced, intuitively corresponding to some greedy maximally concurrent representation.

Consider a MI-trace $[x]_\phi$. From Theorem 1 it follows that for each interval order *po* there is its greedy maximally concurrent extension $gmcr(po)$. Let $SE_{[x]_\phi} = \{gmcr(intord(y)) \mid x \equiv_\phi y\}$ be the set of such extensions generated by the MI-trace $[x]_\phi$, and let $SSEQ_{[x]_\phi} = \{sseq(gmcr(intord(y))) \mid x \equiv_\phi y\}$ be an equivalent set of step sequences.

For example, if we take $x = a_\uparrow c_\downarrow a_\downarrow b_\uparrow b_\downarrow c_\uparrow b_\downarrow d_\uparrow d_\downarrow e_\uparrow e_\downarrow$ from Example 8, then $SE_{[x]_\phi} = \{po_5, po_6, po_9, po_{10}\}$ and $SSEQ_{[x]_\phi}$ comprises the following sequences:

$$
\begin{array}{lll}
a_\uparrow c_\downarrow a_\downarrow c_\downarrow b_\uparrow b_\downarrow d_\uparrow d_\downarrow e_\uparrow e_\downarrow & c_\uparrow a_\downarrow a_\downarrow c_\downarrow b_\uparrow b_\downarrow d_\uparrow d_\downarrow e_\uparrow e_\downarrow & a_\uparrow c_\uparrow c_\downarrow a_\downarrow b_\uparrow b_\downarrow d_\uparrow d_\downarrow e_\uparrow e_\downarrow \\
c_\uparrow a_\uparrow c_\downarrow a_\downarrow b_\uparrow b_\downarrow d_\uparrow d_\downarrow e_\uparrow e_\downarrow & a_\uparrow c_\uparrow a_\downarrow c_\downarrow b_\uparrow b_\downarrow e_\uparrow e_\downarrow d_\uparrow d_\downarrow & c_\uparrow a_\uparrow a_\downarrow c_\downarrow b_\uparrow b_\downarrow e_\uparrow e_\downarrow d_\uparrow d_\downarrow \\
a_\uparrow c_\uparrow c_\downarrow a_\downarrow b_\uparrow b_\downarrow e_\uparrow e_\downarrow d_\uparrow d_\downarrow & c_\uparrow a_\uparrow c_\downarrow a_\downarrow b_\uparrow b_\downarrow e_\uparrow e_\downarrow d_\uparrow d_\downarrow & a_\uparrow a_\downarrow b_\uparrow c_\uparrow b_\downarrow c_\downarrow d_\uparrow d_\downarrow e_\uparrow e_\downarrow \\
a_\uparrow a_\downarrow c_\uparrow b_\uparrow b_\downarrow c_\downarrow d_\uparrow d_\downarrow e_\uparrow e_\downarrow & a_\uparrow a_\downarrow b_\uparrow c_\uparrow c_\downarrow b_\downarrow d_\uparrow d_\downarrow e_\uparrow e_\downarrow & a_\uparrow a_\downarrow c_\uparrow b_\uparrow c_\downarrow b_\downarrow d_\uparrow d_\downarrow e_\uparrow e_\downarrow \\
a_\uparrow a_\downarrow b_\uparrow c_\uparrow b_\downarrow c_\downarrow e_\uparrow e_\downarrow d_\uparrow d_\downarrow & a_\uparrow a_\downarrow c_\uparrow b_\uparrow b_\downarrow c_\downarrow e_\uparrow e_\downarrow d_\uparrow d_\downarrow & a_\uparrow a_\downarrow b_\uparrow c_\uparrow c_\downarrow b_\downarrow e_\uparrow e_\downarrow d_\uparrow d_\downarrow \\
a_\uparrow a_\downarrow c_\uparrow b_\uparrow c_\downarrow b_\downarrow e_\uparrow e_\downarrow d_\uparrow d_\downarrow & & \\
\end{array}
$$

A step sequence $\sigma = A_1 \ldots A_k \in SSEQ_{[x]_\phi}$ is in *greedy maximally concurrent form* if, for every $B_1 \ldots B_m \in SSEQ_{[x]_\phi}$,

- either $k = m$ and $|A_i| = |B_i|$ for $i = 1, \ldots, k$, or
- there is $j \leq k$ such that $|A_i| = |B_i|$ for $i = 1, \ldots, j-1$ and $|A_j| > |B_j|$.

We denote this by $\sigma \in SSEQ_{[x]_\phi}^{gmc}$.

Clearly, $SSEQ_{[x]_\phi}^{gmc} \neq \varnothing$, though in general it may contain more than one step sequence. For example for $x = a_\uparrow c_\uparrow a_\downarrow b_\uparrow c_\downarrow b_\downarrow d_\uparrow d_\downarrow e_\uparrow e_\downarrow$ from Example 8, $SSEQ_{[x]_\phi}^{gmc}$ comprises the following sequences:

$$a_\uparrow c_\uparrow a_\downarrow c_\downarrow b_\uparrow b_\downarrow d_\uparrow d_\downarrow e_\uparrow e_\downarrow \quad c_\uparrow a_\uparrow a_\downarrow c_\downarrow b_\uparrow b_\downarrow d_\uparrow d_\downarrow e_\uparrow e_\downarrow \quad a_\uparrow c_\uparrow c_\downarrow a_\downarrow b_\uparrow b_\downarrow d_\uparrow d_\downarrow e_\uparrow e_\downarrow$$
$$c_\uparrow a_\uparrow c_\downarrow a_\downarrow b_\uparrow b_\downarrow d_\uparrow d_\downarrow e_\uparrow e_\downarrow \quad a_\uparrow c_\uparrow a_\downarrow c_\downarrow b_\uparrow b_\downarrow e_\uparrow e_\downarrow d_\uparrow d_\downarrow \quad c_\uparrow a_\uparrow a_\downarrow c_\downarrow b_\uparrow b_\downarrow e_\uparrow e_\downarrow d_\uparrow d_\downarrow$$
$$a_\uparrow c_\uparrow c_\downarrow a_\downarrow b_\uparrow b_\downarrow e_\uparrow e_\downarrow d_\uparrow d_\downarrow \quad c_\uparrow a_\uparrow c_\downarrow a_\downarrow b_\uparrow b_\downarrow e_\uparrow e_\downarrow d_\uparrow d_\downarrow$$

The first four interval sequences generate po_5 and the last four po_6, which are both stratified orders from Fig. 4.

Let $\ll$ be an arbitrary total ordering of Σ. We extend it to a natural ordering of $\Sigma_{\uparrow\downarrow}$ by $a_\uparrow \ll b_\uparrow \ll a_\downarrow \ll b_\downarrow$ whenever $a \ll b$, and then extend it to a *lexicographical* order of $\Sigma_{\uparrow\downarrow}^*$. Then interval sequence x is in *Foata canonical form* if it is the smallest among those generating step sequences in $SSEQ_{[x]_\phi}^{gmc}$. Note that not every interval trace has a Foata normal form defined this way.

9   Mutex Interval Trace Semantics of Petri Nets

Inhibitor arcs, introduced in [10], allow a transition to check for an absence of a token. In principle they allow 'test for zero', an operator the standard Petri nets do not have (cf. [31]). In this paper, inhibitor nets are just *elementary nets* [37] with inhibitor arcs.

Formally, an *inhibitor* net is a tuple $N = (P, T, F, I, m_0)$, where P is a set of *places*, T is a set of *transitions*, P and T are disjoint, $F \subseteq (P \times T) \cup (T \times P)$ is a *flow relation*, $I \subseteq P \times T$ is a set of *inhibitor arcs* and $m_0 \subseteq P$ is the *initial marking*. An inhibitor arc $(p, e) \in I$ means that e can be enabled only if p *is not marked*. In diagrams (p, e) is indicated by an edge with a small circle at the end. Any set of places $m \subseteq P$ is called a *marking*. The net $\hat{N}$ of Fig. 4 is an inhibitor net with $I = \{(s_3, c_\uparrow), (b, c_\uparrow), (d, e_\uparrow), (e, d_\uparrow)\}$.

A *mutex inhibitor* net (or MI-net) is a tuple $N = (P, T, F, I, M, m_0)$, where (P, T, F, I, m_0) is an inhibitor net and $M \subseteq T \times T$ is a symmetric *mutex relation*. The mutex relation M can only be defined on transitions that can be interpreted as independent, i.e., their neighbourhoods (unions of entries and exits) are disjoint in (P, T, F, I, m_0).

The net N from Fig. 4 is an example of MI-net, where $I = \{(s_3, c)\}$ and $M = \{(d, e), (e, d)\}$. Consider this net N but *without* $M = \{(d, e), (e, d)\}$. Assuming the standard step sequence semantics of inhibitor nets (cf. [1, 15]), we can get from the marking $\{s_4, s_5\}$ to $\{s_6, s_7\}$ either by firing the step $\{e, d\}$, or by firing sequences of singleton steps $\{e\}\{d\}$ or $\{d\}\{e\}$ (i.e., sequences ed or de). The relation $M = \{(d, e), (e, d)\}$ disallows simultaneous execution of d and e, leaving only sequences ed and de.

Inhibitor nets have been introduced in [11] to solve a synchronization problem not expressible in classical Petri nets. Such nets allow 'test for zero', a feature that the standard Petri nets do not have (cf. [4, 31]). Despite their simplicity, basic inhibitor nets [15] can easily express complex behaviours involving weak causality [1, 23, 30], priorities,

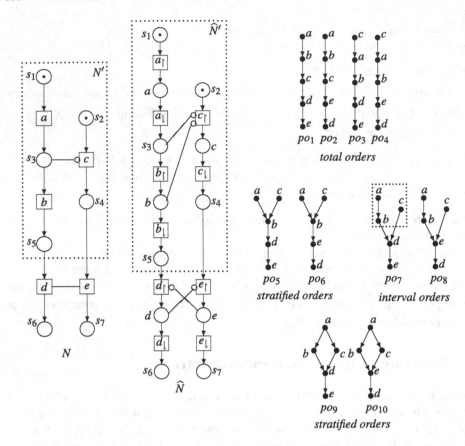

Fig. 4. $\widehat{N}$ is the interval representation of MI-net N. All runs start from $\{s_1, s_2\}$ and end at $\{s_6, s_7\}$. $po_1 -- po_8$ on the right side represent runs of both N and $\widehat{N}$. N' is an inhibitor net and $\widehat{N}'$ is its interval representation. The interval border inside the dotted square is an observation in both N' and $\widehat{N}'$.

various versions of simultaneity, etc. [15,42]. Inhibitor nets used in this paper are extensions of the *elementary net systems* [37] (i.e., we always assume that (P, T, F, m_0) form an elementary net system). The mutex relation (arcs) were introduced in [24] and substantially influenced the development of step traces [12,13].

The mutex relation matters only if net operational semantics allows simultaneous executions, e.g., it is step sequence semantics [1,15], ST-traces semantics [40,42], or interval sequence semantics [16,21]. For the standard firing sequence semantics it is irrelevant. On the other hand, interval sequence semantics for inhibitor nets has been defined as firing sequence semantics of nets that are interpreted as their interval representations. Consider the net N' from Fig. 4 which is N after deleting s_6, s_7, d, and e. The net $\widehat{N}'$ — a subnet of $\widehat{N}$ — is its interval representation.

The basic assumption behind interval executions and interval sequences is that each transition has its beginning and end. In such a case, one cannot adequately describe system states by markings alone. We need to supplement markings with information about transitions that have started, but have not finished yet.

The transformation of an inhibitor net into its interval representation is based on two principles. If inhibitor arcs are not involved, to represent transitions by their beginnings and ends we might just replace each transition $\boxed{t}$ by the net $\boxed{t_\uparrow}\!\!\rightarrow\!\!\bigcirc\!\!t\!\!\rightarrow\!\!\boxed{t_\downarrow}$, as proposed for example in [2] for nets with priorities, or in [44] for timed Petri nets. Each inhibitor arc must be replaced by two when transformation is made, and this construction is explained in detail in [21]. For the nets N' and $\widehat{N}'$ of Fig. 4, the inhibitor arc (s_3, c) in N' is transformed into two inhibitor arcs, $(s_3, c_\uparrow)$ and $(b, c_\uparrow)$, in $\widehat{N}'$.

Consider the net N' in Fig. 4. Assuming that we can 'hold tokens' in executed transitions and holding a token in c overlap with holding tokens in a and b, the net N can generate the interval order from Fig. 4 that is inside dotted square. This interval order can for example be represented by an interval sequence $a_\uparrow c_\uparrow a_\downarrow b_\uparrow b_\downarrow c_\downarrow$ which is a firing sequence of the net $\widehat{N}'$. Now consider the net N which includes the mutex relation $M = \{(e,d), (d,e)\}$. It prevents firing simultaneously e and d. In the net $\widehat{N}$, transition $\boxed{e}$ is replaced by $\boxed{e_\uparrow}\!\!\rightarrow\!\!\bigcirc\!\!e\!\!\rightarrow\!\!\boxed{e_\downarrow}$, transition $\boxed{d}$ is replaced by $\boxed{d_\uparrow}\!\!\rightarrow\!\!\bigcirc\!\!d\!\!\rightarrow\!\!\boxed{d_\downarrow}$, and the simultaneous execution of e and d is modelled either by sequence $e_\uparrow d_\uparrow$ or by $d_\uparrow e_\uparrow$ (followed by $e_\downarrow d_\downarrow$ or $d_\downarrow e_\downarrow$). Inhibitor arcs added by the translation, $(e, d_\uparrow)$ and $(d, e_\uparrow)$, prevent such firing sequences. All this leads to the following definition.

Definition 8. *The* interval representation *of a* MI-*net* $N = (P, T, F, I, M, m_0)$ *is the* inhibitor net $\widehat{N} = (\widehat{P}, \widehat{T}, \widehat{F}, \widehat{I}, m_0)$ *such that* $\widehat{P} = P \cup T$, $\widehat{T} = T_{\uparrow\downarrow}$, $\widehat{I} = \widehat{I}_{iarcs} \cup \widehat{I}_{mut}$, *where:*

$$\widehat{F} = \{(p, t_\uparrow) \mid (p, t) \in F\} \cup \{(t_\downarrow, p) \mid (t, p) \in F\} \cup \{(t_\uparrow, t), (t, t_\downarrow) \mid t \in T\}$$
$$\widehat{I}_{iarcs} = \{(p, t_\uparrow) \mid (p, t) \in I\} \cup \{(r, t_\uparrow) \mid (p, t) \in I \wedge (p, r) \in F\}$$
$$\widehat{I}_{mut} = \{(v, t_\uparrow), (t, v_\uparrow) \mid (t, v) \in M\}.$$

The net $\widehat{N}$ in Fig. 4 is the interval representation of N from the same diagram. Note also that initial markings of N and $\widehat{N}$ are equal (meaning that all new places from $\widehat{P} \setminus P$ are initially empty). In a similar way we can capture states of $\widehat{N}$, where all initiated transitions are also finalised.

We will use the standard black arrowhead notation for the flow arcs, and white dot notation for the inhibitor arcs. Sometimes we write '$\circ^{ia}$' and '$\circ^{m}$' instead of '$\circ$' to indicate which kind of inhibitor arc is involved, i.e., $\widehat{I}_{iarcs}$ or $\widehat{I}_{mut}$.

The interval representation of any MI-net is *always* an inhibitor net and we are interested in interval sequences[2] (that represent interval orders) generated by this inhibitor net. In other words, as for inhibitor nets [21], the interval sequence semantics of N is the firing sequence semantics of $\widehat{N}$.

The standard *firing sequence* semantics for an *inhibitor net* $NI = (P, T, F, I, m_0)$ is defined as follows:

- A transition t is *enabled* at marking m if $^\bullet t \subseteq m$ and $(t^\bullet \cup t^\circ) \cap m = \varnothing$.

---

[2] Defining interval step sequences is mathematically possible but it does not make much sense as $t_\uparrow$ and $t_\downarrow$ are interpreted as event beginning and its end, i.e., they are instantaneous, so their simultaneous occurrence is not observable - when time is continuous, or it can entirely be represented by interleaving - when time is discrete (see [22, 34]).

– An enabled t can *occur* leading to a new marking $m' = (m \setminus {}^\bullet t) \cup t^\bullet$, which is denoted by $m[t\rangle m'$, or $m[t\rangle_N m'$.
– A *firing sequence* from marking m to marking m' is a sequence of transitions $t_1 \ldots t_k$ ($k \geq 0$) for which there are markings $m = m_0, \ldots, m_k = m'$ such that $m_{i-1}[t_i\rangle m_i$, for every $1 \leq i \leq k$. This is denoted by $m[t_1 \ldots t_k\rangle m'$ and $t_1 \ldots t_k \in \mathrm{FS}_{NI}(m \rightsquigarrow m')$. In other words, $\mathrm{FS}_{NI}(m \rightsquigarrow m')$ is the set of *firing sequences* of NI that lead from the marking m to m'.

Definition 9. *Let N and $\widehat{N}$ be as in Definition 8, and $m, m' \subseteq P$. We respectively define the* firing interval sequences *of N from m to m' and the* firing interval orders *of N from m to m', as follows:*

$$\mathrm{FIS}_N(m \rightsquigarrow m') = \mathrm{FS}_{\widehat{N}}(m \rightsquigarrow m')$$
$$\mathrm{FIO}_N(m \rightsquigarrow m') = \mathrm{intord}(\mathrm{FS}_{\widehat{N}}(m \rightsquigarrow m')) .$$

The following result validates the last definition.

Proposition 8. *Let N and $\widehat{N}$ be as in Definition 8, and let $m, m' \subseteq P$.*

1. $\mathrm{FIS}_N(m \rightsquigarrow m') \subseteq IntSeq.$
2. *For all $x \in \mathrm{FIS}_N(m \rightsquigarrow m')$ and $y \in IntSeq$,*

$$\mathrm{intord}(x) = \mathrm{intord}(y) \implies y \in \mathrm{FIS}_N(m \rightsquigarrow m') .$$

Proof. (1) Let $x \in \mathrm{FIS}_N(m \rightsquigarrow m')$. We need to show that $\pi_{\{a_\uparrow, a_\downarrow\}}(x) \in (a_\uparrow a_\downarrow)^*$, for every $a \in \Sigma$. Let $x = z a_\uparrow w$ and $m[z a_\uparrow\rangle m''$. Since $a_\uparrow^\bullet = \{a\}$, we have $a \in m''$. We also have that: (i) for any $m_a \subseteq \widehat{P}$, if $a \in m_a$, then $a_\uparrow$ is not enabled in m_a, and (ii) the only way to remove a from m_a is to fire $a_\downarrow$ (as ${}^\bullet a_\downarrow = \{a\}$). Hence we must have $x = y a_\uparrow w' a_\downarrow v$, where $\pi_{\{a_\uparrow, a_\downarrow\}}(w') = \lambda$. As a result, $\pi_{\{a_\uparrow, a_\downarrow\}}(x) \in (a_\uparrow a_\downarrow)^*$.

(2) Assuming that $\Sigma = T$, let $\simeq$ be a binary relation on $IntSeq$ such that $x \simeq y$ if $x = z a_\uparrow b_\uparrow w$ and $y = z b_\uparrow a_\uparrow w$, for some z, a, b, w. By Theorem 7, $x \simeq^* y \iff \mathrm{intord}(x) = \mathrm{intord}(y)$. Hence it suffices to show that if $x \in \mathrm{FIS}_N(m \rightsquigarrow m')$ and $x \simeq y$, then $y \in \mathrm{FIS}_N(m \rightsquigarrow m')$.

Let $x = z a_\uparrow b_\uparrow w$ and $y = z b_\uparrow a_\uparrow w$. Suppose that $m[z\rangle m_1 [a_\uparrow\rangle m_2 [b_\uparrow\rangle m_3 [w\rangle m'$. This means that both $a_\uparrow$ and $b_\uparrow$ are enabled in m_1, so the statement $m[z\rangle m_1 [b_\uparrow\rangle m_2' [a_\uparrow\rangle m_3 [w\rangle m'$ is also true, i.e., $y \in \mathrm{FIS}_N(m \rightsquigarrow m')$. $\square$

In order to define trace semantics of MI-nets we need to construct suitable relations *wind* and *mut* from a MI-net $N = (P, T, F, I, M, m_0)$. While *mut* is just M, the relation *wind* can only be derived from the structure of the interval representation $\widehat{N}$, similarly as for the interval trace semantics of [21].

Definition 10. *Let N and $\widehat{N}$ be as in Definition 8. We define relations $ind_{\widehat{N}}$, $wind_N$, mut_N, and $mut_{\widehat{N}}$, as follows:*

$$
\begin{aligned}
ind_{\widehat{N}} &= \{(a_\uparrow, b_\uparrow), (a_\downarrow, b_\downarrow) \mid a \neq b \in T\} \cup \\
&\quad \{(a_\uparrow, b_\downarrow) \mid a \neq b \in T \wedge {}^\bullet a_\uparrow^\bullet \cap {}^\bullet b_\downarrow^\bullet = {}^{\circ ia} a_\uparrow \cap {}^\bullet b_\downarrow^\bullet = {}^{\circ ia} b_\downarrow \cap {}^\bullet a_\uparrow^\bullet = \varnothing\} \\
wind_N &= \{(a, b) \mid (b_\uparrow, a_\downarrow) \in ind_{\widehat{N}}\} \\
mut_N &= M \\
mut_{\widehat{N}} &= \{(a_\uparrow, a_\downarrow, b_\uparrow, b_\downarrow) \mid (a, b) \in M\}.
\end{aligned}
$$

Note that the formula for $ind_{\widehat{N}}$ is in the spirit of Mazurkiewicz's original concept and follows from a detailed discussion of similar relations in [21].

It is straightforward to check that $(T, wind_N, mut_N)$ is a MI-trace alphabet. Let $\equiv_N$ be the MI-trace equivalence relation induced by it. The following result shows the soundness of concepts discussed above.

Proposition 9. *Let N and $\widehat{N}$ be as in Definition 8, and $m, m' \subseteq P$. Then:*

$$x \in \text{FIS}_N(m \rightsquigarrow m') \iff [x]_{\equiv_N} \subseteq \text{FIS}_N(m \rightsquigarrow m').$$

Proof. ($\Longleftarrow$) Obvious.
($\Longrightarrow$) It suffices to consider the following two cases.
Case 1: $x = za_\uparrow b_\downarrow w$, $y = zb_\downarrow a_\uparrow w$ and $(a_\uparrow, b_\downarrow) \in ind_{\widehat{N}}$. Suppose that

$$m[z\rangle m_1 [a_\uparrow\rangle m_2 [b_\downarrow\rangle m_3 [w\rangle m'.$$

Since $(a_\uparrow, b_\downarrow) \in ind_{\widehat{N}}$, both $a_\uparrow$ and $b_\downarrow$ are enabled at m_1, so we also have

$$m[z\rangle m_1 [b_\downarrow\rangle m'_2 [a_\uparrow\rangle m_3 [w\rangle m'.$$

Hence $y \in \text{FIS}_N(m \rightsquigarrow m')$.
Case 2: $x = z_1 a_\uparrow a_\downarrow b_\uparrow b_\downarrow z_2$, $y = z_1 b_\uparrow b_\downarrow a_\uparrow a_\downarrow z_2$ and $(a_\uparrow, a_\downarrow, b_\uparrow, b_\downarrow) \in mut_{\widehat{N}}$. Suppose that

$$m[z\rangle m_1 [a_\uparrow\rangle m_2 [a_\downarrow\rangle m_3 [b_\uparrow\rangle m_4 [b_\downarrow\rangle m_5 [w\rangle m'.$$

Since $(a_\uparrow, a_\downarrow, b_\uparrow, b_\downarrow) \in mut_{\widehat{N}}$, both $a_\uparrow$ and $b_\uparrow$ are enabled at m_1. If $a_\uparrow$ is fired then $a \in m_2$ and, since $(a, b_\uparrow) \in \widehat{I}$, $b_\uparrow$ is not enabled in m_2, but $a_\downarrow$ is. However, $b_\uparrow$ is enabled in m_3, etc. Similarly, when we fire $b_\uparrow$ in m_1, so we also have

$$m[z\rangle m_1 [b_\uparrow\rangle m'_2 [b_\downarrow\rangle m_3 [a_\uparrow\rangle m'_4 [a_\downarrow\rangle m_5 [z_2\rangle m'.$$

Hence $y \in \text{FIS}_N(m \rightsquigarrow m')$.   □

10 Concluding Remarks

We have enriched interval traces by the mutex relation. The resulting MI-traces capture not only the 'not later than' relationship, but also the 'no simultaneity' relationship. The concept of mutex relation was adopted from [12], where it was used to extend the comtraces [15] to more powerful step traces. We elaborated on the soundness of the MI-traces definition and their basic properties have been proved. Some new results for interval orders and interval traces have also been provided.

We have also investigated MI-trace semantics for inhibitor Petri nets with mutex arcs. To do this, we have incorporated translations from [12,21,23]. We have also discussed the soundness of the proposed construction.

There are still several open problems in the development of our model, such as the relationship between *wind* and *mut*. They are unconstrained in this paper, but this might not always be a realistic assumption. Few models can handle interval orders as observations. One that can are Higher Dimensional Automata of [7,36], but their relationship to our model is not clear at this moment.

Finally, we would like to point out that interval order semantics (cf. [21,40,42]) and interval semantics (cf. [35]) are incomparable.

Acknowledgment. A partial support by the Discovery NSERC of Canada grant No. 6466-15, and the Leverhulme Trust grant RPG-2022-025 is acknowledged. The authors gratefully acknowledge four anonymous referees, whose comments significantly contributed to the final version of this paper.

References

1. Baldan, P., Busi, N., Corradini, A., Pinna, G.M.: Domain and event structure semantics for Petri nets with read and inhibitor arcs. Inf. Comput. **323**, 129–189 (2004)
2. Best, E., Koutny, M.: Petri net semantics of priority systems. Theor. Comput. Sci. **96**(1), 175–174 (1992)
3. Cartier, P., Foata, D.: Problèmes combinatoires de commutation et réarrangements. LNM, vol. 85. Springer-Verlag, Berlin (1969)
4. Desel, J., Reisig, W.: Place/transition petri nets. In: Reisig, W., Rozenberg, G. (eds.) ACPN 1996. LNCS, vol. 1491, pp. 122–173. Springer, Heidelberg (1998). https://doi.org/10.1007/3-540-65306-6_15
5. Diekert, V., Métivier, Y.: Partial commutation and traces. In: Rozenberg, G., Salomaa, A. (eds.) Handbook of Formal Languages, pp. 457–533. Springer, Heidelberg (1997). https://doi.org/10.1007/978-3-642-59126-6_8
6. Diekert, V., Rozenberg, G., editors. The Book of Traces. World Scientific (1995)
7. Fahrenberg, U., Johansen, C., Struth, G., Ziemiański, K.: Posets with interfaces as a model for concurrency. Inf. Comput. 285 (2022)
8. Fishburn, P.C.: Intransitive indifference with unequal indifference intervals. J. Math. Psychol. **7**, 144–149 (1970)
9. Fishburn, P.C.: Interval Orders and Interval Graphs. John Wiley, New York (1985)
10. Flynn, M.J., Agerwala, T.: Comments on capabilities, limitations and correctness of Petri nets. In: Lipovski, G.J., Szygenda, S.A., editors, Proceedings of the 1st Annual Symposium on Computer Architecture, Gainesville, FL, USA, December 1973, pp. 81–86. ACM (1973)
11. Flynn, M.J., Agerwala, T.: Comments on capabilities, limitations and correctness of Petri nets. In: Lipovski, G.J., Szygenda, S.A., editors, Proceedings of the 1st Annual Symposium on Computer Architecture, Gainesville, FL, USA, December 1973, pp. 81–86. ACM (1973)
12. Janicki, R., Kleijn, J., Koutny, M., Mikulski, Ł: Characterising concurrent histories. Fund. Inform. **139**(1), 21–42 (2015)
13. Janicki, R., Kleijn, J., Koutny, M., Mikulski, Ł: Classifying invariant structures of step traces. J. Comput. Syst. Sci. **104**, 297–322 (2019)
14. Janicki, R., Koutny, M.: Structure of concurrency. Theor. Comput. Sci. **112**(1), 5–52 (1993)
15. Janicki, R., Koutny, M.: Semantics of inhibitor nets. Inf. Comput. **123**(1), 1–16 (1995)
16. Janicki, R., Koutny, M.: Operational semantics, interval orders and sequences of antichains. Fund. Inform. **169**(1–2), 31–55 (2019)
17. Janicki, R., Lauer, P.E.: Specification and Analysis of Concurrent Systems - The COSY Approach, 2nd edn. Springer, EATCS Monographs on Theoretical Computer Science (2012). https://doi.org/10.1007/978-3-642-77337-2
18. Janicki, R., Lauer, P.E., Koutny, M., Devillers, R.: Concurrent and maximally concurrent evolution of nonsequential systems. Theor. Comput. Sci. **43**, 213–238 (1986)
19. Janicki, R., Lê, D.T.M.: Modelling concurrency with comtraces and generalized comtraces. Inf. Comput. **209**(11), 1355–1389 (2011)
20. Janicki, R., Mikulski, Ł: Algebraic structure of step traces and interval traces. Fund. Inform. **175**(1–4), 253–280 (2020)
21. Janicki, R., Yin, X.: Modeling concurrency with interval traces. Inf. Comput. **253**, 78–108 (2017)

22. Janicki, R., Yin, X., Zubkova, N.: Modeling interval order structures with partially commutative monoids. In: Koutny, M., Ulidowski, I. (eds.) CONCUR 2012. LNCS, vol. 7454, pp. 425–439. Springer, Heidelberg (2012). https://doi.org/10.1007/978-3-642-32940-1_30

23. Kleijn, J., Koutny, M.: Formal languages and concurrent behaviours. In: Enguix, G.B., Jiménez-López, M.D., Martín-Vide, C., editors, New Developments in Formal Languages and Applications, volume 113 of Studies in Computational Intelligence, pp. 125–182. Springer, Cham (2008). https://doi.org/10.1007/978-3-540-78291-9_5

24. Kleijn, J., Koutny, M.: Mutex causality in processes and traces of general elementary nets. Fund. Inform. 122(1–2), 119–146 (2013)

25. Laarman, A.: Stubborn transaction reduction. In: Dutle, A., Muñoz, C., Narkawicz, A. (eds.) NFM 2018. LNCS, vol. 10811, pp. 280–298. Springer, Cham (2018). https://doi.org/10.1007/978-3-319-77935-5_20

26. Lengauer, C., Hehner, E.C.R.: A methodolgy for programming with concurrency. In: Händler, W., editor, CONPAR 81: Conference on Analysing Problem Classes and Programming for Parallel Computing, Nürnberg, Germany, June 10–12, 1981, Proceedings, volume 111 of Lecture Notes in Computer Science, pp. 259–270. Springer (1981)

27. Mazurkiewicz, A.: Concurrent program schemes and their interpretations. DAIMI Rep. PB 78, Aarhus University (1977)

28. Mazurkiewicz, A.W.: Introduction to trace theory. In: Diekert, V., Rozenberg, G., editors, The Book of Traces, pp. 3–41. World Scientific, (1995)

29. Mikulski, Ł.: Algebraic structure of combined traces. Log. Methods Comput. Sci., 9(3), (2013)

30. Montanari, U., Rossi, F.: Contextual nets. Acta Infortmatica 32(6), 545–596 (1995)

31. Murata, T.: Petri nets: properties, analysis and applications. Proc. IEEE 77(4), 541–580 (1989)

32. Nagy, B., Akkeleş, A.: Trajectories and traces on non-traditional regular tessellations of the plane. In: Brimkov, V.E., Barneva, R.P. (eds.) IWCIA 2017. LNCS, vol. 10256, pp. 16–29. Springer, Cham (2017). https://doi.org/10.1007/978-3-319-59108-7_2

33. Paulevé, L.: Goal-oriented reduction of automata networks. In: Bartocci, E., Lio, P., Paoletti, N. (eds.) CMSB 2016. LNCS, vol. 9859, pp. 252–272. Springer, Cham (2016). https://doi.org/10.1007/978-3-319-45177-0_16

34. Petri, C.A.: Nets, time and space. Theor. Comput. Sci. 153(1&2), 3–48 (1996)

35. Popova-Zeugmann, L., Pelz, E.: Algebraical characterisation of interval-timed Petri nets with discrete delays. Fundam. Informaticae 120(3–4), 341–357 (2012)

36. Pratt, V.: Modeling concurrency with geometry. In: POPL '91: Proceedings of the 18th ACM SIGPLAN-SIGACT Symposium on Principles of Programming Languages, pp. 311–322. ACM (1991)

37. Rozenberg, G., Engelfriet, J.: Elementary net systems. In: Reisig, W., Rozenberg, G. (eds.) ACPN 1996. LNCS, vol. 1491, pp. 12–121. Springer, Heidelberg (1998). https://doi.org/10.1007/3-540-65306-6_14

38. Shields, M.W.: Adequate path expressions. In: Kahn, G., editor, Semantics of Concurrent Computation, Proceedings of the International Symposium, Evian, France, July 2–4, 1979, volume 70 of Lecture Notes in Computer Science, pp. 249–265. Springer (1979)

39. Shields, M.W.: Concurrent machines. Comput. J. 28(5), 449–465 (1985)

40. van Glabbeek, R., Vaandrager, F.: Petri net models for algebraic theories of concurrency. In: de Bakker, J.W., Nijman, A.J., Treleaven, P.C. (eds.) PARLE 1987. LNCS, vol. 259, pp. 224–242. Springer, Heidelberg (1987). https://doi.org/10.1007/3-540-17945-3_13

41. Vogler, W.: A generalization of trace theory. RAIRO Informatique théorique et Appl. 25(2), 147–156 (1991)

42. Vogler, W.: Partial order semantics and read arcs. Theor. Comput. Sci. 286(1), 33–63 (2002)

43. Wiener, N.: A contribution to the theory of relative position. Proc. Camb. Philos. Soc. **17**, 441–449 (1914)
44. Zuberek, W.M.: Timed Petri nets and preliminary performance evaluation. In: Lenfant, J., Borgerson, B.R., Atkins, D.E., Irani, K.B., Kinniment, D., Aiso, H., editors, Proceedings of the 7th Annual Symposium on Computer Architecture, La Baule, France, May 6–8, 1980, pp. 88–96. ACM (1980)

A Myhill-Nerode Theorem
for Higher-Dimensional Automata

Uli Fahrenberg[1][✉] and Krzysztof Ziemiański[2]

[1] EPITA Research Laboratory (LRE), Le Kremlin-Bicêtre, France
uli@lrde.epita.fr
[2] University of Warsaw, Warsaw, Poland

Abstract. We establish a Myhill-Nerode type theorem for higher-dimensional automata (HDAs), stating that a language is regular precisely if it has finite prefix quotient. HDAs extend standard automata with additional structure, making it possible to distinguish between interleavings and concurrency. We also introduce deterministic HDAs and show that not all HDAs are determinizable, that is, there exist regular languages that cannot be recognised by a deterministic HDA. Using our theorem, we develop an internal characterisation of deterministic languages.

Keywords: higher-dimensional automata · Myhill-Nerode theorem · concurrency theory · determinism

1 Introduction

Higher-dimensional automata (HDAs), introduced by Pratt and van Glabbeek [23,27,28], extend standard automata with additional structure that makes it possible to distinguish between interleavings and concurrency. That puts them in a class with other non-interleaving models for concurrency such as Petri nets [22], event structures [21], configuration structures [31,32], asynchronous transition systems [3,26], and similar approaches [19,24,25,30], while retaining some of the properties and intuition of automata-like models. As an example, Fig. 1 shows Petri net and HDA models for a system with two events, labeled a and b. The Petri net and HDA on the left side model the (mutually exclusive) interleaving of a and b as either $a.b$ or $b.a$; those to the right model concurrent execution of a and b. In the HDA, this independence is indicated by a filled-in square.

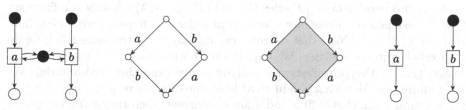

Fig. 1. Petri net and HDA models distinguishing interleaving (left) from non-interleaving (right) concurrency. Left: Petri net and HDA models for $a.b + b.a$; right: HDA and Petri net models for $a \parallel b$.

L. Gomes and R. Lorenz (Eds.): PETRI NETS 2023, LNCS 13929, pp. 167–188, 2023.
https://doi.org/10.1007/978-3-031-33620-1_9

We have recently introduced languages of HDAs [6], which consist of partially ordered multisets with interfaces (ipomsets), and shown a Kleene theorem for them [7]. Here we continue to develop the language theory of HDAs. Our first contribution is a Myhill-Nerode type theorem for HDAs, stating that a language is regular iff it has finite prefix quotient. This provides a necessary and sufficient condition for regularity. Our proof is inspired by the standard proofs of the Myhill-Nerode theorem, but the higher-dimensional structure introduces some difficulties. For example, we cannot use the standard prefix quotient relation but need to develop a stronger one which takes concurrency of events into account.

As a second contribution, we give a precise definition of deterministic HDAs and show that there exist regular languages that cannot be recognised by deterministic HDAs. Our Myhill-Nerode construction will produce a deterministic HDA for such deterministic languages, and a non-deterministic HDA otherwise. (We make no claim as to minimality of our Myhill-Nerode HDAs.) Our definition of determinism is more subtle than for standard automata as it is not always possible to remove non-accessible parts of HDAs. We develop a language-internal characterisation of deterministic languages.

2   Pomsets with Interfaces

HDAs model systems in which labelled events have duration and may happen concurrently. Every event has a time interval during which it is active: it starts at some point, then remains active until its termination and never reappears. Events may be concurrent, that is, their activity intervals may overlap; otherwise, one of the events precedes the other. We also need to consider executions in which some events are already active at the beginning (*source events*) or are still active at the end (*target events*).

At any moment of an execution we observe a list of currently active events (such lists are called *losets* below). The relative position of any two concurrent events on these lists remains the same, regardless of the point in time. This provides a secondary relation between events, which we call *event order*.

To make the above precise, let Σ be a finite alphabet. An *loset*[1] $(U, \dashrightarrow, \lambda)$ is a finite set U with a total order $\dashrightarrow$ called the *event order* and a labelling function $\lambda : U \to \Sigma$. Losets (or rather their isomorphism classes) are effectively strings but consist of concurrent, not subsequent, events.

A *labelled poset with event order* (*lposet*) $(P, <, \dashrightarrow, \lambda)$ consists of a finite set P with two relations: *precedence* $<$ and *event order* $\dashrightarrow$, together with a labelling function $\lambda : P \to \Sigma$. Note that different events may carry the same label: we do *not* exclude autoconcurrency. We require that both $<$ and $\dashrightarrow$ are strict partial orders, that is, they are irreflexive and transitive (and thus asymmetric). We also require that for each $x \neq y$ in P, at least one of $x < y$ or $y < x$ or $x \dashrightarrow y$ or $y \dashrightarrow x$ must hold; that is, if x and y are concurrent, then they must be related by $\dashrightarrow$.

---

[1] Pronunciation: ell-oh-set.

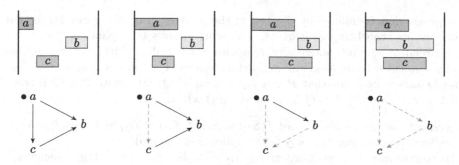

Fig. 2. Activity intervals (top) and corresponding iposets (bottom), see Example 1. Full arrows indicate precedence order; dashed arrows indicate event order; bullets indicate interfaces.

Losets may be regarded as lposets with empty precedence relation; the last condition enforces that their elements are totally ordered by $\dashrightarrow$. A temporary state of an execution is described by an loset, while the whole execution provides an lposet of its events. The precedence order expresses that one event terminates before the other starts. The execution starts at the loset of $<$-minimal elements and finishes with the loset of $<$-maximal elements. The event order of an lposet is generated by the event orders of temporary losets. Hence any two events which are active concurrently are unrelated by $<$ but related by $\dashrightarrow$.

In order to accommodate source and target events, we need to introduce lposets with interfaces (iposets). An *iposet* $(P, <, \dashrightarrow, S, T, \lambda)$ consists of an lposet $(P, <, \dashrightarrow, \lambda)$ together with subsets $S, T \subseteq P$ of source and target *interfaces*. Elements of S must be $<$-minimal and those of T $<$-maximal; hence both S and T are losets. We often denote an iposet as above by $_SP_T$, ignoring the orders and labelling, or use $S_P = S$ and $T_P = T$ if convenient. Source and target events will be marked by "$\bullet$" at the left or right side, and if the event order is not shown, we assume that it goes downwards.

Example 1. Figure 2 shows some simple examples of activity intervals of events and the corresponding iposets. The left iposet consists of three totally ordered events, given that the intervals do not overlap; the event a is already active at the beginning and hence in the source interface. In the other iposets, the activity intervals do overlap and hence the precedence order is partial (and the event order non-trivial).

Given that the precedence relation $<$ of an iposet represents activity intervals of events, it is an *interval order* [12]. In other words, any of the iposets we will encounter admits an *interval representation*: functions b and e from P to real numbers such that $b(x) \leq e(x)$ for all $x \in P$ and $x <_P y$ iff $e(x) < b(y)$ for all $x, y \in P$. We will only consider interval iposets in this paper and omit the qualification "interval". This is *not* a restriction, but rather induced by the semantics.

Iposets may be refined by shortening the activity intervals of events, so that some events stop being concurrent. This corresponds to expanding the precedence relation $<$ (and, potentially, removing event order). The inverse to refinement is called subsumption and defined as follows. For iposets P and Q, we say that Q *subsumes* P (or that P is a *refinement* of Q) and write $P \sqsubseteq Q$ if there exists a bijection $f : P \to Q$ (a *subsumption*) which

- respects interfaces and labels: $f(S_P) = S_Q$, $f(T_P) = T_Q$, and $\lambda_Q \circ f = \lambda_P$;
- reflects precedence: $f(x) <_Q f(y)$ implies $x <_P y$; and
- preserves essential event order: $x \dashrightarrow_P y$ implies $f(x) \dashrightarrow_Q f(y)$ whenever x and y are concurrent (that is, $x \not<_P y$ and $y \not<_P x$).

(Event order is essential for concurrent events, but by transitivity, it also appears between non-concurrent events; subsumptions may ignore such non-essential event order.)

Example 2. In Fig. 2, there is a sequence of refinements from right to left, each time shortening some activity intervals. Conversely, there is a sequence of subsumptions from left to right:

Interfaces need to be preserved across subsumptions, so in our example, the left endpoint of the a-interval must stay at the boundary.

Iposets and subsumptions form a category. The isomorphisms in that category are invertible subsumptions, and isomorphism classes of iposets are called *ipomsets*. Concretely, an isomorphism $f : P \to Q$ of iposets is a bijection which

- respects interfaces and labels: $f(S_P) = S_Q$, $f(T_P) = T_Q$, and $\lambda_Q \circ f = \lambda_P$;
- respects precedence: $x <_P y$ iff $f(x) <_Q f(y)$; and
- respects essential event order: $x \dashrightarrow_P y$ iff $f(x) \dashrightarrow_Q f(y)$ whenever $x \not<_P y$ and $y \not<_P x$.

Isomorphisms between iposets are unique (because of the requirement that all elements be ordered by $<$ or $\dashrightarrow$), hence we may switch freely between ipomsets and concrete representations, see [7] for details. We write $P \cong Q$ if iposets P and Q are isomorphic and let iiPoms denote the set of ipomsets.

Ipomsets may be *glued*, using a generalisation of the standard serial composition of pomsets [13]. For ipomsets P and Q, their *gluing* $P * Q$ is defined if the targets of P match the sources of Q: $T_P \cong S_Q$. In that case, its carrier set is the quotient $(P \sqcup Q)_{/x \equiv f(x)}$, where $f : T_P \to S_Q$ is the unique isomorphism, the interfaces are $S_{P*Q} = S_P$ and $T_{P*Q} = T_Q$, $\dashrightarrow_{P*Q}$ is the transitive closure of $\dashrightarrow_P \cup \dashrightarrow_Q$, and $x <_{P*Q} y$ iff $x <_P y$, or $x <_Q y$, or $x \in P - T_P$ and $y \in Q - S_Q$. We will often omit the "$*$" in gluing compositions. For ipomsets with empty interfaces, $*$ is serial pomset composition; in the general case, matching interface points are glued, see [6,8] or below for examples.

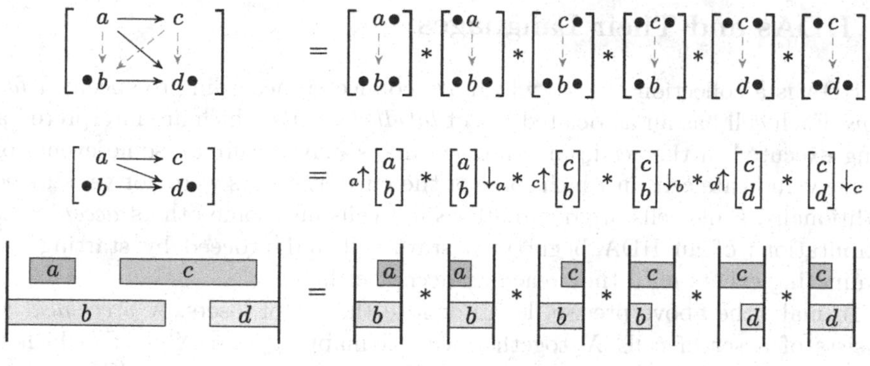

$$
\begin{bmatrix} a \longrightarrow c \\ \llap{\bullet} b \longrightarrow d\bullet \end{bmatrix} = \begin{bmatrix} a\bullet \\ \llap{\bullet} b\bullet \end{bmatrix} * \begin{bmatrix} \bullet a \\ \llap{\bullet} b\bullet \end{bmatrix} * \begin{bmatrix} c\bullet \\ \llap{\bullet} b\bullet \end{bmatrix} * \begin{bmatrix} \bullet c\bullet \\ \llap{\bullet} b \end{bmatrix} * \begin{bmatrix} \bullet c\bullet \\ d\bullet \end{bmatrix} * \begin{bmatrix} \bullet c \\ \llap{\bullet} d\bullet \end{bmatrix}
$$

$$
\begin{bmatrix} a \longrightarrow c \\ \llap{\bullet} b \longrightarrow d\bullet \end{bmatrix} = {}_a\!\uparrow\!\begin{bmatrix} a \\ b \end{bmatrix} * \begin{bmatrix} a \\ b \end{bmatrix}\!\downarrow_a * {}_c\!\uparrow\!\begin{bmatrix} c \\ b \end{bmatrix} * \begin{bmatrix} c \\ b \end{bmatrix}\!\downarrow_b * {}_d\!\uparrow\!\begin{bmatrix} c \\ d \end{bmatrix} * \begin{bmatrix} c \\ d \end{bmatrix}\!\downarrow_c
$$

Fig. 3. Sparse decomposition of ipomset into starters and terminators.

An ipomset P is *discrete* if $<_P$ is empty and $\dashrightarrow_P$ total. Losets are discrete ipomsets with empty interfaces. Discrete ipomsets ${}_U U_U$ are identities for gluing composition and written id_U. A *starter* is an ipomset ${}_{U-A} U_U$, a *terminator* is ${}_U U_{U-A}$; these will be written ${}_A\!\uparrow\!U$ and $U\!\downarrow_A$, respectively.

Any ipomset can be presented as a gluing of starters and terminators [8, Prop. 21]. (This is related to the fact that a partial order is interval iff its antichain order is total, see [12,17,18]). Such a presentation we call a *step decomposition*; if starters and terminators are alternating, the decomposition is *sparse*.

Example 3. Figure 3 shows a sparse decomposition of an ipomset into starters and terminators. The top line shows the graphical representation, in the middle the representation using the notation we have introduced for starters and terminators, and the bottom line shows activity intervals.

Proposition 4. *Every ipomset P has a unique sparse step decomposition.*

A *language* is, a priori, a set of ipomsets $L \subseteq$ iiPoms. However, we will assume that languages are closed under refinement (inverse subsumption), so that refinements of any ipomset in L are also in L:

Definition 5. *A language is a subset $L \subseteq$ iiPoms such that $P \sqsubseteq Q$ and $Q \in L$ imply $P \in L$.*

Using interval representations, this means that languages are closed under shortening activity intervals of events. The set of all languages is denoted $\mathscr{L} \subseteq 2^{\text{iiPoms}}$.

For $X \subseteq$ iiPoms an arbitrary set of ipomsets, we denote by

$$
X\!\downarrow = \{P \in \text{iiPoms} \mid \exists Q \in X : P \sqsubseteq Q\}
$$

its downward subsumption closure, that is, the smallest language which contains X. Then

$$
\mathscr{L} = \{X \subseteq \text{iiPoms} \mid X\!\downarrow = X\}.
$$

3   HDAs and Their Languages

An HDA is a collection of *cells* which are connected according to specified *face maps*. Each cell has an associated list of *labelled events* which are interpreted as being executed in that cell, and the face maps may terminate some events or, inversely, indicate cells in which some of the current events were not yet started. Additionally, some cells are designated *start* cells and some others *accept* cells; computations of an HDA begin in a start cell and proceed by starting and terminating events until they reach an accept cell.

To make the above precise, let $\square$ denote the set of losets. A *precubical set* consists of a set of cells X together with a mapping $\mathsf{ev} : X \to \square$ which to every cell assigns its list of active events. For an loset U we write $X[U] = \{x \in X \mid \mathsf{ev}(x) = U\}$ for the cells of type U. Further, for every $U \in \square$ and subset $A \subseteq U$ there are *face maps* $\delta_A^0, \delta_A^1 : X[U] \to X[U - A]$. The *upper* face maps δ_A^1 terminate the events in A, whereas the *lower* face maps δ_A^0 "unstart" these events: they map cells $x \in X[U]$ to cells $\delta_A^0(x) \in X[U - A]$ where the events in A are not yet active.

If $A, B \subseteq U$ are disjoint, then the order in which events in A and B are terminated or unstarted should not matter, so we require that $\delta_A^\nu \delta_B^\mu = \delta_B^\mu \delta_A^\nu$ for $\nu, \mu \in \{0, 1\}$: the *precubical identities*. A *higher-dimensional automaton* (*HDA*) is a precubical set together with subsets $\bot_X, \top_X \subseteq X$ of *start* and *accept* cells. For a precubical set X and subsets $Y, Z \subseteq X$ we denote by X_Y^Z the HDA with precubical set X, start cells Y and accept cells Z. We do *not* generally assume that precubical sets or HDAs are finite. The *dimension* of an HDA X is $\dim(X) = \sup\{|\mathsf{ev}(x)| \mid x \in X\} \in \mathbb{N} \cup \{\infty\}$.

Example 6. One-dimensional HDAs X are standard automata. Cells in $X[\emptyset]$ are states, cells in $X[a]$ for $a \in \Sigma$ are a-labelled transitions. Face maps δ_a^0 and δ_a^1 attach source and target states to transitions. In contrast to ordinary automata we allow start and accept *transitions* instead of merely states, so languages of such automata may contain not only words but also "words with interfaces". In any case, at most one event is active at any point in time, so the event order is unnecessary.

Example 7. Figure 4 shows an HDA both as a combinatorial object (left) and in a more geometric realisation (right). We write isomorphism classes of losets as lists of labels and omit the set braces in $\delta_{\{a\}}^0$ etc.

An *HDA-map* between HDAs X and Y is a function $f : X \to Y$ that preserves structure: types of cells ($\mathsf{ev}_Y \circ f = \mathsf{ev}_X$), face maps ($f(\delta_A^\nu(x)) = \delta_A^\nu(f(x))$) and start/accept cells ($f(\bot_X) \subseteq \bot_Y$, $f(\top_X) \subseteq \top_Y$). Similarly, a precubical map is a function that preserves the first two of these three. HDAs and HDA-maps form a category, as do precubical sets and precubical maps.

Computations of HDAs are paths: sequences of cells connected by face maps. A *path* in X is, thus, a sequence

$$\alpha = (x_0, \varphi_1, x_1, \ldots, x_{n-1}, \varphi_n, x_n), \tag{1}$$

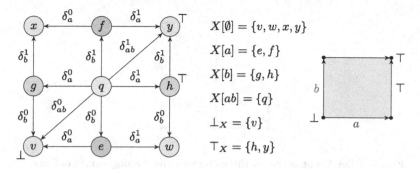

$$X[\emptyset] = \{v, w, x, y\}$$
$$X[a] = \{e, f\}$$
$$X[b] = \{g, h\}$$
$$X[ab] = \{q\}$$
$$\bot_X = \{v\}$$
$$\top_X = \{h, y\}$$

Fig. 4. A two-dimensional HDA X on $\Sigma = \{a, b\}$, see Example 7.

where the x_i are cells of X and the φ_i indicate types of face maps: for every i, $(x_{i-1}, \varphi_i, x_i)$ is either

- $(\delta^0_A(x_i), \nearrow^A, x_i)$ for $A \subseteq \mathsf{ev}(x_i)$ (an *upstep*)
- or $(x_{i-1}, \searrow_B, \delta^1_B(x_{i-1}))$ for $B \subseteq \mathsf{ev}(x_{i-1})$ (a *downstep*).

Upsteps start events in A while downsteps terminate events in B. The *source* and *target* of α as in (1) are $\mathsf{src}(\alpha) = x_0$ and $\mathsf{tgt}(\alpha) = x_n$.

The set of all paths in X starting at $Y \subseteq X$ and terminating in $Z \subseteq X$ is denoted by $\mathsf{Path}(X)^Z_Y$; we write $\mathsf{Path}(X)_Y = \mathsf{Path}(X)^X_Y$, $\mathsf{Path}(X)^Z = \mathsf{Path}(X)^Z_X$, and $\mathsf{Path}(X) = \mathsf{Path}(X)^X_X$. A path α is *accepting* if $\mathsf{src}(\alpha) \in \bot_X$ and $\mathsf{tgt}(\alpha) \in \top_X$. Paths α and β may be concatenated if $\mathsf{tgt}(\alpha) = \mathsf{src}(\beta)$; their concatenation is written $\alpha * \beta$, and we omit the "$*$" in concatenations if convenient.

Path equivalence is the congruence $\simeq$ generated by $(z \nearrow^A y \nearrow^B x) \simeq (z \nearrow^{A \cup B} x)$, $(x \searrow_A y \searrow_B z) \simeq (x \searrow_{A \cup B} z)$, and $\gamma\alpha\delta \simeq \gamma\beta\delta$ whenever $\alpha \simeq \beta$. Intuitively, this relation allows to assemble subsequent upsteps or downsteps into one "bigger" step. A path is *sparse* if its upsteps and downsteps are alternating, so that no more such assembling may take place. Every equivalence class of paths contains a unique sparse path.

Example 8. Paths in one-dimensional HDAs are standard paths, *i.e.*, sequences of transitions connected at states. Path equivalence is a trivial relation, and all paths are sparse.

Example 9. The HDA X of Fig. 4 admits five sparse accepting paths:

$$v \nearrow^a e \searrow_a w \nearrow^b h, \qquad v \nearrow^a e \searrow_a w \nearrow^b h \searrow_b y,$$
$$v \nearrow^{ab} q \searrow_a h, \qquad v \nearrow^{ab} q \searrow_{ab} y, \qquad v \nearrow^b g \searrow_b x \nearrow^a f \searrow_a y.$$

The observable content or *event ipomset* $\mathsf{ev}(\alpha)$ of a path α is defined recursively as follows:

- If $\alpha = (x)$, then $\mathsf{ev}(\alpha) = \mathsf{id}_{\mathsf{ev}(x)}$.
- If $\alpha = (y \nearrow^A x)$, then $\mathsf{ev}(\alpha) = {_A}{\uparrow}\mathsf{ev}(x)$.

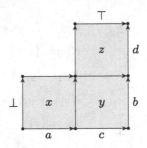

Fig. 5. HDA Y consisting of three squares glued along common faces.

– If $\alpha = (x \searrow_B y)$, then $\mathsf{ev}(\alpha) = \mathsf{ev}(x)\!\downarrow_B$.
– If $\alpha = \alpha_1 * \cdots * \alpha_n$ is a concatenation, then $\mathsf{ev}(\alpha) = \mathsf{ev}(\alpha_1) * \cdots * \mathsf{ev}(\alpha_n)$.

[7, Lemma 8] shows that $\alpha \simeq \beta$ implies $\mathsf{ev}(\alpha) = \mathsf{ev}(\beta)$. Further, if $\alpha = \alpha_1 * \cdots * \alpha_n$ is a sparse path, then $\mathsf{ev}(\alpha) = \mathsf{ev}(\alpha_1) * \cdots * \mathsf{ev}(\alpha_n)$ is a sparse step decomposition.

Example 10. Event ipomsets of paths in one-dimensional HDAs are words, possibly with interfaces. Sparse step decompositions of words are obtained by splitting symbols into starts and terminations, for example, $\bullet\, ab = \bullet\, a * b\, \bullet * \bullet\, b$.

Example 11. The event ipomsets of the five sparse accepting paths in the HDA X of Fig. 4 are $ab\bullet$, ab, $\begin{bmatrix} a \\ b \bullet \end{bmatrix}$, $\begin{bmatrix} a \\ b \end{bmatrix}$, and ba. Figure 5 shows another HDA which admits an accepting path $(\delta_a^0 x \nearrow^a x \searrow_a \delta_a^1 x \nearrow^c y \searrow_b \delta_b^1 y \nearrow^d z \searrow_d \delta_d^1 z)$. Its event ipomset is precisely the ipomset of Fig. 3, with the indicated sparse step decomposition arising directly from the sparse presentation above.

The *language* of an HDA X is

$$\mathsf{Lang}(X) = \{\mathsf{ev}(\alpha) \mid \alpha \text{ accepting path in } X\}.$$

[7, Prop. 10] shows that languages of HDAs are sets of ipomsets which are closed under subsumption, *i.e.*, languages in the sense of Def. 5.

A language is *regular* if it is the language of a finite HDA.

Example 12. The languages of our example HDAs are $\mathsf{Lang}(X) = \left\{ \begin{bmatrix} a \\ b \bullet \end{bmatrix}, \begin{bmatrix} a \\ b \end{bmatrix} \right\}\!\downarrow = \left\{ \begin{bmatrix} a \\ b \bullet \end{bmatrix}, ab\bullet, \begin{bmatrix} a \\ b \end{bmatrix}, ab, ba \right\}$ and

$$\mathsf{Lang}(Y) = \left\{ \begin{bmatrix} a \to c\ \bullet \\ \bullet\, b \to d \end{bmatrix} \right\}\!\downarrow.$$

We say that a cell $x \in X$ in an HDA X is

– *accessible* if $\mathsf{Path}(X)_\perp^x \neq \emptyset$, *i.e.*, x can be reached by a path from a start cell;
– *coaccessible* if $\mathsf{Path}(X)_x^\top \neq \emptyset$, *i.e.*, there is a path from x to an accept cell;
– *essential* if it is both accessible and coaccessible.

A path is *essential* if its source and target cells are essential. This implies that all its cells are essential. Segments of accepting paths are always essential.

The set of essential cells of X is denoted by $\mathsf{ess}(X)$; this is not necessarily a sub-HDA of X given that faces of essential cells may be non-essential. For example, all bottom cells of the HDA Y in Fig. 5 are inaccessible and hence non-essential.

Lemma 13. *Let X be an HDA. There exists a smallest sub-HDA $X^{\mathsf{ess}} \subseteq X$ that contains all essential cells, and $\mathsf{Lang}(X^{\mathsf{ess}}) = \mathsf{Lang}(X)$. If $\mathsf{ess}(X)$ is finite, then X^{ess} is also finite.*

Proof. The set of all faces of essential cells

$$X^{\mathsf{ess}} = \{\delta_A^0 \delta_B^1(x) \mid x \in \mathsf{ess}(X),\ A, B \subseteq \mathsf{ev}(x),\ A \cap B = \emptyset\}$$

is a sub-HDA of X. Clearly every sub-HDA of X that contains $\mathsf{ess}(X)$ must also contain X^{ess}. Since all accepting paths are essential, $\mathsf{Lang}(X^{\mathsf{ess}}) = \mathsf{Lang}(X)$. If $|\mathsf{ess}(X)| = n$ and $|\mathsf{ev}(x)| \leq d$ for all $x \in \mathsf{ess}(X)$, then $|X^{\mathsf{ess}}| \leq n \cdot 3^d$.   □

Track objects, introduced in [6], provide a mapping from ipomsets to HDAs and are a powerful tool for reasoning about languages. We only need some of their properties in proofs, so we do not give a definition here but instead refer to [6, Sect. 5.3]. Let $\square^P$ denote the track object of an ipomset P; this is an HDA with one start cell $c_\perp^P$ and one accept cell $c_P^\top$. Below we list properties of track objects needed in the paper.

Lemma 14. *Let X be an HDA, $x, y \in X$ and $P \in iiPoms$. The following conditions are equivalent:*

1. *There exists a path $\alpha \in \mathsf{Path}(X)_x^y$ such that $\mathsf{ev}(\alpha) = P$.*
2. *There is an HDA-map $f : \square^P \to X_x^y$ (i.e., $f(c_\perp^P) = x$ and $f(c_P^\top) = y$).*

Proof. This is an immediate consequence of [6, Prop. 89].   □

Lemma 15. *Let X be an HDA, $x, y \in X$ and $\gamma \in \mathsf{Path}(X)_x^y$. Assume that $\mathsf{ev}(\gamma) = P * Q$ for ipomsets P and Q. Then there exist paths $\alpha \in \mathsf{Path}(X)_x$ and $\beta \in \mathsf{Path}(X)^y$ such that $\mathsf{ev}(\alpha) = P$, $\mathsf{ev}(\beta) = Q$ and $\mathsf{tgt}(\alpha) = \mathsf{src}(\beta)$.*

Proof. By Lemma 14, there is an HDA-map $f : \square^{PQ} \to X_x^y$. By [6, Lem. 65], there exist precubical maps $j_P : \square^P \to \square^{PQ}$, $j_Q : \square^Q \to \square^{PQ}$ such that $j_P(c_\perp^P) = c_\perp^{PQ}$, $j_P(c_P^\top) = j_Q(c_\perp^Q)$ and $j_Q(c_Q^\top) = c_{PQ}^\top$. Let $z = f(j_P(c_\perp^P))$, then $f \circ j_P : \square^P \to X_x^z$ and $f \circ j_Q : \square^Q \to X_z^y$ are HDA-maps, and by applying Lemma 14 again to j_P and j_Q we obtain α and β.   □

4 Myhill-Nerode Theorem

The *prefix quotient* of a language $L \in \mathscr{L}$ by an ipomset P is the language

$$P \backslash L = \{Q \in iiPoms \mid PQ \in L\}.$$

Similarly, the *suffix quotient* of L by P is $L/P = \{Q \in \text{iiPoms} \mid QP \in L\}$. Denote

$$\text{suff}(L) = \{P \backslash L \mid P \in \text{iiPoms}\}, \qquad \text{pref}(L) = \{L/P \mid P \in \text{iiPoms}\}.$$

We record the following property of quotient languages.

Lemma 16. *If L is a language and $P \sqsubseteq Q$, then $Q \backslash L \subseteq P \backslash L$.*

Proof. If $P \sqsubseteq Q$, then $PR \sqsubseteq QR$. Thus,

$$R \in Q \backslash L \iff QR \in L \implies PR \in L \iff R \in P \backslash L.$$

$\square$

The main goal of this section is to show the following.

Theorem 17. *For a language $L \in \mathscr{L}$ the following conditions are equivalent.*

(a) *L is regular.*
(b) *The set $\text{suff}(L) \subseteq \mathscr{L}$ is finite.*
(c) *The set $\text{pref}(L) \subseteq \mathscr{L}$ is finite.*

We prove only the equivalence between (a) and (b); equivalence between (a) and (c) is symmetric. First we prove the implication (a) $\implies$ (b). Let X be an HDA with $\text{Lang}(X) = L$. For $x \in X$ define languages $\text{Pre}(x) = \text{Lang}(X_\bot^x)$ and $\text{Post}(x) = \text{Lang}(X_x^\top)$.

Lemma 18. *For every $P \in \text{iiPoms}$, $P \backslash L = \bigcup \{\text{Post}(x) \mid x \in X, P \in \text{Pre}(x)\}$.*

Proof. We have

$$
\begin{aligned}
Q \in P \backslash L &\iff PQ \in L \iff \exists\, f : \square^{PQ} \to X = X_\bot^\top \\
&\iff \exists\, x \in X, g : \square^P \to X_\bot^x, \ h : \square^Q \to X_x^\top \\
&\iff \exists\, x \in X : P \in \text{Lang}(X_\bot^x), \ Q \in \text{Lang}(X_x^\top) \\
&\iff \exists\, x \in X : P \in \text{Pre}(x), \ Q \in \text{Post}(x).
\end{aligned}
$$

The last condition says that Q belongs to the right-hand side of the equation. $\square$

Proof. of Theorem. 17, (a) $\implies$ (b). The family of languages $\{P \backslash L \mid P \in \text{iiPoms}\}$ is a subfamily of $\{\bigcup_{x \in Y} \text{Post}(x) \mid Y \subseteq X\}$ which is finite. $\square$

HDA Construction. Now we show that (b) implies (a). Fix a language $L \in \mathscr{L}$, with $\text{suff}(L)$ finite or infinite. We will construct an HDA $\text{MN}(L)$ that recognises L and show that if $\text{suff}(L)$ is finite, then its essential part $\text{MN}(L)^{\text{ess}}$ is finite. The cells of $\text{MN}(L)$ are equivalence classes of ipomsets under a relation $\approx_L$ induced by L which we will introduce below. The relation $\approx_L$ is defined using prefix quotients, but needs to be stronger than prefix quotient equivalence. This is because events may be concurrent and because ipomsets have interfaces. We give examples just after the construction.

For an ipomset $_SP_T$ define its *(target) signature* to be the starter $\text{fin}(P) = {}_{T-S}\uparrow T$. Thus $\text{fin}(P)$ collects all target events of P, and its source interface contains those events that are also in the source interface of P. We also write $\text{rfin}(P) = T - S \subseteq \text{fin}(P)$: the set of all target events of P that are not source events. An important property is that removing elements of $\text{rfin}(P)$ does not change the source interface of P. For example,

$$\text{fin}\left(\left[\begin{smallmatrix}\bullet\, a\,\bullet \\ \bullet\, a \\ c\,\bullet \end{smallmatrix}\right]\right) = [\,{}^{\bullet\, a\,\bullet}_{c\,\bullet}\,], \quad \text{fin}\left([\,{}^{\bullet\ ac\,\bullet}_{\bullet\ b\,\bullet}\,]\right) = [\,{}_{\bullet\, b\,\bullet}^{\ c\,\bullet}\,], \quad \text{fin}\left([\,{}^{ac\,\bullet}_{b\,\bullet}\,]\right) = [\,{}^{c\,\bullet}_{b\,\bullet}\,];$$

rfin is $\{c\}$ in the first two examples and equal to $[\,{}^c_b\,]$ in the last.

We define two equivalence relations on iiPoms induced by L:

- Ipomsets P and Q are *weakly equivalent* ($P \sim_L Q$) if $\text{fin}(P) \cong \text{fin}(Q)$ and $P \backslash L = Q \backslash L$. Obviously, $P \sim_L Q$ implies $T_P \cong T_Q$ and $\text{rfin}(P) \cong \text{rfin}(Q)$.
- Ipomsets P and Q are *strongly equivalent* ($P \approx_L Q$) if $P \sim_L Q$ and for all $A \subseteq \text{rfin}(P) \cong \text{rfin}(Q)$ we have $(P - A)\backslash L = (Q - A)\backslash L$.

Evidently $P \approx_L Q$ implies $P \sim_L Q$, but the inverse does not always hold. We explain in Example 21 below why $\approx_L$, and not $\sim_L$, is the proper relation to use for constructing $\text{MN}(L)$.

Lemma 19. *If* $P \approx_L Q$, *then* $P - A \approx_L Q - A$ *for all* $A \subseteq \text{rfin}(P) \cong \text{rfin}(Q)$.

Proof. For every A we have $(P - A)\backslash L = (Q - A)\backslash L$, and

$$\text{fin}(P - A) = \text{fin}(P) - A \cong \text{fin}(Q) - A = \text{fin}(Q - A),$$

Thus, $P - A \sim_L Q - A$. Further, for every $B \subseteq \text{rfin}(P - A) \cong \text{rfin}(Q - A)$,

$$((P - A) - B)\backslash L = (P - (A \cup B))\backslash L = (Q - (A \cup B))\backslash L = ((Q - A) - B)\backslash L,$$

which shows that $P - A \approx_L Q - A$. □

Now define an HDA $\text{MN}(L)$ as follows. For $U \in \square$,

$$\text{MN}(L)[U] = (\text{iiPoms}_U / \approx_L) \cup \{w_U\},$$

where the w_U are new *subsidiary* cells which are introduced solely to define some lower faces. (They will not affect the language of $\text{MN}(L)$).

The $\approx_L$-equivalence class of P will be denoted by $\langle P \rangle$ (but often just by P in examples). Face maps are defined as follows, for $A \subseteq U \in \square$ and $P \in \text{iiPoms}_U$:

$$\delta_A^0(\langle P \rangle) = \begin{cases} \langle P - A \rangle & \text{if } A \subseteq \text{rfin}(P), \\ w_{U-A} & \text{otherwise,} \end{cases} \qquad \delta_A^1(\langle P \rangle) = \langle P * U{\downarrow}_A \rangle, \qquad (2)$$

$$\delta_A^0(w_U) = \delta_A^1(w_U) = w_{U-A}.$$

In other words, if A has no source events of P, then δ_A^0 removes A from P (the source interface of P is unchanged). If A contains any source event, then $\delta_A^0(P)$ is a subsidiary cell.

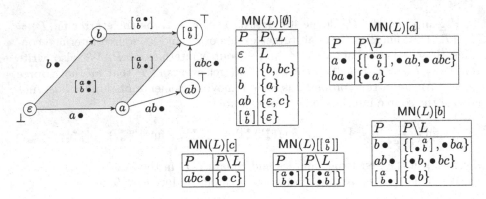

Fig. 6. HDA MN(L) of Example 20, showing names of cells instead of labels (labels are target interfaces of names). Tables show essential cells together with prefix quotients.

Finally, start and accept cells are given by

$$\bot_{\mathsf{MN}(L)} = \{\langle \mathrm{id}_U \rangle\}_{U \in \Box}, \qquad \top_{\mathsf{MN}(L)} = \{\langle P \rangle \mid P \in L\}.$$

The cells $\langle P \rangle$ will be called *regular*. They are $\approx_L$-equivalence classes of ipomsets, lower face maps unstart events, and upper face maps terminate events. All faces of subsidiary cells w_U are subsidiary, and upper faces of regular cells are regular. Below we present several examples, in which we show only the essential part $\mathsf{MN}(L)^{\mathsf{ess}}$ of $\mathsf{MN}(L)$.

Example 20. Let $L = \{[\begin{smallmatrix} a \\ b \end{smallmatrix}], abc\}\!\downarrow = \{[\begin{smallmatrix} a \\ b \end{smallmatrix}], ab, ba, abc\}$. Figure 6 shows the HDA $\mathsf{MN}(L)^{\mathsf{ess}}$ together with a list of essential cells of $M(L)$ and their prefix quotients in L. Note that the state $\langle a \rangle$ has *two* outgoing b-labelled edges: $\langle ab\bullet\rangle$ and $\langle [\begin{smallmatrix} a \\ b \bullet\end{smallmatrix}]\rangle$. The generating ipomsets have different prefix quotients because of $\{[\begin{smallmatrix} a \\ b \end{smallmatrix}], abc\} \subseteq L$ but the same lower face $\langle a \rangle$.

Intuitively, $\mathsf{MN}(L)^{\mathsf{ess}}$ is thus *non-deterministic*; this is interesting because the standard Myhill-Nerode theorem for finite automata constructs deterministic automata. We will give a precise definition of determinism for HDAs in the next section and show in Example 42 that no deterministic HDA X exists with $\mathsf{Lang}(X) = L$.

Example 21. Here we explain why we need to use $\approx_L$-equivalence classes and not $\sim_L$-equivalence classes. Let $L = \{[\begin{smallmatrix} a \\ b \end{smallmatrix}], aa\}\!\downarrow$. Then $\mathsf{MN}(L)^{\mathsf{ess}}$ is as below.

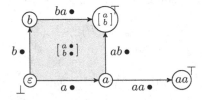

Note that $(aa\bullet)\backslash L = (ba\bullet)\backslash L = \{\bullet a\}$, thus $aa\bullet \sim_L ba\bullet$. Yet $aa\bullet$ and $ba\bullet$ are not strongly equivalent, because $a\backslash L = \{a, b\} \neq \{a\} = b\backslash L$. This provides an example of weakly equivalent ipomsets whose lower faces are not weakly equivalent and shows why we cannot use $\sim_L$ to construct $\mathsf{MN}(L)$.

Example 22. The language $L = \{[\begin{smallmatrix} \bullet\, aa\, \bullet \\ \bullet\, a\, \bullet \end{smallmatrix}]\}$ is recognised by the HDA $\mathsf{MN}(L)^{\mathsf{ess}}$ below:

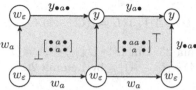

Cells with the same names are identified. Here we see subsidiary cells w_ε and w_a, and regular cells (denoted by y indexed with their signature) that are not coaccessible. The middle vertical edge is $\langle [\begin{smallmatrix} \bullet\, a \\ \bullet\, a\, \bullet \end{smallmatrix}] \rangle$, $y_{\bullet a \bullet} = \langle [\begin{smallmatrix} \bullet\, a\, \bullet \\ \bullet\, a \end{smallmatrix}] \rangle = \langle [\begin{smallmatrix} \bullet\, aa \\ \bullet\, a\, \bullet \end{smallmatrix}] \rangle$, $y_{a\bullet} = \langle [\begin{smallmatrix} \bullet\, aa\, \bullet \\ a \end{smallmatrix}] \rangle$, and $y = \langle [\begin{smallmatrix} \bullet\, a \\ \bullet\, a \end{smallmatrix}] \rangle = \langle [\begin{smallmatrix} \bullet\, aa \\ \bullet\, a \end{smallmatrix}] \rangle$.

$\mathsf{MN}(L)$ is well-defined. We need to show that the formulas (2) do not depend on the choice of a representative in $\langle P \rangle$ and that the precubical identities are satisfied.

Lemma 23. *Let P, Q and R be ipomsets with $T_P = T_Q = S_R$. Then*

$$P\backslash L \subseteq Q\backslash L \implies (PR)\backslash L \subseteq (QR)\backslash L.$$

In particular, $P\backslash L = Q\backslash L$ implies $(PR)\backslash L = (QR)\backslash L$.

Proof. For $N \in \mathsf{iiPoms}$ we have

$$N \in (PR)\backslash L \iff PRN \in L \iff RN \in P\backslash L$$
$$\implies RN \in Q\backslash L \iff QRN \in L \iff N \in (QR)\backslash L. \qquad \square$$

The next lemma shows an operation to "add order" to an ipomset P. This is done by first removing some points $A \subseteq T_P$ and then adding them back in, forcing arrows from all other points in P. The result is obviously subsumed by P.

Lemma 24. *For $P \in \mathsf{iiPoms}$ and $A \subseteq \mathrm{rfin}(P)$, $(P - A) *_A{\uparrow}T_P \sqsubseteq P$.* $\qquad\square$

The next two lemmas, whose proofs are again obvious, state that events may be unstarted or terminated in any order.

Lemma 25. *Let U be an loset and $A, B \subseteq U$ disjoint subsets. Then*

$$U{\downarrow}_B * (U - B){\downarrow}_A = U{\downarrow}_{A\cup B} = U{\downarrow}_A * (U - A){\downarrow}_B. \qquad\square$$

Lemma 26. *Let $P \in \mathsf{iiPoms}$ and $A, B \subseteq T_P$ disjoint subsets. Then*

$$(P * T_P{\downarrow}_B) - A = (P - A) * (T_P - A){\downarrow}_B. \qquad\square$$

Lemma 27. *Assume that $P \approx_L Q$ for $P, Q \in iiPoms_U$. Then $P * U{\downarrow_B} \approx_L Q * U{\downarrow_B}$ for every $B \subseteq U$.*

Proof. Obviously $\mathrm{fin}(P * U{\downarrow_B}) = \mathrm{fin}(P) - B \cong \mathrm{fin}(Q) - B = \mathrm{fin}(Q * U{\downarrow_B})$. For every $A \subseteq \mathrm{rfin}(P) - B \simeq \mathrm{rfin}(Q) - B$ we have

$$((P - A) * (U - A){\downarrow_B})\backslash L = ((Q - A) * (U - A){\downarrow_B})\backslash L$$

by assumption and Lemma 23. But $(P * U{\downarrow_B}) - A = (P - A) * (U - A){\downarrow_B}$ and $(Q * U{\downarrow_B}) - A = (Q - A) * (U - A){\downarrow_B}$ by Lemma 26. □

Proposition 28. *$MN(L)$ is a well-defined HDA.*

Proof. The face maps are well-defined: for δ_A^0 this follows from Lemma 19, for δ_B^1 from Lemma 27. The precubical identities $\delta_A^\nu \delta_B^\mu = \delta_B^\mu \delta_A^\nu$ are clear for $\nu = \mu = 0$, follow from Lemma 25 for $\nu = \mu = 1$, and from Lemma 26 for $\{\nu, \mu\} = \{0, 1\}$. □

Paths and essential cells of $MN(L)$. The next lemma provides paths in $MN(L)$.

Lemma 29. *For every $N, P \in iiPoms$ such that $T_N \cong S_P$ there exists a path $\alpha \in \mathsf{Path}(MN(L))_{\langle N \rangle}^{\langle NP \rangle}$ such that $\mathsf{ev}(\alpha) = P$.*

Proof. Choose a decomposition $P = Q_1 * \cdots * Q_n$ into starters and terminators. Denote $U_k = T_{Q_k} = S_{Q_{k+1}}$ and define

$$x_k = \langle N * Q_1 * \cdots * Q_k \rangle, \qquad \varphi_k = \begin{cases} d_A^0 & \text{if } Q_k = {}_A{\uparrow}U_k, \\ d_B^1 & \text{if } Q_k = U_{k-1}{\downarrow_B} \end{cases}$$

for $k = 1, \ldots, n$. If $\varphi_k = d_A^0$ and $Q_k = {}_A{\uparrow}U_k$, then

$$\delta_A^0(x_k) = \langle N * Q_1 * \cdots * Q_{k-1} * {}_A{\uparrow}U_k - A \rangle$$
$$= \langle N * Q_1 * \cdots * Q_{k-1} * \mathsf{id}_{U_k - A} \rangle = x_{k-1}.$$

If $\varphi_k = d_B^1$ and $Q_k = U_{k-1}{\downarrow_B}$, then

$$\delta_B^1(x_{k-1}) = \langle N * Q_1 * \cdots * Q_{k-1} * U_{k-1}{\downarrow_B} \rangle = x_k.$$

Thus, $\alpha = (x_0, \varphi_1, x_1, \ldots, \varphi_n, x_n)$ is a path with $\mathsf{ev}(\alpha) = P$, $\mathsf{src}(\alpha) = \langle N \rangle$ and $\mathsf{tgt}(\alpha) = \langle N * P \rangle$. □

Our goal is now to describe essential cells of $MN(L)$.

Lemma 30. *All regular cells of $MN(L)$ are accessible. If $P\backslash L \neq \emptyset$, then $\langle P \rangle$ is coaccessible.*

Proof. Both claims follow from Lemma 29. For every P there exists a path from $\langle \mathsf{id}_{S_P} \rangle$ to $\langle \mathsf{id}_{S_P} * P \rangle = \langle P \rangle$. If $Q \in P\backslash L$, then there exists a path $\alpha \in \mathsf{Path}(MN(L))_{\langle P \rangle}^{\langle PQ \rangle}$, and $PQ \in L$ entails that $\langle PQ \rangle \in \top_{MN(L)}$. □

Lemma 31. *Subsidiary cells of* $MN(L)$ *are not accessible. If* $P\backslash L = \emptyset$, *then* $\langle P \rangle$ *is not coaccessible.*

Proof. If $\alpha \in \mathsf{Path}(MN(L))_{\perp}^{wu}$, then it contains a step β from a regular cell to a subsidiary cell (since all start cells are regular). Yet β can be neither an upstep (since lower faces of subsidiary cells are subsidiary) nor a downstep (since upper faces of regular cells are regular). This contradiction proves the first claim.

To prove the second part we use a similar argument. If $P\backslash L = \emptyset$, then a path $\alpha \in \mathsf{Path}(MN(L))_{\langle P \rangle}^{\top}$ contains only regular cells (as shown above). Given that $R\backslash L \neq \emptyset$ for all $\langle R \rangle \in \top_{MN(L)}$, α must contain a step β from $\langle Q \rangle$ to $\langle R \rangle$ such that $Q\backslash L = \emptyset$ and $R\backslash L \neq \emptyset$. If β is a downstep, i.e., $\beta = (\langle Q \rangle \searrow_A \langle Q * U\downarrow_A \rangle)$, and $N \in R\backslash L = (Q * U\downarrow_A)\backslash L$, then $U\downarrow_A * N \in Q\backslash L \neq \emptyset$: a contradiction. If $\beta = (\langle R - A \rangle \nearrow^A \langle R \rangle)$ is an upstep and $N \in R\backslash L$, then, by Lemma 24,

$$(R - A) *_A\uparrow U * N \sqsubseteq R * N \in L,$$

implying that $Q\backslash L = (R - A)\backslash L \neq \emptyset$ by Lemma 16: another contradiction. $\square$

Lemmas 30 and 31 together immediately imply the following.

Proposition 32. $\mathsf{ess}(MN(L)) = \{\langle P \rangle \mid P\backslash L \neq \emptyset\}$. $\square$

$MN(L)$ recognises L. One inclusion follows immediately from Lemma 29:

Lemma 33. $L \subseteq \mathsf{Lang}(MN(L))$.

Proof. For every $P \in \mathsf{iiPoms}$ there exists a path $\alpha \in \mathsf{Path}(MN(L))_{\langle \mathsf{id}_{S_P} \rangle}^{\langle P \rangle}$ such that $\mathsf{ev}(\alpha) = P$. If $P \in L$, then $\varepsilon \in P\backslash L$, i.e., $\langle P \rangle$ is an accept cell. Thus α is accepting and $P = \mathsf{ev}(\alpha) \in \mathsf{Lang}(MN(L))$. $\square$

The converse inclusion requires more work. For a regular cell $\langle P \rangle$ of $MN(L)$ denote $\langle P \rangle \backslash L = P\backslash L$ (this obviously does not depend on the choice of P).

Lemma 34. *If* $S \in \square$ *and* $\alpha \in \mathsf{Path}(MN(L))_{\langle \mathsf{id}_S \rangle}$, *then* $\mathsf{tgt}(\alpha)\backslash L \subseteq \mathsf{ev}(\alpha)\backslash L$.

Proof. By Lemma 31, all cells appearing along α are regular. We proceed by induction on the length of α. For $\alpha = (\langle \mathsf{id}_S \rangle)$ the claim is obvious. If α is non-trivial, we have two cases.

- $\alpha = \beta * (\delta_A^0(\langle P \rangle) \nearrow^A \langle P \rangle)$, where $\langle P \rangle \in MN(L)[U]$ and $A \subseteq \mathsf{rfin}(P) \subseteq U \cong T_P$. By the induction hypothesis,

$$(P - A)\backslash L = \delta_A^0(\langle P \rangle)\backslash L = \mathsf{tgt}(\beta)\backslash L \subseteq \mathsf{ev}(\beta)\backslash L.$$

For $Q \in \mathsf{iiPoms}$ we have

$$Q \in P\backslash L \iff PQ \in L \implies (P - A) *_A\uparrow U * Q \in L \qquad \text{(Lemma 24)}$$
$$\iff {}_A\uparrow U * Q \in (P - A)\backslash L$$
$$\implies {}_A\uparrow U * Q \in \mathsf{ev}(\beta)\backslash L \quad \text{(induction hypothesis)}$$
$$\iff \mathsf{ev}(\beta) *_A\uparrow U * Q \in L$$
$$\iff \mathsf{ev}(\alpha) * Q \in L \iff Q \in \mathsf{ev}(\alpha)\backslash L.$$

Thus, $\langle P \rangle\backslash L = P\backslash L \subseteq \mathsf{ev}(\alpha)\backslash L$.

- $\alpha = \beta * (\langle P \rangle \searrow_B \delta_B^1(\langle P \rangle))$, where $\langle P \rangle \in \mathsf{MN}(L)[U]$ and $B \subseteq U \cong T_P$. By inductive assumption, $P \backslash L = \mathsf{tgt}(\beta) \backslash L \subseteq \mathsf{ev}(\beta) \backslash L$. Thus,

$$\mathsf{tgt}(\alpha) \backslash L = \delta_B^1(\langle P \rangle) \backslash L = \langle P * U \downarrow_B \rangle \backslash L \subseteq (\mathsf{ev}(\beta) * U \downarrow_B) \backslash L = \mathsf{ev}(\alpha) \backslash L.$$

The inclusion above follows from Lemma 23.   □

Proposition 35. $\mathsf{Lang}(MN(L)) = L$.

Proof. The inclusion $L \subseteq \mathsf{Lang}(\mathsf{MN}(L))$ is shown in Lemma 33. For the converse, let $S \in \square$ and $\alpha \in \mathsf{Path}(\mathsf{MN}(L))_{\langle \mathsf{id}_S \rangle}$, then Lemma 34 implies

$$\mathsf{tgt}(\alpha) \in \top_{\mathsf{MN}(L)} \iff \varepsilon \in \mathsf{tgt}(\alpha) \backslash L \implies \varepsilon \in \mathsf{ev}(\alpha) \backslash L \iff \mathsf{ev}(\alpha) \in L,$$

that is, if α is accepting, then $\mathsf{ev}(\alpha) \in L$.   □

Finiteness of MN(L). The HDA $\mathsf{MN}(L)$ is not finite, since it contains infinitely many subsidiary cells w_U. Below we show that its essential part $\mathsf{MN}(L)^{\mathsf{ess}}$ is finite if L has finitely many prefix quotients.

Lemma 36. *If* $\mathsf{suff}(L)$ *is finite, then* $\mathsf{ess}(MN(L))$ *is finite.*

Proof. For $\langle P \rangle, \langle Q \rangle \in \mathsf{ess}(L)$, we have $\langle P \rangle \approx_L \langle Q \rangle$ iff $f(\langle P \rangle) = f(\langle Q \rangle)$, where

$$f(\langle P \rangle) = (P \backslash L, \mathsf{fin}(P), ((P - A) \backslash L)_{A \subseteq \mathsf{rfin}(P)}).$$

We will show that f takes only finitely many values on $\mathsf{ess}(L)$. Indeed, $P \backslash L$ belongs to the finite set $\mathsf{suff}(L)$. Further, all ipomsets in $P \backslash L$ have source interfaces equal to T_P. Since $P \backslash L$ is non-empty, $\mathsf{fin}(P)$ is a starter with T_P as underlying loset. Yet, there are only finitely many starters on any loset. The last coordinate also may take only finitely many values, since $\mathsf{rfin}(P)$ is finite and $(P - A) \backslash L \in \mathsf{suff}(L)$.   □

Proof of Theorem. 17, (b) $\implies$ (a). From Lemma 36 and Lemma 13, $\mathsf{MN}(L)^{\mathsf{ess}}$ is a finite HDA. By Prop. 35 we have $\mathsf{Lang}(\mathsf{MN}(L)^{\mathsf{ess}}) = \mathsf{Lang}(\mathsf{MN}(L)) = L$.   □

Example 37 We finish this section with another example, which shows some subtleties related to higher-dimensional loops. Let L be the language of the HDA shown to the left of Fig. 7 (a looping version of the HDA of Fig. 5), then

$$L = \{\bullet\, a\, \bullet\} \cup \{[\,^{\bullet\,aa\,\bullet}_{\quad b}\,]^n \mid n \geq 1\}\downarrow.$$

Our construction yields $\mathsf{MN}(L)^{\mathsf{ess}}$ as shown on the right of the figure. Here, $e = \langle [\,^{\bullet\,a}_{\,\,b\,\bullet}\,]\rangle$, and cells with the same names are identified. These identifications follow from the fact that $[\,^{\bullet\,aa\,\bullet}_{\,\,bb\,\bullet}\,] \approx_L [\,^{\bullet\,a}_{\,\,b\,\bullet}\,]$, $[\,^{\bullet\,aa}_{\,\,bb\,\bullet}\,] \approx_L [\,^{\bullet\,a}_{\,\,b}\,]$, and $[\,^{\bullet\,aa}_{\,\,b}\,] \approx_L \bullet\, a$. Note that $[\,^{\bullet\,a\,\bullet}_{\,\,b\,\bullet}\,]$ and $[\,^{\bullet\,aa\,\bullet}_{\,\,bb\,\bullet}\,]$ are not strongly equivalent, since they have different signatures: $[\,^{\bullet\,a\,\bullet}_{\,\,b\,\bullet}\,]$ and $[\,^{a\,\bullet}_{\,b\,\bullet}\,]$, respectively.

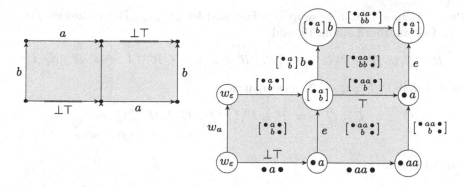

Fig. 7. Two HDAs recognising the language of Example 37. On the left side, start/accept edges are identified.

5   Determinism

We now make precise our notion of determinism and show that not all HDAs may be determinised. Recall that we do not assume finiteness.

Definition 38 *An HDA X is* deterministic *if*

1. *for every $U \in \square$ there is at most one initial cell in $X[U]$, and*
2. *for all $V \in \square$, $A \subseteq V$ and an essential cell $x \in X[V - A]$ there exists at most one essential cell $y \in X[V]$ such that $\delta_A^0(y) = x$.*

That is, in any essential cell x in a deterministic HDA X and for any set A of events, there is at most one way to start A in x and remain in the essential part of X. We allow multiple initial cells because ipomsets in $\mathsf{Lang}(X)$ may have different source interfaces; for each source interface in $\mathsf{Lang}(X)$, there can be at most one matching start cell in X. Note that we must restrict our definition to essential cells as inessential cells may not always be removed (in contrast to the case of standard automata).

A language is deterministic if it is recognised by a deterministic HDA. We develop a language-internal criterion for being deterministic.

Definition 39 *A language L is* swap-invariant *if it holds for all $P, Q, P', Q' \in$ iiPoms that $PP' \in L$, $QQ' \in L$ and $P \sqsubseteq Q$ imply $QP' \in L$.*

That is, if the P prefix of $PP' \in L$ is subsumed by Q (which is, thus, "more concurrent" than P), and if Q itself may be extended to an ipomset in L, then P may be swapped for Q in the ipomset PP' to yield $QP' \in L$.

Lemma 40 *L is swap-invariant iff $P \sqsubseteq Q$ implies $P \backslash L = Q \backslash L$ for all $P, Q \in$ iiPoms, unless $Q \backslash L = \emptyset$.*

Proof. Assume that L is swap-invariant and let $P \sqsubseteq Q$. The inclusion $Q \backslash L \subseteq P \backslash L$ follows from Lemma 16, and

$$R \in Q \backslash L, \ R' \in P \backslash L \iff QR, PR' \in L \implies QR' \in L \iff R' \in Q \backslash L$$

implies that $P \backslash L \subseteq Q \backslash L$. The calculation

$$PP', QQ' \in L, \ P \sqsubseteq Q \iff P' \in P \backslash L, \ Q' \in Q \backslash L, \ P \sqsubseteq Q \implies$$
$$P' \in Q \backslash L \iff QP' \in L$$

shows the converse.

Our main goal is to show the following criterion, which will be implied by Propositions 47 and 49 below.

Theorem 41. *A language L is deterministic iff it is swap-invariant.*

Example 42. The regular language $L = \{[{}^{a}_{b}], ab, ba, abc\}$ from Example 20 is not swap-invariant: using Lemma 40, $ab \bullet \sqsubseteq [{}^{a}_{b}\bullet]$, but $\{ab \bullet\} \backslash L = \{\bullet b, \bullet bc\} \neq \{\bullet b\} = \{[{}^{a}_{b}\bullet]\} \backslash L$. Hence L is not deterministic.

The next examples explain why we need to restrict to essential cells in the definition of determinacy.

Example 43. The HDA in Example 22 is deterministic. There are two different a-labelled edges starting at w_ε (w_a and $\langle[{}^{\bullet a}_{\bullet a}\bullet]\rangle$), yet it does not disturb determinism since w_ε is not accessible.

Example 44. Let $L = \{ab, [{}^{a}_{b}{}^{\bullet}_{\bullet}]\}$. Then $\mathsf{MN}(L)^{\mathsf{ess}}$ is as follows:

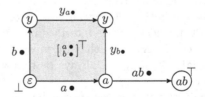

It is deterministic; there are two b-labelled edges leaving a, namely $y_{b\bullet}$ and $ab \bullet$, but only the latter is coaccessible.

Lemma 45. *Let X be a deterministic HDA and $\alpha, \beta \in \mathsf{Path}(X)_\perp$ with $\mathsf{tgt}(\alpha)$, $\mathsf{tgt}(\beta) \in \mathsf{ess}(X)$. If $\mathsf{src}(\alpha) = \mathsf{src}(\beta)$ and $\mathsf{ev}(\alpha) = \mathsf{ev}(\beta)$, then $\mathsf{tgt}(\alpha) = \mathsf{tgt}(\beta)$.*

Proof. We can assume that $\alpha = \alpha_1 * \cdots * \alpha_n$ and $\beta = \beta_1 * \cdots * \beta_m$ are sparse; note that all of these cells are essential. Denote $P = \mathsf{ev}(\alpha) = \mathsf{ev}(\beta)$, then

$$P = \mathsf{ev}(\alpha) = \mathsf{ev}(\alpha_1) * \cdots * \mathsf{ev}(\alpha_n)$$

is a sparse step decomposition of P. Similarly, $P = \mathsf{ev}(\beta_1) * \cdots * \mathsf{ev}(\beta_m)$ is a sparse step decomposition. Yet sparse step decompositions are unique by Prop. 4; hence,

$m = n$ and $\mathsf{ev}(\alpha_k) = \mathsf{ev}(\beta_k)$ for every k. We show by induction that $\alpha_k = \beta_k$. Assume that $\alpha_{k-1} = \beta_{k-1}$. Let $x = \mathsf{src}(\alpha_k) = \mathsf{tgt}(\alpha_{k-1}) = \mathsf{tgt}(\beta_{k-1}) = \mathsf{src}(\beta_k)$. If $P_k = \mathsf{ev}(\alpha_k) = \mathsf{ev}(\beta_k)$ is a terminator $U\!\downarrow_B$, then $\alpha_k = \delta_B^1(x) = \beta_k$. If P_k is a starter ${}_A\!\uparrow U$, then there are $y, z \in X$ such that $\delta_A^0(y) = \delta_A^0(z) = x$. As y and z are essential and X is deterministic, this implies $y = z$ and $\alpha_k = \beta_k$. $\qquad\square$

Lemma 46. *Let α and β be essential paths on a deterministic HDA X. Assume that $\mathsf{src}(\alpha) = \mathsf{src}(\beta)$ and $\mathsf{ev}(\alpha) \sqsubseteq \mathsf{ev}(\beta)$. Then $\mathsf{tgt}(\alpha) = \mathsf{tgt}(\beta)$.*

Proof. Denote $x \in \mathsf{src}(\alpha) = \mathsf{src}(\beta)$ and $y = \mathsf{tgt}(\beta)$. By Lemma 14 there exists an HDA-map $f : \square^{\mathsf{ev}(\beta)} \to X_x^y$. By [6, Lemma 63] there is an HDA-map $i : \square^{\mathsf{ev}(\alpha)} \to \square^{\mathsf{ev}(\beta)}$. We apply Lemma 14 again to the composition $f \circ i$ and obtain that there is a path $\alpha' \in \mathsf{Path}(X)_x^y$ such that $\mathsf{ev}(\alpha') = \mathsf{ev}(\alpha)$. Lemma 45 then implies $\mathsf{tgt}(\alpha) = \mathsf{tgt}(\alpha') = y$. $\qquad\square$

Proposition 47. *If L is deterministic, then L is swap-invariant.*

Proof. Let X be a deterministic automaton that recognises L and fix ipomsets $P \sqsubseteq Q$. From Lemma 16 follows that $Q \backslash L \subseteq P \backslash L$. It remains to prove that if $Q \backslash L \neq \emptyset$, then $P \backslash L \subseteq Q \backslash L$. Denote $U \cong S_P \cong S_Q$.

Let $R \in Q \backslash L$ and let $\omega \in \mathsf{Path}(X)_{\langle \mathrm{id}_U \rangle}^\top$ be an accepting path that recognises QR. By Lemma 15, there exists a path $\beta \in \mathsf{Path}(X)_{\langle \mathrm{id}_U \rangle}$ such that $\mathsf{ev}(\beta) = Q$.

Now assume that $R' \in P \backslash L$, and let $\omega' \in \mathsf{Path}(X)_{\langle \mathrm{id}_U \rangle}^\top$ be a path such that $\mathsf{ev}(\omega') = PR'$. By Lemma 15, there exist paths $\alpha \in \mathsf{Path}(X)_{\langle \mathrm{id}_U \rangle}$ and $\gamma \in \mathsf{Path}(X)^{\mathsf{tgt}(\omega')}$ such that $\mathsf{tgt}(\alpha) = \mathsf{src}(\gamma)$, $\mathsf{ev}(\alpha) = P$ and $\mathsf{ev}(\gamma) = R'$. From Lemma 46 and $P \sqsubseteq Q$ follows that $\mathsf{tgt}(\alpha) = \mathsf{tgt}(\beta)$. Thus, β and γ may be concatenated to an accepting path $\beta * \gamma$. By $\mathsf{ev}(\beta * \gamma) = QR'$ we have $QR' \in L$, i.e., $R' \in Q \backslash L$. $\qquad\square$

Lemma 48. *If $\langle P \rangle \in \mathrm{ess}(MN(L))$ and $A \subseteq \mathrm{rfin}(P)$, then $\langle P - A \rangle \in \mathrm{ess}(MN(L))$.*

Proof. By Lemma 33, $\langle P - A \rangle$ is accessible. By assumption, $\langle P \rangle$ is coaccessible and $(\langle P - A \rangle \nearrow^A \langle P \rangle)$ is a path, so $\langle P - A \rangle$ is also coaccessible. $\qquad\square$

Proposition 49. *If L is swap-invariant, then $MN(L)$ is deterministic.*

Proof. $MN(L)$ contains only one start cell $\langle \mathrm{id}_U \rangle$ for every $U \in \square$.

Fix $U \in \square$, $P, Q \in \mathrm{iiPoms}_U$ and $A \subseteq U$. Assume that $\delta_A^0(\langle P \rangle) = \delta_A^0(\langle Q \rangle)$, i.e., $\langle P - A \rangle = \langle Q - A \rangle$, and $\langle P \rangle, \langle Q \rangle, \langle P - A \rangle \in \mathrm{ess}(MN(L))$. We will prove that $\langle P \rangle = \langle Q \rangle$, or equivalently, $P \approx_L Q$.

We have $\mathrm{fin}(P - A) = \mathrm{fin}(Q - A) =: {}_S\!\uparrow(U - A)$. First, notice that A, regarded as a subset of P (or Q), contains no start events: else, we would have $\delta_A^0(\langle P \rangle) = w_{U-A}$ (or $\delta_A^0(\langle Q \rangle) = w_{U-A}$). As a consequence, $\mathrm{fin}(P) = \mathrm{fin}(Q) = {}_S\!\uparrow U$.

For every $B \subseteq \mathrm{rfin}(P) = \mathrm{rfin}(Q)$ we have

$$(P - A) \approx_L (Q - A) \implies$$
$$(P - (A \cup B)) \backslash L = (Q - (A \cup B)) \backslash L \implies$$
$$((P - (A \cup B)) * {}_{(A-B)}\!\uparrow U) \backslash L = ((Q - (A \cup B)) * {}_{(A-B)}\!\uparrow U) \backslash L.$$

The first implication follows from the definition, and the second from Lemma 23. From Lemma 24 follows that

$$(P - (A \cup B)) *_{(A-B)}{\uparrow}U \sqsubseteq P - B, \quad (Q - (A \cup B)) *_{(A-B)}{\uparrow}U \sqsubseteq Q - B.$$

Thus, by swap-invariance we have $(P - B)\backslash L = (Q - B)\backslash L$; note that Lemma 48 guarantees that neither of these languages is empty. $\square$

6   Conclusion and Further Work

We have proven a Myhill-Nerode type theorem for higher-dimensional automata (HDAs), stating that a language is regular iff it has finite prefix quotient. We have also introduced deterministic HDAs and shown that not all finite HDAs are determinizable.

An obvious follow-up question to ask is whether finite HDAs are *learnable*, that is, whether our Myhill-Nerode construction can be used to introduce a learning procedure for HDAs akin to Angluin's L* algorithm [1] or some other recent approaches [2,15,16]. (See also [33] which introduces learning for pomset automata.)

Our Myhill-Nerode theorem provides a language-internal criterion for whether a language is regular, and we have developed a similar one to distinguish deterministic languages. Another important aspect is the *decidability* of these questions, together with other standard problems such as membership or language equivalence. We believe that membership of an ipomset in a regular language is decidable, but we are less sure about decidability of the other problems.

Given that we have shown that not all regular languages are deterministic, one might ask for the approximation of deterministic languages by other, less restrictive notions. Preliminary results indicate that *ambiguity* does not buy much, given that we seem to have found a language of unbounded ambiguity; an avenue that remains wide open is the one of *history-determinism* [4,14,20].

Lastly, a remark on the fact that we only consider subsumption-closed (or *weak*) languages in this work. While this is quite common in concurrency theory, see for example [10,11,13,34], an extension of our setting to non-weak languages would certainly be interesting. (Note that, for example, languages of Petri nets with inhibitor arcs are non-weak [18].) Such an extension may be obtained by considering *partial* HDAs or HDAs with interfaces, see [5,7,9], but this is subject to future work.

Acknowledgement. We are indebted to Amazigh Amrane, Hugo Bazille, Christian Johansen, and Georg Struth for numerous discussions regarding the subjects of this paper; any errors, however, are exclusively ours.

References

1. Angluin, D.: Learning regular sets from queries and counterexamples. Inf. Comput. **75**(2), 87–106 (1987)
2. Barlocco, S., Kupke, C., Rot, J.: Coalgebra learning via duality. In: Bojańczyk, M., Simpson, A. (eds.) FoSSaCS 2019. LNCS, vol. 11425, pp. 62–79. Springer, Cham (2019). https://doi.org/10.1007/978-3-030-17127-8_4
3. Bednarczyk, M.A.: Categories of asynchronous systems, Ph. D. thesis, University of Sussex, UK (1987)
4. Colcombet, T.: The theory of stabilisation monoids and regular cost functions. In: Albers, S., Marchetti-Spaccamela, A., Matias, Y., Nikoletseas, S., Thomas, W. (eds.) ICALP 2009. LNCS, vol. 5556, pp. 139–150. Springer, Heidelberg (2009). https://doi.org/10.1007/978-3-642-02930-1_12
5. Dubut, J.: Trees in partial higher dimensional automata. In: Bojańczyk, M., Simpson, A. (eds.) FoSSaCS 2019. LNCS, vol. 11425, pp. 224–241. Springer, Cham (2019). https://doi.org/10.1007/978-3-030-17127-8_13
6. Fahrenberg, U., Johansen, C., Struth, G., Ziemiański, K.: Languages of higher-dimensional automata. Math. Struct. Comput. Sci. **31**(5), 575–613 (2021). https://arxiv.org/abs/2103.07557
7. Fahrenberg, U., Johansen, C., Struth, G., Ziemiański, K.: A Kleene theorem for higher-dimensional automata. In: Klin, B., Lasota, S., Muscholl, A. (eds.) CONCUR, volume 243 of Leibniz International Proceedings in Informatics (LIPIcs), pp. 1–18. Schloss Dagstuhl - Leibniz-Zentrum für Informatik (2022). https://arxiv.org/abs/2202.03791
8. Fahrenberg, U., Johansen, C., Struth, G., Ziemiański, K.: Posets with interfaces as a model for concurrency. Inf. Comput. **285**(B), 104914 (2022). https://arxiv.org/abs/2106.10895
9. Fahrenberg, U., Legay, A.: Partial higher-dimensional automata. In: Moss, L.S., Sobociński, P., (eds.) CALCO, volume 35 of Leibniz International Proceedings in Informatics, pp. 101–115. Schloss Dagstuhl - Leibniz-Zentrum für Informatik (2015)
10. Fanchon, J., Morin, R.: Regular sets of pomsets with autoconcurrency. In: Brim, L., Křetínský, M., Kučera, A., Jančar, P. (eds.) CONCUR 2002. LNCS, vol. 2421, pp. 402–417. Springer, Heidelberg (2002). https://doi.org/10.1007/3-540-45694-5_27
11. Fanchon, J., Morin, R.: Pomset languages of finite step transition systems. In: Franceschinis, G., Wolf, K. (eds.) PETRI NETS 2009. LNCS, vol. 5606, pp. 83–102. Springer, Heidelberg (2009). https://doi.org/10.1007/978-3-642-02424-5_7
12. Fishburn, P.C.: Interval orders and interval graphs: a study of partially ordered sets. Wiley (1985)
13. Grabowski, J.: On partial languages. Fundamentae. Informatica **4**(2), 427 (1981)
14. Henzinger, T.A., Piterman, N.: Solving games without determinization. In: Ésik, Z. (ed.) CSL 2006. LNCS, vol. 4207, pp. 395–410. Springer, Heidelberg (2006). https://doi.org/10.1007/11874683_26
15. Howar, F., Steffen, B.: Active automata learning as black-box search and lazy partition refinement. In: Jansen, N., Stoelinga, M., van den Bos, P. (eds.) A Journey from Process Algebra via Timed Automata to Model Learning. LNCS, vol. 13560, pp. 321–338. Springer, Cham (2022). https://doi.org/10.1007/978-3-031-15629-8_17

16. Isberner, M., Howar, F., Steffen, B.: The TTT algorithm: a redundancy-free approach to active automata learning. In: Bonakdarpour, B., Smolka, S.A. (eds.) RV 2014. LNCS, vol. 8734, pp. 307–322. Springer, Cham (2014). https://doi.org/10.1007/978-3-319-11164-3_26

17. Janicki, R., Koutny, M.: Structure of concurrency. Theoret. Comput. Sci. **112**(1), 5–52 (1993)

18. Janicki, R., Koutny, M.: Operational semantics, interval orders and sequences of antichains. Fundamentae Informatica **169**(1–2), 31–55 (2019)

19. Johansen, C.: ST-structures. J. Logic Algeb. Methods Programm. **85**(6), 1201–1233 (2015). https://arxiv.org/abs/1406.0641

20. Kupferman, O., Safra, S., Vardi, M.Y.: Relating word and tree automata. Ann. Pure Appl. Logic **138**(1–3), 126–146 (2006)

21. Nielsen, M., Plotkin, G.D., Winskel, G.: Petri nets, event structures and domains, part I. Theoret. Comput. Sci. **13**, 85–108 (1981)

22. Petri, C.A.: Kommunikation mit Automaten. Number 2 in Schriften des IIM. Institut für Instrumentelle Mathematik, Bonn (1962)

23. Pratt, V.R.: Modeling concurrency with geometry. In: POPL, pp. 311–322, New York City. ACM Press (1991)

24. Pratt, V.: Chu spaces and their interpretation as concurrent objects. In: van Leeuwen, J. (ed.) Computer Science Today. LNCS, vol. 1000, pp. 392–405. Springer, Heidelberg (1995). https://doi.org/10.1007/BFb0015256

25. Pratt, V.R.: Transition and cancellation in concurrency and branching time. Math. Struct. Comput. Sci. **13**(4), 485–529 (2003)

26. Mike, W.: Shields. Concurrent machines. Comput. J. **28**(5), 449–465 (1985)

27. van Glabbeek, R.J.: Bisimulations for higher dimensional automata. Email message, June (1991). http://theory.stanford.edu/rvg/hda

28. van Glabbeek, R.J.: On the expressiveness of higher dimensional automata. Theoret. Comput. Sci. **356**(3), 265–290 (2006)

29. van Glabbeek, R.J.: Erratum to "On the expressiveness of higher dimensional automata". Theoret. Comput. Sci. **368**(1-2), 168–194 (2006)

30. van Glabbeek, R.J., Goltz, U.: Refinement of actions and equivalence notions for concurrent systems. Acta Informatica **37**(4/5), 229–327 (2001)

31. van Glabbeek, R.J., Plotkin, G.D.: Configuration structures. In: LICS, pp. 199–209. IEEE Computer Society (1995)

32. van Glabbeek, R.J., Plotkin, G.D.: Configuration structures, event structures and Petri nets. Theoret. Comput. Sci. **410**(41), 4111–4159 (2009)

33. van Heerdt, G., Kappé, T., Rot, J., Silva, A.: Learning pomset automata. In: FOSSACS 2021. LNCS, vol. 12650, pp. 510–530. Springer, Cham (2021). https://doi.org/10.1007/978-3-030-71995-1_26

34. Vogler, W. (ed.): Modular Construction and Partial Order Semantics of Petri Nets. LNCS, vol. 625. Springer, Heidelberg (1992). https://doi.org/10.1007/3-540-55767-9

Tools

Hippo-CPS: A Tool for Verification and Analysis of Petri Net-Based Cyber-Physical Systems

Remigiusz Wiśniewski[ID], Grzegorz Bazydło[(✉) ID], Marcin Wojnakowski[ID], and Mateusz Popławski[ID]

Institute of Control and Computation Engineering, University of Zielona Góra, Ul. Prof. Z. Szafrana 2, 65-516 Zielona Gora, Poland
g.bazydlo@issi.uz.zgora.pl

Abstract. The paper deals with the verification and analysis techniques offered by the Hippo-CPS system. The presented tool offers alternate examination methods of the Petri net-based cyber-physical system. In particular, the set of proposed modules permits the classification of the system, verification of its main properties (such as liveness, boundedness, and safeness), and the performance of advanced concurrency and sequentiality analysis of the system (including state-space analysis, place invariant analysis, state machine component-based analysis, etc.). Although the paper is focused on the Hippo-CPS application, the presented tools have a strong theoretical background, including adequate algorithms, theorems, and proofs. The functionality of the tools was verified experimentally, by examination of the efficiency and effectiveness of the implemented techniques.

Keywords: Verification · Analysis · Petri net · Cyber-physical system · Boundedness · Safety · Liveness · Invariants · Reachability tree

1 Introduction

A cyber-physical system (CPS) integrates computation and physical processes [1] and focuses on the interactions between the control (cyber) and physical components of the system. Several applications of CPSs can be found in many areas of modern life, such as smart homes, buildings, cities [2], medical systems [3], production systems [4], etc.

One of the most effective technique of CPS modelling are Petri nets [5–9]. Their formal notation and simplicity of use allow for various formal analyses and verifications of the designed concurrent CPS, as well as their automatic translation into models that are very close to the implementation. Basically, a Petri net is a bipartite graph that consists of two types of nodes: places and transitions that are connected by directed arcs [10–12]. Modelling a system using Petri nets has many advantages compared to other approaches [13–15]. The graphical representation of Petri nets makes the modelling relatively simple and intuitive and legible (intuitiveness of graphical representation), and well-developed analysis methods easily detect certain anomalies of system behaviours [10–12, 14, 16].

The design methodology of a Petri net-based CPS consists of several steps, including modelling, verification, and analysis, as well as further prototyping (designing) of the

L. Gomes and R. Lorenz (Eds.): PETRI NETS 2023, LNCS 13929, pp. 191–204, 2023.
https://doi.org/10.1007/978-3-031-33620-1_10

system-oriented for its further implementation within integrated or distributed devices (see [16, 17] for details). Petri net-based techniques allow for the examination of the reliability and robustness of the system at the early specification stage, which may impact significantly the time and costs of the designed system. There are various testing techniques, including analysis of the concurrency and sequentiality relations in the system, as well as verification of its crucial properties, such as boundedness, safeness, and liveness [11–13, 18–23]. In short, examination of such attributes helps, for example, avoid redundancy (unreachable states), and deadlocks in the system [24–28]. On the other hand, verification and analysis of a Petri net-based CPS are not trivial tasks. The main bottleneck refers to the computational complexity of algorithms. Therefore, the existing techniques balance between the optimal results and the reasonable computation time [16].

In the paper, the Hippo-CPS system is presented. The tool is especially dedicated for the Petri net-based CPSs. Such systems are usually strictly oriented on the practical applications, thus they ought to be adequately analysed (verified) and designed. Hippo-CPS consists of several modules that permit to improve such processes. The main aim of the tool was to confront the problem related to the computational complexity of algorithms, by proposing alternative methods of verification and analysis for the modelled CPS. Therefore, the designer is able to select the most suitable technique in order to perform the examination efficiently within the assumed time (in our research it means max. 1 h) and effectively, by gaining the required results. In the paper we focus on the tool, but behind the particular modules of Hippo-CPS there is strong theoretical background, including adequate algorithms, theorems, and proofs that can be found in other Authors' papers (indicated in the description of a particular method).

2 Hippo-CPS

The *Hippo-CPS* system is a set of computer-aided tools (modules) that guide the design, verification, and analysis of the Petri net-based systems. Initially (since 2005) developed as a support for the prototyping of discrete-event systems, it is currently mainly oriented toward CPSs (especially the control part of those systems). Although this paper is focused on the verification and analysis aspects, Hippo-CPS covers much wider design aspects, including decomposition and automatic translation into the destination device description (e.g., Verilog for Field Programmable Gate Arrays, FPGAs). For all modules the input data remain the same, that is, the specification of the system in the Petri Net Hippo (PNH) format (q.v. [17] for details). There is also the possibility of importing systems written in other formats, including known tools and standards, such as *Petri Net Markup Language* (PNML) [29] or *Extensible Markup Language* (XML).

2.1 Architecture

The paper presents seven Hippo-CPS modules related to verifying and analysing the Petri net-based CPS. Although each is considered as separate modules, they all together form one consistent framework. The Hippo-CPS modules are split into three main groups, presented in detail in the next subsections. The first one consists of two modules and refers to the methods of general purposes, i.e., checking of the Petri net class and verification of the liveness, boundedness, and safeness. The second group (two modules) is aimed at the concurrency verification of the Petri net-based CPS. Finally, the third group of three tools is oriented on the sequential verification of the CPS.

2.2 Classification of the Petri Net-Based System

This module permits the classification of the Petri net-based system. Such an arrangement may strongly influence further analysis and design of the CPS, since several classes of the Petri net permit for reduction of the complexity of algorithms. In particular, there are five classes considered, according to [10, 16]:

- State Machine (SM) – each transition has exactly one input and one output place;
- Marked Graph (MG) – each place has exactly one input and one output transition;
- Free-Choice (FC) – every outgoing arc from a place is unique or is a unique incoming arc to a transition;
- Extended Free-Choice (EFC) – every pair of places having a common output transition, has all their output transitions in common;
- Simple net (SN) – every pair of places having a common output transition, one of them has all the output transitions of the other.

The above classes form a hierarchy with respect to expressiveness. In particular, SM and MG belong to the FC, while FC is a part of EFC. Finally, SN includes EFC (and, of course, all the remaining "bottom" classes). Classification of the system is especially important regarding the applied verification and design algorithms. In particular, systems that are classified as EFC have unique properties that may result in the reduction of computational complexity. For example, an optimal decomposition (that is, splitting the Petri net into a minimal number of state machine components, SMCs) is exponential in the general case. However, under certain conditions, such a decomposition can be executed polynomially (see [30] for details). Similarly, verification of the system coverage by sequential components (that is, whether the Petri net is covered by SMCs) can also be bounded by a polynomial in respect of the number of places (cf. [31]). From the technical (programming) point of view, the module was written in C/C++, by examination of the subsequent classes (starting form SM). Its computational complexity is estimated as polynomial (bounded by $O = (|P|^2|T|^2)$, where $|P|$ denotes the number of places, and $|T|$ denotes the number of transitions in the system. The module simply outputs the name of the particular class of a Petri net-based system.

2.3 Reachability Tree Verification (Boundedness, Safeness, Liveness Analysis)

The reachability tree analysis is one of the most popular forms applied to the Petri net-based systems verification. In short, the technique permits the generation of all possible states of the system. However, the total number of states (called *markings*) can be exponential, thus the main bottleneck of this method is related to its exponential complexity. Nevertheless, such a technique can be useful, since it gives very wide opportunities for further verification and analysis of the system. In particular, the main properties of the Petri net-based CPS can be examined, such as boundedness, safeness, and liveness. The module implemented within the Hippo-CPS permits all the important operations related to the reachability tree analysis. There is a capacity for the full reachability tree generation and representation as a figure (picture) for further examination. Furthermore, the tool allows verification of the system. Additional improvements were applied for the examination of the boundedness and safeness of the system. The tool checks boundedness (or safeness, respectively) at each step of the algorithm. If an unbounded (unsafe) place is found, the method terminates, avoiding the computation of the whole reachability tree. Finally, the tool allows for the examination of the liveness property. This operation requires the generation of the complete tree, and further examination of all transitions starting from the initial state (see [32] for details).

Contrary to almost all other Hippo-CPS modules, this tool was written in Java. The output is parameterized according to the user's needs. The tool results in either: the full reachability tree (as a JPG figure), the information about the boundedness, safeness, or liveness of the system. The computational complexity of the method (generation of the reachability tree) is bounded exponentially by the number of places in the system.

2.4 Structural Concurrency Verification (Graph-Based Analysis)

This module permits the computation of the structural concurrency relation of the Petri net-based CPS. Such a relation shows explicitly whether each two pair of places are structurally concurrent. This information is useful in further analysis and design of the system (especially by verification and decomposition techniques, cf. [30, 31]). The main advantage of the method is its polynomial complexity, bounded by $O = (|P| + |T|)^5$ [30] (for EFC systems it can even be computed with a cubic complexity [33]). On the other hand, the tool computes the concurrency relation between at most each pair of places in the system. Therefore, analysis of more complicated models can be difficult or even impossible. Furthermore, the structural concurrency relation does not always coincide with the real one, and may contain redundant pairs of places that are not concurrent (cf. [16]). The algorithm implemented within Hippo-CPS is written in C/C++ and is based on the method initially shown in [34], with further enhancements and modifications (mainly presented in [16]]). In short, the method searches for the structural concurrency relations in the system by consecutive analysis of the transition input and outputs [16]. The resulting concurrency graph holds the structural concurrency relations between every pair of places of the Petri net-based system. Hippo-CPS offers two output formats for this module: the neighbour matrix of the structural concurrency graph, and the structural concurrency relations between every pair of places in the system.

2.5 Concurrency Verification (Concurrency Hypergraph-Based Analysis)

A concurrency hypergraph is an alternative method to the graph-based method for concurrency analysis. Its main advantage stems from the (exact) results it obtains. Each edge (hyperedge) of the obtained structure directly refers to a reachable state (marking) of the system [16, 35]. In other words, a concurrency hypergraph holds complete information about the state space of the examined system. Therefore, it permits for the detailed verification of the concurrency relations in the Petri net-based CPS. Moreover, it is a base for further sequential analysis of the system (q.v. Sect. 2.7). A concurrency hypergraph is closely related to the *reachability set* of the Petri net-based model [10]. However, it is additionally supported by authors' algorithms, definitions (including a new type of *c-exact* hypergraph), theorems, and proofs that permit a much more advanced analysis of the examined system. First of all, the subsequent hyperedges (reachable states) of the system are obtained polynomially [16] (although obtaining the complete set of hyperedges is exponential). Moreover, the structure is a base for further computation of SMCs in the system, which are especially used in the decomposition and analysis of the system [35]. The tool was written in C/C++. Let us underline that Hippo-CPS includes a very wide range of varied hypergraph-based algorithms.

2.6 Structural Sequential Verification (Graph-Based Analysis)

This tool offers two functionalities. The first one permits to obtain the structural sequentiality relation between each pair of places in the Petri net-based system [16]. Such a relation is complementary to the structural concurrency relation described in Sect. 2.4. Furthermore, the module allows for the computation of the *state machine components* (SMCs) in the system. Each SMC forms a sequential component. Calculation of SMCs is essential in the verification, design and analysis of Petri net-based CPS (cf. [16, 18, 21, 30, 31, 36, 37). In particular, the tool examines whether the system is covered by SMCs [31]. If this property holds, the system is bounded and safe [31, 36]. The module was written in C/C++, and it runs in a polynomial time [16, 31]. It is especially applicable in the safeness and boundedness analysis.

2.7 Sequentiality Verification (Sequentiality Hypergraph-Based Analysis)

The sequentiality hypergraph is a structure that preserves the sequential relation between places in the Petri-net based CPS. Its hyperedges strictly refer to the SMCs. Those components are obtained from the concurrency hypergraph (see [16, 35] for details). The SMCs are obtained by computation of exact transversals in the concurrency hypergraph. An exact transversal of a hypergraph is a set of vertices that has exactly one intersection with every edge of a hypergraph [16, 35]. The computation of single exact transversal (single SMC) is polynomial. However, obtaining of the sequentiality hypergraph can be exponential (cf. [16, 35]). Contrary to the graph-based method, the sequentiality hypergraph holds complete information about the sequential relations in the system, since each of its hyperedges may include more than two vertices. Furthermore, there is no need for additional computation of SMCs as it is required in the case of typical graphs. So, it can be directly used in the decomposition of the system [16]. The tool was written in C/C++. The applied algorithms, among others, include the authors' implementation of the DLX technique [38], which operates on the four-way linked matrix. In particular, the method searches for exact transversals in the concurrency hypergraph, which directly refer to the SMCs in the Petri net-based CPS.

2.8 Sequentiality Verification (Linear Algebra Technique)

The last presented tool applies a linear algebra technique. In particular, the method searches for the so-called *place invariants* (p-invariants) in the Petri net-based CPS

[10, 16]. Place invariants have wide application in the design, verification, and analysis of the Petri net models. For example, the system covered by p-invariants is bounded. Furthermore, place invariants are closely related to the sequential components. Under certain assumptions, it is possible to obtain SMCs in the analysed CPS. Therefore (upon additional actions) safeness of the system can be examined [21, 36]. The tool implemented within Hippo-CPS was written in C/C++. It should be noted that obtaining all p-invariants is exponential in the general case [39]. Therefore, the discussed module additionally utilizes the authors' techniques that permit for computation of the reduced set of p-invariants (cf. [21, 36]). The tool is parameterized, and it contains four functionalities. The first computes the complete set of p-invariants, while the second generates all SMCs in the system. Both methods are exact, thus they can be very time-consuming. The remaining two options permit verification of the boundedness and safeness of the system. Those methods utilize the reduced set of invariants; thus, their run-time is usually much faster than the former two.

2.9 Experimental Results

Each of the presented tools was examined in terms of its efficiency (run-time) and effectiveness (proper results). The modules were tested on the dedicated computational server with the use of an Intel® Xeon® Gold 5220 @2.2 GHz processor and 128 GB of RAM. The library of benchmarks contains 242 Petri nets that describe real and hypothetical cyber-physical systems, control systems, and discrete systems. Due to the page limitation here and the huge information of data (242 benchmarks examined by 7 tools that include 13 functionalities), the detailed results can be obtained under the following link: https://hippo-cps.issi.uz.zgora.pl/download/pn2023_results.xlsx.

Let us briefly discuss the obtained data. First of all, it can be observed that classification of the Petri nets was possible for all tested cases. Moreover, it was done in fractions of a second. Moving on to the reachability tree-based tool, the complete structure (as well as liveness verification) was obtained for 223 (92%) benchmarks (14 nets were marked as unbounded and 5 exceeded the assumed time, which was set to 1 h). Furthermore, it was possible to check safeness and boundedness for 237 tests (98%). Several benchmarks require up to several minutes to complete the calculation. Those results were expected since this method is bounded exponentially. Results for structural concurrency and structural sequentiality graphs were achieved in at most a few seconds. Both techniques are bounded by a polynomial in the number of places, thus it was possible to examine all benchmarks. In contrast, the exact methods based on the concurrency and sequentiality hypergraphs were not able to complete all tasks. Concurrency hypergraph was efficient for 225 tests (93%), while sequentiality hypergraph found results for 223 benchmarks (92%). On the other hand, the results gained by those tools were always optimal. Finally, linear algebra-based techniques were able to obtain p-invariants for 237 tests (98%). Let us underline that the experimental results greatly confirmed the theoretical assumptions. Tools based on exact methods (reachability tree, hypergraphs, linear algebra) are effective, but not always efficient. On the other hand, approximate algorithms guarantee the obtaining of results, but they can be insufficient (i.e., graph-based tools).

2.10 A Case-Study Example

This section illustrates the application of the presented Hippo-CPS tools by an example. Figure 1 (left) shows a real model of a crossroads, and Fig. 1 (right) a working model (miniaturization). The purpose of the system is an implementation in the FPGA with the possibility of further dynamic partial reconfiguration (cf. [16, 40, 41] for details).

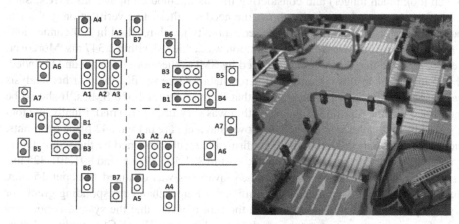

Fig. 1. The idea of the traffic light controller (left) and its miniaturization model (right).

It is assumed that each road contains three independent lanes for cars (right, straight, left), controlled by three traffic lights (*red, yellow, green*). Additionally, there are pedestrian crossings on each road, controlled by two traffic lights (*red, green*). The controller can be configured in several modes (e.g., collision-free, priority for cars, priority for pedestrians, etc.). The Petri net model of the presented CPS is shown in Fig. 2. Each place of the net refers to a particular traffic light (or two lights if controlled in common). The particular lights are numbered and prefixed by the part number and followed by the first letter of a particular signal colour (R-red, Y-yellow, G-green), while letters "A" and "B" refer to the side of the crossroad (cf. Figure 1, left). To clarify the presentation, places are coloured according to the active output signals. The presented Petri net-based CPS consists of 32 places and 12 transitions. Let us now verify and analyse the system with the set of proposed Hippo-CPS tools. The first module classified the system as a *Marked Graph*. It is very useful information since this class may influence the further verification and design steps of the CPS. The run-time of the tool was 0.005ms. The second tool (reachability tree) gave a very important result that the system is live, bounded, and safe. Verifications of those properties were completed within 97.967 ms, 79.008 ms, and 43.831 ms, respectively. This information is essential since it assures that the model is deadlock-free and does not contain unreachable states.

Moving on to the concurrency analysis, there are 233 edges in the structural concurrency graph (obtained within 2.841 ms). This means that there are 233 pairs of concurrent places in the net. Such a fact is essential in the case of the proper functionality of the traffic light controller in order to avoid collisions and unwanted situations where two lights are active simultaneously (i.e., green light for cars and pedestrians at the same line). On the other hand, analysis of each pair of lights can be insufficient and problematic, since there are 233 such pairs. Indeed, the concurrency hypergraph consists of only 12 edges. Moreover, it holds information between all signals that are active simultaneously. It also gives the information that there are in total 12 markings (states) that the traffic light controller may reach. The concurrency hypergraph for this system was obtained in just 0.086 ms. It is surprisingly fast compared to the time used for the generation of the concurrency graph (which took much longer) and considering that the method computes all SMCs. Such a situation clearly and practically shows the need for alternative verification tools. The sequential verification of the CPS by a sequentiality graph resulted in 234 connections between every pair of places. This information was achieved within 1.347 ms. Moreover, the tool reported that the system is covered by SMCs, therefore it is safe and bounded. This confirms the results obtained by the reachability tree verification. Further analysis by the sequentiality hypergraph showed that there are in total 530 SMCs. It should be underlined that the run-time of the algorithm was very fast (2.070 ms). Finally, the tool based on linear algebra was applied. It shows that there are in total 542 place invariants, and 530 SMCs in the system (which confirms the results obtained by the sequentiality hypergraph). These values were computed within 891,400.791 ms and 891,401.325 ms, respectively. It means that the set of place invariants was obtained in about 15 min, thus run-time of those methods was significantly longer than corresponding graph- or hypergraph-based techniques. Moreover, the tool reported that the system is bounded and safe, once more confirming the results gained by other Hippo-CPS modules. Let us underline that the obtained results (liveness, safeness, boundedness, and SMCs) were essential in the verification and further realisation in FPGAs, and the resulting sequential components formed a base for the dynamic partial reconfiguration (according to methods from [16, 41, 42]).

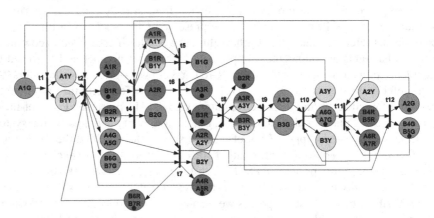

Fig. 2. Specification of the traffic light controller by a Petri net.

3 Installation

The Hippo-CPS tool can be download for free in a form of packed ZIP from http://
hippo-cps.issi.uz.zgora.pl/download/hippo-cps.zip under freeware license. Each module
consists of a *readme* file with the detailed description of the usage instructions.

4 Comparison with Other Tools

There are many noteworthy tools, that are designed to analyse and verify the Petri net-
based system. We have analysed and examined more than 25 available tools that offer
similar functionalities to Hippo-CPS. Let us briefly present a selected few of them.

IOPT-Tools [43] is a very popular and functional web application for editing, simu-
lating, and analysing Petri nets models. The tool allows for the manual decomposition of
the net into synchronised sub-modules. An interesting feature is VHDL or C code generation in terms of FPGA or Arduino implementation. The limitations of the tool are its
analysis of only selected properties of the Petri net (boundedness, safeness, occurrence
of deadlocks) and its inability to analyse place invariants.

PIPE [44] and its successor PIPE2 [45] are Petri net editors and analysis tools. A
big advantage of both programs is open access to the source code and the possibility for
users to develop their own analysis modules or adjust existing ones. PIPE2 enables the
determination of the Petri net class, the calculation of place and transition invariants,
and the analysis of siphons, boundedness, safeness, and deadlocks. The main limitations of the tools refer to the exponential computational complexity of the implemented
algorithms, thus the solution can be not found at all.

CPN Tools [46] is dedicated to Coloured Petri nets (CPNs), but it could also be
applied to other Petri net classes. It offers an analysis module and a graphical editor with
the rare ability to edit the shapes of places, transitions, or arcs. Unfortunately, the tool
is unintuitive, and to use it the study of extensive documentation is needed.

Snoopy [47] is an advanced editor for various classes of Petri nets, including CPNs,
Continuous Petri nets, and Fuzzy Petri nets. The tool has many functions dedicated to
specific Petri net classes and can export models into many formats, e.g., MATLAB,
PNML. A drawback of the tool is the lack of net analysis or simulation. Although there
is an analysis extension [48] of Snoopy called Charlie, it is inconvenient because it
requires the installation of other tools, as well as converting models between them.

Yasper [49] is a simple graphical Petri net editor with a simulation module. Its
interface seems to be user-friendly and accessible. An interesting feature is an export
function to the MS Visio format. Unfortunately, the tool has not been developed since
2005. Moreover, a significant limitation of the program is the lack of Petri net analysis.

JSARP [50, 51] has a very intuitive and friendly user interface. The inbuilt simulator
allows users to graphically indicate the transitions selected for firing. The analysis module generates the Petri net incident matrix and based on the reachability graph, provides
information only about the boundness and liveness of the Petri net.

JARP [52] tool consists of a manual Petri net simulation component, reachability
graph analysis, and invariant computation module. Unfortunately, there is a lack of Petri
net properties analysis, such as boundness, safeness, or liveness.

WOLFGANG [53] is dedicated to two classes of Petri nets: regular P/T (place/transition) Petri nets and CPNs. It offers a simple and intuitive GUI and the possibility of editing shapes or changing the font of labels. Another advantage is compliance with the PNML standard. Unfortunately, the tool has very modest analysis capabilities, limited to verifying the correctness of the Petri net structure, boundness, and soundness.

GreatSPN [54, 55] is a powerful tool for editing and analysing complex Petri nets. Unfortunately, to use the tool there is a need for the manual compilation of sources (with 16 external libraries). The tool is also available on a ready-to-use virtual machine.

TINA [56] has a simple but hardly functional graphical editor. The tool is dedicated to Timed Petri net analysis (similar to the Romeo tool [57]) and is mainly focused on the computation of place and transition invariants, and detailed analysis of the Petri net properties, such as boundedness, liveness, reversibility, and presence of deadlocks.

GPenSIM [58] is a popular toolbox for MATLAB and is used for Petri net-based discrete systems design, while the console tool MIST is devoted to Linux users and it must be compiled from the source. It is dedicated to analysing the safeness of a Petri net (must be entered in MIST format), based on state-space analysis.

A Low Level Petri Net Analyzer (LoLA) [59] is an is a tool for the analysis of Petri nets that is designed particularly for large and complex nets. Unlike IOPT-Tools, PIPE, GreatSPN, LoLA is a console application, which means that it requires calling formulas rather than a graphical user interface. One of the key benefits of LoLA is its speed. It can analyse Petri nets with millions of states and transitions, making it a valuable tool for working with very large and complex Petri net-based models. However, it has also some limitations. As a console application, it can be quite difficult to use, particularly for users who are not familiar with command line interfaces and its own non-standard Petri net format. Additionally, its analysis capabilities are limited to the most basic behavioural properties as boundedness, deadlock occurrence, liveness or soundness. Overall, its console interface and limited analysis capabilities may make it less accessible to some users particularly practitioners.

An interesting tool dedicated to Petri net-based systems analysis is AdamMC [60]. The tool applies model-checking verification techniques, and to reduce the computational complexity it proposes sequential and parallel optimization approaches.

5 Conclusion

Seven Hippo-CPS modules have been presented in the paper. The tools are oriented toward the verification and analysis of the Petri nets. The presented modules offer alternate examination techniques, allowing the user to verify the system in several different ways. The efficiency and effectiveness of the implemented algorithms were examined experimentally. The obtained results confirmed both: theoretical assumptions (in regard to run-time of methods), and proper functionality of modules (achieved results). Future work will include the enhancement of the Hippo-CPS. It is planned to extend the tool with additional verification and design modules. In particular, the verification and design of systems classified as EFC are going to be thoroughly investigated in order to obtain polynomial algorithms (and further implementation within the tool).

Acknowledgements. This work is supported by the National Science Centre, Poland, under Grant number 2019/35/B/ST6/01683.

References

1. Lee, E.A., Seshia, S.A.: Introduction to Embedded Systems: A Cyber-Physical Systems Approach, 2nd edn. The MIT Press, Cambridge (2016)
2. Shih, C.-S., Chou, J.-J., Reijers, N., Kuo, T.-W.: Designing CPS/IoT applications for smart buildings and cities. IET Cyber-Phys. Syst. Theory Appl. **1**(1), 3–12 (2016). https://doi.org/10.1049/iet-cps.2016.0025
3. Dey, N., Ashour, A.S., Shi, F., Fong, S.J., Tavares, J.M.R.S.: Medical cyber-physical systems: a survey. J. Med. Syst. **42**(4), 1–13 (2018). https://doi.org/10.1007/s10916-018-0921-x
4. Patalas-Maliszewska, J., Posdzich, M., Skrzypek, K.: Modelling information for the burnishing process in a cyber-physical production system. Int. J. Appl. Math. Comput. Sci. **32**(3), 345–354 (2022). https://doi.org/10.34768/amcs-2022-0025
5. Zhu, Q., Zhou, M., Qiao, Y., Wu, N.: Petri net modeling and scheduling of a close-down process for time-constrained single-arm cluster tools. IEEE Trans. Syst. Man Cybern. Syst. **48**(3), 389–400 (2018). https://doi.org/10.1109/TSMC.2016.2598303
6. Wiśniewski, R., Bazydło, G., Szcześniak, P.: Low-cost FPGA hardware implementation of matrix converter switch control. IEEE Trans. Circuits Syst. II Express Briefs **66**(7), 1177–1181 (2019). https://doi.org/10.1109/TCSII.2018.2875589
7. Patalas-Maliszewska, J., Wiśniewski, R., Topczak, M., Wojnakowski, M.: Modelling of the effectiveness of integrating additive manufacturing technologies into Petri net-based manufacturing systems. In: 2022 IEEE International Conference on Fuzzy Systems (FUZZ-IEEE), July 2022, pp. 1–9 (2022). https://doi.org/10.1109/FUZZ-IEEE55066.2022.9882766
8. Wiśniewski, R., Wojnakowski, M., Li, Z.: Design and verification of petri-net-based cyber-physical systems oriented toward implementation in field-programmable gate arrays—a case study example. Energies **16**(1), Article no. 1 (2023). https://doi.org/10.3390/en16010067
9. Wojnakowski, M., Wiśniewski, R., Popławski, M., Bazydło, G.: Analysis of control part of cyber-physical systems specified by interpreted Petri nets. In: 2022 IEEE International Conference on Systems, Man, and Cybernetics (SMC), October 2022, pp. 1090–1095 (2022). https://doi.org/10.1109/SMC53654.2022.9945425
10. Murata, T.: Petri nets: properties, analysis and applications. Proc. IEEE **77**(4), 541–580 (1989). https://doi.org/10.1109/5.24143
11. Best, E., Devillers, R., Koutny, M.: Petri Net Algebra. Springer, Heidelberg (2001). https://doi.org/10.1007/978-3-662-04457-5
12. David, R., Alla, H.: Bases of petri nets. In: David, R., Alla, H. (eds.) Discrete, Continuous, and Hybrid Petri Nets, pp. 1–20. Springer, Heidelberg (2010). https://doi.org/10.1007/978-3-642-10669-9_1
13. Reisig, W., Rozenberg, G. (eds.) Lectures on Petri Nets I: Basic Models: Advances in Petri Nets. Springer, Heidelberg (1998). https://doi.org/10.1007/3-540-65306-6
14. Aalst, W.M.P.: Workflow verification: finding control-flow errors using petri-net-based techniques. In: van der Aalst, W., Desel, J., Oberweis, A. (eds.) Business Process Management. LNCS, vol. 1806, pp. 161–183. Springer, Heidelberg (2000). https://doi.org/10.1007/3-540-45594-9_11
15. Patalas-Maliszewska, J., Wiśniewski, R., Topczak, M., Wojnakowski, M.: Design optimization of the Petri net-based production process supported by additive manufacturing technologies. Bull. Pol. Acad. Sci. Tech. Sci. **70**(2), e140693 (2022)

16. Wiśniewski, R.: Prototyping of Concurrent Control Systems Implemented in FPGA Devices. Springer, Heidelberg (2017). https://doi.org/10.1007/978-3-319-45811-3
17. Wisniewski, R., Bazydło, G., Gomes, L., Costa, A., Wojnakowski, M.: Analysis and design automation of cyber-physical system with hippo and IOPT-tools. In: IECON 2019 - 45th Annual Conference of the IEEE Industrial Electronics Society, October 2019, vol. 1, pp. 5843–5848 (2019). https://doi.org/10.1109/IECON.2019.8926692
18. Wojnakowski, M., Wiśniewski, R.: Verification of the boundedness property in a petri net-based specification of the control part of cyber-physical systems. In: Camarinha-Matos, L.M., Ferreira, P., Brito, G. (eds.) DoCEIS 2021. IAICT, vol. 626, pp. 83–91. Springer, Cham (2021). https://doi.org/10.1007/978-3-030-78288-7_8
19. Silva, M., Colom, J.M., Campos, G.C.: Linear algebraic techniques for the analysis of Petri nets. In: Recent Advances in Mathematical Theory of Systems, Control, Networks, and Signal Processing II, pp. 35–42 (1992)
20. Celaya, J.R., Desrochers, A.A., Graves, R.J.: Modeling and analysis of multi-agent systems using Petri nets. In: 2007 IEEE International Conference on Systems, Man and Cybernetics, October 2007, pp. 1439–1444 (2007). https://doi.org/10.1109/ICSMC.2007.4413960
21. Wojnakowski, M., Wiśniewski, R., Bazydło, G., Popławski, M.: Analysis of safeness in a Petri net-based specification of the control part of cyber-physical systems. AMCS 31(4), 647–657 (2021). https://doi.org/10.34768/amcs-2021-0045
22. Esparza, J., Silva, M.: A polynomial-time algorithm to decide liveness of bounded free choice nets. Theor. Comput. Sci. 102(1), 185–205 (1992). https://doi.org/10.1016/0304-3975(92)90299-U
23. Barkaoui, K., Minoux, M.: A polynomial-time graph algorithm to decide liveness of some basic classes of bounded Petri nets. In: Jensen, K. (ed.) ICATPN 1992. LNCS, vol. 616, pp. 62–75. Springer, Heidelberg (1992). https://doi.org/10.1007/3-540-55676-1_4
24. Barkaoui, K., Ben Abdallah, I.: A deadlock prevention method for a class of FMS. In: 1995 IEEE International Conference on Systems, Man and Cybernetics. Intelligent Systems for the 21st Century, October 1995, vol. 5, pp. 4119–4124 (1995). https://doi.org/10.1109/ICSMC.1995.538436
25. Ezpeleta, J., Colom, J.M., Martinez, J.: A Petri net based deadlock prevention policy for flexible manufacturing systems. IEEE Trans. Robot. Autom. 11(2), 173–184 (1995). https://doi.org/10.1109/70.370500
26. Guo, X., Wang, S., You, D., Li, Z., Jiang, X.: A siphon-based deadlock prevention strategy for S3PR. IEEE Access 7, 86863–86873 (2019). https://doi.org/10.1109/ACCESS.2019.2920677
27. Huang, Y., Jeng, M., Xie, X., Chung, S.: Deadlock prevention policy based on Petri nets and siphons. Int. J. Prod. Res. 39(2), 283–305 (2001). https://doi.org/10.1080/00207540010002405
28. Karatkevich, A., Grobelna, I.: Deadlock detection in Petri nets: one trace for one deadlock?. In: 2014 7th International Conference on Human System Interactions (HSI), June 2014, pp. 227–231 (2014). https://doi.org/10.1109/HSI.2014.6860480
29. Gomes, L., Barros, J.P., Costa, A., Nunes, R.: The input-output place-transition petri net class and associated tools. In: 2007 5th IEEE International Conference on Industrial Informatics, June 2007, vol. 1, pp. 509–514 (2007). https://doi.org/10.1109/INDIN.2007.4384809
30. Wiśniewski, R., Karatkevich, A., Adamski, M., Costa, A., Gomes, L.: Prototyping of concurrent control systems with application of Petri nets and comparability graphs. IEEE Trans. Control Syst. Technol. 26(2), 575–586 (2018). https://doi.org/10.1109/TCST.2017.2692204
31. Karatkevich, A.G., Wiśniewski, R.: A polynomial-time algorithm to obtain state machine cover of live and safe Petri nets. IEEE Trans. Syst. Man Cybern. Syst. 50(10), 3592–3597 (2020). https://doi.org/10.1109/TSMC.2019.2894778

32. Popławski, M., Wojnakowski, M., Bazydło, G., Wiśniewski, R.: Reachability tree in live-ness analysis of Petri net-based cyber-physical systems. In: AIP Conference Proceedings, Heraklion, Greece, September 2021
33. Kovalyov, A., Esparza, J.: A polynomial algorithm to compute the concurrency relation of free-choice signal transition graphs. In: Proceedings of International WODES, June 1996
34. Kovalyov, A.V.: Concurrency relations and the safety problem for Petri nets. In: Jensen, K. (ed.) ICATPN 1992. LNCS, vol. 616, pp. 299–309. Springer, Heidelberg (1992). https://doi.org/10.1007/3-540-55676-1_17
35. Wisniewski, R., Wisniewska, M., Jarnut, M.: C-exact hypergraphs in concurrency and sequentiality analyses of cyber-physical systems specified by safe petri nets. IEEE Access **7**, 13510–13522 (2019). https://doi.org/10.1109/ACCESS.2019.2893284
36. Wisniewski, R., Wojnakowski, M., Stefanowicz, Ł.: Safety analysis of Petri nets based on the SM-cover computed with the linear algebra technique. In: AIP Conference Proceedings, vol. 2040, no. 1, p. 080008, November 2018. https://doi.org/10.1063/1.5079142
37. Wiśniewski, R., Karatkevich, A., Adamski, M., Kur, D.: Application of comparability graphs in decomposition of Petri nets. In: 2014 7th International Conference on Human System Interactions (HSI), June 2014, pp. 216–220 (2014). https://doi.org/10.1109/HSI.2014.6860478
38. Knuth, D.E.: Dancing links. arXiv:cs/0011047, November 2000. Accessed 07 Jan 2022. http://arxiv.org/abs/cs/0011047
39. Martínez, J., Silva, M.: A simple and fast algorithm to obtain all invariants of a generalised Petri net. In: Girault, C., Reisig, W. (eds.) Application and Theory of Petri Nets, vol. 52, pp. 301–310. Springer, Heidelberg (1982). https://doi.org/10.1007/978-3-642-68353-4_47
40. Wiśniewski, R., Bazydło, G., Gomes, L., Costa, A.: Dynamic partial reconfiguration of concurrent control systems implemented in FPGA devices. IEEE Trans. Industr. Inf. **13**(4), 1734–1741 (2017). https://doi.org/10.1109/TII.2017.2702564
41. Wiśniewski, R.: Dynamic partial reconfiguration of concurrent control systems specified by Petri nets and implemented in Xilinx FPGA devices. IEEE Access **6**, 32376–32391 (2018). https://doi.org/10.1109/ACCESS.2018.2836858
42. Bazydło, G.: Designing reconfigurable cyber-physical systems using unified modeling language. Energies **16**(3), Article no. 3 (2023). https://doi.org/10.3390/en16031273
43. Gomes, L., Moutinho, F., Pereira, F.: IOPT-tools—a web based tool framework for embedded systems controller development using Petri nets. In: 2013 23rd International Conference on Field Programmable Logic and Applications, September 2013, p. 1 (2013). https://doi.org/10.1109/FPL.2013.6645633
44. Bonet, P., Lladó, C.: PIPE v 2.5: a Petri net tool for performance modelling (2007)
45. Dingle, N.J., Knottenbelt, W.J., Suto, T.: PIPE2: a tool for the performance evaluation of generalised stochastic Petri Nets. SIGMETRICS Perform. Eval. Rev. **36**(4), 34–39 (2009). https://doi.org/10.1145/1530873.1530881
46. Yu, Q., Cai, L., Tan, X.: Airport emergency rescue model establishment and performance analysis using colored Petri nets and CPN tools. Int. J. Aerosp. Eng. **2018**, e2858375 (2018). https://doi.org/10.1155/2018/2858375
47. Heiner, M., Herajy, M., Liu, F., Rohr, C., Schwarick, M.: Snoopy – a unifying Petri net tool. In: Haddad, S., Pomello, L. (eds.) PETRI NETS 2012. LNCS, vol. 7347, pp. 398–407. Springer, Heidelberg (2012). https://doi.org/10.1007/978-3-642-31131-4_22
48. Heiner, M., Schwarick, M., Wegener, J.-T.: Charlie – an extensible petri net analysis tool. In: Devillers, R., Valmari, A. (eds.) PETRI NETS 2015. LNCS, vol. 9115, pp. 200–211. Springer, Cham (2015). https://doi.org/10.1007/978-3-319-19488-2_10
49. van Hee, K., Oanea, O., Post, R., Somers, L., van der Werf, J.M.: Yasper: a tool for workflow modeling and analysis. In: Sixth International Conference on Application of Concurrency to

System Design (ACSD'06), June 2006, pp. 279–282 (2006). https://doi.org/10.1109/ACSD. 2006.37

50. JSARP - Simulador e Analisador de Redes de Petri. http://www.geocities.ws/jsarp_project/ index-2.html. Accessed 21 Dec 2021
51. Oliviera Lino, F.G., Analisador e Simulador de Redes de Petri. Bachelor thesis, University of Rio de Janeiro, Rio de Janeiro (2007). http://www.geocities.ws/jsarp_project/downloads/ monografia2007.pdf. Accessed 11 Dec 2021
52. JARP. http://jarp.sourceforge.net/us/index.html. Accessed 21 Dec 2021
53. WOLFGANG - Petri Net Editor. IIG Telematics, 16 April 2021. https://github.com/iig-uni-freiburg/WOLFGANG. Accessed 21 Dec 2021
54. GreatSPN: The GreatSPN framework version 3.0. 21 December 2021. https://github.com/gre atspn/SOURCES. Accessed 23 December 2021
55. Baarir, S., Beccuti, M., Cerotti, D., De Pierro, M., Donatelli, S., Franceschinis, G.: The GreatSPN tool: recent enhancements. SIGMETRICS Perform. Eval. Rev. **36**(4), 4–9 (2009). https://doi.org/10.1145/1530873.1530876
56. "The TINA toolbox Home Page - TIme petri Net Analyzer - by LAAS/CNRS. http://projects. laas.fr/tina/. Accessed 23 Dec 2021
57. Gardey, G., Lime, D., Magnin, M., Roux, O.(H.): Romeo: a tool for analyzing time Petri nets. In: Etessami, K., Rajamani, S.K. (eds.) CAV 2005. LNCS, vol. 3576, pp. 418–423. Springer, Heidelberg (2005). https://doi.org/10.1007/11513988_41
58. Davidrajuh, R.: Introduction to GPenSIM. In: Davidrajuh, R. (ed.) Petri Nets for Modeling of Large Discrete Systems, pp. 15–27. Springer, Singapore (2021). https://doi.org/10.1007/ 978-981-16-5203-5_2
59. Schmidt, K.: LoLA a low level analyser. In: Nielsen, M., Simpson, D. (eds.) ICATPN 2000. LNCS, vol. 1825, pp. 465–474. Springer, Heidelberg (2000). https://doi.org/10.1007/3-540-44988-4_27
60. Finkbeiner, B., Gieseking, M., Hecking-Harbusch, J., Olderog, E.-R.: AdamMC: a model checker for petri nets with transits against flow-LTL. In: Lahiri, S.K., Wang, C. (eds.) CAV 2020. LNCS, vol. 12225, pp. 64–76. Springer, Cham (2020). https://doi.org/10.1007/978-3-030-53291-8_5

Mochy: A Tool for the Modeling of Concurrent Hybrid Systems

Loïc Hélouët$^{(\boxtimes)}$ (ID) and Antoine Thébault

University Rennes, Inria, CNRS, IRISA, Rennes, France
{loic.helouet,antoine.thebault}@inria.fr

Abstract. This paper introduces MOCHY, a tool designed for the modeling of concurrent systems with variants of stochastic, timed and hybrid Petri nets. Beyond modeling, the tool serves as a platform for fast simulation, and can be used for statistical verification of properties, controller testing, and learning of control rules. The targeted models are variants of stochastic and timed nets where tokens can be continuous quantities depicting trajectories of moving objects. The architecture of the tool is designed to be as adaptive as possible, and allow the redefinition of objects behaviors or transitions firing through the refinement of a few semantic rules. The framework also allows for the integration of controllers. For any model variant, MOCHY can perform fast simulation, and perform statistical verification, evaluate some quantitative properties of a model, or learn control rules for reachability or quantitative objectives.

1 Introduction

This paper introduces a new tool called MOCHY, tailored for the modeling of systems with timed variants of Petri nets, and for fast simulation. The origin of MOCHY stems from the need of fast simulation tools to test traffic management policies for metro networks [12]. We rapidly came to the conclusion that transport networks had so many specificities that time Petri nets, timed Petri nets, or most of their variants were not adapted to the design of such models. First of all, even if models such as time Petri nets (TPNs) are Turing powerful, and can hence simulate most systems, using this expressive power in practice forces to loose the graphical and concurrent nature of nets and results in complex models that are hard to simulate, and cannot be understood by humans. A way to circumvent these issues were to tune existing models to obtain ad-hoc variants of Petri nets. However, this was not satisfactory either, because every transport network comes with its own traffic management policy, i.e. a light form of control that is used to mitigate effects of incidents and delays, that affects the semantics of the model. With this additional constraint, every transport network can have its own ad-hoc semantics, and is hence a new kind of model. An example of variation point for instance is whether a metro network follows a fixed block policy allowing at most one train in each track segment, or a moving block policy that allows several trains in a segment provided they preserve safety headways.

L. Gomes and R. Lorenz (Eds.): PETRI NETS 2023, LNCS 13929, pp. 205–216, 2023.
https://doi.org/10.1007/978-3-031-33620-1_11

The main principles of MOCHY are the following: we consider timed models that can depict trajectories of objects in a bounded environment, such as trains on a track, cars on a road lane, manufactured objects on a conveyor, etc. The simulation scheme of the tool is designed to be as generic as possible. To reach this objective, the semantics of a model is given in a few generic operational rules depicting how the state of a network transforms upon occurrence of a discrete event, or when time elapses. Those rules are repeated within a simulation loop, that may use a controller to select the next actions or delays allowed. Controllers also maintain a memory that can be used for further decisions.

This paper is organized as follows: we first describe the general architecture of the tool, the common features of models that can be simulated by MOCHY, the rules used by its adaptive semantics, and the way runs of MOCHY models are simulated. We then explain how MOCHY has been used to model a metro line in Rennes, and to learn a controller which aim is to help a metro recovering from a bunching situation. These studies show pertinence of our high-level simulation scheme. We conclude with related work, and perspectives on the future development of the tool.

2   Mochy Description

MOCHY is designed to be as modular and adaptable as possible. Its architecture decouples semantics, interfaces, simulation scheme and control. This architecture allow redefinition of a semantics attached to a particular project, modification of a controller to guide choices of delays and actions during simulation, etc. This approach was proved particularly useful when considering models for metro networks: several train management policies have been implemented by simply changing the controller part of the tool.

2.1   Architecture

Mochy's architecture (see Fig. 1) is composed of four main parts: an interface, and three modules describing a Physical Model, a Controller and a Simulator loop. The Physical Model contains the data structures needed to describe the Petri net variant that will be simulated, its initial configuration (mainly initial contents of its places), a net type, which will be used to load a class implementing the semantics rules of the net and a controller if needed. The physical model is loaded from an input project file. The Core part is composed of semantics rules and of a controller. This part is composed of classes that are loaded once the type of net is known from the input project file. The controller and semantic classes define the effects of possible actions and delays on configurations. The Simulator part implements a simulation loop, i.e. it handles semantics rules and control to animate a fixed number of semantics steps set from the interface. The interface allows interaction with these modules, by displaying current configuration of a loaded model, and providing access to the simulation functions of the tool and to standard functions (load, save...).

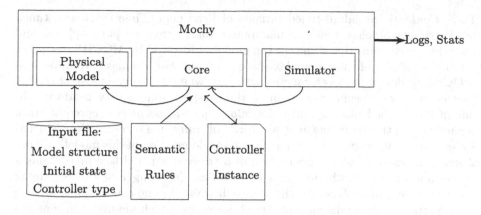

Fig. 1. Mochy architecture

The Physical Model: The goal of MOCHY is to provide a generic tool for the analysis and simulation of Petri nets variants. The models used as input for MOCHY share some common characteristics, namely notions of events represented by transitions of a net, notions of resources represented by places, and the flows of resources consumed/produced by transitions, and timing information. The common elements of MOCHY models are hence close to those of a stochastic timed Petri-Net. The variation points are the contents of places, the way they evolve over time, the firing rules of transitions, and the way a firing of a transition affects place contents. The contents of places can be simple tokens, or more complex objects evolving in a multi dimensional space. We have specialized this generic description to address models for transport networks.

Structure of Nets. The models used by MOCHY are variants of Petri nets with time. They share common features, such as the notions of places, transitions, flows, and time intervals. **Places** are contents holders for quantities that may evolve over time. They can be used as usual as containers for tokens, i.e. integral numbers that are affected by transitions firings, but not by time elapsing. In this case, places represent conditions needed to fire transitions. They can also be used to represent a physical space such as a track portion in a metro network. In this case, contents depict trajectories of objects, i.e. functions depicting evolution of a place contents according to time elapsed. **Transitions** represent classes of events. As usual in Petri nets, they have a preset, i.e. a set of places depicting resources needed for an occurrence of the transition, a postset, i.e. a set of places depicting resources impacted by the firing of a transition. A transition t can be triggered upon conditions that depend on time, and on the contents of places in the preset and in the postset of t. These conditions and the effect of a transition firing vary depending on the semantics rules.

The core structure of a net is hence a tuple $\mathcal{N} = (P, T, A, I)$, where P is a set of places, T is a set of transitions representing events in a system, $A \subseteq (P \times T) \cup (T \times P)$ is a set of arcs connecting places to transitions, and transitions to places. Map $I : T \to \mathbb{Q} \times \mathbb{Q} \times DF$ associates a time interval $[\alpha(t), \beta(t)]$ and a distribution function $f_t : [\alpha(t), \beta(t)] \to [0, 1]$ to each transition $t \in T$.

Place Contents: Standard timed variants of Petri nets (Time Petri nets, timed arc Petri nets, stochastic nets...) manipulate tokens that are put in places, and moved by each transition firing. In the models addressed by MOCHY, we allow for the definition of more complex place contents. For instance, we have used MOCHY to design models for metro networks, where trains can move at several speeds on a track segment as soon as they respect some safety headways. In one of the studied models, called *trajectory nets*, some places represent track segments, and their contents are trajectories of trains in a track portion depicted by *space-time* diagrams (see [12] for a complete description of this model). Figure 2 shows an example of a configuration of a trajectory net. Place p_1 contains a space-time diagram with two train trajectories, depicting how the remaining distance to arrival evolves over time for each train. As one can imagine, as time elapses, the remaining distance to arrival decreases, which modifies the contents of places. When the remaining distance of a trajectory is zero, this trajectory is moved to another place by firing of a transition. The duration of the newly created trajectory is sampled from an interval attached to the transition that will consume it. In the example of Fig. 2, a duration for a new trajectory arriving in p_1 can be sampled between 20 and 25 time units according to a defined probability law, and the initial distance for this trajectory is obviously the size of the space (here a track segment) represented by p_1. The semantics of the model also enforces safety of trains by allowing only trajectories preserving a sufficient headway between trains. This type of model was successfully implemented by instantiating the high-level semantics rules specified below.

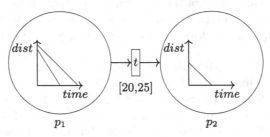

Fig. 2. A simple Trajectory Net.

2.2 Semantics

The current state of a model is called a *configuration*. Configurations are denoted by $C_1, C_2, \ldots$. The semantics of models designed with MOCHY are defined in terms of *timed moves* of the form $C_i \xrightarrow{\delta} C_{i+1}$ which describe how a system evolves when a certain amount of time δ elapses, and *discrete moves* of the form $C_i \xrightarrow{ev} C_{i+1}$, that describe how a system evolves when a discrete event ev (usually the firing of a transition) occurs. We assume a sampling semantics, that is when a transition gets enabled, the time before its urgent firing is chosen according to the current configuration and never changed. In the context of time Petri nets for instance, this corresponds to sampling a duration δ_t within a interval $[\alpha, \beta]$ for every newly enabled transition t when using a discrete move, and considering t as urgent δ_t time units later if it was not disabled before.

In a configuration C, a transition is *urgent* if it **has to** fire or be disabled before some time elapses. In urgent semantics timed moves are forbidden in a configuration C if C has urgent transitions. The notion of urgency may vary from a variant of a model to another. In trajectory nets [12], transitions are urgent if a trajectory of an object has reached a border of the physical space depicted by a place, and a transition is ready to move this object to another part of the net. To allow for the specification of many models, the semantics of a model in MOCHY is given by redefinition of two configuration transformation rules and two functions to test place contents or check values of clocks:

- $R_1(C, \delta)$: depicts the transformation occurring in a configuration C when δ time units elapse
- $R_2(C, ev)$: depicts the transformation occurring in a configuration C when event ev occurs (mainly firing of a transition)
- $F_1(C)$: returns the time that can elapse before a transition becomes urgent
- $F_2(C)$: returns the list of transitions that are firable in C.

2.3   Simulation

Configurations and transformation rules differ for every type of model, but providing functions $R_1(C, \delta), R_2(C, ev), F_1(C), F_2(C)$ is sufficient to implement the simulation loop proposed in Algorithm 1 below. Creating a model with a new semantics in MOCHY hence boils down to coding these rules and functions. Then simulation of a model with MOCHY consists in repeatedly deciding which transition to fire or which delay to elapse.

Algorithm 1. Rule-base Operational semantics for MOCHY

---

set a number of steps n
set an initial configuration C_0; $C = C_0$
$i \leftarrow 1$
while $i \leq n$ **do**
  $L = F_2(C)$
  if $L = \emptyset$ **then**
    $\delta = F_1(C)$
    Use function $R_1(C, \delta)$ for timed moves: $C \xrightarrow{\delta} C' = R_1(C, \delta)$
  else
    Use controller to choose $t \in L$ and update controller's memory
    Use function $R_2(C, t)$ for discrete moves: $C \xrightarrow{t} C' = R_2(C, t)$
  end if
  i++; C=C'
end while

Let us detail the role of controllers during simulation. When several transitions are firable, a controller is used to choose which transition fires. Controllers can be any program making such choice. The most basic controllers chose randomly a transition among firable ones. More involved controllers can be equipped with memory, with a schedule to follow, etc. and can implement complex strategies. We detail in Sect. 3, a controller designed to fix speed and dwell time of metros to recover from bunching situations.

2.4   Inputs-Outputs

MOCHY takes as input **project files** that contain: the class of model used by the project (it is a compiled Java class that implements the semantics rules), a description of the structure of an instance of the loaded model: the places, transitions, and flow relations of the net, the time intervals attached to transitions and the associated distributions. Depending on the specialization of the model, the project file can also provide pointer to additional features: a controller, schedules, etc.). This approach allowed for the modeling of metro networks equipped with a traffic management algorithm designed to adhere to a timetable.

Once a project is loaded, users can play with the specification in an interactive manner, or run a simulation for several steps using the simulation scheme of Algorithm 1. At each stop of the simulator, the contents of places is displayed. The net can be reset to its initial configuration at every stop. During simulation, MOCHY generates **logs**. Each line of a log is of the form `<cdate>;<tname>;<v1>...<vk>` where *cdate* is an occurrence date for an event, *tname* is the name of the transition fired at that date, and $v_1, ...v_k$ are variables values at date *cdate*. The list of variables to save in logs is one of the simulation parameters specified by users. Logs are saved in a file at the end of each simulation, and can then be used by other tools for statistics, process mining...

2.5   User Interface

Figure 3 shows the main window of MOCHY. The top of the window shows the structure of the simulated net, and its current configuration. Below, a control panel for simulation allows launching a simulation for a given number of steps, letting time elapse, or firing a single transition. An additional log part (not shown on the Figure) displays simulation logs, warnings and statistics : mean duration of a simulation after a simulation campaign, performance indicator...

3   Case Studies

MOCHY has been tested successfully on several models of metro networks, with various semantics. In this section, we present a first experiment that successfully used MOCHY to test accuracy of traffic management algorithms, and a second one that allowed to teach a controller to recover from train bunching.

A Metro Line in Rennes: The Metro line A in Rennes has 15 stations and a length of 9 km. The model developed with MOCHY uses a controller that implements a traffic management policy whose goal is to give dwell duration and speed advices to trains in order to adhere to a given timetable. The timetable

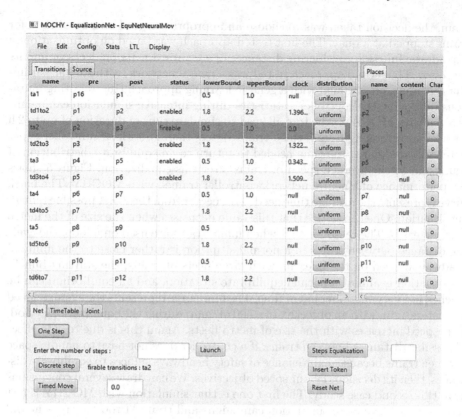

Fig. 3. The main window of MOCHY's graphical interface

describes planned operations for 4 hours, and contains dates for more than 3000 events. The net model is composed of 56 places and 63 transitions. MOCHY showed good performance for the simulation of this metro network: simulation of 10 runs (i.e. sequences of at least 3000 discrete moves) can be performed in 1.5 minutes on an average laptop (HP Elitebook with Intel Core i7). Using the high-level rules of MOCHY, we were able to simulate different traffic management policies, both in a fixed block and in a moving block setting.

Regulation by Equalization. The good performance of the tool allows for its use for applications that require intensive simulation campaigns, such as statistical model-checking or learning techniques. We have used MOCHY to train a neural network in charge of controlling a metro network to recover from a bunching situation. Bunching is a situation where all trains are not well distributed on a network, causing long periods without service in stations, followed by arrival of many trains in a short amount of time. This situation is depicted in Fig. 4.

We have considered a simple network, namely a loop of 20 kilometers, modeled by a net with 60 places and 60 transitions equipped with a controller that aims at equalizing distances between trains in the network (see [10] for more details on the experiment). Several traffic management algorithms have been tested. For each of them, the decision taken was to choose an appropriate speed and dwell time for trains stopped at a quay. The first tested approach was based on optimization of a quadratic function considering distances of a train w.r.t. its predecessor and its successor. The second approach tested was a neural network, with similar parameters as input, and trained with a genetic learning approach [16], selecting mutations improving the controller's statistics during intensive simulation campaigns realized with MOCHY. The tool allowed to simulate runs of duration of up to 2 h for fleet sizes ranging from 5 to 50 trains in less than 15 s. Both controller types were evaluated w.r.t. the time needed to return to an equilibrate distribution of trains and to the average speed of trains after this equalization. Figure 5 shows the performance of a neural network controller trained with MOCHY. The black curve represents the average time needed to recover from the worst possible bunching situation. One can notice that this value decreases when the size of the metro fleet increases. The reason is that when many trains are used on a loop, the average distance between trains in a normal situation is rather close to the minimal headway allowed on the line. So, with large fleets of trains, the starting bunching situation is almost a normal equilibrate situation, and an equilibrium can be reached in a very short amount of time. The blue curve represents the average speed (in meters/min, with an objective of a 500 m/min) during the equalization period. This speed increases with the size of metro fleets. Again this is due to side effect of possible distances between trains: if a controller does not have to handle space between trains because maintenance of safety headways suffices to equilibrate distances, then its decisions favour speed objectives. We can draw several conclusions from this second case study. The first one is that simulation with MOCHY is fast enough to achieve large simulation campaigns and train AI tools such as neural networks. The second conclusion is that the high-level semantics approach is particularly well adapted to the metro setting : changing controllers was done through a minor refinement of a few methods in a class, and could be integrated without changing the description of the physical models of the controlled metro network, nor any other part of the software.

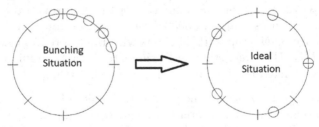

Fig. 4. From a bunching situation to a good distribution of trains in a Metro network.

4   Related Work

Mochy can be seen as a very generic tool allowing for the specification of timed and stochastic variants of nets which semantics can be described as discrete configuration transformations, and timed moves. Such specification is done by easy

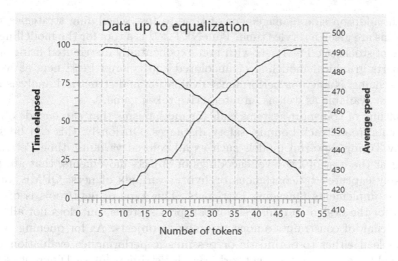

Fig. 5. Time to equalization guided by a neural network for varying fleet sizes.

refinement of configurations definition, and of four methods (the R_1, R_2, F_1, F_2 functions in Sect. 2) from core classes of MOCHY. This vague description encompasses time Petri nets, timed Petri nets, stochastic nets, etc. Once a net variant is specified, MOCHY can load an instance of model, and perform fast simulation based on these rules. Several tools, usually dedicated to a particular variant of Petri nets exist (see [17] for an extensive list). Some of them can handle time, complex firing rules or hybrid variables needed to model trajectories of objects. Many tools are dedicated to time(d) Petri nets and their variants. Romeo [13] is a verification tool for time Petri nets (TPNs). In addition to the standard urgent semantics of TPNs, ROMEO allows for the specification of ad-hoc firing rules, read/inhibitor arcs, and for parameters synthesis. ORIS [5] is close to ROMEO, but is tailored for transient analysis of stochastic timed Petri nets. Tapaal [4] targets verification of timed arcs Petri nets. It allows inhibitor/read arcs. TINA [2] is an analyser for TPNs extended with read arcs, inhibitor arcs, open intervals and data. Several other tools can perform simulation of timed and stochastic variants of nets. ARP analyzes and simulates nets where transitions are attached an interval and a distribution over this interval. Petrisim [19] allows for the simulation of Petri nets with delay between token production and consumption.

Some tools target net variants with colors, a way to introduce variables and data in nets. Alpha/Sim [15] allows for the simulation of stochastic, timed, attributed or colored Petri nets. Great SPN [14] is a tool for generalized (colored) Stochastic Petri nets, that allows for timed simulation. ExSpect [20] is dedicated to the design of business processes and is formally based on colored Petri nets. CPN tools [11] allows for the definition of colored High-level nets and nets with time, and can simulate them to analyse performance of the modeled systems. PnetLab is a tool dedicated to the control of High-level coloured Petri net and allows in addition side management of time to test scheduling strategies when transitions are given a service time. TimeNET [21] is a tool for the modelling and analysis of stochastic Petri nets with non-exponentially distributed firing times. It supports graphical modeling of uncolored and colored Petri nets as well as Markov chains. Numerous performance evaluation and structural analysis algorithms are available as well as an interactive token game.

Modeling of transport systems calls for mechanisms that can encode objects movements, road/track bounds, safety distances. Obviously, this can be simulated by Turing powerful models such as all colored variants, time Petri nets, etc. but at the cost of low-level encodings of objects movements, that are more intuitively captured by continuous or hybrid variants of nets. QPME [3] is a tool that implements Queueing Petri nets. This type of nets/tools is of great interest for the design and analysis of transport systems, but does not allow for the modeling of constraints among the moving objects. As for queuing theory, analyses lead either to optimistic or pessimistic performance evaluation w.r.t. the actual behavior of a train network. Batch Petri nets [6], or Differential Petri nets [8], and the tool Hisim [1] can handle mixed discrete and continuous tokens/ transitions, where places contain quantities that evolve according to differential equations. Discrete transitions firings have the usual semantics, and continuous firings are allowed when place contents exceed some threshold, and moves some quantities of token per time unit. Hisim simulates hybrid nets via a simulation loop that : fires immediate transitions, computes the next event date, progress time to this date, and iterates. Simulation in MOCHY is based on a similar simulation loop.

Fluid-survival-tool [18] considers Hybrid Petri nets to model systems with discrete and continuous quantities, and computes the probability to be in a given state at a certain time, or to verify Stochastic Timed Logic. Time is handled by attaching constant firing times to discrete transitions, distributions on firing times to stochastic transitions, and firing speed to continuous ones.

5   Conclusion and Future Work

This paper has introduced MOCHY, a tool for fast simulation of time Petri nets variants. One advantage of MOCHY is to allow for the specification of new semantics rules as part of the inputs of the model, which gives a huge flexibility to the tool. In particular, the tool was used to develop several variants of nets handling quantitative aspects that are mandatory to design metro networks (e.g. distance between trains, distance to arrival in station...). For all projects handled with MOCHY, the tool showed good performance on average machines, and allowed for extensive simulation campaigns with large case studies modelling existing metro lines. The tool was also successfully used to train neural networks and then use them as controllers.

The MOCHY toolbox can be freely downloaded at the following url: https:// adt-mochy.gitlabpages.inria.fr/mochy/. The available packages contain implementation of semantic rules for several variants of nets, including waiting nets [9] and trajectory nets [12] and examples of models for metro networks. Future distributions will include a statistical model checker for Signal LTL, and Machine Learning techniques to train controllers.

References

1. Amengual, A.: A specification of a hybrid petri net semantics for the HISim simulator. Tech. rep, International Computer Science Institute (2009)
2. Berthomieu, B., Vernadat, F.: Time petri nets analysis with TINA. In: Proceedings of QEST 2006, pp. 123–124. IEEE Computer Society (2006)
3. Buchmann, A., Dutz, C., Kounev, S.: QPME- queueing petri net modeling environment. In: Proceedings of QEST2006, pp. 115–116 (2006)
4. Byg, J., Jørgensen, K.Y., Srba, J.: TAPAAL: editor, simulator and verifier of timed-arc petri nets. In: Liu, Z., Ravn, A.P. (eds.) ATVA 2009. LNCS, vol. 5799, pp. 84–89. Springer, Heidelberg (2009). https://doi.org/10.1007/978-3-642-04761-9_7
5. Carnevali, L., Paolieri, M., Vicario, E.: The ORIS tool: app, library, and toolkit for quantitative evaluation of non-Markovian systems. SIGMETRICS Perform. Evaluation Rev. **49**(4), 81–86 (2022)
6. Demongodin, I.: Generalised batches petri net: Hybrid model for high speed systems with variable delays. Discret. Event Dyn. Syst. **11**(1–2), 137–162 (2001)
7. Demongodin, I., Koussoulas, N.: Differential petri net models for industrial automation and supervisory control. IEEE Trans. Syst. Man Cybern. Syst. **36**(4), 543–553 (2006)
8. Hamdi, F., Messai, N., Manamanni, N.: Design of switched observer using timed differential petri nets: A dwell time approach. In: Proceedings of European Control Conference, ECC 2009, pp. 4641–4646. IEEE (2009)
9. Hélouët, L., Agrawal, P.: Waiting nets. In: Bernardinello, L., Petrucci, L. (eds.) PETRI NETS 2022. LNCS, vol. 13288, pp. 67–89 (2022)
10. Hélouët, L., Fabre, E., Thébault, A.: Optimization of traffic management with learning machines. HAL-03777459 (2022)
11. Jensen, K., Kristensen, L.M.: Coloured Petri Nets. Springer, Heidelberg (2009). https://doi.org/10.1007/b95112
12. Kecir, K.: Performance evaluation of urban rail traffic management techniques. (Évaluation de Performances pour les Techniques de Régulation du Trafic Ferroviaire Urbain), Ph. D. thesis, University of Rennes 1, France (2019)
13. Lime, D., Roux, O.H., Seidner, C., Traonouez, L.-M.: Romeo: a parametric model-checker for petri nets with stopwatches. In: Kowalewski, S., Philippou, A. (eds.) TACAS 2009. LNCS, vol. 5505, pp. 54–57. Springer, Heidelberg (2009). https://doi.org/10.1007/978-3-642-00768-2_6
14. Marsan, M., Balbo, G., Conte, G., Donatelli, S., Franceschinis, G.: Modelling with generalized stochastic petri nets. SIGMETRICS Perform. Evaluation Rev. **26**(2), 2 (1998)
15. Moore, K., Chiang, J.: Alpha/sim: Alpha/sim simulation software tutorial. In: Proceedings of WSC 2000, pp. 259–267 (2000)
16. Palmes, P., Hayasaka, T., Usui, S.: Mutation-based genetic neural network. IEEE Trans. Neural Networks **16**(3), 587–600 (2005)
17. Petri nets tools database: quick overview. https://www.informatik.uni-hamburg.de/TGI/PetriNets/tools/quick.html. Accessed 01 Feb 2023

18. Postema, B., Remke, A., Haverkort, B., Ghasemieh, H.: Fluid survival tool: a model checker for hybrid petri nets. In: Proceedings of Measurement, Modelling, and Evaluation of Computing Systems and Dependability and Fault Tolerance - 17th International GI/ITG Conference, MMB & DFT 2014. LNCS, vol. 8376, pp. 255–259 (2014)
19. Sklenar, J.: Petrisim - environment for simulation of petri networks. In: Proceedings of the 20th Conference of the ASU Object Oriented Modelling and Simulation, pp. 214–221 (1994)
20. van der Aalst, W., et al.: ExSpect 6.4: an executable specification tool for hierarchical colored petri nets. In: Proceedings of ICATPN 2000. LNCS, vol. 1825, pp. 455–464 (2000)
21. Zimmermann, A.: Modelling and performance evaluation with TimeNET 4.4. In: Bertrand, N., Bortolussi, L. (eds.) QEST 2017. LNCS, vol. 10503, pp. 300–303. Springer, Cham (2017). https://doi.org/10.1007/978-3-319-66335-7_19

RENEW: Modularized Architecture and New Features

Daniel Moldt, Jonte Johnsen, Relana Streckenbach, Laif-Oke Clasen(✉),
Michael Haustermann, Alexander Heinze, Marcel Hansson,
Matthias Feldmann, and Karl Ihlenfeldt

University of Hamburg, Faculty of Mathematics, Informatics and Natural Sciences,
Department of Informatics, Hamburg, Germany
laif-oke.clasen@uni-hamburg.de
http://www.paose.de

Abstract. RENEW is an extensible Petri Net IDE that supports the development and execution of high-level Petri Nets and other modeling techniques. Over the past seven years, RENEW has undergone significant development and refinement.

To this end, RENEW's code base has been reworked extensively, and its tool collection has been expanded. The reworking was necessary due to technical debt caused by environmental changes: especially Java's transition from version 9 to 17. Adapting to the latest Java versions enables the modularization of RENEW's architecture through the Java Platform Module System (JPMS) which was introduced with Java 9. Additionally, some new features have been implemented, which were used to test our new architectural design.

One of our main results gave RENEW a cleaner code interface design and a more modern architecture. Examples of the extensions and improvements made are the new P/T-nets with synchronous channels (PTC) formalism and the Modular Model Checker (MoMoC). In addition to the aforementioned changes, the GUI has also been altered and now offers an all-in-one window.

Keywords: Petri Nets · Reference Nets · Synchronous Channels · Tools · Java Platform Module System · Software Architecture

1 Introduction

Long living software requires continuous maintenance, otherwise technical debts are built up [14]. Changes in the environment lead to changed or new requirements with respect to the functionality and technical basis. Java is the main programming language for RENEW[1] [11]. As the development of the software started

---

K. Ihlenfeldt: Supported by all participants of our teaching project classes and many student theses.

[1] **R**eference **N**et **W**orkshop can be installed directly from the website http://renew. de. There is a selection of download options. The license terms can be found on the website mentioned above.

L. Gomes and R. Lorenz (Eds.): PETRI NETS 2023, LNCS 13929, pp. 217–228, 2023.
https://doi.org/10.1007/978-3-031-33620-1_12

at the end of the 20th century, several nowadays deprecated language concepts were used and powerful new ones not yet utilized. Over the years Java was improved which led to some intermediate and, more importantly, to long-term support (LTS) versions[2]. With the introduction of Java 9, the Java Platform Module System (JPMS) was introduced, which provides modules and module layers as abstraction mechanisms that are directly supported by the Java Virtual Machine.

As a result, we started a major refactoring of our whole software system based on our plugin architecture, which has already been presented [4,9,13]. From Java 9 onwards we continuously improved the software in our (teaching) projects and several theses through the JPMS [8]. This was a major transition and severely changed the interfaces and structure of RENEW, leading to the RENEW 4.1 version. Beside the new features since RENEW 2.5 [5,10], which was the last major release, we share our knowledge about the development of a widely used Petri net tool. Main aspects are the insights we gained for the architectural advantages of the JPMS.

In the following we address some formalisms of RENEW (Sect. 2), discuss the objectives of this contribution (Sect. 3), sketch the functionality and some new features of our tool (Sect. 4), present our new architecture (Sect. 5, illustrates some use cases (Sect. 6), and we will finalize with the conclusion (Sect. 7).

2 Formalisms

2.1 Reference Nets

One of the main formalisms of RENEW is the Reference Net formalism [13], which combines the concept of nets-within-nets [20] with reference semantics and the expressiveness of Java code [5].

Every time a simulation is called on a net in RENEW, one instance of it is created and can be observed by the user. This implementation's advantage is the possibility of creating multiple instances of a single net. Another advantage is that a net instance can be created from within a net and its reference is kept as a colored marking [6]. To achieve this goal, users simply need to annotate a transition with the `new` keyword followed by the name of the net they want to use [12].

Synchronous channels call other parts of nets to get a result. They consist of an uplink and a downlink. These two inscriptions are unified by RENEWs runtime for matching identifiers and parameters. That means two transitions are treated as a single one for the purpose of firing [12].

When using Reference Nets in RENEW, Java objects can be used as tokens as well [6]. If a transition is inscribed with `action` followed by a Java expression,

---

[2] Java (Oracle): https://www.oracle.com/java; OpenJDK: https://openjdk.java.net/; Java Community Process: https://jcp.org; background information on Jigsaw the predecessor of the JPMS, starting in 2008 / 2014: https://openjdk.java.net/projects/jigsaw/.

RENEW will execute the instruction upon the firing of the transition [6]. This makes RENEW very expressive, especially in the hands of a skilled Java developer.

Figure 1 depicts two Reference Nets. In the upper part of the window, their models can be seen. The net model to the right takes a token o and calls the Java method toLowerCase() on it. The net model to the left calls an instance of that net, stores it in the token n, and then uses it to convert its other input, the string HELLO, to lowercase. The running simulations are shown in the lower part of the window. The inspection of their tokens reveals that the instance of calledNet is actually in the bottommost place of CallerNet's instance and that the String was converted to lowercase and now is in calledNet's place.

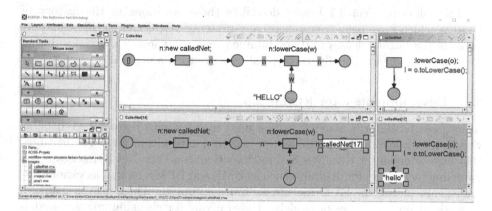

Fig. 1. Screenshot of RENEW running a simulation with Reference Nets

2.2   P/T-Nets with Synchronous Channels

PTC-nets are Place/Transition nets (P/T-nets) with synchronous channels. They were introduced to extend the existing model with synchronous channels as well as with net partitioning, in order to allow for larger models [21].

Figure 2, which is taken from [21], compares the difference between two small nets. The one on the right side uses a transition to synchronize between producer and consumer. The one on the left uses a channel to the same end but the visual separation of the producer and the consumer reflects their status as two separate entities.

3   Objectives

From the outset of RENEW's development, its main purposes have been to draw, edit, and simulate Petri Nets. This mission statement still holds true but is being built upon with each subsequent release adding functionality, improving the IDE, and gaining new application scenarios.

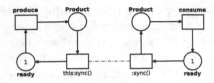
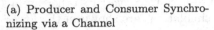
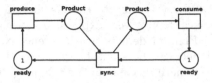

(a) Producer and Consumer Synchronizing via a Channel

(b) Producer and Consumer Synchronizing via a Transition

Fig. 2. Synchronization of a Single Producer and Consumer [21, p. 44]

The following Sect. 3.1 briefly describes those additions in the history of RENEW which have been comprehensively presented [5,6] and clarifies the purposes of the latest releases of its maintained major versions 2.x and 4.x during the past seven years. Finally, the typical user group is described in Sect. 3.2.

3.1   Developments after RENEW 2.5

In its most basic form, RENEW can be used to work with regular P/T-nets, as well as higher level nets including Java nets and nets-within-nets. The features, that were added after RENEW's first introduction as a Petri Net IDE, encompass modeling techniques for diagrams such as UML and BPMN as well as various formalisms including the FA formalism for finite automata. Thus, it even manages to work with other types of models. It also allows for checking created models for desired properties or unwanted side effects.

With the expansion of its capabilities, RENEW also developed into an IDE for Petri Net-based agent-oriented software engineering (PAOSE) [2,15,16]. This approach to software engineering emphasizes on distribution and concurrency. Multi-agent applications can be edited, debugged, and simulated since the introduction of the multi-agent nets framework MULAN [6].

The objective when releasing RENEW 2.6 in April 2022 was to enable the support of new standards and provide useful features in order to improve the user experience. This required for example the reimplementation of the PNML import and export since this standard had changed since its former implementation in RENEW (see Sect. 4.3). The list of improvements also includes an upgrade of automatic net layouts and undo snapshots (see Sects. 4.1 and 4.3) [8].

The introduction of RENEW 4.0 in April 2022 marked the modularization of the software using the Java Platform Module System (JPMS) and the switch from Ant to Gradle as the build tool (see Sect. 5). These changes were made to improve the overall quality of RENEW and to stay current. This new major version also features an enhanced user interface which aims to ease navigating projects developed in RENEW and streamline the overall design. This is achieved by consolidating all functionality to one window, offering the zoom feature and the MiniMap plugin (see Sect. 4.1). Additionally, in order to stay up to date with new Petri-Net-research, RENEW 4.1 introduced the PTC formalism for P/T-nets with synchronous channels which allows for expressive synchronization behavior (see Sect. 4.3) [8]. For future developments, we aim to implement our plugins as HERAKLIT AGENTS [17].

3.2   Users

User groups include both teaching and development on and with RENEW. Within this context, it can be used for teaching theoretical foundations, such as finite automata or Petri nets. In addition, RENEW is utilized by students to model using different modeling techniques like Petri nets or BPMN, which facilitates their understanding.

As already mentioned, RENEW can be used as an IDE for Petri net-based applications. A corresponding approach including framework and toolset was presented in [6]. Thus, applications can be built with RENEW like Settler explained in Sect. 6.2.

Furthermore, there are also developments at RENEW. Here we can distinguish between internal and external developments. In the context of internal developments, RENEW is also used as a development object for teaching and research in our university projects. External developments are easily possible through the plugin system of its open source software in the form of own plugins.

4   Functionality

4.1   Usability

New Look for the UI. The new and improved user interface now consists of only one singular window that includes all functionality. Palettes, drawings, and other helpful tools are all neatly packed into a single frame, the arrangement of which can be customized as desired. Additionally, important windows can still be detached and moved independently of the whole UI.

The new look also includes some quality-of-life features. For example fast access to customization options of graphical components can be found right above the canvas and allows for quick changes. Figure 3 shows the current version of RENEW in use.

Zooming and MiniMap. When working with large nets, it is important to have an overview. Therefore, the zoom feature was introduced, which simplifies editing [11]. This can now either be operated via buttons or, like a large number of RENEW's functions, also has shortcuts for control. If an overview cannot be achieved despite the zoom, a manually activatable MiniMap is also provided. It can be used to simplify navigation in large nets greatly.

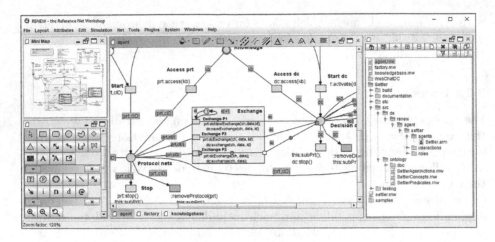

Fig. 3. Screenshot of RENEW in use

Automatic Net Layouts. Previously RENEW provided an option for finetuning an automatic layout processor that slowly moved the diagram around until it matched all criterions as well as possible. Newly implemented is a randomizer that simply skips to a final result [11]. This allows for quick checks of how a good layout might be possible, without taking much time. It also allows the program to come up with new (random) orientations of the diagram that would not be reached by simply spacing some parts differently.

4.2   New Functionality

PTC Formalism. The paper [21] describes the use of synchronous channels in P/T-nets. The implementation provided by these works is now generally available in RENEW with a new plugin. It extends the simulator with two additional formalisms: The Single P/T-net with Channel Compiler allows the use of regular synchronous channels consisting of an up and a downlink in P/T-nets. The P/T nets-within-nets Compiler composes multiple uplinks dynamically with the use of a SystemNet. This allows for a more expressive synchronization behavior.

MoMoC. Modular Model Checker (MoMoC) [22] is a plugin for RENEW that allows for model checking based on reachability graphs. Its goal is to allow for the verification of nets using the CTL model checking algorithm [1,7]. P/T nets as well as higher-level nets can be evaluated. Because RENEW always translates the model into a Petri net representation first when running a simulation, this also allows checking finite automata and other types. Additionally, it allows for visualization of the reachability graph itself as well as quickly checking various net properties such as reversibility, deadlock freedom, and net liveness. It also exposes its framework for further implementation of additional model-checking procedures. Because MoMoC is still in a prerelease state, it is not currently shipped with the 4.1 version. It is, however, available on our websites[3].

4.3 Improvements

PNML Import and Export. Because the PNML standard for saving nets as XML files has been updated since its integration in RENEW, the implementation has been updated too. The import allows for nets created in other software to be loaded into RENEW. Similarly, the export facilitates cross-platform sharing of nets originating in RENEW.

Improved Undo Snapshots. While editing nets it is inevitable to make mistakes. That is why RENEW provides the Undo-option with Ctrl + Z which is commonly used. This feature was recently improved upon where the grouping of movements into a single Snapshot was handled more consistently [11].

5 Architecture

The plugin architecture of RENEW was introduced with the 2.0 version and has already been presented [4,9,13]. This anchored the extensibility of the software as a basic principle. Due to it being based on Java and Java being an object-oriented language, the realization of RENEW is also object-oriented.

Prior attempts to use OSGi as an improvement of our plugin architecture by that library proved to be unsatisfactory. Therefore RENEW 3.0 never became public.

In the fourth major version, we addressed modularization using the Java Platform Module System (JPMS) [8]. One result of this contribution is to provide a deeper insight into the evolved architecture. For this purpose, the modularization of RENEW by the JPMS and the resulting architecture in particular will be discussed next. In addition to this change, RENEW's build tool was changed from Ant to Gradle.

As a starting point for the modularization, we were able to use our plugin architecture. First, we mapped each plugin to exactly one module, which has a module-info file. Within this file, dependencies to other modules, which result from the dependencies of the classes in the plugins, and the provided functionalities are defined.

For a class in module A to use a class in module B, there must be a dependency between module A and module B in the JPMS, expressed by `requires`. However, due to the information hiding principle [18] this is not sufficient. Therefore, module B must additionally release the package, in which the class to be used is located, to the outside. To accomplish this, module B must export the corresponding package, expressed by the keyword `exports`.

---

[3] MoMoC - A Modular Model Checker: https://paose.informatik.uni-hamburg.de/paose/wiki/MoMoC.

The goals to be achieved through modularization are strong encapsulation, high cohesion, loose coupling, and explicit interfaces, which increase the software quality. Here, our initial focus was on stronger encapsulation than before, making interfaces explicit, and increasing cohesion to improve our software architecture. The use of the `exports` and `requires` keywords reinforces the encapsulation concerning the previously existing package structure by introducing an additional layer of access modifiers. Furthermore, the interfaces have been made explicit through the module-info files. In addition, higher cohesion has also been achieved by prohibiting so-called split packages within the JPMS.

This means that Java packages which were distributed over several components before the modularization are now located in a single module. Thus, functionality is now contained in a single component increasing cohesion.

Loose coupling was initially pursued with less focus by us. However, within the module-info, the keywords `uses` and `provides` are suitable for loose coupling. These keywords can be used to express service relations.

But initially, we have deliberately relied exclusively on the requires/exports relation, as described above. This shows that our modularization has still room for improvement, even though we could already achieve some intermediate goals. These include stronger encapsulation, increased cohesion, and the explication of the interfaces.

In JPMS, modules are grouped into module layers at runtime. For this, one or more root modules as well as one or more parents of the module layer are specified for a module layer during creation. All modules which are needed by the root modules including their dependencies and by the dependencies of the dependencies etc. and which are not reachable by the parent relation of the module layers are loaded into the module layer to be created. For our implementation in RENEW, we decided to map each plugin to exactly one module layer. The main reason for this is that module layers can be created and destroyed at runtime. This allows us to load and unload our plugins at runtime. For this, we have implemented a module manager within the loader plugin, which is responsible for the management of the module layers.

From the user's point of view, plugins can now actually be unloaded at runtime through the JPMS in RENEW, which was not possible before. Furthermore, the software quality of RENEW could be increased. Through this, we hope that development times and especially the time-to-market will be reduced. Java itself has so far successfully managed to release a new version every half year after its modularization. We want to get into similar dimensions with RENEW.

During the implementation of modularization using the JPMS, additional new plugins were also developed. The new plugins have been modularized from the beginning. Here, too, the modularization improved the quality right from the start through the aforementioned benefits. These include, for example, the PTChannel plugin. Other internal plugins also benefit from the modularization. This shows that new developments can also benefit from modularization.

In addition to RENEW there is the framework MULAN[4] which was developed at our workspace. MULAN builds on RENEW and implements additional plugins for the agent context. Here we managed to run MULAN both non-modular and modular with the modularized RENEW. This shows that the whole system does not have to be modularized and that modularized and non-modularized components can be run in the same environment.

An overview of the published plugins[5] of RENEW can be seen in Fig. 4. We offer RENEW for Windows, Linux, and Mac. RENEW runs under the Java LTS-Version 11 as well as 17.

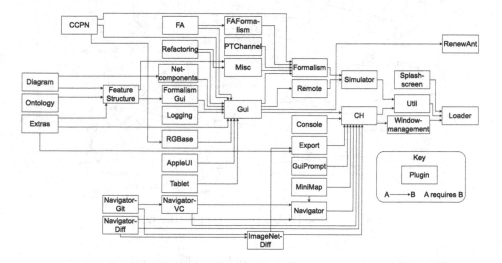

Fig. 4. RENEW 4.1 Plugins

6   Use Cases

6.1   MULAN

MULAN [19] is a framework to develop Petri net-based multi-agent systems. The framework is used in particular in the context of PETRI NET-BASED, AGENT-AND ORGANIZATION-ORIENTED SOFTWARE ENGINEERING (PAOSE, [2]). The PAOSE approach [2] is used to develop multi-agent systems. Cabac comprehensively describes the modeling techniques and the foundations used, which have been developed over the years. Furthermore, MULAN was also developed using nets and Java code, by using the RENEW tool as a basis for MULAN. Due to the modularization of RENEW, we were also able to implement an internal modularized version of the MULAN framework.

---

[4] The source code of MULAN and thus its plugins have not been published yet.
[5] Details about the plugins can be found in the associated READMEs.

6.2 Settler

Another larger example of the possibilities MULAN and RENEW have in developing software is Settler [2]. The famous game Catan (also known as The Settlers of Catan) was developed as a multi-agent system within the context of PAOSE.

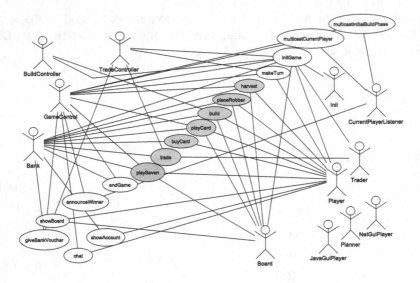

Fig. 5. Coarse Design Diagram of Settler [2,3]

The development time was several years and included several teaching projects. For this purpose, the framework MULAN as well as the tool RENEW were used. By using MULAN and RENEW, Settler was implemented with both nets and Java code. To get an idea of how the larger example consists of interactions between different actors, Fig. 5 shows the coarse design diagram. Due to the modularizing RENEW and MULAN, we were also able to implement an internal modularized version of Settler.

7 Conclusion

RENEW is a tool that has been developed for more than two decades. It is a powerful tool for creating reference nets, but in the meantime, it has also undergone many other developments. Thus, it is now able to support entire programs based on Petri nets and Java. In addition, its extensions can be used not only to control the creation of nets, but also to advance their understanding.

With the old age of the software, it has been necessary to introduce improvements. These are partially visible to the user with the new interface. However, it is much more important that the underlying architecture is kept up to date to support software development in the future. This paper has shown how this challenge has been met.

References

1. Baier, C., Katoen, J.P.: Principles of Model Checking. MIT press, Cambridge (2008)
2. Cabac, L.: Modeling petri net-based multi-agent applications. Dissertation, University of Hamburg, Department of Informatics, Vogt-Kölln Str. 30, D-22527 Hamburg (Apr 2010). https://ediss.sub.uni-hamburg.de/handle/ediss/3691
3. Cabac, L., et al.: PAOSE Settler demo. In: First Workshop on High-Level Petri Nets and Distributed Systems (PNDS) 2005. University of Hamburg, Department of Computer Science, Vogt-Kölln Str. 30, D-22527 Hamburg (Mar 2005). http://www.informatik.uni-hamburg.de/TGI/events/PNDS2005/program_and_abstracts.html
4. Cabac, L., Duvigneau, M., Moldt, D., Rölke, H.: Modeling dynamic architectures using nets-within-nets. In: Ciardo, G., Darondeau, P. (eds.) Applications and Theory of Petri Nets 2005. 26th International Conference, ICATPN 2005, Miami, USA, June 2005. Proceedings. Lecture Notes in Computer Science, vol. 3536, pp. 148–167 (2005), https://doi.org/10.1007/11494744_10
5. Cabac, L., Haustermann, M., Mosteller, D.: Renew 2.5 - towards a comprehensive integrated development environment for petri net-based applications. In: Kordon, F., Moldt, D. (eds.) Application and Theory of Petri Nets and Concurrency - 37th International Conference, PETRI NETS 2016, Toruń, Poland, June 19–24, 2016. Proceedings. Lecture Notes in Computer Science, vol. 9698, pp. 101–112. Springer-Verlag (2016). https://doi.org/10.1007/978-3-319-39086-4_7
6. Cabac, L., Haustermann, M., Mosteller, D.: Software development with Petri nets and agents: approach, frameworks and tool set. Sci. Comput. Program. 157, 56–70 (2018). https://doi.org/10.1016/j.scico.2017.12.003
7. Clarke, E.M., Jr., Grumberg, O., Kroening, D., Peled, D., Veith, H.: Model Checking. MIT Press, Cambridge (2018)
8. Clasen, L.O., Moldt, D., Hansson, M., Willrodt, S., Voß, L.: Enhancement of Renew to version 4.0 using JPMS. In: Köhler-Bußmeier, M., Moldt, D., Rölke, H. (eds.) Proceedings of the International Workshop on Petri Nets and Software Engineering 2022 co-located with the 43rd International Conference on Application and Theory of Petri Nets and Concurrency (PETRI NETS 2022), Bergen, Norway, June 20th, 2022. CEUR Workshop Proceedings, vol. 3170, pp. 165–176. CEUR-WS.org (2022). https://ceur-ws.org/Vol-3170
9. Duvigneau, M.: Konzeptionelle Modellierung von Plugin-Systemen mit Petrinetzen, Agent Technology - Theory and Applications, vol. 4. Logos Verlag, Berlin (2010). http://www.logos-verlag.de/cgi-bin/engbuchmid?isbn=2561&lng=eng&id=
10. Kummer, O., Wienberg, F., Duvigneau, M., Cabac, L., Haustermann, M., Mosteller, D.: Renew - User Guide (Release 2.5). University of Hamburg, Faculty of Informatics, Theoretical Foundations Group, Hamburg (Jun 2016). http://www.renew.de/
11. Kummer, O., Wienberg, F., Duvigneau, M., Cabac, L., Haustermann, M., Mosteller, D.: Renew - the Reference Net Workshop (Feb 2023). http://www.renew.de/, release 4.1
12. Kummer, O., Wienberg, F., Duvigneau, M., Cabac, L., Haustermann, M., Mosteller, D.: Renew - User Guide (Release 4.1). University of Hamburg, Faculty of Informatics, Theoretical Foundations Group, Hamburg (Feb 2023). http://www.renew.de/

13. Kummer, O., Wienberg, F., Duvigneau, M., Schumacher, J., Köhler, M., Moldt, D., Rölke, H., Valk, R.: An extensible editor and simulation engine for Petri nets: renew. In: Cortadella, J., Reisig, W. (eds.) Applications and Theory of Petri Nets 2004. 25th International Conference, ICATPN 2004, Bologna, Italy, June 2004. Proceedings. Lecture Notes in Computer Science, vol. 3099, pp. 484–493. Springer, Berlin Heidelberg New York (Jun 2004). https://doi.org/10.1007/978-3-540-27793-4_29

14. Lilienthal, C.: Komplexität von Softwarearchitekturen, Stile und Strategien. Ph.D. thesis, Staats-und Universitätsbibliothek Hamburg Carl von Ossietzky (2008)

15. Moldt, D.: Petrinetze als Denkzeug. In: Farwer, B., Moldt, D. (eds.) Object Petri Nets, Processes, and Object Calculi, pp. 51–70. No. FBI-HH-B-265/05 in Report of the Department of Informatics, University of Hamburg, Department of Computer Science, Vogt-Kölln Str. 30, D-22527 Hamburg (Aug 2005)

16. Moldt, D.: PAOSE: A way to develop distributed software systems based on Petri nets and agents. In: Barjis, J., Ultes-Nitsche, U., Augusto, J.C. (eds.) Proceedings of The Fourth International Workshop on Modelling, Simulation, Verification and Validation of Enterprise Information Systems (MSVVEIS'06), May 23–24, 2006 - Paphos, Cyprus 2006, pp. 1–2 (2006)

17. Moldt, D., et al.: Enriching heraklit modules by agent interaction diagrams. In: Gomes, L., Lorenz, R. (eds.) Application and Theory of Petri Nets and Concurrency - 44th International Conference, PETRI NETS 2023, Lisboa, Portugal, June 26–30, 2023, Proceedings. Lecture Notes in Computer Science, vol. this volume. Springer (2023)

18. Parnas, D.L., Clements, P.C., Weiss, D.M.: The modular structure of complex systems. IEEE Trans. Softw. Eng. **3**, 259–266 (1985)

19. Rölke, H.: Modellierung von Agenten und Multiagentensystemen - Grundlagen und Anwendungen, Agent Technology - Theory and Applications, vol. 2. Logos Verlag, Berlin (2004). http://logos-verlag.de/cgi-bin/engbuchmid?isbn=0768&lng=eng&id=

20. Valk, R.: Petri nets as token objects - an introduction to elementary object nets. In: Desel, J., Silva, M. (eds.) 19th International Conference on Application and Theory of Petri nets, Lisbon, Portugal. pp. 1–25. No. 1420 in Lecture Notes in Computer Science, Springer-Verlag, Berlin Heidelberg New York (1998). https://doi.org/10.1007/3-540-69108-1_1

21. Voß, L., Willrodt, S., Moldt, D., Haustermann, M.: Between expressiveness and verifiability: P/T-nets with synchronous channels and modular structure. In: Köhler-Bußmeier, M., Moldt, D., Rölke, H. (eds.) Proceedings of the International Workshop on Petri Nets and Software Engineering 2022 co-located with the 43rd International Conference on Application and Theory of Petri Nets and Concurrency (PETRI NETS 2022), Bergen, Norway, June 20th, 2022. CEUR Workshop Proceedings, vol. 3170, pp. 40–59. CEUR-WS.org (2022). https://ceur-ws.org/Vol-3170

22. Willrodt, S., Moldt, D., Simon, M.: Modular model checking of reference nets: MoMoC. In: Köhler-Bußmeier, M., Kindler, E., Rölke, H. (eds.) Proceedings of the International Workshop on Petri Nets and Software Engineering co-locatd with 41st International Conference on Application and Theory of Petri Nets and Concurrency (PETRI NETS 2020), Paris, France, June 24, 2020 (dueto COVID-19: virtual conference). CEUR Workshop Proceedings, vol. 2651, pp. 181–193. CEUR-WS.org (2020). http://ceur-ws.org/Vol-2651/paper12.pdf

Explorative Process Discovery Using Activity Projections

Yisong Zhang(✉) and Wil M. P. van der Aalst(✉)

Chair of Process and Data Science (PADS), RWTH Aachen University, Aachen, Germany
{zhang,wvdaalst}@pads.rwth-aachen.de

Abstract. This paper presents a tool to Explore Process Discovery (EPD) results using activity projection. Our EPD-Tool aims at exploring quality changes after removing activities from an event log. The main idea is to create a projected event log for every non-empty subset of activities and apply process discovery and conformance checking on them. The tool has been implemented as a plugin in ProM. First, EPD-Tool uses a process discovery algorithm to discover Petri net models for each projected event log. Then, EPD-Tool uses a conformance checking technique to compute conformance measures for each projected event log and model pair (L, N), e.g., fitness, precision, and $\mathcal{F}_1$-score. Finally, a dendrogram is generated to visualize the relationship between each log-model pair, thus enabling the systematic exploration of the different models using the dendrogram to find the best-performing node, i.e., a best log-model pair. This method prioritizes activities and detects redundancy in the process, which contributes to process enhancement. Conversely, critical activities are uncovered to help to shorten the processing time or save the process cost. This paper presents the EPD-Tool implementation and some example results.

Keywords: Process mining · Petri nets · Log projection · ProM

1 Introduction

After obtaining event logs from the underlying information systems, stakeholders can use *process mining* techniques to uncover their actual processes, provide insights, diagnose problems, and automatically trigger corrective actions [15,16]. *Process discovery* is a crucial step and the most challenging process mining task, since it aims to learn a process model from example behavior recorded in an event log. Each event in such a log refers to an activity, a well-defined step in some process, a process instance (case), and a timestamp. Process models discovered from event data show the actual process, e.g., the ordering of activities, frequencies, exceptional paths, and bottlenecks.

After obtaining a process model, we evaluate the quality of this model using several measures. Two widely-used control-flow-based *quality criteria* are replay *fitness* and *precision*. Fitness indicates how well the model reflects the behavior of the log. Precision reflects whether the model allows for additional unobserved behavior that is unlikely given the data.

ⓒ The Author(s), under exclusive license to Springer Nature Switzerland AG 2023
L. Gomes and R. Lorenz (Eds.): PETRI NETS 2023, LNCS 13929, pp. 229–239, 2023.
https://doi.org/10.1007/978-3-031-33620-1_13

State-of-the-art process discovery technologies [1–6] focus on the entire event log, and they ignore the impact of individual activity in the whole process. For instance, it is hard to prioritize activities in an event log using a traditional process discovery method, and the method of classifying which activities are redundant or critical is still missing. Therefore, we propose *Explorative Process Discovery using Activity Projections* (EPD) in this paper. This method uses activity-based projection to extract the sub-logs, and discovers the process model after deleting any activity in the event log. Afterward, a conformance checking technique records the changes in process model quality before and after deleting an activity. According to the comparison, we can prioritize activities and judge whether the activity is redundant or critical. We fully implemented the approach using the ProM [7] framework.

The remainder of the paper is structured as follows. In Sect. 2, we introduce basic concepts. Section 3 describes the approach. Section 4 presents the implementation of EPD-Tool and shows how to use this tool. Section 5 evaluates our approach using various data sets. Section 6 concludes the paper.

2   Preliminaries

In this section, we introduce some basic concepts and notations related to our research. The first and most important thing is the event log. Event logs serve as the starting point for any process mining task. An event log is a multiset of traces that describe the life cycle of cases in terms of the activities executed.

Definition 1 (Event Log). *Let $\mathcal{U}_{act}$ be the activity universe, i.e., the set of all possible activity attributes of events. A trace $\sigma = \langle a_1, a_2, \ldots, a_n \rangle \in \mathcal{U}_{act}^*$ is a sequence of activities. An event log $L \in \mathcal{B}(\mathcal{U}_{act}^*)$ is a multiset of traces. $\mathcal{U}_L = \mathcal{B}(\mathcal{U}_{act}^*)$ is the universe of event logs.*

$L_1 = [\langle a, b, c \rangle^5, \langle a, b, c, d, d \rangle^3, \langle a, d \rangle^2]$ and $L_2 = [\langle a, b, c, d \rangle^5, \langle a, b, c, d, d \rangle^3, \langle a, d \rangle^2]$ are two examples of an event log.

Definition 2 (Activity Projection). *Let $L \in \mathcal{U}_L$ be an event log and $\mathcal{A} \subseteq \mathcal{U}_{act}$ be a subset of activities. A projected event log is an event log where all activities not in $\mathcal{A}$ are removed. The projection function is $L{\upharpoonright}_\mathcal{A} = [\sigma{\upharpoonright}_\mathcal{A} \mid \sigma \in L]$ where $\sigma{\upharpoonright}_\mathcal{A}$ is defined recursively: (1) $\langle\rangle{\upharpoonright}_\mathcal{A} = \langle\rangle$ and (2) for $\sigma \in L$:*

$$(\langle a \rangle \cdot \sigma){\upharpoonright}_\mathcal{A} = \begin{cases} \sigma{\upharpoonright}_\mathcal{A} & \text{if } a \notin \mathcal{A} \\ \langle a \rangle \cdot \sigma{\upharpoonright}_\mathcal{A} & \text{if } a \in \mathcal{A} \end{cases} \tag{1}$$

where $\langle a \rangle \cdot \sigma$ appends activity a to trace σ.

Consider L_1 and L_2 introduced before, given $\mathcal{A}_1 = \{b, c\}$ and $\mathcal{A}_2 = \{a, d\}$, then $L_1{\upharpoonright}_{\mathcal{A}_1} = [\langle b, c \rangle^8, \langle\rangle^2]$ and $L_2{\upharpoonright}_{\mathcal{A}_2} = [\langle a, d \rangle^7, \langle a, d, d \rangle^3]$.

Process discovery techniques aim to learn a formal process model based on example behaviors in the event log [15,16]. There is a plethora of process modeling notations, e.g., Business Process Model and Notation (BPMN) [8], Process Trees, etc. Most of these modeling notations can be directly transformed into a Petri net [9], allowing for a

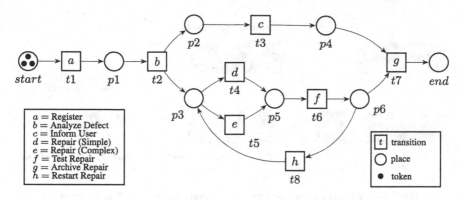

Fig. 1. Labeled Petri net N_1: A product repair process

range of analysis techniques including conformance checking and performance analysis. Therefore, we focus on *labeled accepting Petri nets*, i.e., Petri nets where transitions have activity labels and there is a well-defined initial and final marking. Note that we also allow for silent activities, i.e., transitions that do not have a label and that cannot be observed in the event log. This way, we can also model skipping and handle gateways in BPM and operators in process trees. $\mathcal{U}_N$ is the universe of labeled accepting Petri nets. An example of a labeled Petri net is depicted in Fig. 1.

Our approach does not focus on any specific process discovery algorithm. Instead, *it can use any existing process discovery algorithm provided that it can be converted into a labeled accepting Petri net*. For instance, this paper uses Inductive Miner - infrequent (IMi) [10]. IMi takes an event log as input and discovers a process tree as output which could be transformed into a Petri net directly.

Definition 3 (Exploratory Process Discovery). *disc* : $\mathcal{U}_L \to \mathcal{U}_N$ *is a function that discovers a labeled accepting Petri net for any event log. Given an event log* $L \in \mathcal{U}_L$, *we can discover a model* $N = disc(L)$. (L, N) *is a log-model pair. Given a collection of event log we can create a collection of log-model pairs.*

For any log-model pair (L, N) conformance checking techniques are used to evaluate the quality of process models. In this paper, we use *alignments* [11] to compute replay *fitness* and *precision* [12] of each log-model pair. The $\mathcal{F}_1$-score is based on these.

Definition 4 (Quality Measures). *Let* $(L, N) \in \mathcal{U}_L \times \mathcal{U}_N$ *be a log-model pair.* $\mathcal{F}_{it}(L, N) \in [0, 1]$ *measures fitness (indicating how well the model reproduces the behavior of the log).* $\mathcal{P}_{re}(L, N) \in [0, 1]$ *measures precision (indicating to what degree the model's behavior is likely given the log). The harmonic mean of fitness and precision* $\mathcal{F}_1(L, N) \in [0, 1]$ *is defined as follows:* $\mathcal{F}_1(L, N) = 2\frac{\mathcal{F}_{it}(L,N) \cdot \mathcal{P}_{re}(L,N)}{\mathcal{F}_{it}(L,N) + \mathcal{P}_{re}(L,N)}$.

Here we abstract from the exact computation of fitness and precision and use the alignment-based fitness and precision values implemented in ProM [11, 12].

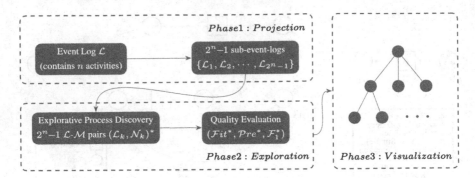

Fig. 2. Overview of explorative process discovery using activity projections.

3   Approach

Since traditional process discovery methods focus on the entire event log and ignore the impact of each activity in the process, they cannot prioritize activities nor classify which activities are redundant or critical. Thus, we propose our "Explorative Process Discovery using Activity Projections" method to address this problem. The approach consists of the following three phases, as Fig. 2 shows:

- **Phase 1**: Projection. Given an event log $L \in \mathcal{U}_L$ containing n activities, for each subset of activities $\mathcal{A}_k$, where $\mathcal{A}_k \subseteq \mathcal{U}_{act}$ and $\mathcal{A}_k \neq \emptyset$, we use activity projection defined in Definition 2 to get a projected event log. Therefore, there are $2^n - 1$ sub-logs $\{L_1, L_2, \cdots, L_{2^n-1}\}$.
- **Phase 2**: Exploration. For each projected event log L_k, we use Explorative Process Discovery as in Definition 3 to discover a Petri net model $disc(L_k) = N_k$, there will be $2^n - 1$ log-model pairs $(L_1, N_1), (L_2, N_2), \cdots, (L_{2^n-1}, N_{2^n-1})$ where $(L_k, N_k) \in \mathcal{U}_L \times \mathcal{U}_N$. Then we use the quality measures described in Definition 4 to compute fitness, precision, and harmonic mean for each pair (L_k, N_k): $\mathcal{F}_{it}(L_k, N_k)$, $\mathcal{P}_{re}(L_k, N_k)$, and $\mathcal{F}_1(L_k, N_k)$.
- **Phase 3**: Visualization. Finally, we visualize the relationship between log-model pairs in a dendrogram and color it using the quality measures to explore the impact of removing a specific activity on the process model.

4   Implementation

The approach has been implemented as a plugin in the ProM framework, named "Explorative Process Discovery using Activity Projections" in the package "ExplorativeProcessDiscovery". To install EPD-Tool, simply download the latest ProM Nightly build from https://promtools.org/, run the PackageManager, select the package "ExplorativeProcessDiscovery", and run ProM. Now the user can import any event log data and apply the plugin "Explorative Process Discovery using Activity Projections". Our tool includes two versions of the function, "Full version" and "Lite version". The main difference between them is that the Full version handles all projected event logs simultaneously, and the Lite version requires users to configure step-by-step to explore

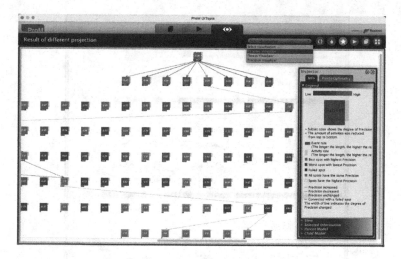

Fig. 3. The main interface of the Full version. Each node corresponds to a log-model pair. Only the connections that improve quality are shown.

the changes in model quality after removing any activity. Moreover, the Full version includes a Pareto optimal model of all the results, while the Lite version does not. We need these two versions because the Full version is time-consuming when processing large [13] event logs, while the Lite version improves efficiency by discarding some insignificant nodes.

4.1 Full Version Discovering Pareto Optimal Models

As Fig. 3 shows, we use a tree structure to show every projected event log simultaneously, and users can choose to color the dendrogram by fitness, precision, or $\mathcal{F}_1$-score through the drop-down list on the upper right. The best route for removal is always shown in this dendrogram. When a node is selected, it will display the connections to its child nodes. For more details, there is a floating "Inspect" window which also includes a "View" panel used to control the zoom function since the generated dendrograms are usually quite large.

- **Square box**: Each node corresponds to each projected event log.
  - **Color**: colors from blue to red; deep blue indicates low quality (fitness, precision, or harmonic mean), while the more red the color is, the better the quality is.
  - **Bottom grey rectangle**: the proportion of events left after projection; the longer the length is, the higher the proportion is.
  - **Right yellow rectangle**: the proportion of activities left after projection, the longer the length, the higher the proportion.
  - **Red box**: best node(s) with the highest quality.

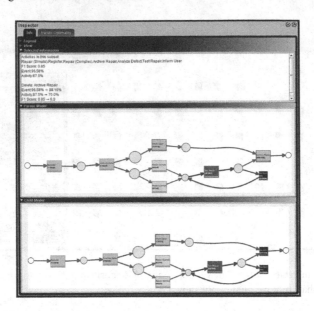

Fig. 4. Detailed information is provided when selecting an edge connecting two models.

- **Blue box**: worst node(s) with the lowest quality.
- **Black box (optional)**: there may exist some log-model pairs that cannot proceed with alignments under limited resources (both RAM and time).
- **Green box (optional)**: all nodes will be colored green if they all have the same best quality.
- **Green border**: the best node(s) will be marked with a green border.
- **Line**: connect related nodes, the width indicates the degree of quality improvement/degradation after deleting the corresponding activity from the upper node to the lower node.
  - **Red line**: quality increased.
  - **Blue line**: quality decreased.
  - **Green line**: quality unchanged.
  - **Black line (optional)**: connected with a "Black box".

By selecting an edge between each pair of nodes, users can collect more information about this node pair, such as which activities are included in this subset, which element was deleted from the parent node, and the performance change between this node pair. Also, users can have an overview of the Petri net models of this node pair, as Fig. 4 shows. We aim to find a "sweet spot" among all projected event logs with the highest fitness and precision. However, such a "sweet spot" is hard to obtain in some cases. Therefore, the concept of Pareto optimality is used to guide the user.

Pareto Optimality. When multiple evaluation indicators exist, an object that is best on all evaluation indicators does not always exist. The concept of Pareto optimality aims to

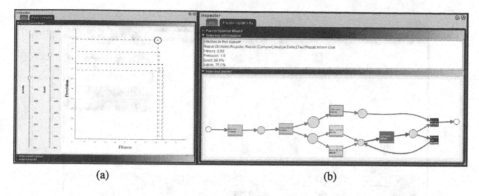

(a)   (b)

Fig. 5. (a) The window of "Pareto optimality". (b) Detailed information of selected "sweet spot".

achieve a trade-off between those indicators, i.e., none of these indicators can be better without making at least one worse.

For most event logs, there is no node with both the highest fitness and the highest precision (except for nodes with only one activity). However, having a process with only one activity makes little sense. Therefore, as Fig. 5(a) shows, in the "Pareto Optimal Model", the user can adjust the sliders to set the ratio of activities and events she wants to keep. After this, the tool will extract and show the "sweet spots" based on the concept of Pareto optimality, which contains the node with the highest fitness, the node with the highest precision, and a set of nodes with a trade-off between fitness and precision. Similarly, the specific information of each projected event log and the discovered Petri net model are visualized by selecting the corresponding node, as Fig. 5(b) shows.

4.2 Lite Version for Guided Exploration

As mentioned above, the Full version will be time-consuming when faced with large event logs, so we also provide a "Lite version" to improve the efficiency of the tool in some cases. As shown in Fig. 6, the dendrogram has only one layer of sub-nodes in this version. The user needs to configure to explore further sub-nodes step by step. Therefore, further operations are introduced in the inspector of the Lite version, such as "Go back", "Go deeper", and "Forward".

- **Go back**: When the interface shows a deep layer, users can select "Go back" to return to the previous layer.
- **Go deeper**: Select "Go deeper" after choosing any child node to explore the deep layer of this child node. Note that it may take a while to display the results after selecting. Because the Lite version calculates each layer of nodes separately by selecting "Go deeper".
- **Forward**: After selecting "Go back", users have a chance to return to the deep layer without recalculation by selecting "Forward". This improves efficiency in some cases because returning to the deep layer through "Go deeper" requires recalculation.

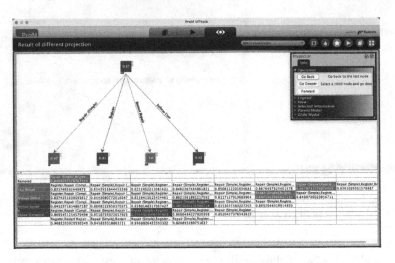

Fig. 6. The main interface of the Lite version.

All other functions in the Lite version are the same as the Full version, including viewing the detailed information of sub-nodes and checking the Petri net model. The only difference is that the Lite version can record the best removals in a table, and the exploration of Pareto optimality models is not supported.

5   Evaluation

In this section, we conduct experiments using two data sets "Repair[1]" and "Road Traffic Fine Management Process" [14] to evaluate our approach. It includes two parts: (1) evaluations of the general functions of our tool and (2) an explanation of why we need a "Lite version" by comparing the time required by the two versions.

5.1   General Functions

First, for a general function of our tool, we can define an activity route or the priority of activities. As shown in Fig. 7 and Fig. 8, we use $\mathcal{F}_1$-score as the evaluation criteria. Here we call the best-performing sub-log the "sweet spot". Each row from top to bottom of these tables records the performance change after removing an activity from the previous sweet spot. Colored cells indicate the sweet spot for the corresponding layer, red means performance increased, and blue means performance decreased. Column "Removed" indicates the activities removed from the previous sweet spot to get the current sweet spot. As a result, the priority of activities in "Repair" is {(Register, Inform User), Restart Repair, Repair (Complex), Repair (Simple), Archive Repair, Analyze Defect, Test Repair}, which means the most frequent (stable) activity in "Repair" is "Register" or "Inform User", and if organizations want to reduce the process or detect

---
[1] https://processmining.org/old-version/files/repairexample.zip.

	Repair (Simp...							
Removed	0.85521833...							
1	Register,Rep...	Repair (Simp...	Repair (Simp...	Repair (Simp...	Repair (Simp...	Repair (Simp...	Repair (Simp...	Repair (Simp...
Test Repair	0.83309870...	0.83466106...	0.82036645...	0.84690495...	0.85340533...	0.86429887...	0.86992081...	0.83610263...
2	Register,Rep...	Repair (Simp...	Repair (Simp...	Repair (Simp...	Repair (Simp...	Repair (Simp...	Repair (Simp...	
Analyze Defect	0.83372240...	0.84230576...	0.81732247...	0.86302009...	0.81004649...	0.88835157...	0.84907260...	
3	Register,Rep...	Repair (Simp...	Repair (Simp...	Repair (Simp...	Repair (Simp...	Repair (Simp...		
Archive Repair	0.84426424...	0.86244819...	0.83553416...	0.88957321...	0.81421830...	0.86926443...		
4	Register,Rep...	Repair (Simp...	Repair (Simp...	Repair (Simp...	Repair (Simp...			
Repair (Complex)	0.87046520...	0.84555529...	0.88118836...	0.86424604...	0.85204273...			
5	Register,Rest...	Repair (Simp...	Repair (Simp...	Repair (Simp...				
Restart Repair	0.96821039...	0.84279947...	0.99688264...	0.77601211...				
6	Register,Info...	Repair (Simp...	Repair (Simp...					
Inform User	0.99954689...	0.99674267...	1.0					
	Register	Repair (Simple)						
	1.0	1.0						

Fig. 7. The best route to remove problematic activities for the data set "Repair".

	Payment,Insert Date...										
Removed	0.7595003188350...										
1	Insert Date Appeal t...	Payment,Receive Result Appea...	Paym...	Paym...	Paym...	Paym...	Paym...	Paym...	Paym...	Paym...	Paym...
Payment	0.8385803343996...	0.6579507118300675	0.72...	0.65...	0.63...	0.77...	0.827...	0.72...	0.79...	0.66...	0.75...
2	Receive Result App...	Insert Date Appeal to Prefectur...	Insert...	Insert...	Insert...	Insert...	Insert ...	Insert...	Insert...	Insert...	
Add penalty	0.7063261808173...	0.7981989401069616	0.63...	0.69...	0.84...	0.87...	0.800...	0.86...	0.76...	0.83...	
3	Receive Result App...	Insert Date Appeal to Prefectur...	Insert...	Insert...	Insert...	Insert...	Insert ...	Insert...	Insert...		
Send for Credit Collection	0.7922438134193...	0.9099752046567058	0.86...	0.77...	0.89...	0.81...	0.932...	0.81...	0.84...		
4	Receive Result App...	Insert Date Appeal to Prefectur...	Insert...	Insert...	Insert...	Insert...	Insert ...	Insert...			
Send Appeal to Prefecture	0.8468958651884...	0.962250806674378	0.97...	0.84...	0.96...	0.84...	0.885...	0.86...			
5	Receive Result App...	Insert Date Appeal to Prefectur...	Insert...	Insert...	Insert...	Insert...	Insert ...				
Appeal to Judge	0.8630499000990...	0.9965471063810547	0.92...	0.99...	0.89...	0.94...	0.939...				
6	Receive Result App...	Insert Date Appeal to Prefectur...	Insert...	Insert...	Insert...	Insert...					
Create Fine	0.7462304133026...	0.8041714947856314	0.88...	0.92...	0.99...	0.99...					
7	Receive Result App...	Insert Date Appeal to Prefectur...	Insert...	Insert...	Insert...						
Send Fine	0.8647719405962...	0.8190476190476191	0.83...	0.83...	0.92...						
8	Receive Result App...	Insert Date Appeal to Prefectur...	Insert...	Insert...							
Insert Date Appeal to Prefecture	0.9997808212315...	0.9087162638412042	0.89...	0.89...							
9	Notify Result Appea...	Receive Result Appeal from Pr...	Recei...								
Receive Result Appeal from Prefecture	1.0	0.9999968695510622	0.98...								
	Insert Fine Notification	Notify Result Appeal to Offender									
	1.0	1.0									

Fig. 8. The best remove route for the data set "Road Traffic Fine Management Process".

problems of this process, they should start from "Test Repair". More specifically, the infrequent (unstable) activities might be {Test Repair, Analyze Defect} because the performance increased after removing these activities, and the performance will decrease if we keep removing any other activities.

Similarly, the priority of activities in "Road Traffic Fine Management Process" is {(Notify Result Appeal to Offender, Insert Fine Notification), Receive Result Appeal from Prefecture, Insert Date Appeal to Prefecture, Send Fine, Create Fine, Appeal to Judge, Send Appeal to Prefecture, Send for Credit Collection, Add penalty, Payment}, that means the most frequent (stable) activity in "Road Traffic Fine Management Process" is "Notify Result Appeal to Offender" or "Insert Fine Notification". If organizations want to reduce the process or detect problems in this process, they should start with "Payment", and the infrequent (unstable) activities might be {Payment, Add penalty, Send for Credit Collection, Send Appeal to Prefecture, Appeal to Judge}.

Moreover, to describe the function of "Pareto optimality" more intuitively, we use the data set "Repair" as a demonstration. As shown in Fig. 5(a), in Pareto optimality, we set the activity and event thresholds to 60% and 80%, respectively, and extracted 4 "sweet spots". Consider the "sweet spot" marked with a red circle, as detailed in Fig. 5(b). After removing the activity Archive Repair, the model's $\mathcal{F}_1$-score changes from 0.85 to 0.9. This result indicates that "Archive Repair" might be an infrequent

Table 1. Comparison of time cost between two versions.

	Data size		Time cost	
	Events	Activities	Full version	Lite version
Repair	11,855	8	38.9s	7.2s
RTFMP	561,470	11	**11,724.9s**	161.0s

(unstable) activity in this process. Therefore, organizations can optimize the whole process by focusing on this activity.

5.2 Scalability

To explain the necessity of the "Lite version" more clearly, we compare the time required by the full and Lite versions to get the same result of priority and classification. As Table 1 shows, for the Repair log, the processing time in the Lite version is around one-tenth that in the Full version, and for RTFMP (Road Traffic Fine Management Process), it decreased from more than 3 h to less than 3 min. (Experiment conducted using an 11th Gen Intel Core i7-1165G7 2.8GHz processor and 16GB RAM.) Although the Lite version does not display all the results at once, it saves much time to observe the impact of removing a specific activity. Additionally, users can still explore the priority of activities step by step instead of just waiting.

6 Conclusion

This paper introduced a new tool for process discovery named the Explorative Process Discovery tool using Activity Projections (EPD-Tool), which is implemented as a ProM plugin. With this plugin, users can explore the impact of removing any activity from the event log on the model. EPD-Tool provides users with insight to identify redundant or critical activities. Therefore, they can optimize processes and reduce process costs based on their expert knowledge. In the future, we plan to refine this tool to improve efficiency and integrate more features to provide users with a deeper insight into the event log incorporating performance related to time and resources (to find problematic resources and time consuming activities). Moreover, there are ways to further improve the scalability of the tool.

Acknowledgments. The authors thank the Alexander von Humboldt (AvH) Stiftung for supporting this research. Funded by the Deutsche Forschungsgemeinschaft (DFG) under Germany's Excellence Strategy, Internet of Production (390621612).

References

1. Augusto, A., Conforti, R., Dumas, M., et al.: Automated discovery of process models from event logs: review and benchmark. IEEE Trans. Knowl. Data Eng. **31**(4), 686–705 (2018)
2. De Weerdt, J., Vanden Broucke, S., Vanthienen, J., et al.: Active trace clustering for improved process discovery. IEEE Trans. Knowl. Data Eng. **25**(12), 2708–2720 (2013)
3. Goedertier, S., Martens, D., Vanthienen, J., et al.: Robust process discovery with artificial negative events. J. Mach. Learn. Res. **10**, 1305–1340 (2009)
4. Vanden Broucke, S., De Weerdt, J.: Fodina: a robust and flexible heuristic process discovery technique. Decis. Support Syst. **100**, 109–118 (2017)
5. Ghose, A., Koliadis, G., Chueng, A.: Process discovery from model and text artifacts. IEEE Congress on Services (Services 2007), 167–174. IEEE (2007)
6. Slaats, T.: Declarative and hybrid process discovery: recent advances and open challenges. J. Data Semant. **9**(1), 3–20 (2020)
7. van Dongen, B.F., de Medeiros, A.K.A., Verbeek, H.M.W., Weijters, A.J.M.M., van der Aalst, W.M.P.: The ProM framework: a new era in process mining tool support. In: Ciardo, G., Darondeau, P. (eds.) ICATPN 2005. LNCS, vol. 3536, pp. 444–454. Springer, Heidelberg (2005). https://doi.org/10.1007/11494744_25
8. Chinosi, M., Trombetta, A.: BPMN: an introduction to the standard. Comput. Stand. Interfaces **34**(1), 124–134 (2012)
9. Murata, T.: Petri nets: properties, analysis and applications. Proc. IEEE **77**(4), 541–580 (1989)
10. Leemans, S.J.J., Fahland, D., van der Aalst, W.M.P.: Discovering block-structured process models from incomplete event logs. In: Ciardo, G., Kindler, E. (eds.) PETRI NETS 2014. LNCS, vol. 8489, pp. 91–110. Springer, Cham (2014). https://doi.org/10.1007/978-3-319-07734-5_6
11. Van der Aalst, W.M.P., Adriansyah, A., Van Dongen, B.F.: Replaying history on process models for conformance checking and performance analysis. Wiley Interdisc. Rev.: Data Min. Knowl. Discov. **2**(2), 182–192 (2012)
12. Adriansyah, A., Munoz-Gama, J., Carmona, J., van Dongen, B.F., van der Aalst, W.M.P.: Alignment based precision checking. In: La Rosa, M., Soffer, P. (eds.) BPM 2012. LNBIP, vol. 132, pp. 137–149. Springer, Heidelberg (2013). https://doi.org/10.1007/978-3-642-36285-9_15
13. Leemans, S.J.J., Fahland, D., Van der Aalst, W.M.P.: Scalable process discovery and conformance checking. Softw. Syst. Mod. **17**(2), 599–631 (2018)
14. de Leoni, M. (Massimiliano); Mannhardt, Felix (2015): Road Traffic Fine Management Process. 4TU.ResearchData Dataset. https://doi.org/10.4121/uuid:270fd440-1057-4fb9-89a9-b699b47990f5
15. Van der Aalst, W.M.P.: Process Mining: Data Science in Action. Springer-Verlag, Berlin (2016). https://doi.org/10.1007/978-3-662-49851-4
16. Van der Aalst, W.M.P., Carmona, J. (eds.): Lecture Notes in Business Information Processing, vol. 448. Springer-Verlag, Berlin (2022)

Verification

Computing Under-approximations of Multivalued Decision Diagrams

Seyedehzahra Hosseini$^{(\boxtimes)}$ and Gianfranco Ciardo

Department of Computer Science, Iowa State University, Ames, IA, USA
{hosseini,ciardo}@iastate.edu

Abstract. Efficient manipulation of binary or multi-valued decision diagrams (BDDs or MDDs) is critical in symbolic verification tools. Despite the applicability of MDDs to real-world tasks such as discovering the reachable states of a model, their large demands on hardware resources, especially memory, limit algorithmic scalability. In this paper, we focus on memory-constrained algorithms that employ a novel $\mathcal{O}(m \log n)$-time under-approximation technique for MDDs, where m and n are the number of MDD edges and nodes, respectively. The effectiveness of our approach is demonstrated experimentally by a reduction in peak memory usage for the symbolic reachability computation of a set of Petri nets.

Keywords: decision diagrams · under-approximation · memory constraints

1 Introduction

Multi-valued decision diagrams (MDDs) are a compact symbolic representation of discrete functions over finite domains, such as those used by verification algorithms to validate a system's intended properties, where we need to manipulate large propositional formulae. To this end, reachability analysis is often the first step in the study of a discrete-state system. The most basic MDD-based method to discover the reachable state space is symbolic breadth-first search (BFS), which finds new reachable states by applying the next-state function to the set of currently-known reachable states, until it finds a fixpoint (until it cannot find any new state). Such approach is highly effective, but it cannot complete reachability analysis for many finite but large systems [8,9]. In other words, "exact" formal methods approaches *always* provide a proof or counterexample for a system property by exhaustively searching through all potential behaviors *given enough resources*, but "under-approximation" approaches *may* deliver the same result *within given resource limitations*.

Ravi et al. proposed three approximations, based on the *density* of each node p (number of minterms for node p divided by number of nodes reachable from p, see Sect. 2) in a fully-reduced binary decision diagram (BDD). The first approach [8] computes each node's density, finds a replacement node (top-down, one of its children, grandchildren, or terminal **0**, in that order), and applies this

---

Work supported in part by National Science Foundation under grant CCF-2212142.

L. Gomes and R. Lorenz (Eds.): PETRI NETS 2023, LNCS 13929, pp. 243–263, 2023.
https://doi.org/10.1007/978-3-031-33620-1_14

replacement based on its impact (number of minterms that would be removed and a lower bound on number of nodes that would be eliminated). Its runtime is quadratic in the BDD size. The second approach [9] (heavy branch subsetting) considers the number of minterms in the node's children, and deletes the child with the fewest minterms until the BDD size drops below a given threshold. Its runtime is linear in the BDD size, but it might create a string of nodes at the top of the BDD, each with one child set to **0**. The third approach [9] (shortest-path) favors short paths, since they encode more minterms, by assigning a path-length to each node v (sum of the length of the shortest paths from root to v and from v to terminal **1**), and deleting nodes with largest path-length. This performs best when the BDD has many paths of various lengths. Up to now, this has been the state of the art on BDD under-approximation.

An important application where this under-approximation can be effectively used to answer questions about the original set is *partial model checking*. For example, suppose we are generating the state space of a system to find out whether it can experience a deadlock. If, at some point, we have generated a (partial) set of reachable states $\mathcal{X}$ encoded in an MDD, but we are running out of memory, we can eliminate some states in $\mathcal{X}$, resulting in a set $\mathcal{X}' \subset \mathcal{X}$, hopefully with much smaller memory requirements (many fewer MDD nodes). Then, we can restart the state space exploration from $\mathcal{X}'$ and, if at any point we find a deadlock state s, we know that the system would be able to reach this deadlock. Furthermore, if each under-approximation $\mathcal{X}' \subset \mathcal{X}$ is chosen with care (e.g., ensuring that the initial state of the system is retained), then the entire reachable state space can still be reached from $\mathcal{X}'$, given enough iterations. If instead we do not find a deadlock, we cannot conclude that the original system is deadlock free, unless we are sure that, upon termination, the set $\mathcal{X}$ of encoded states is not a strict under-approximation, i.e., it is actually the entire state space. This may happen because different reachability algorithms may build the same final set going through different sequences of MDDs, some much more compact than others. This is the reason for the success of the *saturation* algorithm [4].

The rest of this paper is organized as follows. Section 2 gives background on MDD under-approximation and decomposition. Section 3 introduces our MDD under-approximation approach, Sect. 4 shows how to improve its speed, and Sect. 5 uses it for state-space generation. Section 6 reports experimental results on a set of Petri net benchmarks. Finally, Sect. 7 concludes and discusses future work.

2   Preliminaries

An L-level *quasi-reduced multi-valued decision diagram* (MDD) is a directed acyclic edge-labeled multi-graph where:

- Level 0 can only contain the two *terminal* nodes **0** and **1**.
- Each *nonterminal* node p belongs to a level $p.lvl = k \in \{1, ..., L\}$ and has n_k outgoing edges labeled by distinct elements of $\mathcal{S}_k = \{0, ..., n_k - 1\}$, pointing

to nodes at level $k - 1$ or to $\mathbf{0}$, but not all its outgoing edges can point to $\mathbf{0}$. If the edge labeled by i_k points to a node q, we write $p[i_k] = q$.
- There are no *duplicates*: if $p.lvl = q.lvl = k$ and $p[i_k] = q[i_k]$ for all $i_k \in \mathcal{S}_k$, then $p = q$.

(we recall that an alternative canonical version of MDDs, *fully-reduced* MDDs, forbids both duplicate and *redundant* nodes, i.e., any nonterminal node p such that all its outgoing edges point to the same node, $p[0] = p[1] = \cdots = p[n_k - 1]$).

MDDs encode functions of the form $\mathcal{S}_{L:1} = \mathcal{S}_L \times \cdots \times \mathcal{S}_1 \to \mathbb{B}$. Specifically, node p at level k encodes $f_p : \mathcal{S}_{k:1} = \mathcal{S}_k \times \ldots \times \mathcal{S}_1 \to \mathbb{B}$, defined recursively by

$$f_p(i_k, ..., i_1) = \begin{cases} p & \text{if } p.lvl = 0 \\ f_{p[i_k]}(i_{k-1}, ..., i_1) & \text{if } p.lvl > 0. \end{cases}$$

As defined, an MDD can have multiple *roots* at level L, each encoding a "function of interest" (except for the constant function 0, which is encoded by $\mathbf{0}$), but we focus on MDDs encoding a single function, with one root node $r^\star$ at level L, with the understanding that the MDD has no other roots unless stated otherwise, i.e., "MDD $r^\star$" means "the MDD rooted at $r^\star$".

Given node p at level $k > 0$, we recursively define the node reached from p through sequence $\alpha = (i_k, i_{k-1}, ..., i_{h+1}) \in \mathcal{S}_{k:h+1}$, for $L \geq k \geq h \geq 0$, as:

$$p[\alpha] = \begin{cases} p & \text{if } \alpha \text{ is the empty sequence} \\ \mathbf{0} & \text{if } \alpha = (i_k, \beta) \text{ and } p[i_k] = \mathbf{0} \\ q[\beta] & \text{if } \alpha = (i_k, \beta) \text{ and } p[i_k] = q \neq \mathbf{0}. \end{cases}$$

We can use an MDD $r^\star$ to encode a set of *states*. Let the substates reaching node p, or "above" p, and those encoded by p, or "below" p, be respectively $\mathcal{A}(p) = \{\alpha \in \mathcal{S}_{L:k+1} : r^\star[\alpha] = p\}$ and $\mathcal{B}(p) = \{\beta \in \mathcal{S}_{k:1} : p[\beta] = 1\}$. Thus, $\mathcal{A}(p)$ is the set of paths from $r^\star$ to node p and $\mathcal{B}(p)$ the set of paths from p to terminal $\mathbf{1}$. Then, the set of states "traversing" p is $\mathcal{S}(p) = \mathcal{A}(p) \times \mathcal{B}(p)$. As a special case, the set of states encoded by the MDD $r^\star$ is $\mathcal{S}(r^\star) = \mathcal{B}(r^\star) = \mathcal{A}(1) = \mathcal{S}(1)$.

Finally, let $\mathcal{N}(p)$ be the set of nonterminal nodes reachable from p at level k, $\mathcal{N}(p) = \{q : k \geq q.lvl > 0 \land \exists \alpha \in \mathcal{S}_{k:q.lvl+1}, p[\alpha] = q\}$. As a special case, $\mathcal{N}(r^\star)$ is the entire set of nonterminal nodes in the MDD.

Letting $\mathcal{M}(k)$ be the set of MDD nodes at level $k \in \{1, ..., L\}$, we can partition the states encoded by MDD $r^\star$ according to which level-k node they traverse: for any $k \in \{1, ..., L\}$, we have $\mathcal{S}(r^\star) = \bigcup_{p \in \mathcal{M}(k)} \mathcal{S}(p) = \bigcup_{p \in \mathcal{M}(k)} \mathcal{A}(p) \times \mathcal{B}(p)$.

We assume that the nonterminal nodes of the MDD, $\mathcal{N}(r^\star) = \bigcup_{k=1}^{L} \mathcal{M}(k)$, are stored in a *unique table* organized by level. This allows us to access the nodes at specific level efficiently and avoid node duplication.

an MDD rooted at node $r^\star$ can encode a large, even enormous, set of states, but its memory efficiency, measured as the number of states it encodes divided by the number of nodes it uses to encode them, $|\mathcal{S}(r^\star)|/|\mathcal{N}(r^\star)|$, is highly dependent on the specific set being encoded. For example, the *full* set $\mathcal{S}_{L:1}$ requires only a chain of L nonterminal nodes: the node at level k has all its n_k outgoing edges pointing to the node at level $k - 1$, or terminal $\mathbf{1}$ if $k = 1$; this is the same number of nodes required to encode a single state $(i_L, ..., i_1)$: in this MDD, all the outgoing edges of the node at level k point to terminal $\mathbf{0}$, except for the edge labeled with i_k, which points to the node at level $k - 1$, or terminal $\mathbf{1}$ if $k = 1$. Furthermore, it is well-known that the size of the MDD encoding a given set can be highly dependent on the chosen *variable order* [2], that finding the optimal variable order is NP-hard [1], and that some particularly "difficult" subsets of $\mathcal{S}_{L:1}$ require an exponential number of nodes for *any* variable order [2].

One approach explored by researchers to reduce memory consumption (measured in number of nodes) is to under-approximate a set by encoding most of its elements (states), but with substantially fewer nodes. More precisely, we formulate a *threshold* version of the *under-approximation problem* as:

Given MDD $r^\star$ and threshold $T \in \mathbb{N}$, find MDD $s^\star$ such that $|\mathcal{S}(s^\star)|$ is maximum among all MDDs $t^\star$ satisfying $|\mathcal{N}(t^\star)| \leq T$ and $\mathcal{S}(t^\star) \subseteq \mathcal{S}(r^\star)$.

3   Our under-approximation Algorithm

For any nonterminal node p, let its *unique-below-set* be the set of nonterminal nodes that can be reached from the root $r^\star$ only by first traversing p:

$$\mathcal{U}_b(p) = \{q \in \mathcal{N}(p) : \forall \alpha, r^\star[\alpha] = q \Rightarrow \exists \alpha', \alpha'', \alpha = \alpha' \cdot \alpha'' \wedge r^\star[\alpha'] = p\} \subseteq \mathcal{N}(p).$$

$\mathcal{U}_b(p)$ always includes p and has the property that, if we remove p from the MDD (by redirecting to $\mathbf{0}$ any edge pointing to p), the remaining nodes in $\mathcal{U}_b(p)$ become unreachable from $r^\star$, thus they, too, must be removed from the MDD.

Analogously, let the *unique-above-set* of p be the set of nonterminal nodes, at levels strictly above p, that can reach $\mathbf{1}$ only by traversing p:

$$\mathcal{U}_a(p) = \{q \in \mathcal{N}(r^\star) : \forall \alpha, q[\alpha] = \mathbf{1} \Rightarrow \exists \alpha', \alpha'', \alpha = \alpha' \cdot \alpha'' \wedge \alpha' \neq \epsilon \wedge q[\alpha'] = p\}.$$

Again, if we remove p from the MDD, all nodes in $\mathcal{U}_a(p)$ must be removed from the MDD as well, as they encode the empty set.

Intuitively, the key idea in our under-approximation is to select a node $p^\star$ and remove the nodes in $\mathcal{U}(p^\star) = \mathcal{U}_b(p^\star) \cup \mathcal{U}_a(p^\star)$ from the MDD (by redirecting to $\mathbf{0}$ any edge pointing to them). Then, the resulting MDD $s^\star$ satisfies $\mathcal{S}(s^\star) \subset \mathcal{S}(r^\star)$ and $|\mathcal{N}(s^\star)| < |\mathcal{N}(r^\star)|$, since $|\mathcal{N}(r^\star)| \geq |\mathcal{N}(s^\star)| + |\mathcal{U}(p^\star)|$. After this step, we test whether $|\mathcal{N}(s^\star)| \leq T$, and continue removing nodes in this manner if this is not yet the case. It is then essential to devise a good and efficient strategy to pick node $p^\star$ at each iteration. We do so by defining the *density* of node p as

$$Density(p) = |\mathcal{S}(p)| / |\mathcal{U}(p)|,$$

and letting $p^\star$ be a node with the smallest density (it may not be unique).

The approach must ensure that, by eliminating the selected node $p^\star$, the resulting MDD encodes a nonzero function, which would be an obvious but undesirable answer to any under-approximation problem. By checking that $p^\star$ is not

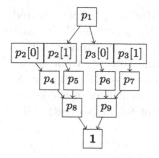

Node p	p_1	p_2	p_3	p_4	p_5	p_6	p_7	p_8	p_9
$|\mathcal{S}(p)|$	4	2	2	1	1	1	1	2	2
$|\mathcal{U}(p^\star)|$	9	4	4	1	1	1	1	4	4
$Density(p)$	4/9	1/2	1/2	1	1	1	1	1/2	1/2

Fig. 1. An MDD where the root is the node with the lowest density.

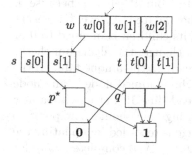

After removing node $p^\star$, edge $s[0]$ points to **0**, thus node s becomes equal to node t, and must be merged with it. This in turn makes node w's pointers $w[0] = w[2]$, and so on (i.e., node w could become equal to some other existing node at its level).

Fig. 2. A portion of an MDD where eliminating $p^\star$ produces a duplicate node.

the only node at its level, the algorithm ensures that the resulting MDD encodes a nonzero function. This is an issue because in some cases, the lowest density nodes in an MDD could those that are the only ones on their level (including $r^\star$), with density $|\mathcal{S}(r^\star)|/|\mathcal{N}(r^\star)$. Figure 1 shows such an MDD, together with the density of each node. The root node p_1 in this MDD is the (only) node with the lowest density. Our algorithm then avoids removing a node if it is the only one at its level (since doing so would remove *all* nodes and *all* states).

We wrote $|\mathcal{N}(r^\star)| \geq |\mathcal{N}(s^\star)| + |\mathcal{U}(p^\star)|$, not $|\mathcal{N}(r^\star)| = |\mathcal{N}(s^\star)| + |\mathcal{U}(p^\star)|$, because, after removing $\mathcal{U}(p^\star)$, some nodes with edges pointing to $p^\star$, once modified to point to **0** instead, might duplicate existing nodes, in which case they are merged with them (this in turn may make nodes pointing to them become duplicates as well, and so on). The MDD of Fig. 2 illustrates this situation.

Ideally, we would compute the number of nodes that become duplicates and are eliminated when removing each candidate node p, so that we could know beforehand the exact size of the resulting MDD if we removed $\mathcal{U}(p)$, but this is too computationally expensive (essentially, it amounts to performing the removal of p and observing its effect on the higher levels). Thus, we instead call a recursive algorithm to eliminate these duplicate nodes *after the fact*, so that $|\mathcal{U}(p^\star)|$ is just a lower bound on the number of nodes actually eliminated by removing $p^\star$, and our under-approximation algorithm is not guaranteed to be optimal.

The algorithm in Fig. 3 detects and removes the duplicate nodes created by the removal of $p^\star$, to ensure that the resulting MDD is canonical. It must be called as RMDUPLICATE($p^\star$), and it removes $p^\star$, $\mathcal{U}(p^\star)$, and any resulting duplicate node. Nodes in $\mathcal{U}_b(p^\star)$ are deleted by disconnecting $p^\star$, while nodes in $\mathcal{U}_a(p^\star)$ and the resulting duplicate nodes are eliminated by RMDUPLICATE. Algorithm RMDUPLICATE uses Map, a mapping of the identifiers of the nodes at level $k^\star = p^\star.lvl$, initialized to $Map(q.id) = q.id$ except for $Map(p^\star.id) = \mathbf{0}$:

$$\forall q \in \mathcal{N}(r^\star), q.lvl = p^\star.lvl, \quad Map(q.id) = \begin{cases} \mathbf{0} & \text{if } q = p^\star \\ q.id & \text{otherwise.} \end{cases}$$

RMDUPLICATE moves through the MDD levels, from $k^\star + 1$ to L because, if node p is mapped to $\mathbf{0}$, all of its ancestors should be updated. At level l, it checks each node q at that level and updates it if any of its children is mapped to $\mathbf{0}$ or any other node (line 12). If the children of node q change, the algorithm checks to see if the modified node q duplicates a node already in the unique table. If the unique table already contains a node q' equal to the modified node q, any node at level $l + 1$ pointing to q should point to q' instead (line 19); otherwise, a new node q^{new} should be created, and any node pointing to q should now point to node q^{new} (line 22). Either way, the required change is recorded by updating the entry for q in the Map for level $l + 1$. After RMDUPLICATE completes, Map for level $k^\star$ contains $p^\star.id$, but the unique table does not include $p^\star$. $Map(p.id)$ is $p.id$ for each nonterminal node $p.id$ if and only if all of its nonterminal children are in the unique table level $k - 1$; otherwise, node p is removed or modified because at least one of its children is removed or modified. Therefore,

$$Map(p.id) = \begin{cases} p.id & \text{if } \forall i \in \mathcal{S}_k : p[i] \in \mathcal{M}(k - 1) \\ \mathbf{0} \text{ or } \{q.id : q \in \mathcal{M}(k)\} & \text{otherwise.} \end{cases}$$

Eliminating duplicate nodes induced by removing $p^\star$ requires $O(\mathcal{N}(r^\star))$ time. RMDUPLICATE is a specialization of Bryant's reduction algorithm [2]: we achieve the same effect, but require a smaller cache (Map) because we proceed by level. The rest of this section describes how to compute node densities. See Table 1 for a summary of the acronyms used in our under-approximation, and their meaning.

3.1 Incoming-edge-count

Algorithm IEC in Fig. 4 counts the number of incoming edges to each nonterminal node. It is called as IEC($r^\star$) after setting the "incoming edge" counter $p.iec$ to 0, for each nonterminal node p. Theorem 1 addresses the correctness of algorithm IEC.

Theorem 1. The call IEC($r^\star$) sets $p.iec$, for any nonterminal node p, to the number of incoming edges to p. Its runtime is $\mathcal{O}(\mathcal{N}(r^\star))$.

Table 1. Acronyms used in our under-approximation and their meaning.

$\mathcal{U}_b(p)$	Unique Below node set of node p
$\mathcal{U}_a(p)$	Unique Above node set of node p
$\mathcal{U}(p)$	Unique node set of node p
$IEC(p)$	Incoming Edge Count of node p
$ASC(p)$	Above State Count of node p
$BSC(p)$	Below State Count of node p
$\mathcal{H}(p)$	Highest-unique-below-set of node p
$\mathcal{L}(p)$	Lowest-unique-above-set of node p

Proof. For a node p at level k, the for-loop at lines 2- 5 is executed $|\mathcal{S}_k|$ times; thus, each outgoing edge from p, if it is an incoming edge for a corresponding nonterminal child of p, is traversed and counted only once. IEC(p) calls itself only on its unvisited children (identified by having their incoming-edge-count equal to 0). Considering the sizes $|\mathcal{S}_k|$ as constants, the runtime is linear in the number of MDD nodes, $\mathcal{O}(\mathcal{N}(r^\star))$.   □

3.2 Above-state-count

Algorithm Asc in Fig. 5 computes the *above-state-counts*, i.e., the number $p.asc = |\mathcal{A}(p)|$ of substates from $r^\star$ to each nonterminal node $p \in \mathcal{N}(r^\star)$. It is called as Asc$(r^\star)$ after initializing $r^\star.asc = 1$, and $p.asc$ to 0 for all other nonterminal nodes p, and after having computed the incoming-edge-counts with the call IEC$(r^\star)$. To keep track of when all edges to a node p have been traversed (implying that counter $p.asc$ has the correct final value and the recursion can proceed downward), algorithm Asc decrements the incoming-edge-count $p.iec$ of node p every time p is reached, so that $p.iec$ will have value 0 after the call Asc$(r^\star)$ completes. Theorem 2 addresses the correctness of algorithm Asc.

Theorem 2. The call Asc$(r^\star)$ sets $p.asc$ to $|\mathcal{A}(p)|$ for any nonterminal node p. Its runtime is $\mathcal{O}(\mathcal{N}(r^\star))$.

Proof. To compute the above-state-count of node p, we need to have computed the correct value $q.iec$ for each node q with a path to p. The algorithm uses the fact that, in each recursive call Asc(p), $p.asc$ has the correct value of above-state-count for node p, and $p.iec = 0$.

Obviously, since IEC sets the incoming-edge-count of the root to 0, $r^\star.iec = 0$.

Similarly, the recursive call Asc$(p[i])$ on line 6 occurs only if $p[i].iec$ is 0, which means that $p[i].asc$ has been updated to take into account the (correct) $q.asc$ value of each parent q of $p[i]$. Then, in each recursive call Asc(p), $p.asc$ has the correct value of the above-state-count. Asc$(r^\star)$ visits each node once, therefore its runtime is $\mathcal{O}(\mathcal{N}(r^\star))$ □

```
 1: procedure RMDUPLICATE(p : node)
 2:    k ← p.lvl;
 3:    for all q ∈ M(k) do                          ▷ Initialization for Map
 4:        if q = p then Map[q] ← 0;
 5:        else Map[q.id] ← q.id;
 6:    for l = k to L − 1 do
 7:        CLEAR(NextMap)
 8:        for all q ∈ M(l + 1) do
 9:            NextMap[q.id] ← q.id;   ▷ Remove node q from unique table level l + 1
10:            NodeChanged ← false
11:            for i = 0 to |S_k| − 1 do
12:                if q[i] ≠ Map[q[i]] then                  ▷ q will change
13:                    q[i] ← Map[q[i]];
14:                    NodeChanged ← true;
15:            if NodeChanged then
16:                if q = 0 then NextMap[q.id] ← 0;
17:                q'.id ← FIND(q, M(l + 1)); ▷ Find node q in unique table level l + 1
18:                if q'.id ≠ NULL then          ▷ Changed node q is a duplicate of q'
19:                    NextMap[q.id] ← q'.id;
20:                else
21:                    q^{new} ← NEWNODE(q);  ▷ Build a node based on changed node q
22:                    NextMap[q.id] ← q^{new}.id;
23:        Map ← NextMap;
```

Fig. 3. Algorithm to remove duplicate nodes.

3.3   Below-state-count

One of the fundamental unary operations for MDDs is to compute the *cardinality* of the set encoded by a node p, i.e., the number of paths from p to **1**. We call this the *below-state-count* of node p. $\text{BSC}(r^\star)$ should be called after setting $p.bsc$ to 0 for all nonterminal nodes p. This algorithm to compute the cardinality is well known, so we include the pseudo-code in Fig. 6, but omit its proof.

3.4   Highest-unique-below-set

The *highest-unique-below-set* of node p is the subset of $\mathcal{U}_b(p) \setminus \{p\}$ containing all the nodes that are in the unique-below-set of p but not in the unique-below-set of a node $q' \neq p$ that is in the unique-below-set of p:

$$\mathcal{H}(p) = \{q \in \mathcal{U}_b(p) \setminus \{p\} : \forall q' \in \mathcal{U}_b(p) \setminus \{p, q\}, q \notin \mathcal{U}_b(q')\}.$$

We define $\mathcal{H}(p)$ because it turns out that every nonterminal node $q \neq r^\star$ belongs to $\mathcal{H}(p)$ for exactly one p, but possibly to $\mathcal{U}_b(p)$ for many nodes p, thus we can store all sets $\mathcal{H}(p)$ using memory linear in the number of MDD nodes, but, in general, (explicitly) storing all sets $\mathcal{U}_b(p)$ would require an overall quadratic memory in the number of MDD nodes. Furthermore, we could still obtain $\mathcal{U}_b(p)$

```
1: procedure IEC(p : node)                          ▷ Node p satisfies p.iec = 0
2:     for i = 0 to |S_{p.lvl}| − 1 do
3:         if p[i].lvl > 0 then
4:             if p[i].iec = 0 then IEC(p[i]);
5:             p[i].iec ← p[i].iec + 1;
```

Fig. 4. Algorithm to compute the incoming-edge-count of each node.

```
1: procedure ASC(p : node)            ▷ Assume r*.asc has been initialized to 1
2:     for i = 0 to |S_{p.lvl}| − 1 do
3:         if p[i].lvl > 0 then
4:             p[i].asc ← p[i].asc + p.asc;
5:             p[i].iec ← p[i].iec − 1;
6:             if p[i].iec=0 then ASC(p[i]);   ▷ Visit p[i] if this was the last edge to it
```

Fig. 5. Algorithm to compute the above-state-count of each node.

as the transitive closure of $\mathcal{H}(p)$, this is explained and proved in Theorem 4, but we do not need to, as we merely need to know its size $|\mathcal{U}_b(p)|$ to compute our under-approximation, not its actual elements.

Theorem 3. The set $\{\mathcal{H}(q) : q \in \mathcal{U}_b(p), \mathcal{H}(q) \neq \emptyset\}$ is a partition of $\mathcal{U}_b(p) \setminus \{p\}$.

Proof. To prove the proposition we must verify that $\mathcal{H}(p)$ satisfies the following:

1. $\bigcup_{q \in \mathcal{U}_b(p)} \mathcal{H}(q) = \mathcal{U}_b(p) \setminus \{p\}$. First, it is easy to show that, if $t \in \bigcup_{q \in \mathcal{U}_b(p)} \mathcal{H}(q)$ then $t \in \mathcal{U}_b(p) \setminus \{p\}$. If $t \in \mathcal{H}(q)$ for some $q \in \mathcal{U}_b(p)$, then, all paths from r^* to q pass through p, and all paths from r^* to t pass through q. Therefore, all paths from r^* to t pass through p, so $t \in \mathcal{U}_b(p) \setminus \{p\}$. To prove containment in the other direction, consider the lowest node $q \in \mathcal{U}_b(p)$ such that $t \in \mathcal{U}_b(q)$; there must be such a node since, at the very least, we could have $q = p$. But since q is the lowest node satisfying $t \in \mathcal{U}_b(q)$, then no other node q' between q and t can satisfy $t \in \mathcal{U}_b(q')$, thus $t \in \mathcal{H}(q)$, by definition.
2. For any given pair of nodes $q, q' \in \mathcal{U}_b(p)$, $\mathcal{H}(q)$ and $\mathcal{H}(q')$ are disjoint, i.e., $\mathcal{H}(q) \cap \mathcal{H}(q') = \emptyset$. By contradiction, assume that $\exists q, s, t \in \mathcal{U}_b(p), s \neq t$ and $q \in \mathcal{H}(s) \cap \mathcal{H}(t)$, i.e., $q \in \mathcal{H}(s)$ and $q \in \mathcal{H}(t)$, therefore $q \in (\mathcal{U}_b(s) \setminus \{s\}) \cap (\mathcal{U}_b(t) \setminus \{t\})$, thus $q \in \mathcal{U}_b(s)$ and $q \in \mathcal{U}_b(t)$. This means that any path from r^* to q must pass through both s and t and, since $s \neq t$, nodes s and t must be at different levels. Without loss of generality, assume that s is above t, then q cannot be in $\mathcal{H}(s)$, thus we have a contradiction. □

Theorem 4. Let the reflexive and transitive closure of $\mathcal{H}(p)$ for a given node p be defined as $\mathcal{H}^*(p) = \mathcal{H}(p) \cup \mathcal{H}(\mathcal{H}(p)) \cup \cdots$, where $\mathcal{H}(\{p_1, ..., p_c\}) = \bigcup_{d=1}^{c} \mathcal{H}(p_d)$. If $\mathcal{U}_b(p)$ contains nodes beyond p, then $\{\mathcal{H}^n(p) : n \in \mathbb{N}, \mathcal{H}^n(p) \neq \emptyset\}$ is a coarser partition than $\{\mathcal{H}(q) : q \in \mathcal{U}_b(p), \mathcal{H}(q) \neq \emptyset\}$. Thus, $\mathcal{H}^*(p) = \mathcal{U}_b(p) \setminus \{p\}$.

Proof. We need to prove that, for any $q \in \mathcal{U}_b(p)$, there is a minimum n such that $\mathcal{H}(q) \subseteq \mathcal{H}^n(p)$. Consider $t \in \mathcal{U}_b(p) \setminus \{p\}$. Since Theorem 3 states that

```
1: function BSC(p : node)
2:     if p = 1 ∨ p = 0 then return p;
3:     if p.bsc = 0 then
4:         for i = 0 to |S_{p.lvl}| − 1 do p.bsc ← p.bsc + BSC(p[i]);
5:     return p.bsc;
```

Fig. 6. Algorithm to compute the below-state-count of each node.

```
1: function UBC(p : node)                                    ▷ ∀p ∈ N(r*) : p.ubc ← 1
2:     if p.vst = 1 then return p.ubc;
3:     for all q ∈ H(p) do p.ubc ← p.ubc + UBC(q);           ▷ H(p) : Sec. 3.6.
4:     p.vst ← 1;
5:     return p.ubc;
```

Fig. 7. Algorithm to compute the unique-below-count.

$\{H(q) : q \in U_b(p), H(q) \neq \emptyset\}$ is a partition of $U_b(p) \setminus \{p\}$, there exists a q such that $t \in H(q)$. If $q = p$, then $t \in H^n(p)$ for $n = 1$. If $q \neq p$, then we know that $U_b(p)$ contains p, q, and t, and that there exists a node q_1 such that $q \in H(q_1)$. If $q_1 = p$, then $q \in H(p)$, $t \in H(q)$ which means that $t \in H^2(p)$; otherwise, we can repeat the reasoning and eventually, since $U_b(p)$ is a finite set, we must eventually find a $q_n = p$, implying that $t \in H^{n+1}(p)$.   □

Procedure UBC(r^*) computes the size $|U_b(p)|$ of the unique-below-set for any nonterminal node $p \in N(r^*)$, using the information in $H(p)$.

3.5 Lowest-unique-above-set

The *lowest-unique-above* set of node p is the subset of $U_a(p)$ containing all nodes that are in the unique-above-set of p but not in the unique-above-set of a node in the unique-above-set of p:

$$\mathcal{L}(p) = \{q \in U_a(p) : \forall q' \in U_a(p) \setminus \{q\}, q \notin U_a(q')\}.$$

As for $H(p)$, we define $\mathcal{L}(p)$ because it turns out that every node belongs to $\mathcal{L}(p)$ for exactly one p, but possibly to $U_a(p)$ for many nodes p, thus we can store all sets $\mathcal{L}(p)$ using memory linear in the number of MDD nodes, but, in general, we cannot (explicitly) store all sets $U_a(p)$ in linear memory. Furthermore, again, we could obtain $U_a(p)$ as the transitive closure of $\mathcal{L}(p)$, as stated in Theorem 6, but we do not need to, we only need to compute its size $|U_a(p)|$.

Theorem 5. The set $\{\mathcal{L}(q) : q \in U_a(p)\}$ is a partition of $U_a(p)$.

Proof. Similar to that of Theorem 3.   □

Theorem 6. Let the reflexive and transitive closure of $\mathcal{L}(p)$ for a given node p be defined as $\mathcal{L}^*(p) = \mathcal{L}(p) \cup \mathcal{L}(\mathcal{L}(p)) \cup \cdots$, where $\mathcal{L}(\{p_1, ..., p_c\}) = \bigcup_{d=1}^{c} \mathcal{L}(p_d)$. If $U_a(p)$ contains nodes beyond p, then $\{\mathcal{L}^n(p) : n \in \mathbb{N}\}$ is a coarser partition than $\{\mathcal{L}(q) : q \in U_a(p)\}$. Thus, $\mathcal{L}^*(p) = U_a(p)$.

1: **function** $\text{UAC}(p : node)$ $\triangleright \forall p \in \mathcal{N}(r^\star) : p.uac \leftarrow 0$
2:     **if** $p.vst = 1$ **then return** $p.uac$;
3:         **for all** $q \in \mathcal{L}(p)$ **do** $p.uac \leftarrow p.uac + 1 + \text{UAC}(q)$; $\triangleright \mathcal{L}(p)$: Sec. 3.6
4:         $p.vst \leftarrow 1$;
5:         **return** $p.uac$

Fig. 8. Algorithm to compute the unique-above-count

Proof. Similar to that of Theorem 4.   □

Procedure $\text{UAC}(r^\star)$ computes the size of the unique-above-set $|\mathcal{U}_a(p)|$, for any nonterminal node $p \in \mathcal{N}(r^\star)$, by recursively using the information in $\mathcal{L}(p)$.

3.6 Dominator and Post-dominator

A simplistic iterative algorithm to calculate $\mathcal{H}(p)$ and $\mathcal{L}(p)$ for all nodes p has quadratic complexity in the number of MDD nodes [5]; to reduce this complexity, we use a *dominator* algorithm. Given a flow graph with a single source and sink (in our case, $r^\star$ and **1**), a node v *dominates* another node w, if every path from the $r^\star$ to w must traverse v Every node $w \neq r^\star$ has at least one dominator. A node v is the *immediate dominator* of w, denoted by $idom(w) = v$, if v dominates w and every other dominator of w also dominates v. Every node $w \neq r^\star$ has a unique $idom(w)$. Importantly, q is in $\mathcal{U}_b(p)$ iff q is in $dom(p)$, and is the immediate dominator of p iff it is the only node in p's highest-unique-below-set $\mathcal{H}(p)$.

The dominator algorithm builds a *dominator tree* whose nodes $\mathcal{V}$ are the MDD nodes and whose edges $\{(idom(w), w) : w \in \mathcal{V} \setminus \{r^\star\}\}$ form a direct tree rooted at $r^\star$. It performs a depth-first search and assigns the visit time to each node, effectively defining a total order, where $v > w$ means that the visit time of node v is larger than that of node w. The dominator algorithm uses the visit time of the node instead of the original node label in the following steps. Next, for each node $w \neq r^\star$, it defines the "semidominator" $sdom(w) \in \mathbb{N}$ as:

$$sdom(w) = \min\{v : \exists \text{ path } v = v_0, v_1, ..., v_j = w \text{ s.t. } v_i > w \text{ for } 1 \leq i \leq j - 1\}.$$

The algorithm uses Theorem 7 to compute $sdom(w)$ for any $w \neq r^\star$:

Theorem 7. (from [6]) For any node $w \neq r^\star$:

$$sdom(w) = \min(\{v \mid (v, w) \text{ is an edge and } v < w\} \cup \{sdom(u) \mid u > w \text{ and}$$

$$\text{there is an edge } (v, w) \text{ such that } u \text{ is an ancestor of } v, u \xrightarrow{*} v\}).$$

Then the algorithm uses Corollary 1 to compute the immediate dominator of all nodes using the semidominator information.

1: **procedure** UNDERAPPROXONE($r^\star$: $node, T_{min}$: int, T_{max} : int)
2:     **if** $|\mathcal{N}(r^\star)| < T_{max}$ **then return**;
3:     **while** $|\mathcal{N}(r^\star)| > T_{min}$ **do**
4:         BSC($r^\star$);
5:         IEC($r^\star$);   ▷ Initialize $r^\star.iec$ to 0
6:         ASC($r^\star$);   ▷ Initialize $r^\star.asc$ to 1
7:         UAC($r^\star$); UBC($r^\star$);
8:         **for all** $p \in \mathcal{N}(r^\star)$ **do** $Density(p) \leftarrow (p.ac \cdot p.bc)/p.uac + p.ubc$;
9:         $p^\star \leftarrow$ pick one of the nodes in $\{p \mid \forall p' \in \mathcal{N}(r^\star), Density(p) \leq Density(p')\}$;
10:         RMDUPLICATE($p^\star$);

Fig. 9. Algorithm to under-approximate MDD $r^\star$ by selecting one node at a time.

Corollary 1. (from [6]) Let $w \neq r^\star$ and u be a node for which $sdom(u)$ is the minimum among nodes u satisfying $sdom(w) \xrightarrow{+} u \xrightarrow{+} w$, i.e., $sdom(w)$ is a proper ancestor of u and u is a proper ancestor of w, then

$$idom(w) = \begin{cases} sdom(w) & \text{if } sdom(w) = sdom(u) \\ idom(u) & \text{otherwise.} \end{cases}$$

A node w *post-dominates* another node v, if every path from v to $\mathbf{1}$ traverse w. A node v is the immediate post-dominator of w, if v post-dominates w and every other post-dominator of w also post-dominates v. Again, node q is in $\mathcal{U}_a(p)$ iff q is in $postdom(p)$ and node q is the post-dominator for node p iff node q is the only node in p's lowest-unique-above-set.

$$q \in postdom(p) \iff q \in \mathcal{U}_a(p) \qquad ipostdom(p) = q \iff \mathcal{L}(p) = \{q\}$$

The post-dominator algorithm applies the dominator algorithm to the reverse MDD (with source $\mathbf{1}$ and sink $r^\star$) to compute the lowest-unique-above-sets.

3.7   Under-approximation (one Node at a Time)

Given MDD $r^\star$, the call UNDERAPPROXONE($r^\star, T_{min}, T_{max}$) computes an under-approximation for $r^\star$ if the size of the MDD $r^\star$ is greater than T_{max}. The algorithm reduces the size of the MDD so that it does not exceed T_{min} (T_{min} must be at least the number of MDD levels). T_{max} and T_{min} introduce hysteresis to avoid calling the under-approximation too frequently. In practice, T_{max} should be as large as possible and T_{min} a fraction of T_{max} (in our experiments, it is $0.6 \cdot T_{max}$). The algorithm computes the below-state-count, above-state-count, unique-count, and density (lines 4, 6, 7, and 8) for each node at each iteration of the while-loop. Then, it selects a single node $p^\star$ with lowest density (line 9) and removes from the MDD the nodes in $\mathcal{U}(p^\star)$ and any resulting duplicate using RMDUPLICATE, until the number of MDD nodes is at most T_{min}.

As the algorithm recomputes the information after each deletion, the selected node $p^\star$ is the "quasi-optimal" choice: it is optimal based on density information, but it ignores the effect of removing duplicate nodes.

4   Speeding up the under-approximation

In large models, recomputing the above-state-count, below-state-count, incoming-edge-count, and unique-count after deleting each set of nodes $\mathcal{U}(p^\star)$ can be costly. UNDERAPPROXMANY selects instead *a set* of nodes $\mathcal{P}^\star$, and deletes all nodes in $\bigcup_{p \in \mathcal{P}^\star} \mathcal{U}(p)$ before recomputing all node densities.

While eliminating duplicate nodes caused by deleting just *one* set $\mathcal{U}(p^\star)$ is slightly simpler, identifying and removing *all* duplicate nodes created by removing the set of nodes $\bigcup_{p \in \mathcal{P}^\star} \mathcal{U}(p)$ has the same time complexity, thus its cost can be better amortized. The call RMDUPLICATESET($\mathcal{P}^\star$) in Fig. 10 finds and removes the duplicate nodes created by eliminating the nodes in $\bigcup_{p \in \mathcal{P}^\star} \mathcal{U}(p)$, to ensure MDD canonicity. $\mathcal{K} = \{p^\star.lvl : p^\star \in \mathcal{P}^\star\}$ stores the MDD levels of the selected nodes (line 2), and *Map* maps the identifiers of nodes at level $k^\star = \min\{\mathcal{K}\}$, initialized as

$$\forall q \in \mathcal{N}(r^\star), q.lvl = p^\star.lvl, \quad Map(q.id) = \begin{cases} q.id, & \text{if } q \notin \mathcal{P}^\star \\ \mathbf{0}, & \text{otherwise.} \end{cases}$$

RMDUPLICATESET traverses the MDD from level $\min(\mathcal{K}) + 1$ to L. Like RMDUPLICATE, it starts at level $k^\star + 1$; if q's child $q[i]$ is mapped to another node (line 15), node q must change (line 16), and the algorithm checks if the new node is a duplicate of a node q'. Any edge pointing to q from higher-level nodes must be changed.

UNDERAPPROXONE selects node $p^\star$ with lowest density, deletes $\mathcal{U}(p^\star)$, and recomputes the density ($\mathcal{U}$, $\mathcal{B}$, and $\mathcal{A}$), while UNDERAPPROXMANY selects a set of nodes $\mathcal{P}^\star$, one after another, but it does not update the density information after selecting each node. This reduces execution time since calculating density is a heavy duty operation, but uses increasingly stale, thus less precise, information. This is because not only the number of nodes eliminated or merged after calling RMDUPLICATESET is not taken into account, but also because, by selecting a sequence of nodes $(p_1, p_2, ..., p_k)$, the selection of any node except for p_1 uses an approximation of the correct values for $\mathcal{U}$, $\mathcal{A}$, and $\mathcal{B}$.

When the MDD size $\mathcal{N}(r^\star)$ exceeds T_{max}, a call to UNDERAPPROXMANY reduces the size to T_{min} or less. Selecting a set of nodes $\mathcal{P}^\star$ instead of just node $p^\star$ increases the chances of deleting more nodes than necessary. This is because, ideally, every time the algorithm selects node $p_i \in \mathcal{P}^\star$, $\mathcal{S}(p_i)$ should be disjoint from $\mathcal{S}(p_j)$, for any other $p_j \in \mathcal{P}^\star$, but this is not necessarily the case. Before adding p_i to $\mathcal{P}^\star$, the algorithm checks that $|\mathcal{S}(p_i)| + \sum_{p_j \in \mathcal{P}^\star} |\mathcal{S}(p_j)| < |\mathcal{S}(r^\star)|$, to ensure that removing the set of nodes $\mathcal{P}^\star$ does not produce an empty MDD. UNDERAPPROXMANY uses a (lower bound) on the percentage ψ of $|\mathcal{S}(r^\star)|$ that must be kept as a constraint: if $|\mathcal{S}(p_i)| + \sum_{p_j \in \mathcal{P}^\star} |\mathcal{S}(p_j)| > \psi \cdot |\mathcal{S}(r^\star)|$, the algorithm does not add p_i to $\mathcal{P}^\star$; instead, it removes just $\mathcal{U}(\mathcal{P}^\star)$ and any duplicate nodes created by removing $\mathcal{U}(\mathcal{P}^\star)$ from the MDD, then it recomputes the new densities of the nodes of the resulting MDD.

The exact call is UNDERAPPROXMANY($r^\star, T_{min}, T_{max}, \psi$), where T_{min} is the selected T_{min}, T_{max} is the maximum (triggering) threshold, and ψ is the

1: **procedure** RMDUPLICATESET($\mathcal{P}^*$)
2:    $\mathcal{K} \leftarrow \{l : p.lvl = l, p \in \mathcal{P}^*\};$ ▷ $\mathcal{K}$ is a set of levels
3:    **for** $p \in \mathcal{M}(\min(\mathcal{K}))$ **do**   ▷ Initialization loop
4:        **if** $p \in \mathcal{P}^*$ **then** $Map[p] \leftarrow 0;$
5:        **else** $Map[p] \leftarrow p.id;$
6:    **for** $l = \min(\mathcal{K}) + 1$ to L **do**
7:        $NextMap.$CLEAR;
8:        **if** $l \in \mathcal{K}$ **then** $\mathcal{K} \leftarrow \mathcal{K} \setminus \{l\};$
9:        **for all** $q \in \mathcal{M}(l+1)$ **do**
10:            $NodeChanged \leftarrow false;$
11:            **if** $q \notin \mathcal{P}^*$ **then**   ▷ If node q is not marked for deletion
12:                $NextMap[q.id] \leftarrow q.id;$
13:                $\mathcal{M}(l+1).$REMOVE(q); ▷ Remove node q from unique table level $l + 1$
14:                **for** $i = 0$ to $|\mathcal{S}_k| - 1$ **do**
15:                    **if** $q[i] \neq Map[q[i]]$ **then**   ▷ q should change
16:                        $q[i] \leftarrow Map[q[i]];$
17:                        $NodeChanged \leftarrow true;$
18:                **if** $NodeChanged$ **then**
19:                    $q'.id \leftarrow \mathcal{M}(l+1).$FIND($q$);   ▷ Find q from unique table level $l+1$
20:                    **if** $q'.id \neq NULL$ **then**   ▷ Node q is a duplicate of q'
21:                        $NextMap[q.id] \leftarrow q'.id;$
22:                    **else**
23:                        $q^{new} \leftarrow$ NEWNODE(q); ▷ Build q^{new} based on changed node q
24:                        $NextMap[q.id] \leftarrow q^{new}.id;$
25:            $NextMap[q] \leftarrow 0;$
26:        $Map \leftarrow NextMap;$

Fig. 10. Algorithm to remove a set of duplicate nodes.

maximum percentage of removed states (required to be strictly less than 100%). The greater ψ is, the less frequently the algorithm needs to recompute node densities. line 11 ensures that UNDERAPPROXMANY deletes at least one node at a time, even when ψ is near zero (in which case UNDERAPPROXMANY behaves like UNDERAPPROXONE).

5 Application

The first step in the study of a discrete system is often reachability analysis, i.e., the computation of its reachable states. Given an initial state set $\mathcal{S}_{init} \subseteq \mathcal{S}_{L:1}$ and a *next-state* function of the form $\mathcal{T} : \mathcal{S}_{L:1} \rightarrow 2^{\mathcal{S}_{L:1}}$, the reachability set $\mathcal{S}_{rch}$ is the smallest set $\mathcal{X}$ satisfying $\mathcal{X} = \mathcal{X} \cup \mathcal{T}(\mathcal{X}) \cup \mathcal{S}_{init}$. The breadth-first (BF) method is a common exploration approach for MDD-based reachability analysis, as it naturally implements this definition of $\mathcal{S}_{rch}$ as a fixpoint. It starts by initializing $\mathcal{S}_{rch}$ to $\mathcal{S}_{init}$, and repeatedly adds to it the states reachable from it in one application of $\mathcal{T}$, until no more new states are found. At the ith iteration, $\mathcal{S}_{rch}$ contains all states at distance up to i from $\mathcal{S}_{init}$. Thus, it builds $\mathcal{S}_{rch}$ as $\mathcal{S}_{init} \cup \mathcal{T}(\mathcal{S}_{init}) \cup \mathcal{T}^2(\mathcal{S}_{init}) \cup \cdots$.

```
1: procedure UNDERAPPROXMANY(r* : node, Tmin : int, Tmax : int, ψ : float)
2:     if |N(r*)| < Tmax then return ;
3:     while |N(r*)| < Tmin do
4:         BSC(r*);
5:         IEC(r*);                                          ▷ Initialize r*.iec to 0
6:         ASC(r*);                                          ▷ Initialize r*.asc to 1
7:         UAC(r*); UBC(r*);
8:         for all p ∈ N(r*) do Density(p) ← (p.ac × p.bc)/p.uc;
9:         RemovedNodes ← ∅;
10:        repeat                              ▷ Select a set of nodes with lowest density.
11:            select node p ∉ RemovedNodes with lowest Density;
12:            if ∑_{p∈P*} |S(p)| > ψ × |S(r*)| then break;
13:            P* ← P* ∪ {p};
14:            RemovedNodes ← RemovedNodes ∪ U(p);
15:        until |N(r*)| − |RemovedNodes| > Tmin
16:        RMDUPLICATESET(P*);
```

Fig. 11. Under-approximating MDD r^* by selecting many nodes at a time.

```
1: procedure BF(S_init, T_min, T_max)
2:     S_rch ← S_init;
3:     S_pre ← ∅;
4:     while S_pre ≠ S_rch do
5:         S_pre ← S_rch;
6:         S_rch ← S_rch ∪ T(S_pre);
7:     return S_rch;
```

Fig. 12. Algorithm to compute the reachable state space using breadth-first.

The chained BF (ChBF) approach [7] observes that, if $\mathcal{T}$ is partitioned as $\mathcal{T} = \bigcup_{\alpha \in \mathcal{E}} \mathcal{T}_\alpha$, where $\mathcal{E}$ is a set of (asynchronous) *events*, runtime and memory requirements may be reduced by using a different iteration: if $\mathcal{E} = \{\alpha, \beta, \gamma\}$, the generic i^{th} ChBF iteration updates $\mathcal{S}_{rch}$ using three sequential steps: (1) $\mathcal{S}_{rch} \leftarrow \mathcal{S}_{rch} \cup \mathcal{T}_\alpha(\mathcal{S}_{rch})$; (2) $\mathcal{S}_{rch} \leftarrow \mathcal{S}_{rch} \cup \mathcal{T}_\beta(\mathcal{S}_{rch})$; (3) $\mathcal{S}_{rch} \leftarrow \mathcal{S}_{rch} \cup \mathcal{T}_\gamma(\mathcal{S}_{rch})$. This has the effect of potentially accelerating convergence to the fixpoint, as the ith iteration discovers states reachable not just through one of the three single events, but also through one of the sequences of events $\alpha\beta$, $\alpha\gamma$, $\beta\gamma$, or $\alpha\beta\gamma$.

ChBF was proposed in conjunction to Petri net models [7], the formalism we use for our experiments. In this case the events are the Petri net transitions, which are by definition asynchronous, and $\mathcal{T}_\alpha(i) = \emptyset$ if Petri net transition α is not enabled in marking i, while $\mathcal{T}_\alpha(i) = \{j\}$ if transition α is enabled in marking i and its firing in i leads (deterministically) to marking j.

```
1: procedure CHBFUA(S_init, T_min, T_max, ψ)
2:     y ← S_init;
3:     S_rch ← S_init;
4:     while true do
5:         for α ∈ ℰ do
6:             y ← S_rch;
7:             S_rch ← S_rch ∪ T_α;
8:             if S_rch ≠ y then              ▷ Applying α added some states
9:                 UNDERAPPROX(S_rch, T_min, T_max, ψ);
10:                S_rch ← S_rch ∪ S_init;
11:                lastNonZeroIteration ← α;
12:            else if lastNonZeroIteration = α then  ▷ No event in ℰ added states
13:                return S_rch;
```

Fig. 13. Chained breadth-first algorithm with under-approximation.

Figure 13 shows our ChBF approach invoking "UNDERAPPROX", i.e., either UNDERAPPROXONE or UNDERAPPROXMANY, whenever the MDD size exceeds T_{max} (line 9). Either under-approximation may delete the initial state(s), which would then make it impossible to generate the entire state space. Thus, CHBFUA adds back the initial state(s) after each under-approximation (line 10). CHBF is exactly the same as CHBFUA, except it does not have lines 9, and 10.

As shown, algorithm CHBFUA might not halt because, after calling under-approximation, the set of states could be exactly the same as after the previous under-approximation: the algorithm is in a cycle where it adds and removes the same set of states. To recognize this situation, we should keep the old set of states, but this would require storing two MDDs; we use instead the old number of states and nodes as a proxy to (conservatively) detect this problem and let CHBFUA output a partial state space S_{part} instead of the full state space S_{rch}.

Assume that, before calling under-approximation, the number of MDD nodes is $n_{old} > T_{max}$ and the number of states is rs_{old}, and that, after calling under-approximation once and firing one or more events, the numbers are $n_{new} > T_{max}$ and rs_{new}. If $n_{new} = n_{old}$ and $rs_{new} = rs_{old}$, the algorithm applies one more event resulting in n'_{new} MDD nodes and rs'_{new} states. Then, three cases may arise:

1. If $n'_{new} > n_{new}$ and $rs'_{new2} > rs_{new}$, the MDD with n_{new} nodes is not a fixpoint; the algorithm conservatively decides that the MDD with n_{new} nodes is the same as that with n_{old} nodes, it refrains from calling under-approximation, and returns the partial state space S_{part} encoded by the MDD with n'_{new} nodes.
2. If $n'_{new} \leq n_{new}$ and $rs'_{new} > rs_{new}$, CHBFUA continues its normal execution.
3. If $n'_{new} = n_{new}$ and $rs'_{new} = rs_{new}$, the algorithm applies a new event and repeats the check for cases 1, 2, and 3, until either case 1 or 2 happens, or all events have been applied once without discovering new states (in which case it reached a fixpoint and returns S_{rch}, encoded by $n_{old} = n_{new}$ nodes).

6   Results

We designed a set of experiments and ran them on a Linux workstation with 16GB of RAM. We implemented CHBFUA, chained breadth-first reachability with under approximation, within the model checker SMART [3]. Our benchmark is a subset of the bounded models from the Model Checking Contest (MCC) 2021 (https://mcc.lip6.fr/2021/). Models are described as Petri nets, and most of them have one or more scaling parameters that affect their state space size. 799 models in the MCC benchmark are bounded, 499 of which generate the next-state function within 60 seconds, and 259 of which generate the entire state space using CHBF within one hour. Of these, we eliminated 72 models because they have the same peak and final number of nodes using CHBF (the under-approximation algorithm does not make sense for such models; admittedly this cannot be determined a priori). Thus, we considered the remaining 187 models. For our experiments, we selected $T_{min} = 10,000$ and $T_{max} = 15,000$, the percentage ψ of the minimum number of states to be kept was set to 0.5, and the maximum execution time for each run was set to 24 hours. The peak number of nodes for 123 of the 187 models is less than 15,000, therefore the under-approximation is not triggered on those models (thus CHBFUA behaves exactly like CHBF on them). Using UNDERAPROXMANY, 19 of the remaining 64 models generate the complete state space using under-approximation in less than 24 hours; in these models, whenever the number of node exceeds T_{max}, the algorithm selects a set of nodes $\mathcal{P}^*$ and deletes $\bigcup_{p \in \mathcal{P}^*} \mathcal{U}(p)$, until the number of nodes is less than T_{min}. The algorithm adds $\mathcal{S}_{init}$ back and finally generates the complete state space $\mathcal{S}_{rch}$. 15 models out of remaining 64 generate only a partial state space $\mathcal{S}_{part}$.

For the other 30 models out of the remaining 64 models, our algorithm is unable to generate either $\mathcal{S}_{part}$ or $\mathcal{S}_{rch}$ in 24 h. Given enough time, it would always terminate and generate the complete state space or a partial state space. For example, if we increase the running time from 24 to 48 hours, 4 of these 30 models can generate a partial state space. If the model is run indefinitely and the final number of nodes is greater than T_{max}, our approach would in principle eventually generate a partial state space because the number of increasing possible state space sequences is bounded.

6.1   Experimental Results

We compare CHBFUA with CHBF in terms of both memory and time. The more frequently the under-approximation calculates node densities, the slower our algorithm will be. Thus, UNDERAPPROXONE is slower than UNDERAPPROXMANY, and we report only the results for the latter.

Figure 14 compares the peak node and time ratios for CHBF and CHBFUA ($Peak_{\text{CHBFUA}}/Peak_{\text{CHBF}}$ and $Time_{\text{CHBFUA}}/Time_{\text{CHBF}}$ respectively) for models where CHBFUA generates $\mathcal{S}_{rch}$. For these models, the final number of nodes is less than T_{max}, otherwise the CHBFUA would not be able to generate the entire state space (whenever the number of nodes is greater than T_{max}, CHBFUA

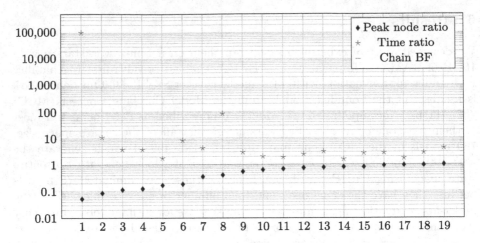

Fig. 14. Time and peak node ratios for the 19 models where CHBFUA generates the entire state space (sorted by increasing peak node ratio).

calls under-approximation to reduce the number of nodes to no more than T_{min}, thus this would eventually result in finding only a partial state space). This experiment shows that:

- The smaller the peak node ratio, the more the under-approximation algorithm is applicable to the model. The peak number of nodes generated by CHBFUA in most cases (except model 19) is less than the peak number of nodes generated by CHBF.
- The peak number of nodes for model 19 in CHBFUA is slightly higher than CHBF, i.e., the peak node ratio is greater than one. This can happen because, after deleting a set of nodes, even just applying the transition relation T_α for one transition α may result in an MDD with more nodes than the peak number of nodes needed by the CHBF algorithm.
- The runtime ratio in all cases is greater than one, because once the number of nodes reaches T_{max} and CHBFUA invokes UNDERAPPROXMANY, it calculates node's density to select and delete nodes until the number of nodes is less than or equal to T_{min}. Calculating the density information and adding back removed states causes CHBFUA to have a higher runtime than CHBF.

Table 2 shows detailed experimental results for models where CHBFUA generates S_{rch}. The more CHBFUA invokes the under-approximation (row "#UA calls"), the larger its runtime is than that of CHBF. Also, in most cases, the fewer times the under-approximation algorithm is invoked, the closer the peak of CHBFUA and peak of CHBF are; this is because it is more likely that T_{max} is close to the peak number of nodes in CHBF.

Figure 15 reports instead the final node and state space ratios for CHBFUA and CHBF ($FinalNode_{\text{CHBFUA}}/FinalNode_{\text{CHBF}}$ and $|S_{\text{CHBFUA}}|/|S_{\text{CHBF}}|$, where $FinalNode_{\text{CHBFUA}}$ is the final number of nodes generated by CHBFUA and

Table 2. Results for models where CHBFUA generates the complete state space.

Model#	1	2	3	4	5	6	7	8	9	10
#UA calls	533	137	71	101	27	217	59	423	43	15
Peak nodes CHBF	845,847	348,203	222,344	293,111	130,602	105,738	46,549	70,858	30,065	26,251
Peak nodes CHBFUA)	43,970	29,683	25,162	36,320	21,887	19,379	16,247	28,642	16,330	16,587
runtime CHBF (sec)	1,692	275	443	308	250	1,268	237	392	196	206
runtime CHBFUA (sec)	100,345	2,952	1,649	1,126	434	10,243	978	32,482	567	411
Model#	11	12	13	14	15	16	17	18	19	
#UA calls	7	11	5	9	4	6	4	6	28	
Peak nodes CHBF	26,917	24,983	20,213	20,078	18,873	23,875	21,813	17,235	18,773	
Peak nodes CHBFUA	18,282	18,674	15,559	16,106	15,251	21,748	20,491	16,245	18,899	
runtime CHBF (sec)	233	31	32	240	61	12	22	41	133	
runtime CHBFUA (sec)	433	76	99	379	165	33	37	115	557	

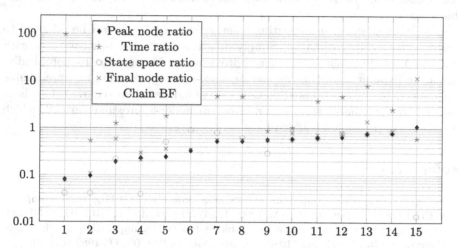

Fig. 15. Runtime, peak node, state space, and final node ratios for models where CHBFUA generates a partial state space (sorted by increasing peak node ratio).

$|\mathcal{S}_{\text{CHBFUA}}|$ is the size of the state space generated by CHBFUA), for models where CHBFUA generates $\mathcal{S}_{part}$. The final number of nodes for most of these models (except model 13 and 15) is greater than T_{max}. The state space ratio is always less than one, since the CHBFUA does generate the complete state space. The final number of nodes generated by CHBFUA in most models is less than the final number of nodes for CHBF, however CHBFUA encodes only a portion of the entire state space. In some models, e.g., 15, the algorithm detects a partial state space faster (time ratio less than one), but the final node ratio is greater than one, indicating that the algorithm is unable to merge nodes to obtain a denser MDD. In these cases, a self-adjusting heuristic could be beneficial.

7 Conclusions and Future Work

We presented a new algorithm for MDDs under-approximation that uses a more precise density than in previously-proposed techniques for BDDs. We demonstrated the soundness of our approach by applying it to the symbolic Petri net state-space generation, where it can compute the entire state space, or possibly a subset of it, with lower memory requirements, at the price of longer runtimes.

Further work is needed towards reducing the number of user-provided parameters. Specifically, we envision a self-adjusting heuristic that automatically chooses and updates (upward or downward) the percentage ψ parameter and the minimum threshold for under-approximation, by self-monitoring the algorithm's own performance (in practical applications, the maximum threshold would instead be likely set to a large value dictated by the amount of available RAM).

Finally, it is worth investigating whether our approach can be adapted to compute an over-approximation. Simply substituting a highest-density node with terminal 1 would result in an over-approximation but, for the monotonically-increasing fixpoint algorithm we use for state-space generation, an unreachable state added by an over-approximation call would never be removed; this is in contrast to a reachable state removed by an under-approximation call, which can always in principle be added back.

References

1. Bollig, B., Wegener, I.: Improving the variable ordering of OBDDs is NP-complete. IEEE Trans. Comput. **45**(9), 993–1002 (1996). https://doi.org/10.1109/12.537122
2. Bryant: graph-based algorithms for Boolean function manipulation. IEEE Trans. Comput. **C-35**(8), 677–691 (1986). https://doi.org/10.1109/TC.1986.1676819
3. Ciardo, G., Jones, R., Marmorstein, R., Miner, A., Siminiceanu, R.: SMART: stochastic model-checking analyzer for reliability and timing. In: Proceedings International Conference on Dependable Systems and Networks, pp. 545- (2002). https://doi.org/10.1109/DSN.2002.1028976
4. Ciardo, G., Marmorstein, R., Siminiceanu, R.: The saturation algorithm for symbolic state-space exploration. Int. J. Softw. Tools Technol. Transfer **8**(1), 4–25 (2006)
5. Hosseini, S.: memory constrained algorithms for multi-valued decision diagrams. Master's thesis, Iowa State University (2021). https://www.proquest.com/dissertations-theses/memory-constrained-algorithms-multi-valued/docview/2628162662/se-2
6. Lengauer, T., Tarjan, R.E.: A fast algorithm for finding dominators in a flowgraph. ACM Trans. Program. Lang. Syst. **1**(1), 121–141 (1979). https://doi.org/10.1145/357062.357071
7. Pastor, E., Roig, O., Cortadella, J., Badia, R.M.: Petri net analysis using Boolean manipulation. In: Valette, R. (ed.) ICATPN 1994. LNCS, vol. 815, pp. 416–435. Springer, Heidelberg (1994). https://doi.org/10.1007/3-540-58152-9_23

8. Ravi, K., McMillan, K.L., Shiple, T.R., Somenzi, F.: Approximation and decomposition of binary decision diagrams. In: Proceedings of the 35th Annual Design Automation Conference, pp. 445–450. DAC 1998, Association for Computing Machinery, New York, NY, USA (1998). https://doi.org/10.1145/277044.277168
9. Ravi, K., Somenzi, F.: High-density reachability analysis. In: Proceedings of the 1995 IEEE/ACM International Conference on Computer-Aided Design, pp. 154–158. ICCAD 1995, IEEE Computer Society, USA (1995)

Stochastic Decision Petri Nets

Florian Wittbold[1]([✉]) [iD], Rebecca Bernemann[1] [iD], Reiko Heckel[2] [iD],
Tobias Heindel[3] [iD], and Barbara König[1] [iD]

[1] Universität Duisburg-Essen, Duisburg, Germany
florian.wittbold@uni-due.de
[2] University of Leicester, Leicester, UK
[3] Heliax Technologies GmbH, Berlin, Germany

Abstract. We introduce stochastic decision Petri nets (SDPNs), which
are a form of stochastic Petri nets equipped with rewards and a control
mechanism via the deactivation of controllable transitions. Such nets can
be translated into Markov decision processes (MDPs), potentially leading
to a combinatorial explosion in the number of states due to concurrency.
Hence we restrict ourselves to instances where nets are either safe, free-
choice and acyclic nets (SAFC nets) or even occurrence nets and policies
are defined by a constant deactivation pattern. We obtain complexity-
theoretic results for such cases via a close connection to Bayesian networks, in particular we show that for SAFC nets the question whether
there is a policy guaranteeing a reward above a certain threshold is NP^{PP}-
complete. We also introduce a partial-order procedure which uses an
SMT solver to address this problem.

1 Introduction

State-based probabilistic systems are typically modelled as Markov chains [28],
i.e., transition systems where transitions are annotated with probabilities. This
admits an intuitive graphical visualization and efficient analysis techniques [17].
By introducing additional non-determinism, one can model a system where a
player can make decisions, enriched with randomized choices. This leads to the
well-studied model of Markov decision processes (MDPs) [6,15] and the challenge
is to synthesize strategies that maximize the reward of the player.

In this paper we study stochastic systems enriched with a mechanism for
decision making in the setting of concurrent systems. Whenever a system exhibits
a substantial amount of concurrency, i.e., events that may potentially happen
in parallel, compiling it down to a state-based system – such as an MDP – can
result in a combinatorial state explosion and a loss in efficiency of MDP-based
methods. We base our models on stochastic Petri nets [21], where Petri nets are
a standard formalism for modelling concurrent systems, especially such systems
where resources are generated and consumed. When considering the discrete-
time semantics of such stochastic nets, it is conceptually easy to transform them
into Markov chains, but this typically leads to a state space explosion.

There exist successful partial order methods for analyzing concurrent systems that avoid explicit interleavings and the enumeration of all reachable states.

L. Gomes and R. Lorenz (Eds.): PETRI NETS 2023, LNCS 13929, pp. 264–285, 2023.
https://doi.org/10.1007/978-3-031-33620-1_15

Instead, they work with partial orders – instead of total orders – of events. While such techniques are well understood in the absence of random choices, leading for instance to methods such as unfoldings [14], there are considerable difficulties to reconcile probability and partial order. Progress has been made by the introduction of the concept of branching cells [1] that encapsulate independent choices, but to our knowledge there is no encompassing theory that provides off-the-shelf partial order methods for computing the probability of reaching a certain goal (e.g. marking a certain place) in a stochastic net.

The contributions of this paper are the introduction of a new model: stochastic decision Petri nets (SDPNs) and its connection to Markov decision processes (MDPs). The transformation of SDPNs into MDPs is relatively straightforward, but may lead to state space explosion, i.e., exponentially many markings, due to the concurrency inherent in the Petri net. This can make the computation of the optimal policy infeasible. We restrict ourselves to a subclass of nets which are safe, acyclic and free-choice (SAFC) and to constant policies and study the problem of determining a policy that guarantees a payoff above some bound. Our result is that the problem SAFC-POL of determining such a policy, despite the restrictions, is still NP^{PP}-complete. We reduce from the D-MAP problem for Bayesian networks [24] (in fact the two problems are interreducible under mild restrictions) and show the close connection of reasoning about stochastic Petri nets and Bayesian networks. Furthermore, for SAFC nets, there is a partial-order solution procedure via an SMT solver, for which we obtain encouraging runtime results. For the simpler free-choice occurrence nets, we obtain an NP-completeness result.

Note that the paper contains some proof sketches, while full proofs and an additional example can be found in [29].

2 Preliminaries

By $\mathbb{N}$ we denote the natural numbers without 0, while $\mathbb{N}_0$ includes 0.

Given two sets X, Y we denote by $(X \to Y)$ the set of all functions from X to Y. Given a function $f \colon X \to \mathbb{N}_0$ or $f \colon X \to \mathbb{R}$ with X finite, we define $\|f\|_\infty = \max_{x \in X} f(x)$ and $\mathrm{supp}(f) = \{x \in X \mid f(x) \neq 0\}$.

Complexity Classes: In addition to well-known complexity classes such as P and NP, our results also refer to PP (see [23]). This class is based on the notion of a probabilistic Turing machine, i.e., a non-deterministic Turing machine whose transition function is enriched with probabilities, which means that the acceptance function becomes a random variable. A language L lies in PP if there exists a probabilistic Turing machine M with polynomial runtime on all inputs such that a word $w \in L$ iff it is accepted with probability strictly greater than $1/2$. As probabilities we only allow numbers ρ that are efficiently computable, meaning that the i-th bit of ρ is computable in a time polynomial in i. (See [2] for a discussion on why such probabilistic Turing machines have equal expressivity with those based on fair coins, which is not the case if we allow arbitrary numbers.)

Given two complexity classes A, B and their corresponding machine models, by A^B we denote the class of languages that are solved by a machine of class

A, which is allowed to use an oracle answering yes/no-questions for a language $L \in B$ at no extra cost in terms of time or space complexity. In particular $\mathsf{NP^{PP}}$ denotes the class of languages that can be accepted by a non-deterministic Turing machine running in polynomial time that can query a black box oracle solving a problem in PP.

By Toda's theorem [27], a polynomial time Turing machine with a PP oracle ($\mathsf{P^{PP}}$) can solve all problems in the polynomial hierarchy.

In order to prove hardness results we use the standard polynomial-time many-one reductions, denoted by $A \leq_p B$ for problems A, B (see [16]).

Stochastic Petri Nets: A stochastic Petri net [21] is given by a tuple $N = (P, T, {}^\bullet(), (){}^\bullet, \Lambda, m_0)$ where P and T are finite sets of places and transitions, ${}^\bullet( ), ( ){}^\bullet : T \to (P \to \mathbb{N}_0)$ determine for each transition its pre-set and post-set including multiplicities, $\Lambda : T \to \mathbb{R}_{>0}$ defines the firing rates and $m_0 : P \to \mathbb{N}_0$ is the initial marking. By $\mathcal{M}(N)$ we denote the set of all markings of N, i.e., $\mathcal{M}(N) = (P \to \mathbb{N}_0)$.

We will only consider the discrete-time semantics of such nets. The firing rates determine stochastically which transition is fired in a marking where multiple transitions are enabled: When transitions $t_1, \ldots, t_n \in T$ are enabled in a marking $m \in \mathcal{M}(N)$ (i.e., ${}^\bullet t_i \leq m$ pointwise), then transition t_i fires with probability $\Lambda(t_i)/\sum_{j=1}^{n} \Lambda(t_j)$, resulting in a discrete step $m \to_{t_i} m' := m - {}^\bullet t_i + t_i{}^\bullet$. In particular, the firing rates have no influence on the reachability set $\mathcal{R}(N) := \{m \in \mathcal{M}(N) \mid m_0 \to^* m\}$ but only define the probability of reaching certain places or markings. Defining "empty" transitions $m \to_\varepsilon m$ for markings $m \in \mathcal{R}(N)$ where no transition is enabled, such a stochastic Petri net can be interpreted as a Markov chain on the set of markings $\mathcal{M}(N)$.

This Markov chain thus generates a (continuous) probability space over sequences $(m_0, m_1, \ldots) \in \mathcal{M}(N)^\omega$ where a sequence is called valid if m_0 is the initial marking of the Petri net and for a prefix $(m_0, \ldots, m_n)$ all cones $\{(m'_0, m'_1, \ldots) \in \mathcal{M}(N)^\omega \mid \forall k = 0, \ldots, n : m'_k = m_k\}$ have non-zero probability. We write $\mathcal{FS}(N) := \{\mu \in \mathcal{M}(N)^\omega \mid \mu \text{ is valid}\}$ to denote the set of valid sequences. We assume that no two transitions have the same pre- and postconditions to have a one-to-one-correspondence between valid sequences and firing sequences $\mu : (m_0 \to_{t_1} m_1 \to_{t_2} \ldots)$.

For a firing sequence μ, we write $\mu^k : m_0 \to_{t_1} m_1 \to_{t_2} \cdots \to_{t_k} m_k$ to denote the finite subsequence of the first k steps, $\operatorname{len}(\mu) := \min\{k \in \mathbb{N} \mid t_k = \varepsilon\} - 1$, for its length, as well as

$$pl(\mu) := \bigcup_{n=0}^{\infty} \operatorname{supp}(m_n) \qquad tr(\mu) := \{t_n \mid n \in \mathbb{N}\} \setminus \{\varepsilon\}$$

to denote the set of places reached in μ (or, analogously, μ^k), and the set of fired transitions in μ (independent of their firing order), respectively.

We are, furthermore, interested in the following properties of Petri nets: A Petri net N as above is called

- *ordinary* iff all transitions require and produce at most one token in each place ($\| {}^{\bullet}t \|_{\infty}, \| t^{\bullet} \|_{\infty} \le 1$ for all $t \in T$);
- *safe* iff it is ordinary and all reachable markings also only have at most one token in each place ($\| m \|_{\infty} \le 1$ for all $m \in \mathcal{R}(N)$);
- *acyclic* iff the transitive closure $\prec_N^+$ of the causal relation $\prec_N$ (with $p \prec_N t$ if ${}^{\bullet}t(p) > 0$ and $t \prec_N p$ if $t^{\bullet}(p) > 0$) is irreflexive;
- an *occurrence net* iff it is safe, acyclic, free of backward conflicts (all places have at most one predecessor transition, i.e., $|\{t \mid t^{\bullet}(p) > 0| \le 1$ for all $p \in P$) and self-conflicts (for $x \in P \cup T$, there exist no two distinct conflicting transitions $t, t' \in T$, i.e., transitions sharing preconditions, on which x is causally dependent, i.e., $t, t' \prec_N^+ x$), and the initial marking has no causal predecessors (for all $p \in P$ with $m_0(p) = 1$, we have $t^{\bullet}(p) = 0$ for all $t \in T$);
- *free-choice* [13] iff it is ordinary and all transitions $t, t' \in T$ are either both enabled or disabled in all markings (i.e., ${}^{\bullet}t = {}^{\bullet}t'$ or $\mathrm{supp}({}^{\bullet}t) \cap \mathrm{supp}({}^{\bullet}t') = \emptyset$);
- φ-*bounded* (for $\varphi \colon \mathbb{N}_0 \to \mathbb{N}_0$) iff all its runs, starting from m_0, have at most length $\varphi(|P| + |T|)$, i.e., iff $\mathrm{len}(\mu) \le \varphi(|P| + |T|)$ for all firing sequences $\mu \in \mathcal{FS}(N)$.

We will abbreviate the class of free-choice occurrence Petri nets as FCON, safe and acyclic free-choice nets as SAFC nets, and the class of φ-bounded Petri nets as $[\varphi]$BPN. Note that FCON $\subseteq$ SAFC and also SAFC $\subseteq$ $[id]$BPN for the identity id.[1]

We also introduce some notation specifically for SAFC nets: As common in the analysis of safe Petri nets, we will interpret markings as well as pre- and postconditions of transitions as subsets of the set P of places rather than functions $P \to \{0, 1\} \subseteq \mathbb{N}_0$.

The set of maximal configurations will be denoted by $\mathcal{C}^{\omega}(N) := \{tr(\mu) \mid \mu \in \mathcal{FS}(N)\}$ and configurations by $\mathcal{C}(N) := \{tr(\mu^k) \mid \mu \in \mathcal{FS}(N), k \in \mathbb{N}_0\}$.

An important notion in the analysis of a (free-choice) net are branching cells (see also [1,8]). We will define a cell to be a subset of transitions $\mathbb{C} \subseteq T$ where all transitions $t \in \mathbb{C}$ share their preconditions and all $t' \in T \setminus \mathbb{C}$ share no preconditions with $t \in \mathbb{C}$. In other words, $\mathbb{C}$ is an equivalence class of a relation $\leftrightarrow$ on T defined by

$$\forall t, t' \in T : t \leftrightarrow t' \iff {}^{\bullet}t = {}^{\bullet}t'.$$

We will write $\mathbb{C}_t := [t]^{\leftrightarrow}$ to denote the equivalence class of transition $t \in T$ and ${}^{\bullet}\mathbb{C} := \bigcup_{t \in \mathbb{C}} {}^{\bullet}t$ as well as $\mathbb{C}^{\bullet} := \bigcup_{t \in \mathbb{C}} t^{\bullet}$ to denote the sets of pre- and postplaces of $\mathbb{C}$, respectively. The set of all cells of a net N is denoted by $BC(N)$.

Markov Decision Processes: A Markov decision process (MDP) is a tuple (S, A, δ, r, s_0) consisting of finite sets S, A of states and actions, a function $\delta \colon S \times A \to \mathcal{D}(S)$ of probabilistic transitions (where $\mathcal{D}(S)$ is the set of probability distributions on S), a reward function $r \colon S \times A \times S \to \mathbb{R}$ of rewards and an initial state $s_0 \in S$ (see also [6,15]).

---

[1] Indeed, $[id]$BPN contains any safe and acyclic Petri net, omitting the free-choice constraint.

A policy (or strategy) for an MDP is some function $\pi \colon S \to A$. It has been shown that such stationary deterministic policies can act optimally in such an (infinite-horizon) MDP setting (see also [15]). A policy gives rise to a Markov chain on the set of states with transitions $s \mapsto \delta(s, \pi(s)) \in \mathcal{D}(S)$. The associated probability space is $s_0 S^\omega$, the set of all infinite paths on S starting with s_0, which – due to its uncountable nature – has to be dealt with using measure-theoretic concepts. As before we equip the probability space with a σ-algebra generated by all cones, i.e., all sets of words sharing a common prefix.

The value (or payoff) of a policy π is then given as the expectation of the (undiscounted) total reward (where s_i, $i \in \mathbb{N}_0$ are random variables, mapping an infinite path to the i-th state, i.e., they represent the underlying Markov chain):

$$\mathbb{E}\left[\sum_{n \in \mathbb{N}_0} r(s_n, \pi(s_n), s_{n+1})\right].$$

To avoid infinite values, we have to assume that the sum is bounded.

The problem of finding an optimal policy $\pi \colon S \to A$ for a given MDP (S, A, δ, r, s_0) with finite state and action space is known to be solvable in polynomial time using linear programming [15,19].

Bayesian Networks: Bayesian networks are graphical models that give compact representations of discrete probability distributions, exploiting the (conditional) independence of random variables.

A (finite) probability space $(\Omega, \mathbb{P})$ consists of a finite set Ω and a probability function $\mathbb{P} \colon \Omega \to [0,1]$ such that $\sum_{\omega \in \Omega} \mathbb{P}(\omega) = 1$. A Bayesian network [25] is a tuple (X, Δ, P) where

- $X = (X_i)_{i=1,\ldots,n}$ is a (finite) family of random variables $X_i \colon \Omega \to V_i$, where V_i is finite.
- $\Delta \subseteq \{1, \ldots, n\} \times \{1, \ldots, n\}$ is an acyclic relation that describes dependencies between the variables, i.e., its transitive closure Δ^+ is irreflexive. By $\Delta^i = \{j \mid (j, i) \in \Delta\}$ we denote the parents of node i according to Δ.
- $P = (P_i)_{i=1,\ldots,n}$ is a family of probability matrices $P_i \colon \prod_{j \in \Delta^i} V_j \to \mathcal{D}(V_i)$, whose entries are given by $P_i(v_i \mid (v_j)_{j \in \Delta^i})$.

A probability function $\mathbb{P}$ is consistent with such a Bayesian network whenever for $v = (v_i)_{i=1,\ldots,n} \in \prod_{i=1}^n V_i$ we have

$$\mathbb{P}(X = v) = \prod_{i=1}^n P_i(v_i \mid (v_j)_{j \in \Delta^i}).$$

The size of a Bayesian network is not just the size of the graph, but the sum of the size of all its matrices (where the size of an $m \times n$-matrix is $m \cdot n$). In particular, note that a node with k parents in a binary Bayesian network (i.e., with $|V_i| = 2$ for all i) is associated with a 2×2^k probability matrix.

Example 2.1. *An example Bayesian network is given in Fig. 1. There are four random variables (a, b, c, d) with codomain $\{0, 1\}$. The tables in the figure denote the conditional probabilities, for instance $P_d(0 \mid 01) = \mathbb{P}(X_d = 0 \mid X_a = 0, X_b = 1) = 1/6$, i.e., one records the probability that a random variable has a certain value, dependent on the value of its parents in the graph. The probability $\mathbb{P}(X = 0100) = \mathbb{P}(X_a = 0, X_b = 1, X_c = 0, X_d = 0)$ is obtained by multiplying $P_a(0) \cdot P_b(1) \cdot P_c(0 \mid 0) \cdot P_d(0 \mid 01) = 1/3 \cdot 1/2 \cdot 2/3 \cdot 1/6 = 1/54$.*

We are interested in the following two problems for Bayesian networks (see also [24]):

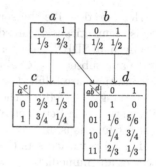

Fig. 1. A Bayesian Network.

- D-PR: Given the Bayesian network (X, Δ, P) and $E = \{X_{i_1}, \ldots, X_{i_\ell}\} \subseteq X$, $e \in V_E := \prod_{j=1}^{\ell} V_{i_j}$ (the evidence) and a rational $p > 0$, does it hold that $\mathbb{P}(E = e) > p$? This problem is known to be PP-complete [20].
- D-MAP: Given a Bayesian network (X, Δ, P), a rational number $p > 0$, disjoint subsets $E, F \subseteq X,^2$ and evidence $e \in V_E$, does there exist $f \in V_F$ such that $\mathbb{P}(F = f, E = e) > p$, or, if $\mathbb{P}(E = e) \neq \emptyset$, equivalently, $\mathbb{P}(F = f \mid E = e) > p$ (by adapting the bound p). It is known that this problem, also known as maximum a-posteriori problem, is $\mathsf{NP}^{\mathsf{PP}}$-complete (see [11, 20]).

The corresponding proof in [24] also shows that the D-MAP problem remains $\mathsf{NP}^{\mathsf{PP}}$-complete if F only contains uniformly distributed 'input' nodes, i.e., nodes X_i with $\Delta^i = \emptyset$ and $P_i(x_i) = 1/|V_i|$, as well as $V_i = \{0, 1\}$ for all $i = 1, \ldots, n$.

In particular, the following problem (where E, F are switched!) is still $\mathsf{NP}^{\mathsf{PP}}$-complete: Given a binary Bayesian network (X, Δ, P) (i.e., $V_i = \{0, 1\}$ for all i), a rational $p > 0$, disjoint subsets $E, F \subseteq X$ where F only contains uniformly distributed input nodes, as well as evidence $e \in V_E$, does there exist $f \in V_F$ such that $\mathbb{P}(E = e \mid F = f) > p$ (as $\mathbb{P}(F = f) = 1/2^{|F|}$ is independent of f and known due to uniformity)? We will, in the rest of this paper, refer to this modified problem as D-MAP instead of the original problem above.

Example 2.2 *(D-MAP). Given the Bayesian Network in Fig. 1 with $F = \{X_a\}$ (MAP variable), $E = \{X_c, X_d\}$, $e = (0, 1) \in V_c \times V_d$ (evidence) and $p = 1/3$, we ask whether $\exists f \in \{0, 1\}$: $\mathbb{P}(X_c = 0, X_d = 1 \mid X_a = f) > 1/3$. When choosing $f = 1 \in V_a$, the probability $\mathbb{P}(X_c = 0, X_d = 1 \mid X_a = 1) = 3/4 \cdot (1/2 \cdot 3/4 + 1/2 \cdot 1/3) = 13/32 > 1/3$ exceeds the bound. Note that to compute the value in this way, one has to sum up over all possible valuations of those variables that are neither evidence nor MAP variables, indicating that this is not a trivial task.*

---

^{2} The variables contained in F are called MAP variables.

3   Stochastic Decision Petri Nets

We will enrich the definition of stochastic Petri nets to allow for interactivity, similar to how MDPs [6] extend the definition of Markov chains.

Definition 3.1. *A stochastic decision Petri net (SDPN) is a tuple $(P, T, {}^\bullet(), ()^\bullet, \Lambda, m_0, C, R)$ where $(P, T, {}^\bullet(), ()^\bullet, \Lambda, m_0)$ is a stochastic Petri net; $C \subseteq T$ is a set of controllable transitions; $R\colon \mathcal{P}(P) \to \mathbb{R}$ is a reward function.*

Here we describe the semantics of such SDPNs in a semi-formal way. The precise semantics is obtained by the encoding of SDPNs into MDPs in Sect. 4.

Given an SDPN, an external agent may in each step choose to manually deactivate any subset $D \subseteq C$ of controllable transitions (regardless of whether their preconditions are fulfilled or not). As such, if transitions $D \subseteq C$ are deactivated in marking $m \in \mathcal{M}(N)$, the SDPN executes a step according to the semantics of the stochastic Petri net $N_D = (P, T \setminus D, {}^\bullet(), ()^\bullet, \Lambda_D, m_0)$ where the pre- and post-set functions and Λ_D are restricted accordingly.

For all rewarded sets $Q \in \text{supp}(R)$, the agent receives an "immediate" reward $R(Q)$ once all the places $p \in Q$ are reached at one point in the execution of the Petri net (although not necessarily simultaneously). In particular, any reward is only received once. Note that this differs from the usual definition of rewards as in MDPs, where a reward is received each time certain actions is taken in given states. However, logical formulae

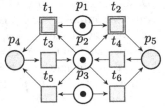

Fig. 2. Example SDPN

over reached places (such as "places p_1 and p_2 are reached without reaching place q") are more natural to represent by such one-time rewards instead of cumulative rewards.[3] The framework can be extended to reward markings instead of places but at the cost of an exponential explosion, since to be able to compute the one-time step-wise rewards not only already reached places but already reached markings would have to be memorized. Note that a reward need not be positive.

More formally, given a firing sequence $\mu : m_0 \to_{t_1} m_1 \to_{t_2} \dots$, the agent receives a value or payoff of $V(pl(\mu))$ where $V(M) := \sum_{Q \subseteq M} R(Q)$.

Example 3.2. *As an example consider the SDPN in Fig. 2. The objective is to mark both places coloured in yellow at some point in time (not necessarily at the same time). This can be described by a reward function R which assigns 1 to the set $\{p_4, p_5\}$ containing both yellow places and 0 to all other sets.*

The transitions with double borders (t_1, t_2) are controllable and it turns out that the optimal strategy is to deactivate both t_1 and t_2 first, in order to let t_5 or t_6 mark either of the two goal places before reaching the marking $(1, 1, 0, 0, 0)$ from which no information can be gained which of the two goal places have been marked. An optimal strategy thus has to have knowledge of already achieved sub-goals in terms of visited places. In this case, the strategy can deactivate one of the transitions (t_1, t_2) leading to the place already visited.

---

[3] Firings of transitions can also easily be rewarded by adding an additional place.

Policies may be dependent on the current marking and the places accumulated so far. Now, for a given policy $\pi : \mathcal{M}(N) \times \mathcal{P}(P) \to \mathcal{P}(C)$, determining the set $\pi(m, Q) \subseteq C$ of deactivated transitions in marking m for the set Q of places seen so far, we consider the (continuous) probability space $m_0 \mathcal{M}(N)^\omega$, describing the infinite sequence $m_0 \to_{t_1} m_1 \to_{t_2} \ldots$ of markings generated by the Petri net under the policy π (i.e., if in step n the transitions $D_n := \pi(m_{n-1}, \bigcup_{k=0}^{n-2} \operatorname{supp}(m_k))$ are deactivated).

Then we can consider the expectation of the random variable $V \circ pl$, i.e.,

$$\mathbb{V}^\pi := \mathbb{E}^\pi [V \circ pl],$$

over the probability space $m_0 \mathcal{M}(N)^\omega$. We will call this the value of π and, if $\pi \equiv D \subseteq C$ is constant, simply write $\mathbb{V}^D$ which we will call the value of D.

For the complexity analyses we assume that R is only stored on its support, e.g., as a set $R \subseteq \mathcal{P}(P) \times \mathbb{R}$ which we will interpret as a dictionary with entries $[Q : R(Q)]$ for some $Q \subseteq P$, as for many problems of interest the size of the support of the reward function can be assumed to be polynomially bounded w.r.t. to the set of places and transitions.

We consider the following problems for stochastic Petri nets, where we parameterize over a class $\mathcal{N}$ of SDPNs and (for the second problem) over a class $\Psi \subseteq (\mathcal{M}(N) \times \mathcal{P}(P) \to \mathcal{P}(C))$ of policies:

- $\mathcal{N}$-VAL: Given a rational $p > 0$, a net $N \in \mathcal{N}$ and a policy $\pi \in \Psi$ for N, decide whether $\mathbb{V}^\pi > p$.
- $\mathcal{N}$-POL: Given a rational $p > 0$ and a net $N \in \mathcal{N}$, decide whether there exist a policy $\pi \in \Psi$ such that $\mathbb{V}^\pi > p$.

  Although paramterized over sets of policies, we will omit Ψ if is clear from the context (in fact we will restrict to constant policies from Sect. 5 onwards).

4 Stochastic Decision Petri Nets as Markov Decision Processes

We now describe how to transform an SDPN into an MDP, thus fixing the semantics of such nets. For unbounded Petri nets, the resulting MDP has an infinite state space, but we will restrict to the finite case later.

Definition 4.1. *Given an SDPN $N = (P, T, F, \Lambda, C, R, m_0)$ where m_0 is not the constant zero function, the MDP for N is defined as the tuple (S, A, δ, r, s_0) where*

- $S = \mathcal{R}(N) \times \mathcal{P}(P)$ *(product of reachable markings and places collected),*
- $A = \mathcal{P}(C)$ *(sets of deactivated transition as actions),*
- $\delta : (\mathcal{R}(N) \times \mathcal{P}(P)) \times \mathcal{P}(C) \to \mathcal{D}(\mathcal{R}(N) \times \mathcal{P}(P))$*, with*

$$\delta((m, Q), D)((m', Q')) := \begin{cases} p(m' \mid m, D) & \text{if } Q' = Q \cup \operatorname{supp}(m), \\ 0 & \text{otherwise,} \end{cases}$$

where

$$p(m' \mid m, D) = \frac{\sum_{t \in En(m,D), m \to_t m'} \Lambda(t)}{\sum_{t \in En(m,D)} \Lambda(t)}$$

whenever $En(m, D) := \{t \in T \backslash D \mid {}^{\bullet}t \leq m\} \neq \emptyset$. *If* $En(m, D) = \emptyset$, *we set* $p(m' \mid m, D) = 1$ *if* $m = m'$ *and* 0 *if* $m \neq m'$. *That is,* $p(m' \mid m, D)$ *is the probability of reaching* m' *from* m *when transitions* D *are deactivated.*

- $r \colon S \times A \times S \to \mathbb{R}$ *(reward function) with*

$$r((m, Q), D, (m', Q')) := \begin{cases} \sum_{Q \subseteq Y \subseteq Q'} R(Y) & \text{if } Q = \emptyset, \\ \sum_{Q \subsetneq Y \subseteq Q'} R(Y) & \text{if } Q \neq \emptyset. \end{cases}$$

- $s_0 = (m_0, \emptyset)$

The transition probabilities are determined as for regular stochastic Petri nets where we consider only the rates of those transitions that have not been deactivated and that can be fired for the given marking. If no transition is enabled, we stay at the current marking with probability 1.

Note that the reward for the places reached in a marking m is only collected when we fire a transition leaving m. This is necessary as in the very first step we also obtain the reward for the empty set, which might be non-zero, and due to the fact that the initial marking is assumed to be non-empty, this reward for the empty set is only collected once.

The following result shows that the values of policies $\pi \colon S \to A$ (note that these are exactly the policies for the underlying SDPN) over the MDP are equal to the ones over the corresponding SDPN.

Proposition 4.2. *Let* $N = (P, T, F, \Lambda, C, R, m_0)$ *be an SDPN and* $M = (S, A, \delta, r, s_0)$ *the corresponding MDP. For any policy* $\pi \colon S \to A$, *we have*

$$(\mathbb{V}^{\pi} =) \mathbb{E}^{\pi} [V \circ pl] = \mathbb{E}^{\pi} \left[\sum_{n \in \mathbb{N}_0} r(\mathbf{s}_n, \pi(\mathbf{s}_n), \mathbf{s}_{n+1}) \right]$$

where $(\mathbf{s}_n)_n$ *is the Markov chain resulting from following policy* π *in* M.

This provides an exact semantic for SDPNs via MDPs. Note, however, that for analysis purposes, even for safe Petri nets, the reachability set $\mathcal{R}(N)$ (as a subset of $\mathcal{P}(P)$) is generally of exponential size whence the transformation into an MDP can at best generally only yield algorithms of exponential worst-case-time. Hence, we will now restrict to specific subproblems and it will turn out that even with fairly severe restrictions to the type of net and the policies allowed, we obtain completeness results for complexity classes high in the polynomial hierarchy.

5   Complexity Analysis for Specific Classes of Petri Nets

For the remainder of this paper, we will consider the problem of finding optimal *constant* policies for certain classes of nets. In other words, the agent chooses *before* the execution of the Petri net which transitions to deactivate for its *entire* execution. For a net N, the policy space is thus given by

$$\Psi(N) = \{\pi : \mathcal{M}(N) \to \mathcal{P}(C) \mid \pi \equiv D \subseteq C\} \doteq \mathcal{P}(C).$$

Since one can non-deterministically guess the maximizing policy (there are only exponentially many) and compute its value, it is clear that the complexity of the policy optimization problem $\mathcal{N}$-POL is bounded by the complexity of the corresponding value problem $\mathcal{N}$-VAL as follows: If, for a given class $\mathcal{N}$ of Petri nets, $\mathcal{N}$-VAL lies in the complexity class C, then $\mathcal{N}$-POL lies in NP^{C}.

We will now show the complexity of these problems for the three Petri net classes FCON, SAFC, and $[\varphi]$BPN and work out the connection to Bayesian networks. In the following we will assume that all probabilities are efficiently computable, allowing us to simulate all probabilistic choices with fair coins.

5.1   Complexity of Safe and Acyclic Free-Choice Decision Nets

We will first consider the case of Petri nets where the length of runs is bounded.

Proposition 5.1. *For any polynomial* φ, *the problem* $[\varphi]$BPN-VAL *is in* PP. *In particular,* $[\varphi]$BPN-POL *is in* NP^{PP}.

Proof (sketch). Given a Petri net N, a policy π and a bound p, a PP-algorithm for $[\varphi]$BPN-VAL can simulate the execution of the Petri net and calculate the resulting value, checking whether the expected value for π is greater than the pre-defined bound p. For this, we have to suitably adapt the threshold (with an affine function ψ) so that the probabilistic Turing machine accepts with probability greater than 1/2 iff the reward for the given policy is strictly greater than p.

As the execution of the Petri net takes only polynomial time in the size of the Petri net (φ), this can be performed by a probabilistic Turing machine in polynomial time whence $[\varphi]$BPN-VAL lies in PP.

Since a policy can be guessed in polynomial time, we can also infer that $[\varphi]$BPN-POL is in NP^{PP}.   □

This easily gives us the following corollary for SAFC nets.

Corollary 5.2. *The problem* SAFC-VAL *is in* PP *and* SAFC-POL *in* NP^{PP}.

Proof. This follows directly from Proposition 5.1 and the fact that SAFC $\subseteq$ $[id]$BPN.   □

Proposition 5.3. *The problem* SAFC-POL *is* NP^{PP}-*hard and, therefore, also* NP^{PP}-*complete.*

Proof (sketch). This can be proven via a reduction D-MAP $\leq_p$ SAFC-POL, i.e., representing the modified D-MAP problem for Bayesian networks as a decision problem in safe and acyclic free-choice nets. NP^{PP}-completeness then follows together with Corollary 5.2. Note that we are using the restricted version of the D-MAP problem as explained in Sect. 2 (uniformly distributed input nodes, binary values).

We sketch the reduction via an example: we take the Bayesian network in Fig. 1 and consider a D-MAP instance where $E = \{X_c, X_d\}$ (evidence, where we fix the values of c, d to be $0, 1$), $F = \{X_a\}$ (MAP variables) and p is a threshold. That is, the question being asked for the Bayesian network is whether there exists a value x such that $\mathbb{P}(X_c = 0, X_d = 1 \mid X_a = x) > p$.

This Bayesian network is encoded into the SAFC net in Fig. 3, where transitions with double borders are controllable and the yellow places give a reward of 1 when both are reached (not necessarily at the same time). Transitions either have an already indicated rate of 1 or the rate can be looked up in the corresponding matrix of the BN. The rate of a transition $t^i_{x_1 x_2 \to x_3}$ is the probability value $P_i(x_3 \mid x_1 x_2)$, where P_i is the probability matrix for $i \in \{a, b, c, d\}$.

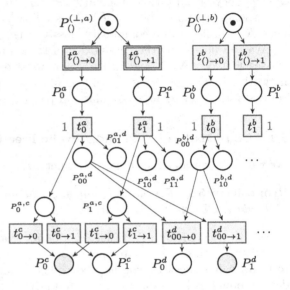

Fig. 3. SAFC net corresponding to BN in Fig. 1.

Intuitively the first level of transitions simulates the probability tables of P^a, P^b, the nodes without predecessors in the Bayesian network, where for instance the question of whether P^a_0 or P^a_1 are marked corresponds to the value of the random variable X_a associated with node a. Since X_a is a MAP variable, its two transitions are controllable. Note that enabling both transitions will never give a higher reward than enabling only one of them. (This is due to the fact that $\max\{x, y\} \geq p_1 \cdot x + p_2 \cdot y$ for $p_1, p_2 \geq 0$ with $p_1 + p_2 = 1$.)

The second level of transitions (each with rate 1) is inserted only to obtain a free-choice net by creating sufficiently many copies of the places in order to make all conflicts free-choice.

The third level of transitions simulates the probability tables of P^c, P^d, only to ensure the net being free-choice we need several copies. For instance, transition $t^c_{0 \to 0}$ consumes a token from place $P^{a,c}_0$, a place specifically created for the entry $P^c(c = 0 \mid a = 0)$ in the probability table of node c.

In the end the aim is to mark the places P^c_0 and P^d_1, and we can find a policy (deactivating either $t^a_{() \to 0}$ or $t^a_{() \to 0}$) such that the probability of reaching both places exceeds p if and only if the D-MAP instance specified above has a solution.

This proof idea can be extended to more complex Bayesian networks, for a more formal proof see [29]. □

In fact, a reduction in the opposite direction (from Petri nets to Bayesian networks) is possible as well under mild restrictions, which shows that these problems are closely related.

Proposition 5.4. *For two given constants k, ℓ, consider the following problem: let N be a SAFC decision Petri net, where for each branching cell the number of controllable transitions is bounded by some constant k. Furthermore, given its reward function R, we assume that $|\cup_{Q \in \text{supp}(R)} Q| \le \ell$. Given a rational number p, does there exist a constant policy π such that $\mathbb{V}^\pi > p$?*
This problem can be polynomially reduced to D-MAP.

Proof (sketch). We sketch the reduction, which is inspired by [8], via an example: consider the SAFC net in Fig. 5, where the problem is to find a deactivation pattern such that the payoff exceeds p. We encode the net into a Bayesian network (Fig. 4), resulting in an instance of the D-MAP problem.

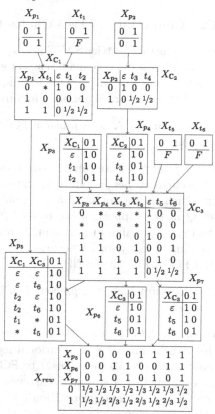

We have four types of random variables: place variables (X_p, $p \in P$), which record which place is marked; transition variables ($X_{t_1}, X_{t_5}, X_{t_6}$), one for each controllable transition, which are the MAP variables; cell variables ($X_{\mathbb{C}_i}$ for $\mathbb{C}_1 = \{t_1, t_2\}$, $\mathbb{C}_2 = \{t_3, t_4\}$, $\mathbb{C}_3 = \{t_5, t_6\}$) which are non-binary and which record which transition in the cell was fired or whether no transition was fired (ε); a reward variable (X_{rew}) such that $\mathbb{P}(X_{rew} = 1)$ equals the function ψ applied to the payoff. Note that we use the affine function ψ from the proof of Proposition 5.1 to represent rewards as probabilities in the interval $[0, 1]$. The threshold for the D-MAP instance is $\psi(p)$. Dependencies are based on the structure of the given SAFC net. For instance, $X_{\mathbb{C}_3}$ is dependent on X_{p_3}, X_{p_4} (since $\bullet\mathbb{C}_3 = \{p_3, p_4\}$) and X_{t_5}, X_{t_6} (since t_5, t_6 are the controllable transitions in $\mathbb{C}_3$).

Both the matrices of cell and place variables could become exponentially large, however this problem can be resolved easily by dividing the matrices into smaller ones and cascading them. Since the number of controllable transitions is bounded by k and the number of rewarded places by ℓ, they will not cause an exponential blowup of the corresponding matrix. □

Fig. 4. Bayesian network obtained from the SAFC net in Fig. 5 below. Entries $*$ are 'don't-care' values.

Corollary 5.5. *The problem* SAFC-VAL *is* PP-*hard and, therefore, also* PP-*complete.*

Proof. We note that using the construction in the proof of Proposition 5.3 with the set F of MAP variables being empty, we can reduce the D-PR problem for Bayesian networks to the SAFC-VAL problem, showing that SAFC-VAL is PP-hard. Using Corollary 5.2, this yields that SAFC-VAL is PP-complete.   □

Corollary 5.6. *For any polynomial* $\varphi : \mathbb{N}_0 \to \mathbb{N}_0$ *fulfilling* $\varphi(n) \geq n$ *for all* $n \in \mathbb{N}_0$, *the problem* $[\varphi]$BPN-VAL *is* PP-*complete and* $[\varphi]$BPN-POL *is* NP^{PP}-*complete.*

Proof. As any safe and acyclic free-choice net is an id-bounded net, it is, in particular, a φ-bounded net with φ as above, and we have SAFC-VAL $\leq_p$ $[\varphi]$BPN-VAL and SAFC-POL $\leq_p$ $[\varphi]$BPN-POL. Propositions 5.1 and 5.3 as well as Corollary 5.5, therefore show that $[\varphi]$BPN-VAL is PP-complete and $[\varphi]$BPN-POL is NP^{PP}-complete.   □

5.2   Complexity of Free-Choice Occurrence Decision Nets

Now we further restrict SAFC nets to occurrence nets, which leads to a substantial simplification. The main reason for this is the absence of backwards-conflicts, which means that each place is uniquely generated, making it easier to trace causality, i.e., there is a unique minimal configuration that generates each place.

Proposition 5.7. *The problem* FCON-VAL *is in* P. *In particular,* FCON-POL *is in* NP.

Proof (sketch). Determining the probability of reaching a set of places Q in an occurrence net amounts to multiplying the probabilities of the transitions on which the places in Q are causally dependent. This can be done for every set Q in the support of the reward function R, which enables us to determine the expected value in polynomial time, implying that FCON-VAL lies in P. By guessing a policy for an occurrence net with controllable transitions, we obtain that FCON-POL lies in NP.   □

Proposition 5.8. *The problem* FCON-POL *is* NP-*hard and, therefore, also* NP-*complete.*

Proof (sketch). To show NP-hardness we reduce 3-SAT (the problem of deciding the satisfiability of a propositional formula in conjunctive normal form with at most three literals per clause) to FCON-POL. Given a formula ψ, this is done by constructing a simple occurrence net with parallel controllable transitions, one for each atomic proposition ℓ in ψ. Then we define a reward function with polynomial support in such a way that the expected reward for the constructed net is larger or equal than the number of clauses iff the formula has a model. The correspondence between the model and the policy is such that transitions whose atomic propositions are evaluated as true are deactivated.   □

6   An Algorithm for SAFC Decision Nets

Here we present a partial-order algorithm for solving the policy problem for SAFC (decision) nets. It takes such a net and converts it into a formula for an SMT solver. We will assume the following, which is also a requirement for occurrence nets:

Assumption 6.1. For all places $p \in m_0$: $^\bullet p := \{t \in T \mid p \in t^\bullet\} = \emptyset$.

This is a mild assumption since any transition $t \in {}^\bullet p$ for a place $p \in m_0$ in a safe and acyclic net has to be dead as all places can only be marked once.

We are now using the notion of (branching) cells, introduced in Sect. 2: The fact that the SDPN is safe, acyclic and free-choice ensures that choices in different cells are taken independently from another, so that the probability of a configuration $\tau \in \mathcal{C}(N)$ under a specific deactivation pattern $D \subseteq C$ is given by

$$\mathbb{P}^D(tr \supseteq \tau) = \prod_{t \in \tau} \frac{\chi_{T \setminus D}(t) \cdot \Lambda(t)}{\sum_{t \in \mathcal{C}_t \setminus D} \Lambda(t)} = \begin{cases} 0 & \text{if } \tau \cap D \neq \emptyset \\ \prod_{t \in \tau} \frac{\Lambda(t)}{\sum_{t' \in \mathcal{C}_t \setminus D} \Lambda(t')} & \text{otherwise} \end{cases}$$

where $\chi_{T \setminus D}$ is the characteristic function of $T \setminus D$ and $0/0$ is defined to yield 0.

The general idea of the algorithm is to rewrite the reward function $R : \mathcal{P}(P) \to \mathbb{R}$ on sets of places to a reward function on sets of transitions that yields a compact formula for computing the value $\mathbb{V}^D$ for specific sets D (i.e., solving SAFC-VAL), that we can also use to solve the policy problem SAFC-POL via an SMT solver.

We first need some definitions:

Definition 6.2. *For a maximal configuration $\tau \in \mathcal{C}^\omega(N_D)$ for a given deactivation pattern $D \subseteq C$, we define its set of prefixes in $\mathcal{C}(N_D)$ to be*

$$\text{pre}^D(\tau) := \{\tau' \in \mathcal{C}(N_D) \mid \tau' \subseteq \tau\}$$

which corresponds to all configurations that can lead to the configuration τ. We also define the set of extensions of a configuration $\tau \in \mathcal{C}(N_D)$ in $\mathcal{C}^\omega(N_D)$, which corresponds to all maximal configurations that τ can lead to, as

$$\text{ext}^D(\tau) := \{\tau' \in \mathcal{C}^\omega(N_D) \mid \tau \subseteq \tau'\}.$$

Definition 6.3. *Let N be a Petri net with a reward function $R: \mathcal{P}(P) \to \mathbb{R}$ on places and a deactivation pattern D. A reward function $[R]: \mathcal{P}(T) \to \mathbb{R}$ on transitions is called consistent with R if for each firing sequence $\mu \in \mathcal{FS}(N_D)$:*

$$V(pl(\mu)) = \sum_{Q \subseteq pl(\mu)} R(Q) = \sum_{\tau \in \text{pre}^D(tr(\mu))} [R](\tau).$$

This gives us the following alternative method to determine the expected value for a net (with given policy D):

Lemma 6.4. *Using the setting of Definition 6.3, whenever $[R]$ is consistent with the reward function R and $[R](\tau) = 0$ for all $\tau \notin \mathcal{C}(N)$, the expected value for the net N under the constant policy D is:*

$$\mathbb{V}^D = \sum_{\tau \subseteq T} \mathbb{P}^D(tr \supseteq \tau) \cdot [R](\tau).$$

Note that $[R](tr(\mu)):=V(pl(\mu))$ for $\mu \in \mathcal{FS}(N)$ fulfills these properties trivially. However, rewarding only maximal configurations can lead, already in occurrence nets with some concurrency, to an exponential support (w.r.t. the size of the net and its reward function). The goal of our algorithm is to instead make use of the sum over the configurations by rewarding reached places immediately in the corresponding configuration, generating a function $[R]$ that fulfills the properties above and whose support remains of polynomial size in occurrence nets. Hence, we have some form of partial-order technique, in particular concurrent transitions receive the reward independently of each other (if the reward is not dependent on firing both of them).

The rewriting process is performed by iteratively 'removing maximal cells' and resembles a form of backward-search algorithm. First of all, $\preceq_N^*$ (the reflexive and transitive closure of causality $\prec_N$) induces a partial order $\sqsubseteq$ on the set $BC(N)$ of cells via

$$\forall \mathbb{C}, \mathbb{C}' \in BC(N) : \mathbb{C} \sqsubseteq \mathbb{C}' \iff \exists t \in \mathbb{C}, t' \in \mathbb{C}' : t \preceq_N^* t'.$$

Let all cells $(\mathbb{C}_1, \ldots, \mathbb{C}_m)$ with $m = |BC(N)|$ be ordered conforming to $\sqsubseteq$, then we let N_k denote the Petri net consisting of places $P_k := P \setminus (\bigcup_{l>k} \mathbb{C}_l^\bullet) \cup (\bigcup_{l \leq k} \mathbb{C}_l^\bullet)$ (where the union with the post-sets is only necessary if backward-conflicts exist) and transitions $T_k := \bigcup_{l \leq k} \mathbb{C}_l$, the remaining components being accordingly restricted (note that the initial marking m_0 is still contained in P_k by Assumption 6.1). In particular, it holds that $N = N_m$ as well as $T_0 = \emptyset$ and $P_0 = \{p \in P \mid \forall t \in T : p \notin t^\bullet\}$.

Let N be a Petri net with deactivation pattern D, $\mu \in \mathcal{FS}(N_D)$ be a firing sequence and $k \in \{1, \ldots, |BC(N)|\}$. We write $tr_{\leq k}(\mu) := tr(\mu) \cap T_k$ for the transitions in the first k cells and $tr_{>k}(\mu) := tr(\mu) \setminus T_k$ for the transitions in the cells after the k-th cell as well as $pl_{\leq k}(\mu) := m_0 \cup (\bigcup_{t \in tr_{\leq k}(\mu)} t^\bullet)$ for the places reached after all transitions in the first k cells were fired.

We will now construct auxiliary reward functions $R[k]$ that take pairs of a set of places ($U \subseteq P_k$) and of transitions ($V \subseteq T \setminus T_k$) as input and return a reward. Intuitively, $R[k](U, V)$ corresponds to the reward for reaching all places in U and then firing all transitions in V afterwards where reaching U ensures that all transitions in V can fire.

Starting with the reward function $R[m] : \mathcal{P}(P) \times \{\emptyset\} \to \mathbb{R}, (M, \emptyset) \mapsto R(M)$, we iteratively compute reward functions $R[k] : \mathcal{P}(P_k) \times \mathcal{P}(T \setminus T_k) \to \mathbb{R}$ for $k \geq 0$:

$$R[k](U, V) := \begin{cases} R[k+1](U, V) & \text{if } \mathbb{C}_{k+1} \cap V = \emptyset \\ \displaystyle\sum_{\substack{U' \cap t^\bullet \neq \emptyset \\ U = U' \setminus t^\bullet \cup \,^\bullet t}} R[k+1](U', V \setminus \{t\}) & \text{if } \mathbb{C}_{k+1} \cap V = \{t\} \\ 0 & \text{otherwise} \end{cases}$$

The first case thus describes a scenario where no transition from the $(k + 1)$-th cell is involved while the second case sums up all rewards that are reached when some transition t in the cell has to be fired (that is, all rewards that are given when some of the places in $t^\bullet$ are reached). We give non-zero values only to sets V that contain at most one transition of each cell and U has to contain the full pre-set of t of the transition t removed from V. This is done in order to ensure that in subsequent steps those transitions that generate $^\bullet t$ are in the set to which we assign the reward. This guarantees that V is always a configuration of N after marking U while $R[k](U, V)$ is zero if the transitions in V cannot be fired after U. In this way, rewards are ultimately only given to configurations and to no other sets of transitions, enabling us later to compute the probabilities of those configurations.

And if N is an occurrence net, every entry in $R[k+1]$ produces at most one entry in $R[k]$, meaning that $\text{supp}(R[k]) \leq \text{supp}(R[k+1])$.

Now we can prove that the value of a firing sequence is invariant when rewriting the auxiliary reward functions as described above.

Proposition 6.5. *The auxiliary reward functions satisfy*

$$\sum_{V \subseteq tr_{>k}(\mu)} \sum_{U \subseteq pl_{\leq k}(\mu)} R[k](U, V) = \sum_{V \subseteq tr_{>k+1}(\mu)} \sum_{U \subseteq pl_{\leq k+1}(\mu)} R[k+1](U, V),$$

for $k \in \{0, \ldots, |BC(N)| - 1\}$.

Hence, for every $\mu \in \mathcal{FS}(N)$

$$V(pl(\mu)) = \sum_{U \subseteq pl(\mu)} R[|BC(N)|](U, \emptyset) = \sum_{V \subseteq tr_{>k}(\mu)} \sum_{U \subseteq pl_{\leq k}(\mu)} R[k](U, V),$$

which means that we obtain a reward function on transitions consistent with R by defining $[R] : \mathcal{P}(T) \to \mathbb{R}$ as

$$[R](V) := \sum_{U \subseteq m_0} R[0](U, V).$$

This leads to the following corollary:

Corollary 6.6. *Given a net N and a deactivation pattern D, we can calculate the expected value*

$$\mathbb{V}^D = \mathbb{E}[V \circ pl] = \sum_{\tau \subseteq T} \prod_{t \in \tau} \frac{\chi_{T \setminus D}(t) \cdot \Lambda(t)}{\sum_{t' \in \mathbb{C}_t \setminus D} \Lambda(t')} [R](\tau).$$

Checking whether some deactivation pattern D exists such that this term is greater than some bound p can be checked by an SMT solver.

Note that, in contrast to the naive definition of $[R]$ only on maximal configurations, this algorithm constructs a reward function on configurations that, for occurrence nets, has a support with at most $\text{supp}(R)$ elements. For arbitrary SAFC nets, the support of $[R]$ might be of exponential size.

Example 6.7. *We take the Petri net from Fig. 5 as an example (where all transitions have firing rate 1). The reward function R is given in the table below. By using the inclusion-exclusion principle we ensure that one obtains reward 1 if one or both of the yellow places are marked at some point without ever marking the red place.*

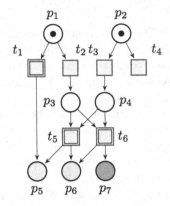

Fig. 5. A SAFC decision net. The goal is to mark one or both of the yellow places at some point without ever marking the red place. (Color figure online)

The optimal strategy is obviously to only deactivate the one transition (t_6) which would mark the red place.

The net has three cells $\mathbb{C}_1 = \{t_1, t_2\}, \mathbb{C}_2 = \{t_3, t_4\}$, and $\mathbb{C}_3 = \{t_5, t_6\}$ where $\mathbb{C}_1, \mathbb{C}_2 \sqsubseteq \mathbb{C}_3$. As such, $R[3] = R$ with R below and obtain $R[2]$ (due to $P_2 = \{p_1, p_2, p_3, p_4, p_5\}$). In the next step, we get (by removing t_3 and t_4) $R[1]$ and finally $R[0]$, from which we can derive $[R]$, the reward function on transitions, as described above.

This allows us to write the value for a set D of deactivated transitions as follows (where if both $t_5, t_6 \in D$, we assume the last quotient to be zero)

$$\mathbb{V}^D = \frac{\chi_{T \setminus D}(t_1)}{\chi_{T \setminus D}(t_1) + 1} + \frac{1}{\chi_{T \setminus D}(t_1) + 1} \frac{1}{2} \frac{\chi_{T \setminus D}(t_5)}{\chi_{T \setminus D}(t_5) + \chi_{T \setminus D}(t_6)}$$

$R = [\{p_5\} : 1, \{p_6\} : 1, \{p_5, p_6\} : -1, \{p_5, p_7\} : -1, \{p_6, p_7\} : -1, \{p_5, p_6, p_7\} : 1]$

$R[2] = [(\{p_5\}, \emptyset) : 1, (\{p_3, p_4\}, \{t_5\}) : 1, (\{p_3, p_4, p_5\}, \{t_6\}) : -1]$

$R[1] = [(\{p_5\}, \emptyset) : 1, (\{p_2, p_3\}, \{t_3, t_5\}) : 1, (\{p_2, p_3, p_5\}, \{t_3, t_6\}) : -1]$

$R[0] = [(\{p_1\}, \{t_1\}) : 1, (\{p_1, p_2\}, \{t_2, t_3, t_5\}) : 1]$

$[R] = [\{t_1\} : 1, \{t_2, t_3, t_5\} : 1]$

Writing $x_i := \chi_{T \setminus D}(t_i) \in \{0, 1\}, i = 1, 5, 6$, the resulting inequality

$$\frac{x_1}{x_1 + 1} + \frac{1}{2} \frac{1}{x_1 + 1} \frac{x_5}{x_5 + x_6} > p$$

can now be solved by an SMT solver with Boolean variables x_1, x_5, and x_6 (i.e., $x_1, x_5, x_6 \in \{0, 1\}$).

Runtime Results: To test the performance of our algorithm, we performed runtime tests on specific families of simple stochastic decision Petri nets, focussing on the impact of concurrency and backward-conflicts on its runtime. All families are based on a series of simple branching cells each containing two transitions, one controllable and one non-controllable, reliant on one place as a precondition. Each non-controllable transition marks a place to which we randomly assigned a reward according to a normal distribution (in particular, it can be negative). The families differ in how these cells are connected, testing performance with concurrency, backward-conflicts, and sequential problems, respectively (for a detailed overview of the experiments see [29]).

Rewriting the reward function (and, thus, solving the value problem) produced expected results: Runtimes on nets with many backward-conflicts are exponential while the rewriting of reward functions of occurrence nets exhibits a much better performance, reflecting its polynomial complexity.

To solve the policy problem based on the rewritten reward function, we compared the performances of naively calculating the values of each possible deactivation pattern with using an SMT solver (Microsoft's z3, see also [12]). Tests showed a clear impact on the representation of the control variables (describing the deactivation set D) as booleans or as integers bounded by 0 and 1 with the latter showing a better performance. Furthermore, the runtime of solving the rewritten formula with an SMT solver showed a high variance on random reward values. Nonetheless, the results show the clear benefit of using the SMT solver on the rewritten formula in scenarios with a high amount of concurrency, with much faster runtimes than the brute force approach. In scenarios without concurrency, this benefit vanishes, and in scenarios with many backward-conflicts, the brute force approach is considerably faster than solving the rewritten function with an SMT solver. The latter effect can be explained by the rewritten reward function $[R]$ having an exponential support in this scenario.

All in all, the runtime results reflect the well-known drawbacks and benefits of most partial-order techniques, excelling in scenarios with high concurrency while having a reduced performance if there are backward- and self-conflicts.

7   Conclusion

We have introduced the formalism of stochastic decision Petri nets and defined its semantics via an encoding into Markov decision processes. It turns out that finding optimal policies for a model that incorporates concurrency, probability and decisions, is a non-trivial task. It is computationally hard even for restricted classes of nets and constant policies. However, we remark that workflow nets are often SAFC nets and a constant deactivation policy is not unreasonable, given that one cannot monitor and control a system all the time. We have also presented an algorithm for the studied subproblem, which we view as a step towards efficient partial-order techniques for stochastic (decision) Petri nets.

Related Work: Petri nets [26] are a well-known and widely studied model of concurrent systems based on consumption and generation of resources. Several subclasses of Petri nets have received attention, among them free-choice nets [13] and occurrence nets, where the latter are obtained by unfolding Petri nets for verification purposes [14].

Our notion of stochastic decision Petri nets is an extension of the well-known model of stochastic Petri nets [21]. This model and a variety of generalizations are used for the quantitative analyses of concurrent systems. Stochastic Petri nets come in a continuous-time and in a discrete-time variant, as treated in this paper. That is, using the terminology of [28], we consider the corresponding Markov chain of jumps, while in the continuous-time case, firing rates determine not only the probability which transition fires next, but also how fast a transition will fire dependent on the marking. These firing times are exponentially distributed, a distribution that is memoryless, meaning that the probability of a transition firing is independent on its waiting time.

Our approach was motivated by extending the probabilistic model of stochastic Petri nets by a mechanism for decision making, as in the extension of Markov chains [28] to Markov decision processes (MDPs) [6]. Since the size of a stochastic Petri net might be exponentially smaller than the Markov chain that it generates, the challenge is to provide efficient methods for determining optimal strategies, preferably partial order methods that avoid the explicit representation of concurrent events in an interleaving semantics. Our complexity results show that the quest for such methods is non-trivial, but some results can be achieved by suitably restricting the considered Petri nets.

A different approach to include decision-making in Petri nets was described by Beccuti et al. as Markov decision Petri nets [4,5]. Their approach, based on a notion of well-formed Petri nets, distinguishes explicitly between a probabilistic part and a non-deterministic part of the Petri net as well as a set of components that control the transitions. They use such nets to model concurrent systems and obtain experimental results. In a similar vein, graph transformation systems – another model of concurrent systems into which Petri nets can be encoded – have been extended to probabilistic graph transformation systems, including decisions in the MDP sense [18]. The decision is to choose a set of rules with the same left-hand side graph and a match, then a randomized choice is made among these rules. Again, the focus is on modelling and to our knowledge neither of these approaches provides complexity results.

Another problem related to the ones considered in this paper is the computation of the expected execution time of a timed probabilistic Petri net as described in [22]. The authors treated timed probabilistic workflow nets (TPWNs) which assumes that every transition requires a fixed duration to fire, separate from the firing probability. They showed that approximating the expected time of a sound SAFC TPWN is #P-hard which is the functional complexity class corresponding to PP. While the problems studied in their paper and in our paper are different, the fact that both papers consider SAFC nets and obtain a #P- respectively PP-hardness result seems interesting and deserves further study.

Our complexity results are closely connected with the analysis of Bayesian networks [25], which are a well-known graphical formalism to represent conditional dependencies among random variables and can be employed to reason about and compactly represent probability distributions. The close relation between Bayesian networks and occurrence nets was observed in [8], which gives a Bayesian network semantics for occurrence nets, based on the notion of branching cells from [1] that were introduced in order to reconcile partial order methods – such as unfoldings – and probability theory. We took inspiration from this reduction in Proposition 3 and another of our reductions (Proposition 5.3) – encoding Petri nets as Bayesian networks – is a transformation going into the other direction, from Bayesian networks to SAFC nets.

In our own work [7,9] we considered a technique for uncertainty reasoning, combining both Petri nets and Bayesian networks, albeit in a rather different setting. There we considered Petri nets with uncertainty, where one has only probabilistic knowledge about the current marking of the net. In this setting Bayesian networks are used to compactly store this probabilistic knowledge and the main challenge is to update respectively rewrite Bayesian networks representing such knowledge whenever the Petri net fires.

Future Work: As future work we plan to consider more general classes of Petri nets, lifting some of the restrictions imposed in this paper. In particular, it would be interesting to extend the method from Sect. 6 to nets that allow infinite runs. Furthermore, dropping the free-choice requirement is desirable, but problematic. While the notion of branching cells does exist for stochastic nets (see [1,8]), it does not accurately reflect the semantics of stochastic nets (see e.g. the discussion on confusion in the introduction of [8]).

As already detailed in the introduction, partial-order methods for analyzing probabilistic systems, modelled for instance by stochastic Petri nets, are in general poorly understood. Hence, it would already be a major result to obtain scalable methods for computing payoffs values for a stochastic net without decisions, but with a high degree of concurrency.

In addition we plan to use the encoding of Petri nets into Bayesian networks from [8] (on which we based the proof of Proposition 5.4) and exploit it to analyze such nets by using dedicated methods for reasoning on Bayesian networks.

Naturally, it would be interesting to extend analysis techniques in such a way that they can deal with uncertainty and derive policies when we have only partial knowledge, as in partially observable Markov decision process (POMDPs), first studied in [3]. However, this seems complex, given the fact that determining the best strategy for POMDPs is a non-trivial problem in itself [10].

Similarly, it is interesting to introduce a notion of time as in continuous-time Markov chains [28], enabling us to compute expected execution times as in [22].

Last but not least, our complexity analysis and algorithm focus on finding optimal constant policies. A natural step would be to instead consider the problem of finding optimal positional strategies as defined in Sect. 3, which is the focus of most works on Markov decision processes (see for example [10]).

References

1. Abbes, S., Benveniste, A.: True-concurrency probabilistic models: branching cells and distributed probabilities for event structures. Inf. Comput. **204**(2), 231–274 (2006)
2. Arora, S., Barak, B.: Computational complexity: a modern approach. Cambridge University Press (2009)
3. Astrom, K.J.: Optimal control of Markov decision processes with incomplete state estimation. J. Math. Anal. Appl. **10**, 174–205 (1965)
4. Beccuti, M., Amparore, E.G., Donatelli, S., Scheftelowitsch, D., Buchholz, P., Franceschinis, G.: Markov decision Petri nets with uncertainty. In: Beltrán, M., Knottenbelt, W., Bradley, J. (eds.) EPEW 2015. LNCS, vol. 9272, pp. 177–192. Springer, Cham (2015). https://doi.org/10.1007/978-3-319-23267-6_12
5. Beccuti, M., Franceschinis, G., Haddad, S.: Markov decision Petri net and Markov decision well-formed net formalisms. In: Kleijn, J., Yakovlev, A. (eds.) ICATPN 2007. LNCS, vol. 4546, pp. 43–62. Springer, Heidelberg (2007). https://doi.org/10.1007/978-3-540-73094-1_6
6. Bellman, R.: A Markovian decision process. J. Math. Mech. **6**(5), 679–684 (1957)
7. Bernemann, R., Cabrera, B., Heckel, R., König, B.: Uncertainty reasoning for probabilistic Petri nets via Bayesian networks. In Proceedings of FSTTCS 2020, vol. 182 of LIPIcs, pp. 1–17. Schloss Dagstuhl - Leibniz Center for Informatics (2020)
8. Bruni, R., Melgratti, H.C., Montanari, U.: Bayesian network semantics for Petri nets. Theor. Comput. Sci. **807**, 95–113 (2020)
9. Cabrera, B., Heindel, T., Heckel, R., König, B.: Updating probabilistic knowledge on condition/event nets using Bayesian networks. In Proceedings of CONCUR 2018, vol. 118 of LIPIcs, pp. 1–17. Schloss Dagstuhl - Leibniz Center for Informatics (2018)
10. Cassandra, A.R.: Exact and approximate algorithms for Markov decision processes, Ph. D. thesis, Brown University, USA (1998)
11. de Campos, C.P.: New complexity results for MAP in Bayesian networks. In: Proceedings of IJCAI 2011, pp. 2100–2106. IJCAI/AAAI (2011)
12. de Moura, L., Bjørner, N.: Z3: an efficient SMT solver. In: Ramakrishnan, C.R., Rehof, J. (eds.) TACAS 2008. LNCS, vol. 4963, pp. 337–340. Springer, Heidelberg (2008). https://doi.org/10.1007/978-3-540-78800-3_24
13. Desel, J., Esparza, J.: Free choice Petri nets. Number 40 in Cambridge Tracts in Theoretical Computer Science. Cambridge University Press (1995)
14. Esparza, J., Heljanko, K.: Unfoldings: a partial order approach to model checking. Springer (2008). https://doi.org/10.1007/978-3-540-77426-6
15. Feinberg, E.A., Shwartz, A. (eds.): Handbook of Markov Decision Processes. Kluwer, Boston, MA (2002)
16. Garey, M.R., Johnson, D.S.: Computers and Intractability. Freeman (1979)
17. Grinstead, C., Snell, L.: Markov chains. In Introduction to Probability, chapter 11, pp. 405–470. American Mathematical Society, second edition (1997)
18. Krause, C., Giese, H.: Probabilistic graph transformation systems. In: Ehrig, H., Engels, G., Kreowski, H.-J., Rozenberg, G. (eds.) ICGT 2012. LNCS, vol. 7562, pp. 311–325. Springer, Heidelberg (2012). https://doi.org/10.1007/978-3-642-33654-6_21
19. Littman, M.L., Dean, T.L., Kaelbling, L.P.: On the complexity of solving Markov decision problems. In: Proceedings of UAI 1995, pp. 394–402. Morgan Kaufmann (1995)

20. Littman, M.L., Majercik, S.M., Pitassi, T.: Stochastic Boolean satisfiability. J. Autom. Reason. **27**(3), 251–296 (2001)
21. Marsan, M.A.: Stochastic Petri nets: an elementary introduction. In: Rozenberg, G. (ed.) APN 1988. LNCS, vol. 424, pp. 1–29. Springer, Heidelberg (1990). https://doi.org/10.1007/3-540-52494-0_23
22. Meyer, P.J., Esparza, J., Offtermatt, P.: Computing the expected execution time of probabilistic workflow nets. In: Vojnar, T., Zhang, L. (eds.) TACAS 2019. LNCS, vol. 11428, pp. 154–171. Springer, Cham (2019). https://doi.org/10.1007/978-3-030-17465-1_9
23. Papadimitriou, C.H.: Computational Complexity. Addison-Wesley (1994)
24. Park, J.D., Darwiche, A.: Complexity results and approximation strategies for MAP explanations. J. Artif. Intell. Res. **21**, 101–133 (2004)
25. Pearl, J.: Causality: models, reasoning, and inference. Cambridge University Press (2000)
26. Reisig, W.: Petri Nets: An Introduction. EATCS Monographs on Theoretical Computer Science. Springer-Verlag, Berlin, Germany (1985)
27. Toda, S.: PP is as hard as the polynomial-time hierarchy. SIAM J. Comput. **20**(5), 865–877 (1991)
28. Tolver, A.: An introduction to Markov chains. University of Copenhagen, Department of Mathematical Sciences (2016)
29. Wittbold, F., Bernemann, R., Heckel, R., Heindel, T., König, B.: Stochastic decision Petri nets (2023). arXiv:2303.13344

Token Trail Semantics – Modeling Behavior of Petri Nets with Labeled Petri Nets

Robin Bergenthum[1]($\boxtimes$), Sabine Folz-Weinstein[2], and Jakub Kovář[3]

[1] Fakultät für Mathematik und Informatik, FernUniversität in Hagen, Hagen, Germany
`robin.bergenthum@fernuni-hagen.de`
[2] Lehrgebiet Softwaretechnik und Theorie der Programmierung, FernUniversität in Hagen, Hagen, Germany
`sabine.folz-weinstein@fernuni-hagen.de`
[3] Lehrgebiet Programmiersysteme, FernUniversität in Hagen, Hagen, Germany
`jakub.kovar@fernuni-hagen.de`

Abstract. There are different semantics for Petri nets. The behavior of a Petri net is either its set of enabled firing sequences, the reachability graph, a set of process nets, a valid partial language, its branching process, or any other known semantics taken from the literature. Every semantics has different advantages in different applications. Some focus on the set of reachable states and can model conflicts well. Other focus on the control flow of actions and can directly specify concurrency. Yet, every semantics has its drawbacks. State graphs explode in size when there is concurrency. Sequential and partial languages explode in size if there is conflict. Furthermore, all semantics use different concepts, definitions, graphical representations, and related algorithms. In this paper, we introduce token trails to define whether a labeled Petri net is in the language of another Petri net. Using labeled Petri nets as a specification language, we show how to faithfully model behavior including conflict and concurrency. Furthermore, we prove that token trail semantics faithfully covers all other semantics of Petri nets and, thus, serves as a kind of meta semantics.

Keywords: Petri nets · Labeled nets · Token trails · Semantics · Compact tokenflows · Modeling behavior · Conflict · Concurrency

1 Introduction

Petri nets [1, 9, 10, 18, 20] have formal semantics, an intuitive graphical representation, and can express conflict and concurrency among the occurrences of actions of a system. Petri nets model actions by transitions, local states by places, and the relations between actions and local states by arcs. We model a state of a Petri net by putting tokens in places. A marked Petri net can change its state by firing a transition. A transition can fire if every place in its pre-set is marked. If a transition fires, it consumes tokens from its pre-set and produces tokens in its post-set. This firing rule is very intuitive and easy to formalize. This surely is a big part of the success of Petri nets as a modeling language and why we love Petri nets.

© The Author(s), under exclusive license to Springer Nature Switzerland AG 2023
L. Gomes and R. Lorenz (Eds.): PETRI NETS 2023, LNCS 13929, pp. 286–306, 2023.
https://doi.org/10.1007/978-3-031-33620-1_16

The firing rule is the core of every Petri net semantics. Although the firing rule itself is simple, there are a lot of different semantics for Petri nets in the literature. There is the sequential language of a Petri net, there are state graphs, partially ordered runs, process nets, branching processes, prime event structures, and so on and so forth. Every semantics has its own advantages and disadvantages in different applications. On the one hand, every semantics is specialized, and we can choose the best fit for every application. On the other hand, it is a mess of different definitions to choose from and even if we choose the correct semantics there are drawbacks inherent to that choice.

For example, repeatedly processing the firing rule creates so called firing sequences. The set of enabled firing sequences is the language of a marked Petri net. This language is very easy to handle but it is not able to specify concurrency. It is easy to come up with a Petri net where we can fire two transitions in any order, but not concurrently. Furthermore, a firing sequence cannot directly specify conflict. When there is conflict, we need one firing sequence for every combination of options in every conflict. Thus, even for a simple Petri net, the size of the language may be huge.

The set of all reachable states, together with the set of all transitions from one state to another, is called the reachability graph of a Petri net. This semantics is still relatively easy to handle. In contrast to firing sequences, these state graphs can very conveniently express conflict, merging, and looping of sequences of actions. But again, state graphs cannot express concurrency. Even worse, if there is concurrency there is the so-called state space explosion where the number of global states grows exponentially in the number of local states.

There are step sequences and state graphs based on multisets of transitions. These semantics can specify concurrent sets of transitions but still, the number of global states explodes just like in every other state graph. Furthermore, sequences of steps are rather technical. Thus, it is neither easy nor intuitive to specify behavior using combinations of sequences of steps.

To model concurrency, there are partially ordered runs [7, 15, 19, 22]. A run is a firing sequence, but the sequence is a partial, not a total order. Thus, it is not sufficient to a have sequences of global markings enabling transitions anymore. We need partially ordered sets of local markings. These sets of markings are called compact tokenflows [3]. Using runs we can easily model concurrent behavior, but just like for firing sequences, it is not possible to directly specify conflict. Again, if there is conflict, we need a run for every possible combination of options.

Compact tokenflows in runs abstract from the history of tokens. We can consider labeled partial orders and regular tokenflows to include the history [16]. We can use so called process nets to include the history and even distinguish individual tokens. Yet, labeled partial orders and process nets have the same disadvantages as runs and compact tokenflows.

We can extend runs and process nets with an additional conflict relation and get prime event structures and branching processes [22]. These semantics can specify concurrency but also merge identical prefixes of runs or process nets. Remark, we can branch but not merge so that these structures fan-out and it is hard to keep track of the relations between the different conflict-free sets of partially ordered nodes. Furthermore, they are not able to directly define looping behavior.

Altogether, we identify two major problems: Firstly, although the definition of a Petri net is easy and clean, the different semantics are all over the place. Just to give one example, the concepts of valid regular tokenflows, enabled cuts in runs, sets of enabled step sequences, process nets, valid compact tokenflows, valid prime event structures, and branching processes all define the same partial language for every Petri net. Yet, there are these different definitions, proofs of their equivalence, different graphical representations, and different algorithms in the literature. Wouldn't it be nice to have some easy to understand meta semantic covering them all? Secondly, it is still not possible to come up with an intuitive and compact graphical representation of the behavior of a Petri net if there is conflict and concurrency.

In this paper, to tackle these problems, we refer the reader to the first sentence of this section. If there is conflict and concurrency, use Petri nets. We introduce token trail semantics for Petri nets. Using token trails, we define whether a labeled Petri net is in the language of another Petri net. To show that this is a valid and useful definition, we prove that if a labeled net models a firing sequence, a state graph, or a run, the labeled net is in the net language of a Petri net if and only if, the firing sequence, the state graph, or the run is in the language of this Petri net as well. Thus, the language defined by token trails will respect and cover all the above-mentioned semantics. Furthermore, we prove that every Petri net is in its own net language. We show examples of how to faithfully model behavior truly specifying conflict and concurrency generating readable graphical representations of executions using general labeled nets. We show how to calculate token trails and introduce a web-tool to demonstrate that token trails are a simple yet very powerful semantics for Petri nets.

2   Preliminaries

Let $\mathbb{N}$ be the non-negative integers. Let f be a function and B be a subset of the domain of f. We write $f|_B$ to denote the restriction of f to B. As usual, we call a function $m \colon A \to \mathbb{N}$ a multiset and write $m = \sum_{a \in A} m(a) \cdot a$ to denote multiplicities of elements in m. Let $m' \colon A \to \mathbb{N}$ be another multiset. We write $m \leq m'$ if $\forall a \in A : m(a) \leq m'(a)$ holds. We denote the transitive closure of an acyclic and finite relation $<$ by $<^*$. We denote the skeleton of $<$ by $<^\diamond$. The skeleton of $<$ is the smallest relation $\vartriangleleft$ so that $\vartriangleleft^* = <^*$ holds. Let $(V, <)$ be some acyclic and finite graph, $(V, <^\diamond)$ is called its Hasse diagram.

We model distributed systems by Petri nets [5, 9, 18, 20].

Definition 1. A Petri net is a tuple (P, T, W) where P is a finite set of places, T is a finite set of transitions so that $P \cap T = \emptyset$ holds, and $W \colon (P \times T) \cup (T \times P) \to \mathbb{N}$ is a multiset of arcs. A marking of (P, T, W) is a multiset $m \colon P \to \mathbb{N}$. Let m_0 be a marking, we call $N = (P, T, W, m_0)$ a marked Petri net and m_0 the initial marking of N.

Figure 1 depicts a marked Petri net. We show transitions as rectangles, places as circles, the multiset of arcs as a set of weighted arcs, and the initial marking as a set of black dots called tokens.

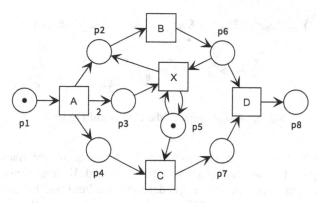

Fig. 1. A marked Petri net.

For Petri nets there is a firing rule. Let t be a transition of a marked Petri net (P, T, W, m_0). We denote $°t = \sum_{p \in P} W(p,t) \cdot p$ the weighted pre-set of t. We denote $t° = \sum_{p \in P} W(t,p) \cdot p$ the weighted post-set of t. A transition t can fire in marking m if $m \geq °t$ holds. Once transition t fires, the marking of the Petri net changes from m to $m' = m - °t + t°$.

In our example marked Petri net, transition A can fire in the initial marking. If A fires, this removes one token from p_1. Additionally, firing A produces a new token in p_2, two new tokens in p_3, and a new token in p_4. In this new marking transitions B and C can fire. A is not enabled anymore, because there are no more tokens in p_1. Firing transition B will enable transition X and transition D. Firing transition C will disable transition X.

Repeatedly processing the firing rule of a Petri net produces so-called firing sequences. These firing sequences are the most basic behavioral model of Petri nets. For example, the sequence $ABXCBD$ is enabled in the marked Petri net of Figure 1. The sequence $ACBD$ is another example. Let N be a marked Petri net, the set of all enabled firing sequences of N is the sequential language of N.

Another formalism to model the behavior of a Petri net is the reachability graph. A marking is reachable if there is a firing sequence that produces this marking. The reachability graph of a marked Petri net $N = (P, T, W, m_0)$ is a tuple (R, T, X) where R is the set of reachable markings of N, T is the set of transitions of N, and X (called transitions as well) is a set of triples in $R \times T \times R$ so that (m, t, m') is in X if and only if t is enabled in (P, T, W, m), and firing t in m leads to the marking m'.

We call a tuple (R', T', X', i) a state graph enabled in N if there is an injective function $g : R' \to R$, $g(i) = m_0$, $T' \subseteq T$, $\forall (m, t, m') \in X' : (g(m), t, g(m')) \in X$, and for every $m' \in R'$ there is a directed path from i to m' using the elements of X' as arcs. Roughly speaking, every node of a state graph relates to a reachable state, we don't have to include all transitions and states if every node can be reached from the initial node. Thus, a state graph is kind of a prefix of a reachability graph. We call the set of enabled state graphs the state language of N.

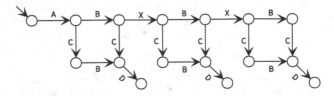

Fig. 2. A state graph of the Petri net of Figure 1.

Figure 2 depicts a state graph modeling the behavior of the marked Petri net depicted in Figure 1. The state graph has 16 states and 18 transitions labeled with transitions of the Petri net. The state graph describes the Petri nets behavior as follows. At first, we must fire transition A. Then, we have some choices. We can execute the loop $BXBXB$ until place p_3 of the Petri net is empty, or we can fire transition C at any time during this loop. As soon as we fire C, we disable transition X by removing a token from place p_5. If there is an occurrence of transition B after the last occurrence of transition X, we can fire transition D once.

The state language includes firing sequences as the set of all paths through the graphs. In this sense, state graphs can merge firing sequences on shared states and can contain loops. Yet, these graphs are not able to directly express concurrency.

Firing A in the initial marking depicted in Figure 1 leads to the marking $p_2 + 2 \cdot p_3 + p_4$. In this marking, transitions B and C can fire concurrently because they don't share tokens. Neither firing sequences nor state graphs can express this concurrency. Two transitions can occur in any order but cannot be executed at the same time. Therefore, there are additional semantics of Petri nets in the literature, able to explicitly express concurrency. There are step semantics of Petri nets [14], process net semantics of Petri nets [13], tokenflow semantics of Petri nets [16], and compact tokenflow semantics of Petri nets [3]. Fortunately, these semantics are equivalent [3, 16, 17, 21] and all define the same partial language. In a partial language, every so-called run is a partially ordered set of events. Obviously, runs can express concurrency and are a very intuitive approach to model behavior of a distributed system.

Definition 2. Let T be a set of labels. A labeled partial order is a triple $(V, \ll, l)$ where V is a finite set of events, $\ll \subseteq V \times V$ is a transitive and irreflexive relation, and the labeling function $l : V \to T$ assigns a label to every event. A run is a triple $(V, <, l)$ iff $(V, <^*, l)$ is a labeled partial order. A run $(V, <, l)$ is also called a labeled Hasse diagram iff $<^\diamond = <$ holds.

Using tokenflow semantics or compact tokenflow semantics, we can decide if a run is in the partial language of a Petri net in polynomial time. Tokenflows, just like branching processes, track the history of tokens. Compact tokenflows define a partially ordered set of local states and thus, abstract from this history. Compact tokenflows are more efficient [4]. Roughly speaking, a compact tokenflow is a distribution of tokens on the arcs of a run so that every event receives enough tokens, no event must pass too many tokens, and all events share tokens from the initial marking.

Definition 3. Let $N = (P, T, W, m_0)$ be a marked Petri net and $run = (V, <, l)$ be a run so that $l(V) \subseteq T$ holds. A compact tokenflow is a function $x : (V \cup <) \to \mathbb{N}$. Let $v \in V$ be an event. We denote $in(v) := x(v) + \sum_{v' < v} x(v', v)$ the inflow of v, and $out(v) = \sum_{v < v'} x(v, v')$ the outflow of v. We define, x is valid for $p \in P$ iff the following conditions hold:

   (i) $\forall v \in V : in(v) \geq W(p, l(v))$,
   (ii) $\forall v \in V : out(v) \leq in(v) + W(l(v), p) - W(p, l(v))$, and
   (iii) $\sum_{v \in V} x(v) \leq m_0(p)$.

run is enabled in N iff there is a valid compact tokenflow for every $p \in P$. The set of all enabled runs of N is the partial language of N.

Figure 3 depicts three different runs modeling the behavior of the marked Petri net depicted in Figure 1. Every run starts with executing transition A. The first run models the concurrent execution of transitions B and C before firing D. The second run models the execution of the loop BXB concurrently to transition C. But transition C can only occur after X because of place p_5. The third run models two times the loop, and again modeling that there is no occurrence of X after the occurrence of C.

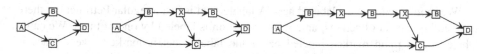

Fig. 3. Three runs of the Petri net of Figure 1.

Figure 4 depicts the branching process of the Petri net of Figure 1. For formal definitions we refer the reader to [13, 20]. Just note, that the maximal conflict-free sets of events of Figure 4 are the three runs of Figure 3. Using the branching process, we merge identical prefixes of the three runs and directly model conflict. But we can only branch and not merge, which is weird when we model a sequence of choices. Only the first choice will be modeled directly, any other following choice will be copied and distributed over the consistency sets of the branching process. Adding such upwards-closed conflict relation comes at a high cost of readability of the modeled behavior.

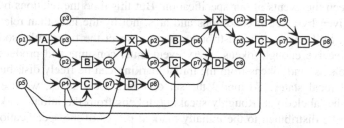

Fig. 4. The branching process of the Petri net of Figure 1.

Figure 2, Figure 3, and Figure 4 all model the behavior of the Petri net of Figure 1. Figure 2 is unable to express the concurrency of transitions B and C. Figure 3 needs three separate runs, because runs cannot contain conflict. Figure 4 is troublesome to read because it cannot merge states as state graphs can. Thus, there will always be some tradeoff choosing one semantics over the others.

3   Token Trails

In this section, we introduce token trails for Petri nets. Using token trails, we define whether a labeled Petri net is in the language of another Petri net.

We define the rise of a transition as the difference between the number of tokens in the pre-set and the number of tokens in the post-set of a transition. Whenever there are arc weights, the rise is the difference between the weighted sums.

Definition 4. Let $N = (P, T, W, m)$ be a marked Petri net, let $t \in T$ be a transition. We denote the weighted sum of tokens $t^{\blacktriangle}$ as $t^{\blacktriangle} := \sum_{(p,t) \in W} W(p,t) \cdot m(p)$, the weighted sum of tokens $t^{\blacktriangle}$ as $t^{\blacktriangle} := \sum_{(p,t) \in W} W(p,t) \cdot m(p)$ and define the rise as $t^{\triangle}$ of transition t as $t^{\triangle} := t^{\blacktriangle} - t^{\blacktriangle}$.

We model behavior by labeled nets. A labeled net is just a regular Petri net but there is an additional set of actions, and every transition is labeled by one of them. We call a labeled net a plain marked labeled net if the labeled net is marked, every place is carrying at most one token in the initial marking, and there are no arc weights.

Definition 5. A labeled net is a tuple (C, E, F, A, l) where (C, E, F) is a Petri net, A is a finite set of actions, and $l : E \rightarrow A$ is an injective labeling function. A marking m of (C, E, F, A, l) is a marking of (C, E, F).
We call (C, E, F, A, l, m) a plain marked labeled net if (C, E, F, A, l, m) is a marked labeled net, $F \leq \sum_{f \in ((C \times E) \cup (E \times C))} f$, and $m \leq \sum_{p \in C} p$ holds.

If a labeled net models behavior, similarly to process nets and branching processes, every transition models an event and every place models a condition. Thus, we also call them events and conditions. The arcs between events and conditions form the control flow of the behavior. Now, we need to define when a labeled net is in the language of a Petri net. We follow the ideas of compact tokenflows and model valid distributions of tokens between the events of our specification. But this time the relations between the events are given by a set of conditions and arcs, not by the later-than relation of the partial order. Still, every valid distribution must respect the firing rule, so that every event must receive enough tokens, every event must consume and produce the right number of tokens, and tokens from the initial marking can be freely distributed over a set of initial local states. To model the set of initial local states, we use an initial marking of the labeled net. Roughly speaking, tokens from the initial marking of the Petri net can be distributed to the initially marked places of the specification.

We call a valid distribution of tokens over the local states of a labeled net a token trail. Such a distribution is a multiset of conditions. Thus, it is straight forward to just formalize token trails as markings of a marked labeled net.

Definition 6. Let $S = (C, E, F, A, l, m_x)$ be a marked labeled net and let m be a marking of S. Let $N = (P, T, W, m_0)$ be a marked Petri net, $A \subseteq T$, and $p \in P$ be a place. The marking m is a token trail for p iff

(I) $\forall e \in E \colon e^{\blacktriangle} \geq W(p, l(e))$,

(II) $\forall e \in E \colon e^{\triangle} = W(l(e), p) - W(p, l(e))$, and

(III) $\sum_{c \in C} m_x(c) \cdot m(c) = m_0(p)$.

S is enabled in N iff there is a token trail for every $p \in P$. The set of all enabled labeled nets of N is the net language of N.

We just kind of brute-forced the definition of a token trail and the definition of the net language of a Petri net. Remark, although conditions (I) and (II) look just like the regular firing rule of Petri nets, they are not. They are derived from conditions (i) and (ii) of compact tokenflows. For example, in the firing rule, we require that there are enough tokens in every place in the pre-set of a transition. Fix a place p and a transition t, we need at least $W(p, t)$ tokens in p. In Definition 6, t needs $W(p, l(t))$ tokens as well, but these tokens can arrive at t over different paths through the labeled net. Tokens could have been produced earlier and then be distributed over the dependencies defined by the conditions of the specification until they arrive at t. Thus, the sum of all tokens distributed over all conditions in the prefix of t count towards the number $W(p, l(t))$ of tokens needed for $l(t)$ to be enabled. A token trail does not model a state of the Petri net, it models a distribution of tokens where tokens can travel using conditions of the specification. Obviously, this idea is directly taken from compact tokenflows, where tokens travel along the defined later-than relation.

In this paper, we lift the idea of compact tokenflows to arbitrary labeled nets. Tokenflows only travel in one direction, can synchronize, but not loop nor merge. Token trails can utilize every condition of a specification, thus, can split, branch, merge, synchronize, and loop. In the remainder of the paper, we show examples and prove that Definition 6 is well-defined. We argue that the net language of a Petri net covers, unifies, and extends existing semantics for Petri nets faithfully.

4 Token Trails for Transition Systems and Partial Languages

In this section, we prove that the net language covers the state language and the partial language of a Petri net. If we model state graphs and runs as marked labeled nets, we only need plain nets.

Let e be a transition of a plain marked labeled net. We denote ${}^{\bullet}e$ and $e^{\bullet}$ the pre-set and the post-set of e. Using plain nets only, we can simply calculate $e^{\blacktriangle} = \sum_{c \in e^{\bullet}} m(c)$ and $e^{\triangle} = \sum_{c \in {}^{\bullet}e} m(c)$. Furthermore, in a plain marked labeled net, the initial marking is one-bound. We simplify condition (III) of Definition 6 to $\sum_{c \in m_x} m(c) = m_0(p)$.

Token Trails for Firing Sequences. A firing sequence specifies sequential behavior of a Petri net. The biggest application dealing with this type of semantics is of course process mining [2]. In process mining, observations of a running workflow system are recorded in (sequential) event logs. This behavioral specification is used to mine, evaluate, operate, and optimize business processes.

It is easy to model a firing sequence as a labeled net. Every event relates to exactly one transition of the firing sequence. Every condition relates to one element of the total order relation. Arcs connect events and conditions to form a sequence. Consequently, there is one place with an empty pre-set, and we mark this place as the starting point by one token in the initial marking of the labeled net.

Fig. 5. A plain marked labeled net modeling the firing sequence *ABXCBD*.

Figure 5 depicts a marked labeled net modeling a sequence of transitions of a firing sequence. In figures of labeled nets, we show the labels of the transitions, not their name. In this example, there are two different transitions in the labeled net both labeled *B*. The token marks the first condition as the initial local state of the sequence.

Definition 8. We call a plain marked labeled net a **sequence net** if there is exactly one place i with an empty prefix, the initial marking is i, there is exactly one place o with an empty postfix, every other place and transition has exactly one predecessor and one successor, and there is a path from i to o visiting all transitions and places.

To construct a token trail for p_1 of Figure 1 in the labeled net depicted in Figure 5, we start by looking at condition (III). There is a token in the initial marking of p_1 of Figure 1 and therefore, we must put this token in the first condition of Figure 5. According to condition (I), the event labeled *A* gets enough tokens and, according to condition (II), this event must consume this token because its rise needs to be -1. Now, every other condition is unmarked, because, according to condition (II), the rise for all other labels must be 0. Remark, we mentioned that the difference between the regular firing rule and the conditions of a token trail is that we sum and weight all tokens in the prefix of an event. In a sequence net there are no weights and there is only one condition in every prefix, thus, there is no actual difference between the regular firing rule and the rules of token trails.

Figure 6 depicts eight copies of the sequence net of Figure 5. The first copy shows the token trail for place $p1$ of Figure 1. The second copy shows the token trail for $p2$ of Figure 1 and so forth. There is a token trail for every place of Figure 1 and thus, the sequence net depicted in Figure 5 is in the net language of the Petri net of Figure 1.

We look at the columns of conditions in Figure 6 to directly see the one-to-one relation between a set of token trails of a sequence net and the set of markings enabling a firing sequence. This comes down to the fact that transitions neither branch nor merge, so we can prove the following theorem.

Theorem 1. Let S be a sequence net. S models a firing sequence of a Petri net N iff there is a token trail in S for every place of N. Furthermore, the token trail in S for a place p is the p-component of markings generated by the firing sequence.

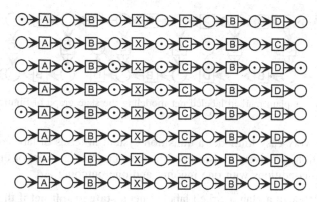

Fig. 6. Eight token trails for the places of Figure 1 in copies of the labeled net of Figure 5.

Proof. Let $t_0 t_1 \ldots t_n$ be a firing sequence of $N = (P, T, W, m_0)$. There is a unique sequence of markings $m_1 m_2 \ldots m_{n+1}$ so that t_i is enabled in m_i, and firing t_i in m_i yields m_{i+1}. If a marked sequence net $S = (C, E, F, A, l, m_x)$ models the firing sequence $t_0 t_1 \ldots t_n$, we can rename the elements of the net as follows: There is one place c_0 with an empty prefix carrying one token, transitions $e_0, e_1, \ldots, e_n$ labeled $l(e_i) = t_i$ and places $c_1, c_2, \ldots, c_{n+1}$ so that the multiset of arcs is $\sum_i ((c_i, e_i) + (e_i, c_{i+1}))$, $A = T$, and $m_x = c_0$.

Fix a place $p \in P$. The marking $m = \sum_i m_i(p) \cdot c_i$ of S is the token trail for p because $\sum_{c \in m_x} m(c) = m(c_0) = m_0(p)$ and thus condition (III) holds. For all e_i, firing t_i in m_i yields m_{i+1}. Thus, $m_{i+1} = m_i - {}^\circ t_i + t_i^\circ$ holds. We look at the p-component of this equation to get $m_{i+1}(p) = m_i(p) - W(p, t_i) + W(t_i, p)$ and thus, $e_i^\Delta = m(c_{i+1}) - m(c_i) = m_{i+1}(p) - m_i(p) = W(l(e_i), p) - W(p, l(e_i))$. This is condition (II). For all t_i, t_i is enabled in m_i. Thus, $m_i \geq {}^\circ t_i$ holds. Again, $m_i(p) \geq W(p, t_i)$ and $\sum_{c \in {}^\bullet e_i} m(c) = m(c_i) = m_i(p) \geq W(p, l(e_i))$ and thus condition (I) holds as well. Furthermore, for every p, m is completely defined by the sequence of markings $m_1 m_2 \ldots m_{n+1}$.

Let S be a sequence net and for every $p \in P$ let m_p be a token trail for p. Conditions (I), (II), and (III) hold and we use the same arguments as above backwards to construct the p-components of a sequence of markings $m_1 m_2 \ldots m_{n+1}$ enabling $t_0 t_1 \ldots t_n$ in m_0. ∎

Theorem 1 shows that token trails respect the definitions of firing sequences. A firing sequence is in the sequential language of a Petri net if and only if the related sequence net is in the net language.

Token Trails for Transition Systems. A firing sequence models behavior as a sequence of actions. We use transition systems to model behavior focused on states. The biggest application dealing with this type of semantics is asynchronous circuit design [8]. In circuit design, we specify behavior as a transition system and use region-based approaches to synthesize a Petri net to be implemented.

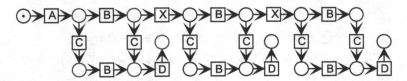

Fig. 7. A plain marked labeled net modeling the state graph of Figure 2.

Again, it is very easy to model a transition system in terms of labeled nets. Every state is a condition, and every transition of the state graph is an event of the labeled net connecting two conditions with one ingoing and one outgoing arc.

Definition 9. We call a plain marked labeled net a **state graph net** if there is exactly one place i so that there is a path from i to any other place, the initial marking is i, and every transition has exactly one predecessor and one successor.

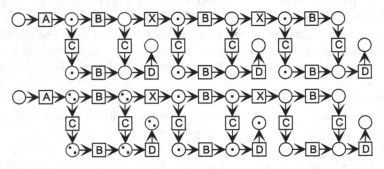

Fig. 8. The token trails for the place p_2 and p_3 of Figure 1 in the labeled net of Figure 7.

Figure 8 depicts two copies of the state graph net of Figure 7 with two token trails. The marking of the first state graph net is a token trail for p_2 of Figure 1. The initial marking of p_2 is 0, thus, the marking of the initial place of Figure 8 must be 0 as well. The rise of the event labeled A is 1, all events labeled B have a rise of -1, all events labeled C or D have a rise of 0, and all events labeled X have a rise of 1. The marking of the second state graph net of Figure 8 is a token trail for place p_3 of Figure 1.

Like in sequence nets, events of a state graph net do not branch. Thus, there is just one place in the pre-set of every event where tokens can arrive, and there is only one place where an event can pass the tokens to. We build a token trail as follows: we put the number of initial tokens in the initial place of the labeled net. Then we start a breadth-first search at the initial place. For every visited event, we have the number of ingoing tokens and just calculate the tokens to put in the successor place, using the rise of the label. This will construct a token trail if merging paths agree on the number of tokens and if no marking must be negative. Obviously, this is the case if the state graph net models reachable states of the original Petri net, only considering the place p. Like for sequence nets, there is a one-to-one relation between a set of token trails of a state graph net and the set of markings of a state graph in the language of the Petri net.

Theorem 2. Let S be a state graph net. S models a state graph of a Petri net N iff there is a token trail in S for every place of N. Furthermore, the token trail in S for a place p is the p-component of every state of the state graph.

Proof. Let $G = (R, T, X, i)$ be a state graph of $N = (P, T, W, m_0)$. For every transition $(m', t, m'') \in X$, t is enabled in m' and firing t in m' yields m''. If a marked state graph net $S = (C, E, F, A, l, m_x)$ models G, we can rename this net as follows: there is a transition $e_{(m',t,m'')}$ labeled $l\left(e_{(m',t,m'')}\right) = t$ for every transition, and a places p_r for every state $r \in R$ so that the multiset of arcs is $\sum_{(m',t,m'') \in X} ((p_{m'}, t) + (t, p_{m''}))$, $A = T$, and $m_x = p_i$.

For every state s of G there is a cycle free path of transitions leading from the initial state to s. This path is a firing sequence. $S = (C, E, F, A, l, m_x)$ models G and all transitions of S have exactly one predecessor and exactly one successor. The path in G relates to a subnet in S so that this net is a sequence net. Fix a place $p \in P$, we apply Theorem 1 to get $m = \sum_{r \in R} r(p) \cdot p_r$ is the only candidate for a token trail in S for p. Again, there is a one-to-one relation between the p-component of every state of the state graph and the token trail in S for p. Even if a transition (m', t, m'') in G is part of a cycle, $e_{(m',t,m'')}$ is unbranched so that, $e_{(m',t,m'')}^{\triangle} = m_{m''}(p) - m_{m'}(p) = m''(p) - m'(p) = W(t, p) - W(p, t)$ and $\sum_{c \in {}^{\bullet}e_{(m',t,m'')}} m(c) = m_{m'}(p) = m'(p) \geq W(p, t)$ holds. m is a token trail for p. Again, the p-components of the set of states define the token trail for every place. Using the same arguments backwards, we also get the other direction. ∎

Theorem 2 shows that token trails respect the definitions of state graphs. A transition system is a state graph of a Petri net if and only if the related state graph net is in the net language.

Token Trails for Partial Languages. A partial language models behavior as partially ordered sequences of actions. Using partial languages, we can model concurrency of action occurrences. The biggest application dealing with this kind of semantics is business process management [12]. Modern business processes, like for example Order-to-Cash, Quote-to-Order workflow processes, are distributed over different people, departments, and systems of a company, so that we model these processes using partially ordered runs.

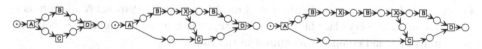

Fig. 9. Three labeled nets modeling the partially ordered sequences of actions of Figure 3.

Modeling a partial order of events in terms of labeled nets is easy. Every event is a labeled transition, and we model the skeleton of the partial order by a set of conditions. We add a condition between two transitions whenever there is a later-than relation between the two related events. Thus, every condition has at most one ingoing and one outgoing arc. There is concurrency but no conflict. Figure 9 depicts three labeled nets modeling three runs.

Definition 10. We call a plain marked labeled net a **partial order net** if the net is acyclic, every transition has at least one ingoing and at least one outgoing arc, every place has at most one ingoing and at most one outgoing arc, and the initial marking is the sum of places with an empty prefix.

Figure 10 depicts three copies of the three partial order nets of Figure 9 with token trail markings. For partial order nets, for the first time in this paper, we actually have to sum-up ingoing and outgoing tokens in order to calculate the rise of a transition. For example, the rise of transition A is 1 in the first row of examples, 2 in the second row, and 0 in the last row.

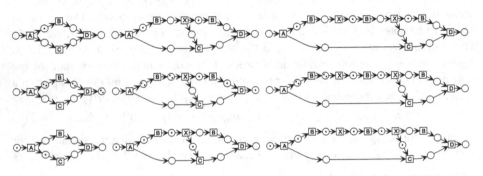

Fig. 10. Token trails for places p_2, p_3, and p_5 of Figure 1 in the labeled nets of Figure 9.

In Figure 10, the token trails of the first row all relate to place p_2 of Figure 1. The token trails of the second row relate to place p_3, and the token trails of the last row relate to p_5. Remark, in partial order nets there is not a one-to-one relation between the rises of events to token trails anymore. For example, the token trail in the middle of Figure 10 relates to place p_3. One token travels from X to the final marking. This token could also travel via C and thus lead to another token trail related to p_3. For partial order nets, there is a one-to-one relation between a token trail in the partial order net and a compact tokenflow in the labeled Hasse diagram.

Theorem 3. Let S be a partial order net. S models a run of a Petri net N iff there is a token trail in S for every place of N. Furthermore, there is a one-to-one relation between token trails in S and compact tokenflows in the run.

Proof. Let $run = (V, <, l)$ be a run of $N = (P, T, W, m_0)$. For every $p \in P$ there is a compact tokenflow x in run for p. If a marked partial order net $S = (C, E, F, A, l, m_x)$ models run, we can rename this net as follows: for every event $v \in V$ there is a transition $e_v \in E$ with $l(e_v) = l(v)$, for every arc $(v, v') \in {<}$ there is a place $c_{(v,v')} \in C$ so that ${}^\bullet c_{(v,v')} = \{e_v\}$ and $c_{(v,v')}{}^\bullet = \{e_{v'}\}$. For every $v \in V$ with an empty prefix there is a place c_v^i in the pre-set of v, for every $v \in V$ with an empty postfix there is a place c_v^f in the post-set of v, $A = T$, and $m_x = \sum_{c \in C, {}^\bullet c = \emptyset} c$.

Fix a place $p \in P$ and its compact tokenflow x. Conditions (i), (ii), (iii) hold. The main idea is to construct another valid compact tokenflow x so that (b) and (c) hold as well.

(b) $\forall v \in V, v^\bullet \neq \emptyset : out(v) = in(v) + W(l(v), p) - W(p, l(v))$,
(c) $\sum_{v \in V, {}^\bullet v \neq \emptyset} x(c) = m_0(p)$.

As long as there is an event v with a non-empty prefix so that its outflow is not yet as big as the inflow plus $W(l(v), p) - W(p, l(v))$, there is a path from v to an event with an empty post-set. We can add the missing tokens to every arc of this path to fix the outflow of v. For every event on this path, inflow and outflow will be increased by the same number, for the last event only the inflow will be increased so that (i), (ii), (iii) still hold. We repeat until (b) holds. If there is an event v with a non-empty prefix so that $x(v) > 0$, there is a path from an event v' with an empty prefix to v. Again, we add $x(v)$ to $x(v')$, add $x(v)$ tokens to every event on the path and set $x(v)$ to 0 to move tokens consumed from the initial marking to the initial events, only increasing inflow and outflow of every event by the same amount. Now, the value of the compact tokenflow is 0 on non-initial events. If the sum of this tokenflow is not yet as big as the number of tokens in the initial marking, we add tokens on a path from an event with an empty prefix to an event with an empty postfix. Altogether, we construct a compact tokenflow x so that (i), (ii), (iii), (b), and (c) hold.

We construct a token trail m in S for p as follows. $m := \sum_{(v,v') \in <} x(v, v') \cdot c_{(v,v')} + \sum_{v \in V, {}^\bullet v = \emptyset} x(v) \cdot c_v^i + \sum_{v \in V, v^\bullet = \emptyset} (in(v) + W(l(v), p) - W(p, l(v))) \cdot c_v^f$.

m is a token trail for p because the only difference between x and m is that we moved tokens from the initial marking to the initial places and added the right number of tokens to the final places so that (b) implies (II). Obviously, (i) implies (I), and (c) implies (III) because all initial places of S are marked with one token in m_x.

This time the other direction is even simpler because we build a valid compact tokenflow from a token trail by just ignoring tokens in the final places and moving tokens from the initial places to the initial events. Now, (I), (II), and (III) imply (i), (ii), and (iii) directly without even adapting tokenflow on the edges of *run*. ∎

Theorem 3 shows that token trails cover the definitions of compact tokenflows. A run is in the partial language of a Petri net if and only if the related partial order net is in the net language.

Like in the previous subsection, going from firing sequences to state graphs, we can add conflict to a run of a Petri net. To add conflict, we use the formalism of labeled prime event structures of a Petri net. The main idea is that there is an additional conflict relation so that sets of conflict-free events are runs. The conflict relation is upwards closed so that runs branch, but do not merge. Thus, every prime event structure is covered by runs and is enabled in a Petri net if every run of the prime event structure is enabled with a compact token flow so that the flows match on shared prefixes of the prime event structure. We add this idea to the proof of Theorem 3, like going from proof of Theorem 1 to the proof of Theorem 2, to get that there are also matching token trails.

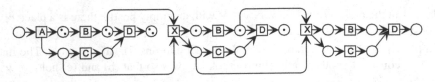

Fig. 11. Labeled net modeling the prime event structure for the three runs of Figure 9 marked with a token trail for place p_3 of Figure 1.

Figure 11 depicts a labeled net modeling a prime event structure. The conflict free sets of events of this net are the runs of Figure 9. In this prime event structure net we don't show its initial marking (just the first condition), but depict a token trail for place p_3 of Figure 1 as a composition of the three token trails for p_3 depicted in the second row of Figure 10.

A labeled net is a branching process of a Petri net if there is an additional one-to-one relation between tokens in the Petri net and tokens in the token trails. This must be examined in future work. Up to this point, the net language covers runs, as well as prime event structures of a Petri net.

5   Token Trails for Labeled Petri Nets

In Section 4, we proved that token trails cover firing sequences, state graphs, and partial languages using plain marked labeled nets only. In this section, we show that the token trail semantics is a well-defined generalization of existing semantics, and we show how to model behavior using general marked labeled nets. We show examples of labeled nets and Petri nets to argue that some labeled nets model behavior of these Petri nets and some do not.

Token Trails for Arbitrary Initial Markings. In the definition of a token trail, we consider arbitrary initial markings. Up to this point, only modeling firing sequences, state graphs, and partial languages there is no need to put multiple tokens in conditions.

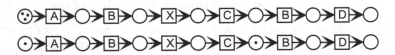

Fig. 12. Two marked labeled nets.

The first labeled net in Figure 12 depicts our running example sequentially ordered net, but this time with an initial marking of three tokens in the first place. Thus, it is neither a plain marked labeled net, nor a sequence net anymore. However, if we consider the token trails of Figure 6, trails number 2, 3, 4, 6, 7, and 8 are still token trails for the related places p_2, p_3, p_4, p_6, p_7, and p_8 of Figure 1. If we change the initial marking of Figure 1 to $3 \cdot p_1 + 3 \cdot p_5$, then token trails 1 and 5 would relate to p_1 and p_5. The first labeled net in Figure 12 is only in the net language of a Petri net if the sequence $ABXCBD$ can be executed concurrently three times to itself. This perfectly matches our intuition looking at the first net of Figure 12.

The second labeled net in Figure 12 depicts an even more sophisticated initial marking. We specify that we start the sequence at the beginning and simultaneously before the last two actions B and D. In this example, the first token trail of Figure 6 is still a trail for p_1 of Figure 1 because the fifth place of the sequence is not marked so that the sum defined by condition (III) is 1 and thus, is the initial marking of p_1 in Figure 1. Similarly, the token trails 4, 5, 6, and 8 still hold for their related places. If we consider places p_2 and p_7, we must add one token each to the initial marking of the Petri net of Figure 1 to fix trails number 2 and 7. This is consistent with the intuition behind the specification in Figure 12 because the marked Petri net should be able to execute BD from the initial marking. However, something is strange about the third token trail of Figure 6 because p_3 is not connected to transitions B and D but still, we must add one token to the initial marking of p_3 so that the third token trail of Figure 6 is a token trail for p_3. But if we really think about the behavior we specify, we see that in Figure 12, using the first initial token, the sequence of events $ABXCB$ reaches the same condition as using the second initial token and just execute the second event labeled B. In other words, we specify that we want to have a Petri net so that firing $ABXCB$ and firing B leads to the same state. This is only satisfied if we add one additional token to p_3.

We can add an initial marking to any kind of labeled net to model a multiset of local states as starting points. This is the main idea of condition (III) of Definition 6.

Token Trails for Merged Local States. Figure 13 depicts labeled nets modeling the behavior of our running example in terms of marked labeled nets. There is conflict, a merge, a split, and synchronization. Obviously, they are neither a state graph net nor a partial order net. The upper net of Figure 13 specifies behavior where we start by an action A. Then there is a split. In the top part of the split, we have the choice between the sequence just B, or the sequence of actions BXB. Executing either of these two sequences will lead to the merge at the local state c_6. In parallel to this choice, we have action C followed by a synchronization using action D. This kind of control flow seems to fit the Petri net depicted in Figure 1.

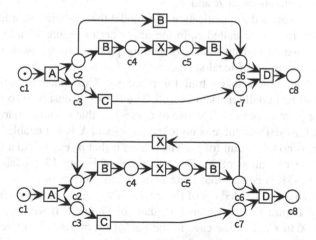

Fig. 13. A marked labeled net with a merge and a marked labeled net with a loop.

Now, we check if the upper labeled net of Figure 13 is in the net language of the Petri net of Figure 1 by constructing a token trail for every place. Figure 14 depicts the token trails for places p_1, p_2, p_6, p_4, p_7, and p_8.

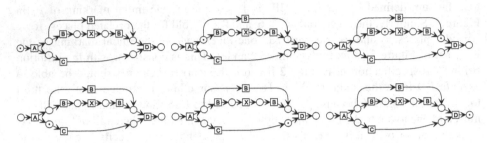

Fig. 14. Token trails in copies of the labeled net of Figure 13 for some places of Figure 1.

We still miss a token trail for the other places. For the place p_3, the rise of X must be -1 and thus, we need at least one token in c_4. Assume there are n tokens in c_4. The rise of B must be 0 and thus, there must be n tokens in c_2 as well. With the same argument there must be n tokens in c_6, and there must be n tokens in c_5, all because B must have a rise of 0. Consequently, the number of tokens in c_4 is the number of tokens in c_5, therefore, the rise of X cannot be -1. There is no token trail for the place p_3 in the upper labeled net of Figure 13. Remark, this is perfectly fine with our intuition! In the first labeled net of Figure 13, we specify that executing the loop, and not executing the loop, leads to the same local state c_6. Thus, if the upper net of Figure 13 is in the net language of some Petri net, executing or skipping the loop must lead to the same state. This holds for places p_1, p_2, p_6, p_4, p_7, and p_8, but not for place p_3. If we skip the loop, there are two tokens left in p_3. If we execute the loop once, there is one token left in p_3. Altogether, the upper labeled net in Figure 13 specifies that counting iterations is not allowed. It is only in the net language of Figure 1 if we were to delete p_3 and thus, allow for arbitrary iterations of B and X.

If we want to change the specification and model that counting should be possible, we must split the local state called c_6 in the upper net of Figure 13 into two separate states. Thus, we would end up with a branching process. Here, token trail semantic perfectly handles merging of local states.

We are still missing a token trail for place p_5. This place is initially marked, therefore, c_1 must be initially marked as well. The rise of A must be 0 so that the initial token can either go to c_2 or to c_3. The rise of C is -1 so that we must mark c_3 and there is no token left for c_2. Thus, there is no token for c_4 and X is not enabled according to (I). Again, there is no token trail for p_5. The reason is that p_5 ensures that C can only be executed after the execution of X. The upper net of Figure 13 models C and X as independent. Thus, it is correct that there is no token trail for p_5.

Summing up the first labeled net of Figure 13, if we delete p_3 and p_5 from Figure 1, the upper net of Figure 13 is in the net language of Figure 1. If we keep p_5 and add a condition from X to C, as is the case in the partial order nets of Figure 9, the initial token can go to X and another token from X to C. This labeled net would be in the net language of Figure 1 without p_3. If we want to count loops, we must split c_6 to allow a branched local state. These examples highlight how token trails deal with merging of alternative executions and shared local states.

Token Trails for Loops. We just indirectly specified looping behavior using a shared local state. What if the specification directly models a loop? The second labeled net of Figure 13 depicts an example where we specify a similar behavior but this time using a loop. Again, there is a loop of B and X in parallel to the action C. Here, we directly specify that there is at least one execution of X and at least two executions of B before another X can return to the local state c_2. With the same arguments as before, we can't construct a token trail for place p_3 where X (or B) is producing, and D is consuming tokens because we must always be able to go back to c_2. Again, it is not possible to count Bs and Xs.

We can copy the token trails of Figure 14 into the second labeled net of Figure 13 to directly see that there are token trails for the places p_1, p_2, p_6, p_4, p_7, and p_8. There is no token trail for p_3 because of the merged local state c_2 and there is no token trail for p_5 because C and X are modeled as independent again. Thus, again token trails match our intuition and handle Figure 13 correctly, although it is a specification with a loop.

To really put token trails to the test we introduce the first Petri net of Figure 15. This example might look strange at first, but we want to have a bounded Petri net where we must execute the loop of our running example at least once. Here, after the first execution of B the place p_5 is not marked yet. We execute X to mark p_5 and enable B again. After firing B a second time p_5 and p_6 are marked, enabling D. We can fire Y to move a token from p_5 to p_3 to reset the marking to the state already visited after firing the first B.

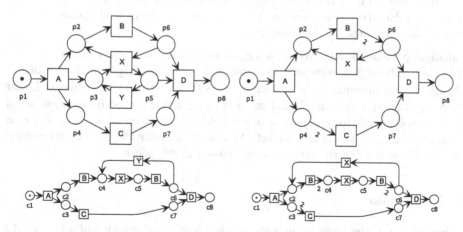

Fig. 15. Two Petri nets and two labeled nets of their languages.

In Figure 15, the first labeled net is in the net language of the first Petri net. Intuitively, Figure 15 specifies that neither the local states c_2 and c_5 nor the local states c_4 and c_6 are merged. Using an additional action Y, we can distinguish entering and re-entering the loop. Here, the token trail for place p_5 is simply $c_5 + c_6$. The token trail for place p_3 is simply $c_2 + c_4$. Token trails faithfully handle merges and loops.

Token Trails for Weighted Arcs. The second Petri net of Figure 15 depicts our running example net with additional arc weights. The second labeled net of the same figure is in its net language. We need condition c_3 twice to execute the event labeled C. Whenever we execute an event labeled B, we produce condition c_4 or c_6 twice. The marking c_3 in the labeled net is a token trail of p_4. The rise of the event labeled A is still 1, the rise of the event labeled C is -2. The marking $c_4 + c_6$ is a token trail for p_6. Remark, in the second Petri net of Figure 15, C is not enabled after firing A in the initial marking, but the same holds for the depicted labeled net. Token trails just respect condition c_3 and the related arc weights.

At the end of this section, we show one more strong argument highlighting that the net language is well-defined. We prove that every labeled net without duplicate labels is in its own net language.

Theorem 4. Let S be a marked labeled net without duplicate labels, so that S models a Petri net N. S is in the net language of N.

Proof. Let $N = (P, T, W, m_0)$ be a Petri net and $S = (C, E, F, A, l, m_x)$ be a labeled net without duplicate labels. S models N, so we rename all elements of S so that $S = (P, T, W, T, id, m_0)$ holds. Fix a place $p \in P$ in N, $m = p$ is a token trail for p in S because $m(c)$ is only 1 for $c = p$ and 0 for any other condition.
Conditions (I), (II), and (III) hold because $\forall e \in T : \sum_{c \in \bullet e} W(c, e) \cdot m(c) = W(p, t)$, $\forall e \in T : \sum_{c \in P}(W(e, c) - W(c, e)) \cdot m(c) = (W(e, p) - W(p, e)) \cdot m(p) = W(e, p) - W(p, e)$, and $\sum_{c \in C} m_0(c) \cdot m(c) = m_0(p)$. ∎

Every Petri net is in its own net language. Token trails work perfectly fine for any kind of state-based or event-based specification, as well as for general labeled nets with loops, initial markings, and arc weights.

Calculating Token Trails. We implemented token trails as a new module of the I ♥ Petri Nets website. The website is available at www.fernuni-hagen.de/ilovepetrinets/. The 🦊 module implements the conditions of Definition 6 as a simple Integer Linear Program. We can drag a labeled net and a Petri net to the related Buttons and see if there are token trails for every place of the Petri net. Click on some place to see an example token trail in the labeled net. At www.fernuni-hagen.de/ilovepetrinets/fox/ please find the webtool and all the examples used in this paper.

6   Conclusion

In this paper, we introduced token trails. A token trail is a distribution of tokens on the set of conditions of a labeled net respecting the consumption and production of tokens of the labels of the events. Whenever there is conflict, a token trail must agree on the number of tokens in every condition of a labeled net. Whenever there is concurrency, a token trail can distribute tokens to local states. We have proven that token trails cover firing sequences, state graphs, and partial languages of Petri nets. Furthermore, they faithfully extend Petri net semantics to labeled nets. Token trails have a very intuitive graphical representation and are very easy to calculate. In addition to Petri nets, we only need the additional concept of labels to directly define token trail semantics. For all these reasons, we see token trails as a kind of meta semantics for Petri nets.

Besides the rather formal stuff, we see a lot of applications for token trails. In future work, we will define synthesis based on token trails. The goal is to come up with a framework so that we specify behavior of a system in terms of labeled nets and get the resulting Petri net for free. Using token trails, we can unify the definitions of state-based and event-based regions. We refer the reader to the workshop paper [6] for a first glimpse at this new region definition. Roughly speaking, we will specify labeled nets and calculate a set of places for the set of labels so that for every place there is a token trail using the ILP defined in Definition 6. We use the concept of minimal token trails (i.e., minimal number of tokens) to get a finite result. A prototype of the approach is already implemented in the 🏠 module of the I ♥ Petri Nets website.

References

1. van der Aalst, W.M.P., van Dongen, B.F.: Discovering Petri Nets from Event Logs. In: Jensen, K., van der Aalst, W.M.P., Balbo, G., Koutny, M., Wolf, K. (eds.) Transactions on Petri Nets and Other Models of Concurrency VII. LNCS, vol. 7480, pp. 372–422. Springer, Heidelberg (2013). https://doi.org/10.1007/978-3-642-38143-0_10
2. van der Aalst, W.M.P., Carmona, J.: Process Mining Handbook. Springer (2022). https://doi.org/10.1007/978-3-031-08848-3
3. Bergenthum, R., Lorenz, R.: Verification of Scenarios in Petri Nets Using Compact Tokenflows. In: Fundamenta Informaticae, vol. 137, no. 1, pp. 117–142. IOS Press (2015)
4. Bergenthum, R.: Firing Partial Orders in a Petri Net. In: Buchs, D., Carmona, J. (eds.) PETRI NETS 2021. LNCS, vol. 12734, pp. 399–419. Springer, Cham (2021). https://doi.org/10.1007/978-3-030-76983-3_20
5. Bergenthum, R.: Petrinetze: Grundlagen der Formalen Prozessanalyse. In: Prozessmanagement und Process-Mining, De Gruyter Studium, pp. 125–152. De Gruyter (2021)
6. Bergenthum, R., Kovar, J.: A First Glimpse at Petri Net Regions. In: Proceedings of Application and Theory of Petri Nets 2022, CEUR Workshop Proceedings 3167, pp. 60–68 (2022)
7. Best, E., Devillers, R.: Sequential and Concurrent Behaviour in Petri Net Theory. In: Theoretical Computer Science 55, nr. 1, pp. 87–136. Elsevier (1987)
8. Cortadella, J., Kishinevsky, M., Kondratyev, A., Lavagno, L., Yakovlev, A.: Hardware and Petri Nets Application to Asynchronous Circuit Design. In: Nielsen, M., Simpson, D. (eds.) ICATPN 2000. LNCS, vol. 1825, pp. 1–15. Springer, Heidelberg (2000). https://doi.org/10.1007/3-540-44988-4_1
9. Desel, J., Reisig, W.: Place/Transition Petri Nets. In: Reisig, W., Rozenberg, G. (eds.) ACPN 1996. LNCS, vol. 1491, pp. 122–173. Springer, Heidelberg (1998). https://doi.org/10.1007/3-540-65306-6_15
10. Desel, J., Juhás, G.: What is a Petri Net? In: Ehrig, H., Juhás, G., Padberg, J., Rozenberg, G. (eds.) Unifying Petri Nets, Advances in Petri Nets, LNCS 2128, pp. 1–25. Springer, Cham (2001). https://doi.org/10.1007/3-540-45541-8_1

11. van Dongen, B.F., de Medeiros, A.K.A., Verbeek, H.M.W., Weijters, A.J.M.M., van der Aalst, W.M.P.: The ProM Framework: A New Era in Process Mining Tool Support. In: Ciardo, G., Darondeau, P. (eds.) ICATPN 2005. LNCS, vol. 3536, pp. 444–454. Springer, Heidelberg (2005). https://doi.org/10.1007/11494744_25

12. Dumas, M., La Rosa, M., Mendling, J., Reijers, H.A.: Fundamentals of Business Process Management. Springer, Heidelberg (2013). https://doi.org/10.1007/978-3-642-33143-5

13. Goltz, U., Reisig, W.: Processes of Place/Transition-Nets. In: Diaz, J. (eds.) Automata Languages and Programming, vol. 154, pp. 264–277. Springer, Heidelberg (1983). https://doi.org/10.1007/BFb0036914

14. Grabowski, J.: On Partial Languages. In: Fundamenta Informaticae, vol. 4, no. 2, pp. 427–498. IOS Press (1981)

15. Janicki, R., Koutny, M.: Structure of Concurrency. In: Theoretical Computer Science 112, no. 1, pp. 5–52. Elsevier (1993)

16. Juhás, G., Lorenz, R., Desel, J.: Can I Execute My Scenario in Your Net? In: Ciardo, G., Darondeau, P. (eds.) Proceedings of Application and Theory of Petri Nets 2005, LNCS 3536, pp. 289–308. Springer, Heidelberg (2005). https://doi.org/10.1007/11494744_17

17. Kiehn, A.: On the Interrelation Between Synchronized and Non-Synchronized Behavior of Petri Nets. In: Elektronische Informationsverarbeitung und Kybernetik, vol. 24, no. 1–2, pp. 3–18 (1988)

18. Peterson, J.L.: Petri Net Theory and the Modeling of Systems. Prentice-Hall, Englewood Cliffs (1981)

19. Pratt, V.: Modelling Concurrency with Partial Orders. In: International Journal of ParallelProgramming 15, pp. 33–71 (1986)

20. Reisig, W.: Understanding Petri Nets - Modeling Techniques, Analysis Methods, Case Studies. Springer, Heidelberg (2013). https://doi.org/10.1007/978-3-642-33278-4

21. Vogler, W. (ed.): Modular Construction and Partial Order Semantics of Petri Nets. LNCS, vol. 625. Springer, Heidelberg (1992). https://doi.org/10.1007/3-540-55767-9

22. Winskel, G.: Event Structures. In: Brauer, W., Reisig, W., Rozenberg, G. (eds.) ACPN 1986. LNCS, vol. 255, pp. 325–392. Springer, Heidelberg (1987). https://doi.org/10.1007/3-540-17906-2_31

On the Reversibility of Circular Conservative Petri Nets

Raymond Devillers[(⊠)]

Université Libre de Bruxelles, Boulevard du Triomphe C.P. 212,
1050 Bruxelles, Belgium
raymond.devillers@ulb.be

Abstract. The paper examines how to decide if a given (initially marked) Petri net is reversible, i.e., may always return to the initial situation. In particular, it concentrates on a very specific subclass of weighted circuits where the total number of tokens is constant, for which the worst case complexity is not known. Various ways to tackle the problem are considered, and some subcases are derived for which the problem is more or less easy.

Keywords: Petri net · weighted circuit · reversibility · polynomial complexity

1 Introduction

A specific feature of P/T Petri nets is that many interesting problems are decidable, but their (worst case) complexity may be huge and hard to analyse. For instance, it is known since the eigthies that the reachability problem is decidable [9,12,13] (while it becomes undecidable if we add at least two inhibitor arcs [15]), but only from 2018 that the worst case complexity is ackermanian [4,5,10]. As a consequence several other decidable properties, like the existence of home states [1], the reversibility (is the initial marking a home state?) [7],... are also (at least) ackermanian. An idea is then to consider structural subclasses to determine if some of these properties become easier to check. In this respect, a colleague[1] incidentally mentioned to me that it was not even known if the reversibility for a very special subclass of weighted circuits, where the total number of tokens is constant, is polynomial or exponential.

That is the problem we shall tackle in this paper, considering various approaches, exhibiting necessary and sufficient conditions, and special subcases where the problem may be solved (more or less) easily.

The structure of the paper is as follows: after recalling the context, Sect. 2 concentrates on invariant weighted circular Petri Nets. Some algorithms are developed in Sect. 3 and necessary and sufficient conditions in Sect. 4. Reduction to potential reachability is exploited in the next Section and the search for the largest dead number of tokens is considered in Sect. 6. The last Section, as usual concludes and suggests possible future developments.

---

[1] Pr. Eike Best, retired from Carl von Ossietzky Universität Oldenburg, Germany.

© The Author(s), under exclusive license to Springer Nature Switzerland AG 2023
L. Gomes and R. Lorenz (Eds.): PETRI NETS 2023, LNCS 13929, pp. 307–323, 2023.
https://doi.org/10.1007/978-3-031-33620-1_17

2   Invariant Weighted Circular Petri Nets

Definition 1. PETRI NETS

A *(weighted, initially marked, place-transition or P/T) Petri net* (PN for short) is denoted as $N = (P, T, F, M_0)$ where P is a set of places, T is a disjoint set of transitions $(P \cap T = \emptyset)$, F is the flow function $F \colon ((P \times T) \cup (T \times P)) \to \mathbb{N}$ specifying the arc weights (in graphical representations, arcs with a null weight are usually omitted), and M_0 is the initial marking (where a marking is a mapping $M \colon P \to \mathbb{N}$, indicating the number of tokens in each place). We shall only consider here finite nets, where P and T are finite sets. Its *incidence matrix* is the matrix $C \colon (P \times T) \to \mathbb{Z}$ defined as $C(p, t) = F(t, p) - F(p, t)$, i.e., the difference between the number of tokens produced and absorbed by t on p. For each node $x \in P \cup T$, its post-set is $x^\bullet = \{y \in P \cup T | F(x, y) > 0\}$ and its preset is $^\bullet x = \{y \in P \cup T | F(y, x) > 0\}$.

A transition $t \in T$ is *enabled* at a marking M, denoted by $M[t\rangle$, if $\forall p \in P \colon M(p) \geq F(p, t)$. The firing of t (when enabled) leads from M to M', denoted by $M[t\rangle M'$, if $M[t\rangle$ and $M'(p) = M(p) - F(p, t) + F(t, p)$. This can be extended, as usual, to $M[\sigma\rangle M'$ for *(firing) sequences* $\sigma \in T^*$, and $[M\rangle$ denotes the set of markings reachable from M. The net is *bounded* if there is $k \in \mathbb{N}$ such that $\forall M \in [M_0\rangle, p \in P \colon M(p) \leq k$. The *reachability graph* $RG(N)$ of N is the labelled transition system with the set of vertices $[M_0\rangle$, initial state M_0, label set T, and set of edges $\{(M, t, M') \mid M, M' \in [M_0\rangle \wedge M[t\rangle M'\}$. There is a (non-trivial) cycle in this reachability graph if $M[\sigma\rangle M$ for some $M \in [M_0\rangle$ and $\sigma \neq \varepsilon$ (the empty sequence).

If $\sigma \in T^*$ is a firing sequence, its *Parikh vector* $\Psi(\sigma)$ is the (column) vector in $\mathbb{N}^T$ counting for each transition $t \in T$ the number of its occurrences in σ. The *state equation* says that, if $M[\sigma\rangle M'$, i.e., if M' is reachable from M, then $M' = M + C \cdot \Psi(\sigma)$. Conversely, one says that a marking M' is *potentially reachable* from M if, for some column vector $\nu \in \mathbb{N}^T$, we have $M' = M + C \cdot \nu$. We shall also say that a column vector $\nu \in \mathbb{N}^T$ is *fireable* at a marking M if there is a firing sequence $M[\sigma\rangle$ with $\Psi(\sigma) = \nu$; the *support* of ν is the set of transitions with a non-null component: $sup(\nu) = \{t \in T | \nu(t) > 0\}$; if[2] $C \cdot \nu = \mathbf{0}$ and $\nu \neq \mathbf{0}$, ν is said to be a *semiflow* of N; if a semiflow is fireable at some marking, it defines a non-trivial cycle at M. A (line) vector $\boldsymbol{\mu} \in \mathbb{N}^S$ is said to be an S-invariant (also called a *P-invariant* or a *place-invariant* in the literature) if $\boldsymbol{\mu} \cdot C = \mathbf{0}$.

Moreover, we shall define $maxex_N \in (\mathbb{N} \cup \{\infty\})^T$ as the extended T-vector satisfying: $\forall t \in T$, $maxex_N(t)$ is the maximal number of times t may be executed in firing sequences of N from M_0, allowing the case $maxex_N(t) = \infty$ when t may be executed an unbounded number of times in firing sequences.

N is *persistent* if for all reachable markings $M, M', M'' \in [M_0\rangle$ and transitions $t', t'' \in T$ such that $M[t'\rangle M'$ and $M[t''\rangle M''$ with $t' \neq t''$, there exists a (reachable) marking M''' such that $M'[t''\rangle M'''$ and $M''[t'\rangle M'''$. It is *backward persistent* if for all reachable markings $M, M', M'' \in [M_0\rangle$ and transitions $t', t'' \in T$ such that $M'[t'\rangle M$ and $M''[t''\rangle M$ with $t' \neq t''$, there exists a marking $M''' \in [M_0\rangle$

---

[2] In general, $\mathbf{0}$ denotes a null vector of the adequate size and shape (line or column).

such that $M'''[t''\rangle M'$ and $M'''[t'\rangle M''$.

N is reversible if $\forall M \in [M_0\rangle : M_0 \in [M\rangle$, i.e., whatever marking is reached from the initial situation, it is always possible to go back to this initial situation. N deadlocks in a marking $M \in [M_0\rangle$ if $\nexists t \in T : M[t\rangle$. It is *trivially reversible* if N deadlocks in M_0. It is *live* if $\forall M \in [M_0\rangle, t \in T : \exists M' \in [M\rangle$ such that $M'[t\rangle$.
A marking M is a *home state* if $\forall M' \in [M_0\rangle : M \in [M'\rangle$ (M is always reachable).
Note that a net is reversible iff its initial state (marking) is a home state.   □1

In the following, we shall assume that N does not deadlock in M_0 (easy to check), so that trivial reversibility is avoided. From the state equation, it occurs that if μ is an S-invariant, for any reachable marking $M \in [M_0\rangle$ we have $\mu \cdot M = \mu \cdot M_0$, i.e., the weighted number of tokens $\mu \cdot M$ is invariant. The same is true if M is potentially reachable from M_0.

Definition 2. SOME SUBCLASSES
A P/T net N is a *weighted marked graph* [6] (WMG, also called weighted T-nets in [16]) if $\forall p \in P, |p^\bullet| \leq 1 \wedge |^\bullet p| \leq 1$.
It is a *weighted circuit* [16] if it has the shape of a circle, i.e., (up to some renaming) for some $n \in \mathbb{N}$, $P = \{p_1, p_2, \ldots, p_n\}$, $T = \{t_1, t_2, \ldots, t_n\}$, $\forall i \in [1, n] : t_i^\bullet = \{p_i\} \wedge {}^\bullet t_i = \{p_{i\ominus 1}\}$ (with $i \ominus 1 = i - 1$ if $i > 1$ and n if $i = 1$; symmetrically, we shall define $i \oplus 1 = i + 1$ if $i < n$ and 1 if $i = n$).
It is moreover *invariant* (IWC for short) if $\forall i \in [1, n] : F(t_i, p_i) = w_i = F(p_{i\ominus 1}, t_i)$.   □2

Weighted circuits, and in particular the ones with an S-invariant, have been examined in [2,14]. Here we shall concentrate on invariant ones (where the unit vector $\mu = (1, \ldots, 1)$ is an S-invariant), for which more specific results may be obtained.

Corollary 1. INVARIANCE PROPERTY AND PERSISTENCE
In any IWC, whatever the initial marking, the total number of tokens in the system remains constant: $M' \in [M\rangle \Rightarrow \sum_{i=0}^{n} M(p_i) = \sum_{i=0}^{n} M'(p_i)$.
Any IWC, like any weighted circuit and any WMG, is persistent and backward persistent, whatever the initial marking.
In any IWC (like in any weighted circuit and any WMG), if $M[\tau\rangle \wedge M[\sigma\rangle$, then $M[\tau(\sigma \overset{\bullet}{-} \tau)\rangle$ and $M[\sigma(\tau \overset{\bullet}{-} \sigma)\rangle$ are both enabled firing sequences (leading to the same marking), where $\sigma \overset{\bullet}{-} \tau$ is the residue of σ by τ, i.e., the sequence σ where one drops the first $\min(\Psi(\tau)(t), \Psi(\sigma)(t))$ occurrences of each transition $t \in T$ (Keller's theorem [8]).   □1

An IWC is pictured in Fig. 1.

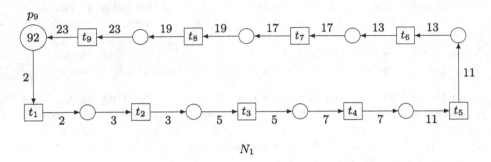

Fig. 1. An IWC with 9 places and transitions, and 92 tokens permanently in the system.

Since IWCs are WMGs, we may apply results from [6,16], for instance:

Proposition 1. POTENTIAL AND EFFECTIVE REACHABILITY (TH 3 OF [6])
If N is an IWC with incidence matrix C and $M = M_0 + C \cdot \nu \geq 0$ with $\nu \leq maxex_N$, then there exists a firing sequence $M_0[\sigma\rangle M$ with $\Psi(\sigma) = \nu$.   □1

Proposition 2. CYCLES IMPLY REVERSIBILITY (LEM 4(3) OF [6])
If N is an IWC, it is neutral in the sense of [11,16]; it has a unique minimal semiflow π, with support T, and each semiflow is a multiple of π.
If its reachability graph has a non-trivial cycle, then there is a cycle with Parikh vector π around each arc of the reachability graph (hence also around each reachable marking) and the net is reversible.   □2

Corollary 2. REVERSIBILITY AND LIVENESS
An IWC is live iff it is non-trivially reversible.   □2

And since adding initial tokens may not destroy an existing cycle, we also get:

Corollary 3. MONOTONICITY
If an IWC is (non-trivially) reversible, adding initial tokens preserves reversibility.
Reversely, if an IWC is not reversible, dropping initial tokens preserves non-reversibility.   □3

In the following, we shall consider IWCs with $n > 1$. Indeed, if $n = 0$ the system disappears, and if $n = 1$, when $M_0(p_1) < w_1$ the system is deadlocked and thus trivially reversible, while if $M_0(p_1) \geq w_1$ we have $M_0[t_1\rangle M_0$ and the system is always reversible, so that the reversibility problem is not interesting.

3 Some Easy Algorithms

Let us first determine the minimal semiflow π of an IWC N (we shall usually denote $\pi(t_i)$ by π_i).

Lemma 1. MINIMAL SEMIFLOW
$\forall i : \pi_i = \text{lcm}\{w_1, w_2, \ldots, w_n\}/w_i$.

Proof: By definition, π is a semiflow iff $w_1 \cdot \pi_1 = w_2 \cdot \pi_2 = \ldots = w_n \cdot \pi_n$, hence the formula yields a semiflow. Moreover, since lcm is the smallest common multiple of its arguments, it is known that, for each prime number p, if p occurs in w_i with a maximal exponent, it does not occur at all in π_i. Hence there is no sub-multiple of π, hence no smaller semiflow.   □1

In order to determine if N is reversible, we must thus check if $M_0[\pi\rangle$ (meaning that there is a firing sequence $M_0[\sigma\rangle$ such that $\Psi(\sigma) = \pi$, see Definition 1). To do that, we simply have to check that, if $M_0[\nu\rangle M$ with $\nu \leq \pi$ and $\nu \neq \pi$, it is possible to perform from M some transition in the support of $\pi - \nu$, hence elongating ν towards π. Indeed, from the fact that $M_0[\mathbf{0}\rangle$, if this is true we shall be able to exhibit that $M_0[\pi\rangle$; and from Keller's theorem (see Cor 1), if $M_0[\pi\rangle \wedge M_0[\nu\rangle$, it is possible to elongate ν to π.

This may be done by the following (non-deterministic) algorithm:

```
procedure REVERS1(N)
    if ∄i : M₀(p_{i⊖1}) ≥ w_i then
        return(trivial non-reversibility)
    end if
    compute π; M = M₀
    while ∃i : π(t_i) > 0 ∧ M(p_{i⊖1}) ≥ w_i do
        π(t_i) = π(t_i) − 1; M(p_{i⊖1}) = M(p_{i⊖1}) − w_i; M(p_i) = M(p_i) + w_i
    end while
    if π = 0 then
        return(reversible)
    else
        return(not reversible)
    end if
end procedure
```

The computation of π is quite simple, but its size may be huge and the worst case complexity of the algorithm is exponential. Indeed, in case of reversibility, the number of iterations of the while-loop is $\sum_i \pi(t_i) = \text{lcm}\{w_1, w_2, \ldots, w_n\} \cdot (\sum_i 1/w_i)$. For instance, if w_i is the ith prime number, all the w_i's are pairwise prime and $\text{lcm}\{w_1, w_2, \ldots, w_n\} = \prod_i w_i$. Then, it is known that, asymptotically, $w_n \sim n \cdot \log(n)$, $\sum_i 1/w_i \sim \log\log(n)$, $\prod_i w_i \sim e^{n \cdot \log(n)}$, hence the worst case complexity of Revers1 is exponential. It should thus only be used when $\text{lcm}\{w_1, w_2, \ldots, w_n\}$ is not too high, or when there are very few tokens in the system, leading to a quick deadlock.

For instance, Revers1 shows that N_1 in Fig. 1 is reversible, but the needed number of steps is 334406399 and takes around 6 s on a standard processor.

It is possible to speed up algorithm Revers1 by scanning systematically the transitions and firing them as much as possible in each update. This leads to the following deterministic algorithm:

procedure REVERS2(N)

    if $\nexists i : M_0(p_{i\ominus 1}) \geq w_i$ **then**

        return(trivial non-reversibility)

    end if

    compute π; $M = M_0$; continue:=true

    while continue **do**

        continue:=false

        for $i \leftarrow 1$ to n **do**

            if $\pi(t_i) > 0 \wedge M(p_{i\ominus 1}) \geq w_i$ **then**

                continue:=true

                $k = \min(\pi(t_i), \lfloor M(p_{i\ominus 1})/w_i \rfloor)$

                $\pi(t_i) = \pi(t_i) - k$

                $M(p_{i\ominus 1}) = M(p_{i\ominus 1}) - k \cdot w_i$

                $M(p_i) = M(p_i) + k \cdot w_i$

            end if

        end for

    end while

    if $\pi = 0$ **then**

        return(reversible)

    else

        return(not reversible)

    end if

end procedure

This may be especially efficient if there are many tokens circulating in the system. For instance, if there are initially at least $\mathrm{lcm}\{w_1, w_2, \ldots, w_n\}$ tokens in p_1, a single scan of each place will lead to the conclusion that the given net is reversible. By contrast, if there are few tokens, the speed up will be polynomial, and the worst case remains exponential. For instance, Revers2 still shows that N_1 in Fig. 1 is reversible, but the number of needed steps in the while loop is 4738072 and it takes around 2 s on a standard processor.

4   Some Necessary and Sufficient Conditions

Instead of trying to build progressively a reproducing firing sequence, like in the previous section, we may exploit the fact that the total number of tokens is invariant for the special class of nets we consider (see Corollary 1). In the following we shall denote by $m = \sum_i M_0(p_i)$ this number: in the terminology of [2,14], this is the *weight* of the initial, hence of each reachable, marking.

A first easy property says that, if m is large enough, the net is reversible, whatever the distribution of the tokens in the various places:

Proposition 3. LARGE MARKINGS ARE REVERSIBLE

If $m > (\sum_i w_i) - n$, then the net is reversible.

Proof: If $m > (\sum_i w_i) - n = \sum_i (w_i - 1)$, from the fact that for each reachable marking M we have $\sum_i M(p_i) = m$, by the pigeon hole principle in each reachable marking at least one place p_i has at least $w_{i \oplus 1}$ tokens, so that $t_{i \oplus 1}$ is enabled. As a consequence, no reachable marking is dead, it is always possible to extend a firing sequence, and since the net is bounded, at some point we shall form a cycle. But then we know that the net is reversible (see Proposition 2). □3

We thus have an easy sufficient condition of reversibility, and its negation yields a necessary condition for non-reversibility. Still with the terminology of [2, 14], $(\sum_i w_i) - n + 1$ is the least live marking weight, i.e., the least marking weight such that all markings with at least that weight are reversible (and live). Indeed, from that weight all the markings are reversible from Proposition 3 and the marking such that $M(p_i) = w_{i \oplus 1} - 1$, with weight $(\sum_i w_i) - n$ immediately deadlocks. With respect to [2,14], there is no need here (i.e., for the class of IWCs) to solve a (non-easy) Frobenius diophantine equation to find the least live marking weight.

For instance, this condition immediately tells us that the net N_1 in Fig. 1 is reversible, without performing the long computation corresponding to the algorithms in the previous section, since in this case $\sum_i (w_i - 1) = 1 + 2 + 4 + 6 + 10 + 12 + 16 + 18 + 22 = 91$. In fact, we here have a "limit" initial marking allowing to be able to apply this result, and indeed, with 91 tokens instead of 92 in the place p_9, Algorithm Revers2 tells us that the net becomes non-reversible (after 614322 executions of the while loop).

There is, symmetrically, an easy property saying that if we do not have enough tokens, the net is non-reversible, whatever the distribution of the tokens in the various places:

Proposition 4. SMALL MARKINGS ARE NON-REVERSIBLE

If $\frac{(m+n-1)!}{m! \cdot (n-1)!} < \text{lcm}\{w_1, w_2, \ldots, w_n\} \cdot (\sum 1/w_i)$, then the net is non-reversible.

Proof: Combinatory analysis tells us that $\frac{(m+n-1)!}{m! \cdot (n-1)!}$ yields the number of ways to distribute m tokens between n places. Since a minimal cycle returning to the initial state (like one returning to any other state) has a length $\text{lcm}\{w_1, w_2, \ldots, w_n\} \cdot (\sum 1/w_i)$ (see Lemma 1), we also need that number of different markings to be visited during that cycle. Hence the property.   □4

We thus have an easy sufficient condition of non-reversibility, and its negation yields a necessary condition for reversibility. These conditions characterise situations which are or are not reversible whatever the initial distribution of the tokens. Between those bounds, in general (but not always) the reversibility will rely on the initial marking (we shall soon see such examples). However, it is not sure that some marking just above the limit case given by Proposition 4 is not reversible, hence we know some weights such that no marking with that weight is reversible, but not the largest one.

We may also mention an easy condition ensuring reversibility for specific markings:

Proposition 5. LARGE MARKINGS IN SOME PLACE IMPLY REVERSIBILITY
If, for some place p_i, $M_0(p_i) \geq \text{lcm}\{w_1, w_2, \ldots, w_n\}$, then the net is reversible.

Proof: It is easy to see that, in this case, Algorithm Revers2 starting at place p_i will generate a reproducing firing sequence in one execution of the while loop.
□5

Very often, Proposition 3 will imply Proposition 5 (like for net N_1 in Fig. 1), but not always, as illustrated by Fig. 2: with $6 = \text{lcm}\{2, 3\}$ tokens in p_9, Proposition 5 tells us that the net is reversible, while $\sum_i (w_i - 1) = 10$ so that Proposition 3 may not be applied. But distributing those tokens by putting 3 of them in p_7 and a single one in p_1, p_2 and p_3, yields non-reversibility, with the same total number of tokens.

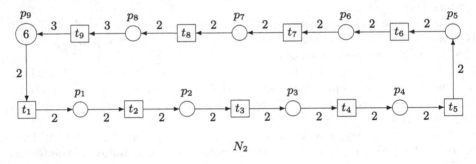

N_2

Fig. 2. An IWC with 9 places and transitions, and 6 tokens permanently in the system.

As mentioned in the proof of Proposition 5, instead of applying the latter, we may perform Revers2 while limiting the number of executions of the while loop to 2: one to gather tokens in a single place and one to observe that a reproducing firing sequence was obtained (or that the system deadlocked before). More generally, it may be interesting to perform Revers2 while limiting the number of executions of the while loop to a polynomial number (in the size of the net), like n: in practice, in many cases (but not all) we shall conclude the reversibility or the non-reversibility of the given IWC net.

An important point to observe is that all the checks considered in this section (as well as the trivial reversibility check) are clearly polynomial. However, they do not cover all the cases. The next section will go a bit further.

5   Potential Reachability Checks

From the previous remark, we shall now assume that the considered IWC N is not trivially reversible, and that no check from the previous section is conclusive.
An important point to observe for the class of IWC nets is the following:

Proposition 6. POTENTIAL AND EFFECTIVE REACHABILITY *In a reversible IWC, if a marking is potentially reachable, it is effectively reachable.*

Proof: If the net N is reversible, since π has support T, $maxex_N = \infty^T$ so that, from Proposition 1, any potentially reachable marking is effectively reachable.

$$\square 6$$

Corollary 4. POTENTIAL REACHABILITY AND REVERSIBILITY
If a deadlocking marking is potentially reachable in an IWC, the latter is non-reversible. $\square 4$

Note, that this does not mean that the potentially reachable deadlocking marking is effectively reachable, since it may happen that another deadlocking marking is reached instead. But we cannot progress forever, otherwise at some point we shall close a (non-empty) cycle, with support T, and by Keller's theorem the deadlocking sequence may be elongated, hence cannot be deadlocking.

5.1   The Limit Case

We shall here assume that $m = \sum_i M_0(p_i) = \sum_i (w_i - 1)$, so that Proposition 3 "just" fails. Since the number m of tokens in the system does not change, the only deadlocking marking that could possibly be reached is M_D such that $\forall i : M_D(p_i) = w_{i \oplus 1} - 1$. A similar remark is that the only marking which cannot be preceded by another one is M^D such that $\forall i : M^D(p_i) = w_i - 1$, so that starting from M^D will lead to the longest firing sequences built by algorithms Revers1 and Revers2. From Corollary 4, we thus have:

Proposition 7. POTENTIAL REACHABILITY OF M_D
When $\sum_i M_0(p_i) = \sum_i (w_i - 1)$, an IWC is reversible if and only if M_D is not potentially reachable.

Proof: If M_D is not potentially reachable, it is not reachable either and any firing sequence may be elongated, leading finally to a cycle (the IWC is bounded), hence to reversibility (see Proposition 2).
If M_D is potentially reachable, Corollary 4 applies.   □7

We thus have to determine if the equation $M_0 + C \cdot x = M_D$ has a solution $x \in \mathbb{N}^T$. In fact, it is equivalent to determine if there is a solution in $\mathbb{Z}^T$ since it is always possible to add any multiple of π to it. If we denote $M = M_D - M_0$, we thus have to solve $C \cdot x = M$ in the integers, with

$$
C = \begin{bmatrix}
w_1 & -w2 & 0 & \ldots & 0 \\
0 & w_2 & -w_3 & \ldots & 0 \\
\ldots & & & & \\
\ldots & & & & \\
0 & 0 & 0 & \ldots & -w_n \\
-w_1 & 0 & 0 & \ldots & w_n
\end{bmatrix}
$$

It is known that solving an integer linear system is polynomial, for instance by constructing the Hermite normal form of the matrix [3], or by a direct examination of the system when it has a particular form. We shall here illustrate the latter. Note first that C has rank $n-1$, and that $\sum_i M(p_i) = 0$, so that we only have to solve $n-1$ lines. Let us consider the $n-1$ first ones.

1. $w_1 \cdot x_1 - w_2 \cdot x_2 = M(p_1)$ is a classical linear diophantine equation. It has no solution if $\gcd(w_1, w_2)$ does not divide $M(p_1)$; then there is no solution to the whole system of equations, M_D is not potentially reachable, and the given IWC is reversible; this may be checked in polynomial time. Otherwise it has infinitely many solutions. Let us thus assume $\gcd(w_1, w_2) = g_1$, $w_1 = g_1 \cdot w_1'$, $w_2 = g_1 \cdot w_2'$ and $M(p_1) = g_1 \cdot m_1$. Then w_1' and w_2' are pairwise prime and we may use the (polynomial) extended Euclide algorithm to find the Bezout coefficients $b_1, b_2 \in \mathbb{Z}$ such that $w_1' \cdot b_1 - w_2' \cdot b_2 = 1$. Then the solutions are $x_1 = m_1 \cdot b_1 + w_2' \cdot z_2$ and $x_2 = m_1 \cdot b_2 + w_1' \cdot z_2$, where z_2 is any integer in $\mathbb{Z}$.

2. $w_2 \cdot x_2 - w_3 \cdot x_3 = M(p_2)$ is a similar linear diophantine equation. If reversibility was not deduced in the previous step, from the previous relations, it may be rewritten as $w_2 \cdot (m_1 \cdot b_2 + w_1' \cdot z_2) - w_3 \cdot x_3 = M(p_2)$, or $(w_2 \cdot w_1') \cdot z_2 - w_3 \cdot x_3 = M(p_2) - w_2 \cdot m_1 \cdot b_2$, or $(w_2 \cdot w_1') \cdot z_2 - w_3 \cdot x_3 = m_2'$, with $m_2' = M(p_2) - w_2 \cdot m_1 \cdot b_2$. It has no solution if $\gcd(w_2 \cdot w_1', w_3)$ does not divide m_2', then there is no solution to the whole system of equations, M_D is not potentially reachable, and the given IWC is reversible; otherwise it has infinitely many solutions. Let us thus assume $\gcd(w_2 \cdot w_1', w_3) = g_2$, $w_2 \cdot w_1' = g_2 \cdot w_2''$, $w_3 = g_2 \cdot w_3'$ and $m_2' = g_2 \cdot m_2$. Then w_2'' and w_3' are pairwise prime and we may use the extended Euclide algorithm to find the Bezout coefficients $b_2', b_3 \in \mathbb{Z}$ such that $w_2'' \cdot b_2' - w_3' \cdot b_3 = 1$. Then the solutions are $z_2 = m_2 \cdot b_2' + w_3' \cdot z_3$ and $x_3 = m_2 \cdot b_3 + w_2'' \cdot z_3$, where z_3 is any integer in $\mathbb{Z}$.

3..n-2 We may then continue and, at each stage, there is a simple condition specifying if there is a solution. If not, we may stop with the answer that the IWC is reversible. Otherwise we construct two linear expressions depending on a new integer variable: the first one yields the general form for an auxiliary variable introduced in the previous stage and the second one drives the general expression for a variable allowing to build the diophantine equation to solve at the next stage.

n-1 When we arrive at the last stage, it is not necessary to build the general form of the present diophantine equation: if it is solvable (otherwise, as before we may deduce the given IWC is reversible), we may choose any particular solution (for instance the one originating directly from the Bezout coefficients) and go back to progressively build the components of a possible solution x of $M_0 + C \cdot x = M_D$. We may deduce that the given IWC is non-reversible from Proposition 7.

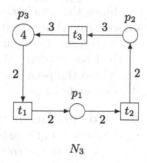

$$N_3$$

Fig. 3. An IWC with 3 places and transitions, and 4 tokens permanently in the system.

For instance, in the limit system N_3 of Fig. 3 (with $4 = 2 + 2 + 3 - 3$, so that we are indeed in the limit case), the constraint on p_1 is $2 \cdot x_1 - 2 \cdot x_2 = w_2 - 1 = 1$ which is not solvable. Hence the system is reversible. On the contrary, with 3 tokens in p_3 and one in p_1, it is not.

Other cases may be solved rather quickly. We may observe that, adding constraints i to j (we allow $j < i$ since the constraints may be considered circularly), we get $w_i \cdot x_i - w_{j \oplus 1} \cdot w_{j \oplus 1} = \sum_{k=i}^{j} (w_k - 1 - M_0(p_k))$; hence, if $\gcd(w_i, w_{j \oplus 1})$ does not divide $\sum_{k=i}^{j} (w_k - 1 - M_0(p_k))$, we may conclude there is no solution, hence the given IWC is reversible.

The Relatively Prime Subcase. Let us now consider the special subcase where all the w_i's are pairwise prime (like in Fig. 1).
In the first step, $g_1 = 1$, $w_1' = w_1$ and there is always a solution.
Then, in the second step, we have to solve the equation $(w_2 \cdot w_1) \cdot z_2 - w_3 \cdot x_3 = m_2'$. Since w_3 is prime with both w_1 and w_2, it is known it is also prime with $w_1 \cdot w_2$ (otherwise, there would be a prime divisor of w_3 and $w_1 \cdot w_2$, which then should divide either w_1 or w_2, contradicting the pairwise prime hypothesis). Thus, $g_2 = 1$, $w_2'' = w_1 \cdot w_2$ and there is always a solution.

We may then proceed similarly for the next steps. For instance, in step i, we shall get an equation of the form $(w_i \cdots w_2 \cdot w_1) \cdot z_i - w_{i+1} \cdot x_{i+1} = m'_i$, and since w_{i+1} is prime with $(w_i \cdots w_2 \cdot w_1)$, we shall get that $g_i = 1$ and there is always a solution.

Hence, the given IWC is non-reversible.

And since a deadlocking IWC remains deadlocking if we drop some initial tokens (see Corollary 3), this amounts to a proof of the following:

Theorem 1. PAIRWISE PRIME CASE
If the weights w_i of an IWC are pairwise prime, it is reversible iff its (constant) number of tokens is greater than $\sum_i (w_i - 1)$.   □1

5.2   The Non-limit Case

We shall here assume[3] that $0 < m = \sum_i M_0(p_i) < \sum_i (w_i - 1)$, and that the weights are not pairwise prime. Since the number m of tokens in the system does not change, the only deadlocking markings that could possible be reached are characterised by the fact that they are between 0 and M_D. In fact, those limits are not included, but we shall not exploit this explicitly, because it will be implied by the hypothesis on m and the preservation of the total number of tokens. Corollary 4 may then be reformulated as:

Corollary 5. LINEAR EXPRESSION OF REVERSIBILITY
An IWC is reversible if and only if the system of constraints $0 \leq M_0 + C \cdot x \leq M_D$, or equivalently $-M_0 \leq C \cdot x \leq M_D - M_0$, is not solvable in $x \in \mathbb{N}^T$.   □5

In fact, it is again equivalent to determine if there is a solution in $\mathbb{Z}^T$ since it is always possible to add any multiple of π to it to reach $\mathbb{N}^T$. This result also shows that reversibility in our case is in co-NP since, if we are given a possible solution x, we may check in polynomial time if $0 \leq M_0 + C \cdot x \leq M_D$, i.e., if the IWC is not reversible. Moreover, if $\sum_i (w_i - 1) - \sum_i M_0(p_i)$ is not too high, there is a limited number of possible solutions of the system $M_0 + C \cdot x = M$ with $0 \leq M \leq M_D$ and $\sum_i M(p_i) = \sum_i M_0(p_i)$ to be considered, and each of them may be solved polynomially. For instance, if $\sum_i (w_i - 1) - \sum_i M_0(p_i) = 1$, we only have to check at most n vectors M (M_D minus one token in some place). But there is no certainty that we shall reach a conclusion in a polynomial time that way in general.

Like for the limit case, let us consider each linear constraint in turn. A difference with the limit case is that here we shall need to consider all the lines, and not only the first $(n - 1)$ ones, since $\sum_i M_0(p_i) \neq 0$ and $\sum_i (M_D(p_i) - M_0(p_i)) \neq 0$ while $\forall j : \sum_i C_{j,i} = 0$.

---

[3] The case 0 is excluded from Proposition 4, since we assumed the latter may not be applied, and by the fact we assumed the initial marking does not deadlock.

For each i, the ith constraint is $-M_0(p_i) \leq w_i \cdot x_i - w_{i \oplus 1} \cdot x_{i \oplus 1} \leq M(p_i)$, with as before $M = M_D - M_0$. Let $g_i = \gcd(w_i, w_{i \oplus 1})$. For any $y \in \mathbb{Z}, k \in \mathbb{N}$, we shall denote by $\lfloor y \rfloor_k$ the largest multiple of k not greater than y, and by $\lceil y \rceil_k$ the smallest multiple of k not smaller than y, i.e., $\lfloor y \rfloor_k = k \cdot \lfloor y/k \rfloor$ and $\lceil y \rceil_k = k \cdot \lceil y/k \rceil$. The constraint is then equivalent to

$$\lceil -M_0(p_i) \rceil_{g_i} = -\lfloor M_0(p_i) \rfloor_{g_i} \leq w_i \cdot x_i - w_{i \oplus 1} \cdot x_{i \oplus 1} \leq \lfloor M(p_i) \rfloor_{g_i}. \tag{1}$$

Adding all the constraints (1), we get

$$-\sum_i \lfloor M_0(p_i) \rfloor_{g_i} \leq 0 \leq \sum_i \lfloor M(p_i) \rfloor_{g_i} \tag{2}$$

Since $M_0 \geq 0$, the first part of the constraint (2) is trivial, but the second one may be more interesting.

Let us consider the net N_1 from Fig. 2, but with only 4 token in p_9 (and none elsewhere). The first seven constraints are $0 \leq 2 \cdot x_i - 2 \cdot x_{i \oplus 1} \leq \lfloor 1 \rfloor_2 = 0$, since $w_i = 2 = w_{i \oplus 1}$, $g_i = 2$, $M_0(p_i) = 0$ and $M_D(p_i) = 1$; the eighth one is $0 \leq 2 \cdot x_8 - 3 \cdot x_9 \leq 2$, since $w_8 = 2$, $w_9 = 3$, $g_8 = 1$, $M_0(p_8) = 0$ and $M_D(p_8) = 2$; the ninth one is $-4 \leq 3 \cdot x_9 - 2 \cdot x_1 \leq -3$, since $w_9 = 3$, $w_1 = 2$, $g_9 = 1$, $M_0(p_9) = 4$ and $M_D(p_9) = 1$. The sum yields $-4 \leq 0 \leq -1$, which means there is no solution, hence the net is reversible; the same arises of course if one adds tokens to the initial marking, either in p_9 or elsewhere; on the contrary, if we drop one initial token in p_9, the sum-constraint is non-conclusive, and an execution of Revers2 indeed shows we have non-reversibility; the same arises with some redistributions of the 4 tokens between the various places.

Unfortunately, this does not cover all the cases. Let us consider for example the system on Fig. 4, where the weights are the first prime numbers, but with 2's added in between. The pivoting number of tokens is 38 and the g_i's are all 1, so that whenever we start with $m \leq 38$ tokens, the summing constraint is $-m \leq 0 \leq 38 - m$, which potentially allows a solution to the linear system of constraints above. However, down to 36 tokens in place p_9, the system is reversible (after 85406 steps of Revers1, or 1776 scans of Revers2), i.e., there is no solution.

It is however possible to refine the reasoning above to get more conclusive cases. Let $g_{i,j} = \gcd(w_i, w_{j \oplus 1})$ (so that $g_{i,i} = g_i$). Since $\sum_{k=i}^{j}(w_k \cdot x_k - w_{k \oplus 1} \cdot x_{k \oplus 1}) = w_i \cdot x_i - w_{j \oplus 1} \cdot x_{j \oplus 1}$, we have $w_i \cdot x_i - w_{j \oplus 1} \cdot x_{j \oplus 1} \leq \sum_{k=i}^{j} M(p_k)$, so that if $g_{i,j} > 1$, we must have $w_i \cdot x_i - w_{j \oplus 1} \cdot x_{j \oplus 1} \leq \lfloor \sum_{k=i}^{j} M(p_k) \rfloor_{g_{i,j}}$, i.e., we may replace $\sum_{k=i}^{j} M(p_k)$ by $\lfloor \sum_{k=i}^{j} M(p_k) \rfloor_{g_{i,j}}$. We may do it for any interval $[i,j]$, even circularly (i.e., going from n to 1), as well as for disjoint intervals. But also, if disjoint intervals are included in another one, we may do the same while replacing the partial sums corresponding to included intervals by floor expressions as done above. And finally, summing the constraints we shall get that 0 must not be greater than the sum of all those partial floor expressions (including for intervals $[i,i]$, even if $g_i = 1$): if this is not the case, we may conclude there is no solution and the IWC is reversible.

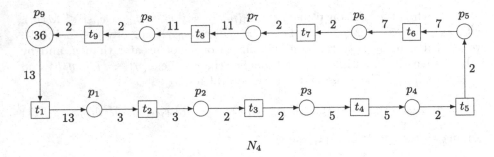

N_4

Fig. 4. An IWC with 9 places and transitions, and 6 tokens in permanence in the system.

For instance, in Fig. 4, the interesting intervals are $[3, 4]$, $[5, 6]$, $[7, 8]$ and $[9, 2]$, as well as $[3, 6]$, $[5, 8]$, $[7, 2]$, $[9, 4]$, $[3, 8]$, $[5, 2]$, $[7, 4]$ (all with gcd $= 2$), and the whole set (for instance $[3, 2]$). Since $M = (2, 1, 4, 1, 6, 1, 10, 1, -24 = 12 - 36)$, we may observe that $M(p_3) + M(p_4) = 5$ may be replaced by 4, $M(p_5) + M(p_6) = 7$ may be replaced by 6, $M(p_7) + M(p_8) = 11$ may be replaced by 10 and $M(p_9) + M(p_1) + M(p_2) = -21$ may be replaced by -22. No more reinforcement may be applied, and for the whole set we thus obtain: $0 \le 4 + 6 + 10 - 22 = -2$; hence there is no solution and we may deduce the IWC is reversible without relying to the algorithm Revers2 (or Revers1).

Unfortunately, we may not obtain certainty in all cases with this method. Let us thus consider the various constraints individually.

For the individual constraint for place p_1, we get:
$-\lfloor M_0(p_1) \rfloor_{g_1} \le w_1 \cdot x_1 - w_2 \cdot x_2 \le \lfloor M(p_1) \rfloor_{g_1}$, or equivalently $m_{1,1} \le w_{1,1} \cdot x_1 - w_{1,2} \cdot x_2 \le m_{1,2}$, with $m_{1,1} = -\lfloor M_0(p1)/g_1 \rfloor$, $m_{1,2} = \lfloor (M_D(p_1) - M_0(p_1))/g_1 \rfloor$, $w_{1,1} = w_1/g_1$ and $w_{1,2} = w_2/g_1$. As before, let $b_{1,1}, b_{1,2}$ be the Bezout coefficients of $w_{1,1}, w_{1,2}$, so that $w_{1,1} \cdot b_{1,1} - w_{1,2} \cdot b_{12} = 1$ (note that this also means that $w_{1,2}$ and $b_{1,1}$ are relatively prime, as well as $w_{1,2}$ and $b_{1,1}$). The general form of the solutions is then $x_1 = y_1 \cdot b_{1,1} + w_{1,2} \cdot z_1$ and $x_2 = y_1 \cdot b_{1,2} + w_{1,1} \cdot z_1$, with $z_1 \in \mathbb{Z}$ and $y_1 \in \mathbb{Z} \cap [m_{1,1}, m_{1,2}]$. It may happen that $m_{1,1} = m_{1,2}$, in which case there is a single possible value for y_1 and the general expression only relies on a single variable z_1 instead of two. It is not possible that $m_{1,1} > m_{1,2}$ since $M_D(p_1) = w_2 - 1$ and g_1 divides w_2.

The individual constraint for place p_2 is:
$-\lfloor M_0(p_2) \rfloor_{g_2} \le w_2 \cdot x_2 - w_3 \cdot x_3 \le \lfloor M(p_2) \rfloor_{g_2}$. Replacing x_2 by its expression above, this yields $-\lfloor M_0(p_2) \rfloor_{g_2} \le w_2 \cdot b_{1,2} \cdot y_1 + w_2 \cdot w_{1,1} \cdot z_1 - w_3 \cdot x_3 \le \lfloor M(p_2) \rfloor_{g_2}$. Since $w_{1,1}$ and $b_{1,2}$ are relatively prime, $\gcd\{w_2 \cdot b_{1,2}, w_2 \cdot w_{1,1}, w_3\} = \gcd(w_2, w_3) = g_2$. Again, it may happen that $-\lfloor M_0(p_2) \rfloor_{g_2} = \lfloor M(p_2) \rfloor_{g_2}$, but not $-\lfloor M_0(p_2) \rfloor_{g_2} > \lfloor M(p_2) \rfloor_{g_2}$ since $M_D(p3) = w_3 - 1$ and g_2 divides w_3. The general solution may be written (in polynomial time) in the form $y_1 = a_{2,1,1} \cdot y_{2,1} + a_{2,1,2} \cdot y_{2,2}$, $z_1 = a_{2,2,1} \cdot y_{2,1} + a_{2,2,2} \cdot y_{2,2} + a_{2,2,3} \cdot y_{2,3}$, $x_3 = a_{2,3,1} \cdot y_{2,1} + a_{2,3,2} \cdot y_{2,2} + a_{2,3,3} \cdot y_{2,3}$, where all the coefficients are in $\mathbb{Z}$ (we may even assume $a_{2,1,2} \in \mathbb{N}$), $y_{2,1} \in \mathbb{Z} \cap [-\lfloor M_0(p_2)/g_2 \rfloor, \lfloor (M_D(p_2) - M_0(p_2))/g_2 \rfloor]$

and $y_{2,2}, y_{2,3} \in \mathbb{Z}$. But since $y_1 \in \mathbb{Z} \cap [m_{1,1}, m_{1,2}]$, this introduces additional constraints on $y_{2,1}$ and $y_{2,2}$.

For instance, if $w_1 = 4, w_2 = 3, w_4 = 3$, $M_0(p1) = 0$ and $M_0(p_2) = 3$, we get $4 \cdot x_1 - 3 \cdot x_2 = d_1 \in [0, 2]$ and $3 \cdot x_2 - 4 \cdot x_3 \in [-3, 0]$. This leads to $d_1 = 3 \cdot d_2 + 4 \cdot y$ in general and to $d_1 = -d_2$ for $d_2 \in [-2, 0]$ if we take the constraints into account, which is easy and reduces the range to explore to 3 consecutive values instead of 4. But if $M_0(p_2) = 0$, we get the constraint $d_2 \in [0, 3]$, which leads to $d_1 = 0$ or $d_1 = -d_2 + 4$ for $d_2 \in [2, 3]$, breaking the range for d_2 into two small but non-consecutive ranges, hence potentially leading to an exponential number of ranges to explore later.

This does not mean there is no polynomial way to obtain the solution, but presently we do not see how to do it, nor how to prove that the problem remains NP-complete.

6 Largest Dead Number of Tokens

In the same way, we may be interested in the least live number of tokens, such that all initial markings with at least this number of tokens are live, and we may search for the largest dead number of tokens, such that all initial markings with at most this number of tokens deadlock. However, while the least live number of tokens is easy to obtain ($(\sum_i w_i) - n + 1$) and only relies on the multi-set of arc weights, not on their sequence, the same is not true for the largest dead number of tokens.

Presently, we do not know an efficient way to compute this number. We may for instance apply Algorithm Revers2 on all the initial markings with a certain number of tokens and increase the latter until we find a live marking, or decrease the latter until we find no live marking, or proceed by dichotomy. It is however possible to reduce the number of markings to consider, for instance by only considering the ones where all the places but one have less tokens than the corresponding arc weights.

For instance, for the weights $(10, 14, 15, 21)$, the largest dead number of tokens is 42, while for the weights $(14, 10, 15, 21)$, it is 41. The reason of this feature is probably the fact that the graphs of the gcd-relationship between the successive arc weights are rather different, as illustrated in Fig. 5.

This is in some sense a least counter-example, with 4 transitions and places, since this phenomenon does not occur with 3 (or 2) transitions and places. To see this we may observe the following.

Lemma 2. INVARIANCE BY ROTATION
If an IWC with arc weights $w_1, w_2, \ldots, w_n$ is live for the marking $m_1, m_2, \ldots, m_n$, then the IWC with arc weights $w_2, \ldots, w_n, w_1$ is live for the marking $m_2, \ldots, m_n, m_1$.

322   R. Devillers

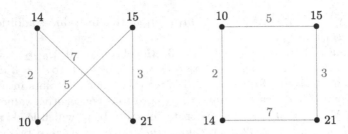

Fig. 5. On the left, the gcd-graph between the weights $(10, 14, 15, 21)$, on the right the same for the weights $(14, 10, 15, 21)$.

Proof: Obvious: the enabled transition sequences are the same.   □2

Lemma 3. INVARIANCE BY REFLECTION
If an IWC with arc weights $w_1, w_2, \ldots, w_n$ is live for the marking $m_1, m_2, \ldots, m_n$, then the IWC with arc weights $w_n, \ldots, w_2, w_1$ is live for the marking $m_n, \ldots, m_2, m_1$.

Proof: Obvious: a transition sequence $M[t_1 t_2 \ldots t_k \rangle M'$ is enabled in the first IWC iff so is the sequence $M'[t_k \ldots t_2 t_1 \rangle M$ in the second one. As a consequence $M[t_1 t_2 \ldots t_k \rangle M$ is enabled in the first IWC iff so is the sequence $M[t_k \ldots t_2 t_1 \rangle M$ in the second one. The cycles are thus the same.   □3

Corollary 6. INVARIANCE BY ISOMETRY
If the sequence of weights of an IWC is the same as the sequence of another one up to rotations and reflections, the live markings are the same up to the same rotations and reflections. In particular there is a live marking of the first IWC with m tokens iff the same is true for the second IWC. As a consequence, the largest dead number of tokens is the same for these two IWCs.   □6

Corollary 7. IWC WITH 2 OR 3 TRANSITIONS
The largest dead number of tokens of two IWCs with the same multisets of arc weights are the same.

Proof: When the multisets of IWCs with two or three transitions are the same, these IWCs are isometric.   □7

7   Conclusions and Perspectives

We have shown that even for the very restricted family of IWC Petri nets, determining if a system is reversible (or live) remains hard to tackle, but that some subcases may be handled more or less easily. It remains of course to determine the exact worst case complexity of the general class IWC. It could also be interesting to consider other subclasses of structurally persistent nets.

Acknowledgements. The author appreciated the remarks and encouragements of the anonymous referees.

References

1. Best, E., Esparza, J.: Existence of home states in petri nets is decidable. Inf. Process. Lett. **116**(6), 423–427 (2016). https://doi.org/10.1016/j.ipl.2016.01.011
2. Chrzastowski-Wachtel, P., Raczunas, M.: Liveness of weighted circuits and the Diophantine problem of Frobenius. In: Ésik, Z. (ed.) FCT 1993. LNCS, vol. 710, pp. 171–180. Springer, Heidelberg (1993). https://doi.org/10.1007/3-540-57163-9_13
3. Cohen, H.: A course in computational algebraic number theory. Graduate Texts in Mathematics, vol. 138. Springer, Heidelberg (1993). https://doi.org/10.1007/978-3-662-02945-9. http://www.worldcat.org/oclc/27810276
4. Czerwinski, W., Lasota, S., Lazic, R., Leroux, J., Mazowiecki, F.: The reachability problem for petri nets is not elementary. J. ACM **68**(1), 1–28 (2021). https://doi.org/10.1145/3422822
5. Czerwinski, W., Orlikowski, L.: Reachability in vector addition systems is Ackermann-complete. In: 62nd IEEE Annual Symposium on Foundations of Computer Science, FOCS 2021, Denver, CO, USA, 7-10 February 2022, pp. 1229–1240 (2021). https://doi.org/10.1109/FOCS52979.2021.00120
6. Devillers, R., Hujsa, T.: Analysis and synthesis of weighted marked graph Petri nets. In: Application and Theory of Petri Nets and Concurrency - 39th International Conference, PETRI NETS 2018, Bratislava, Slovakia, 24–29 June 2018, Proceedings, pp. 19–39 (2018)
7. de Frutos-Escrig, D., Johnen, C.: Decidability of home space property. Tech. Rep. 503, Laboratoire de Recherche en Informatique, Université de Paris-Sud (1989)
8. Keller, R.M.: A fundamental theorem of asynchronous parallel computation. In: Sagamore Computer Conference, 20-23 August 1974, LNCS, vol. 24, pp. 102–112 (1975). https://doi.org/10.1007/3-540-07135-0_113
9. Kosaraju, S.R.: Decidability of reachability in vector addition systems (preliminary version). In: Proceedings of the 14th Annual ACM Symposium on Theory of Computing, 5–7 May 1982, San Francisco, California, USA, pp. 267–281 (1982)
10. Lasota, S.: Improved Ackermannian lower bound for the petri nets reachability problem. In: Berenbrink, P., Monmege, B. (eds.) 39th International Symposium on Theoretical Aspects of Computer Science (STACS 2022). Leibniz International Proceedings in Informatics (LIPIcs), vol. 219, pp. 1–15. Schloss Dagstuhl – Leibniz-Zentrum für Informatik, Dagstuhl, Germany (2022). https://drops.dagstuhl.de/opus/volltexte/2022/15856
11. Lien, Y.E.: Termination properties of generalized Petri nets. SIAM J. Comput. **5**(2), 251–265 (1976)
12. Mayr, E.W.: An algorithm for the general Petri net reachability problem. In: Proceedings of the 13th Annual ACM Symposium on Theory of Computing, 11–13 May 1981, Milwaukee, Wisconsin, USA, pp. 238–246 (1981)
13. Mayr, E.W.: An algorithm for the general Petri net reachability problem. SIAM J. Comput. **13**(3), 441–460 (1984)
14. Raczunas, M., Chrzastowski-Wachtel, P.: A Diophantine problem of Frobenius in terms of the least common multiple. Discrete Math. **150**(1-3), 347–357 (1996). https://doi.org/10.1016/0012-365X(95)00199-7
15. Reinhardt, K.: Reachability in Petri nets with inhibitor arcs. Electr. Notes Theor. Comput. Sci. **223**, 239–264 (2008)
16. Teruel, E., Chrzastowski-Wachtel, P., Colom, J.M., Silva, M.: On weighted T-systems. In: Jensen, K. (ed.) ICATPN 1992. LNCS, vol. 616, pp. 348–367. Springer, Heidelberg (1992). https://doi.org/10.1007/3-540-55676-1_20

Automated Polyhedral Abstraction Proving

Nicolas Amat$^{(\boxtimes)}$, Silvano Dal Zilio , and Didier Le Botlan

LAAS-CNRS, Université de Toulouse, INSA, CNRS, Toulouse, France
nicolas.amat@laas.fr

Abstract. We propose an automated procedure to prove polyhedral abstractions for Petri nets. Polyhedral abstraction is a new type of state-space equivalence based on the use of linear integer constraints. Our approach relies on an encoding into a set of SMT formulas whose satisfaction implies that the equivalence holds. The difficulty, in this context, arises from the fact that we need to handle infinite-state systems. For completeness, we exploit a connection with a class of Petri nets that have Presburger-definable reachability sets. We have implemented our procedure, and we illustrate its use on several examples.

Keywords: Automated reasoning · Abstraction techniques · Reachability problems · Petri nets

1 Introduction

We describe a procedure to automatically prove *polyhedral abstractions* between pairs of parametric Petri nets. Polyhedral abstraction [2,6] is a new type of equivalence that can be used to establish a "linear relation" between the reachable markings of two Petri nets. Basically, an abstraction is a triplet of the form (N_1, E, N_2), where E is a system of linear constraints between the places of two nets N_1 and N_2. The idea is to preserve enough information in E so that we can rebuild the reachable markings of N_1 knowing only the ones of N_2, and vice versa.

In this context, we use the term *parametric* to stress the fact that we manipulate semilinear sets of markings, meaning sets that can be defined using a Presburger arithmetic formula C. In particular, we reason about parametric nets (N, C), instead of marked nets (N, m_0), with the intended meaning that all markings satisfying C are potential initial markings of N. We also define an extended notion of polyhedral equivalence between parametric nets, denoted $(N_1, C_1) \approx_E (N_2, C_2)$, whereas our original definition [1] was between marked nets only (see Definition 1).

We show that given a valid equivalence statement $(N_1, C_1) \approx_E (N_2, C_2)$, it is possible to derive a Presburger formula, in a constructive way, whose satisfaction implies that the equivalence holds. We implemented this procedure on top of an

L. Gomes and R. Lorenz (Eds.): PETRI NETS 2023, LNCS 13929, pp. 324–345, 2023.
https://doi.org/10.1007/978-3-031-33620-1_18

SMT-solver for Linear Integer Arithmetic (LIA) and show that our approach is applicable in practice (Sect. 7). Our method is only a semi-procedure though, since there are two possible outcomes when the equivalence does not hold: either we can generate a formula that is unsound, or our procedure does not terminate, and each of these outcomes provide useful information on why the equivalence does not hold.

This decidability result is not surprising, since most equivalence problems on Petri nets are undecidable [15,16]. If anything, it makes the fact that we may often translate our problem into Presburger arithmetic quite remarkable. Indeed, polyhedral abstraction is by essence related with the *marking equivalence* problem, which amounts to decide if two Petri nets with the same set of places have the same reachable markings; a problem proved undecidable by Hack [17]. Also, polyhedral equivalence entails trace equivalence, another well-known undecidable equivalence problem when we consider general Petri nets [17,18].

Description of Our Approach and Related Works. We introduced the concept of polyhedral abstraction as a way to solve reachability problems more efficiently. We applied this approach to several problems: originally for model-counting, that is to count the number of reachable markings of a net [12,13]; then to check reachability formulas and to find inductive invariants [1,2]; and finally to speed-up the computation of concurrent places (places that can be marked simultaneously in a reachable marking) [5,6]. We implemented our approach in two symbolic model-checkers developed by our team: Tedd, a tool based on Hierarchical Set Decision Diagrams (SDD) [25], part of the Tina toolbox [22]; and SMPT [3,21], an SMT-based model-checker focused on reachability problems [4].

In each case our approach can be summarized as follows. We start from an initial net (N_1, C_1) and derive a polyhedral abstraction $(N_1, C_1) \approx_E (N_2, C_2)$ by applying a set of *abstraction laws* in an iterative and compositional way. Finally, we solve a reachability problem about N_1 by transforming it into a reachability problem on net N_2, which should hopefully be easier to check. A large number of the laws we implement in our tools derive from structural reduction rules [11], or are based on the elimination of redundant places and transitions, with the goal to obtain a "reduced" net N_2 that is smaller than N_1.

We also implement several other kinds of abstraction rules—often subtler to use and harder to prove correct—which explains why we want machine checkable proofs of equivalence. For instance, some of our rules are based on the identification of Petri nets subclasses in which the set of reachable markings equals the set of potentially reachable ones, a property we call the PR-R equality in [19,20]. We use this kind of rules in the example of the "SwimmingPool" model of Fig. 8, a classical example of Petri net often used in case studies (see e.g. [10]).

We give an example of a basic abstraction law in Fig. 1, with an instance of rule (CONCAT) that allows us to fuse two places connected by a direct, silent transition. We give another example with (MAGIC), in Fig. 2, which illustrates a more complex agglomeration rule, and refer to other examples in Sect. 7.

The parametric net (N_1, C_1) (left of Fig. 1) has a condition which entails that place y_2 should be empty initially ($y_2 = 0$), whereas net (N_2, C_2) has a

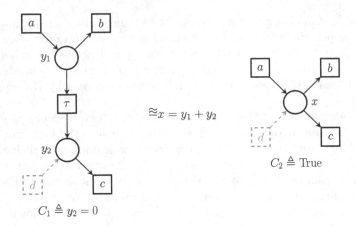

Fig. 1. Equivalence rule (CONCAT), $(N_1, C_1) \approxeq_E (N_2, C_2)$, between nets N_1 (left) and N_2 (right), for the relation $E \triangleq (x = y_1 + y_2)$.

trivial constraint, which can be interpreted as simply $x \geqslant 0$. We can show (see Sect. 3) that nets N_1 and N_2 are E-equivalent, which amounts to prove that any marking $(y_1 : k_1, y_2 : k_2)$ of N_1, reachable by firing a transition sequence σ, can be associated with the marking $(x : k_1 + k_2)$ of N_2, also reachable by the same firing sequence. Actually, we prove that this equivalence is sound when no transition can input a token directly into place y_2 of N_1. This means that the rule is correct in the absence of the dashed transition (with label d), but that our procedure should flag the rule as unsound when transition d is present.

The results presented in this paper provide an automated technique for proving the correctness of polyhedral abstraction laws. This helps us gain more confidence on the correctness of our tools and is also useful if we want to add new abstraction rules. Indeed, up until now, all our rules where proven using "manual theorem proving", which can be tedious and error-prone.

Incidentally, the theory we developed for this paper also helped us gain a better understanding of the constraints necessary when designing new abstraction laws. A critical part of our approach relies on the ability, given a Presburger predicate C, to encode the set of markings reachable from C by firing only silent transitions, that we denote $\tau_C^\star$ in the following. Our approach draws a connection with previous works [7,8,24] that study the class of Petri nets that have Presburger-definable reachability sets; also called *flat nets*. We should also make use of a tool implemented by the same authors, called FAST, which provides a method for representing the reachable set of flat nets. Basically, we gain the insight that polyhedral abstraction provides a way to abstract away (or collapse) the sub-parts of a net that are flattable. Note that our approach may work even though the reachability set of the whole net is not semilinear, since only the part that is abstracted must be flattable. We also prove that when $(N_1, C_1) \approxeq_E (N_2, C_2)$ then necessarily the sets $\tau_{C_1}^\star$ and $\tau_{C_2}^\star$ are semilinear.

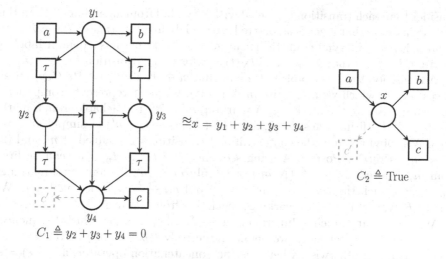

$$\approx x = y_1 + y_2 + y_3 + y_4$$

$$C_1 \triangleq y_2 + y_3 + y_4 = 0$$

$$C_2 \triangleq \text{True}$$

Fig. 2. Equivalence rule (MAGIC).

Outline and Contributions. The paper is organized as follows. We define our central notion of *parametric polyhedral abstraction* in Sect. 3 and prove several of its properties in Sect. 6. In particular, we prove that polyhedral abstraction is a congruence, and that it is preserved when "duplicating labeled transitions". These properties mean that every abstraction law we prove can be safely applied in every context, and that each law can be used as a "rule schema". Our definition relies on a former notion of polyhedral equivalence, that we recall in Sect. 2, together with a quick overview of our notations. We describe our proof procedure in Sect. 4, which is defined as the construction of a set of four *core requirements*, each expressed as separate quantified LIA formulas. A key ingredient in this translation is to build a predicate, $\tau_C^\star$, which encodes the markings reachable by firing only the silent transitions of a net. We defer the definition of this predicate until Sect. 5, where we show how it can be obtained using the output of the FAST tool. We also describe a method for automatically certifying that the resulting predicate is sound, which means that we do not have to trust the soundness of any outside software component, except SMT solvers. We conclude by presenting the results obtained with a new tool implementing our approach, called Reductron, on some concrete examples.

2   Petri Nets and Polyhedral Abstraction

A Petri net is a tuple $(P, T, \text{Pre}, \text{Post})$, where $P = p_1, \ldots, p_n$ is a finite set of places, $T = t_1, \ldots, t_k$ is a finite set of transitions (disjoint from P), and $\text{Pre} : T \to (P \to \mathbb{N})$ and $\text{Post} : T \to (P \to \mathbb{N})$ are the pre- and post-condition functions (also known as the flow functions of the net). A state of a net, also called a marking, is a mapping $m : P \to \mathbb{N}$ (also denoted $\mathbb{N}^P$) that assigns a number of tokens, $m(p)$, to each place p in P. A marked net (N, m_0) is a pair consisting of a net, N, and an initial marking, m_0. In the following, we will often consider that each transition is labeled with a symbol from an alphabet Σ. In this case, we assume that a net is associated with a labeling function $l : T \to \Sigma \cup \{\tau\}$, where τ is a special symbol for the silent action. Every net has a default labeling function, l_N, such that $\Sigma = T$ and $l_N(t) = t$ for every transition $t \in T$.

A transition $t \in T$ is enabled at a marking $m \in \mathbb{N}^P$ if $m(p) \geqslant \mathrm{Pre}(t, p)$ for all places $p \in P$, which we also write $m \geqslant \mathrm{Pre}(t)$, where $\geqslant$ represents component-wise comparison of the markings. A marking $m' \in \mathbb{N}^P$ is reachable from a marking $m \in \mathbb{N}^P$ by firing transition t, denoted $(N, m) \xrightarrow{t} (N, m')$ or simply $m \xrightarrow{t} m'$ when N is obvious from the context, if: (1) transition t is enabled at m, and (2) $m' = m - \mathrm{Pre}(t) + \mathrm{Post}(t)$. A firing sequence $\varrho = t_1, \ldots, t_n \in T^*$ can be fired from m, denoted $(N, m) \xRightarrow{\varrho} (N, m')$ or simply $m \xRightarrow{\varrho} m'$, if there exist markings $m_0, \ldots, m_n$ such that $m = m_0$, $m' = m_n$, and $m_i \xrightarrow{t_{i+1}} m_{i+1}$ for all $i < n$. We denote $R(N, m_0)$ the set of markings reachable from m_0 in N.

We can lift any labeling function $l : T \to \Sigma \cup \{\tau\}$ to a mapping of sequences from T^* to Σ^*. Specifically, we define inductively $l(\varrho.t) = l(\varrho)$ if $l(t) = \tau$ and $l(\varrho.t) = l(\varrho).l(t)$ otherwise, where . is the concatenation operator, and $l(\epsilon) = \epsilon$, where ϵ is the empty sequence, verifying $\epsilon.\sigma = \sigma.\epsilon = \sigma$ for any $\sigma \in \Sigma^*$. Given a sequence of labels $\sigma \in \Sigma^*$, we write $(N, m) \xRightarrow{\sigma} (N, m')$ if there exists a firing sequence $\varrho \in T^*$ such that $(N, m) \xRightarrow{\varrho} (N, m')$ and $\sigma = l(\varrho)$. In this case, σ is referred to as an *observable sequence* of the marked net (N, m). In some cases, we have to consider firing sequences that must not finish with τ transitions. Hence, we define a relation $(N, m) \xRightarrow{\sigma\rangle} (N, m')$, written simply $m \xRightarrow{\sigma\rangle} m'$, as follows:

- $(N, m) \xRightarrow{\epsilon\rangle} (N, m)$ holds for all marking m.
- $(N, m) \xRightarrow{\sigma.a\rangle} (N, m')$ holds for any markings m, m' and $a, \sigma \in \Sigma \times \Sigma^*$, if there exists a marking m'' and a transition t such that $l(t) = a$ and $(N, m) \xRightarrow{\sigma} (N, m'') \xrightarrow{t} (N, m')$.

It is immediate that $m \xRightarrow{\sigma\rangle} m'$ implies $m \xRightarrow{\sigma} m'$. Note the difference between $m \xRightarrow{\epsilon} m'$, which stands for any sequence of τ transitions, and $m \xRightarrow{\epsilon\rangle} m'$, which implies $m = m'$ (the sequence is empty).

We use the standard graphical notation for nets, where places are depicted as circles and transitions as squares such as the nets displayed in Fig. 1.

Polyhedral Abstraction. We define an equivalence relation that can be used to describe a linear dependence between the markings of two different nets, N_1 and N_2. Assume V is a set of places $p_1, \ldots, p_n$, considered as variables, and let m be a mapping in $V \to \mathbb{N}$. We define $\underline{m}$ as a linear formula, whose unique model in $\mathbb{N}^V$ is m, defined as $\underline{m} \triangleq \bigwedge \{x = m(x) \mid x \in V\}$. By extension, given a Presburger formula E, we say that m is a (partial) solution of E if the formula $E \wedge \underline{m}$ is consistent. Equivalently, we can view $\underline{m}$ as a substitution, where each variable $x \in V$ is substituted by $m(x)$. Indeed, the formula $F\{m\}$ (the substitution $\underline{m}$ applied to F) and $F \wedge \underline{m}$ admit the same models. Given two mappings $m_1 \in \mathbb{N}^{V_1}$ and $m_2 \in \mathbb{N}^{V_2}$, we say that m_1 and m_2 are *compatible* when they have equal values on their shared domain: $m_1(x) = m_2(x)$ for all x in $V_1 \cap V_2$. This is a necessary and sufficient condition for the system $\underline{m_1} \wedge \underline{m_2}$ to be consistent. Finally, if V is the set of free variables of $\underline{m_1}, \underline{m_2}$, and the free variables of E are included in V, we say that m_1 and m_2 are related up-to E, denoted $m_1 \equiv_E m_2$, when $E \wedge \underline{m_1} \wedge \underline{m_2}$ is consistent.

$$m_1 \equiv_E m_2 \quad \Leftrightarrow \quad \exists m \in \mathbb{N}^V . \; m \models E \wedge \underline{m_1} \wedge \underline{m_2} \tag{1}$$

This relation defines an equivalence between markings of two different nets ($\equiv_E \; \subseteq \mathbb{N}^{P_1} \times \mathbb{N}^{P_2}$) and, by extension, can be used to define an equivalence between nets themselves, that is called *polyhedral equivalence* in [2,5], where all reachable markings of N_1 are related to reachable markings of N_2 (and conversely), as explained next.

Definition 1 (E-abstraction). *Assume $N_1 = (P_1, T_1, \mathrm{Pre}_1, \mathrm{Post}_1)$ and $N_2 = (P_2, T_2, \mathrm{Pre}_2, \mathrm{Post}_2)$ are two Petri nets, and E a Presburger formula whose free variables are included in $P_1 \cup P_2$. We say that the marked net (N_2, m_2) is an E-abstraction of (N_1, m_1), denoted $(N_1, m_1) \sqsubseteq_E (N_2, m_2)$, if and only if:*

(A1) *The initial markings are compatible with E, meaning $m_1 \equiv_E m_2$.*
(A2) *For all observable sequences $(N_1, m_1) \overset{\sigma}{\Rightarrow} (N_1, m_1')$ in N_1, there is at least one marking m_2' over P_2 such that $m_1' \equiv_E m_2'$, and for all markings m_2' over P_2 such that $m_1' \equiv_E m_2'$ we have $(N_2, m_2) \overset{\sigma}{\Rightarrow} (N_2, m_2')$.*

We say that (N_1, m_1) is *E-equivalent* to (N_2, m_2), denoted $(N_1, m_1) \equiv_E (N_2, m_2)$, when we have both $(N_1, m_1) \sqsubseteq_E (N_2, m_2)$ and $(N_2, m_2) \sqsubseteq_E (N_1, m_1)$.

By definition, given an equivalence statement $(N_1, m_1) \equiv_E (N_2, m_2)$, then for every marking m_2' reachable in N_2, the set of markings of N_1 consistent with $E \wedge \underline{m_2'}$ is non-empty (condition (A2)). This defines a partition of the reachable markings of (N_1, m_1) into a union of "convex sets"—hence the name polyhedral abstraction—each associated to one (at least) reachable marking in N_2.

Although E-abstraction looks like a simulation, it is not, since the pair of reachable markings m_1', m_2' from the definition does not satisfy $(N_1, m_1') \sqsubseteq_E (N_2, m_2')$ in general. This relation $\sqsubseteq_E$ is therefore broader than a simulation, but suffices for our primary goal, that is Petri net reduction. Of course, $\equiv_E$ is not a bisimulation either. It is also quite simple to show that checking E-abstraction equivalence is undecidable in general.

Theorem 1 (Undecidability of E-equivalence). *The problem of checking whether a statement $(N_1, m_1) \equiv_E (N_2, m_2)$ is valid is undecidable.*

Proof. Assume that N_1 and N_2 are two nets with the same set of places, such that all transitions are silent. Then $(N_1, m_1) \equiv_{\mathrm{True}} (N_2, m_2)$, an E-abstraction for the trivial constraint $E \triangleq \mathrm{True}$, entails that (N_1, m_1) and (N_2, m_2) must have the same reachability set. This property is known as *marking equivalence* and is undecidable [17]. $\qquad\square$

3   Parametric Reduction Rules and Equivalence

E-abstraction is defined on marked nets (Definition 1), thus the reduction rules defined in [1,2], which are E-abstraction equivalences, mention marked nets as well. Their soundness was proven manually, using constrained parameters for initial markings. Such constraints on markings are called *coherency constraints*.

Coherency Constraints. We define a notion of *coherency constraint*, C, that must hold not only in the initial state, but also in a sufficiently large subset of reachable markings. We have already seen an example with the constraint $C_1 \triangleq y_2 = 0$ used in rule (CONCAT). Without the use of C_1, rule (CONCAT) would be unsound since net N_2 (right of Fig. 1) could fire transition b more often than its counterpart, N_1.

Since C is a predicate on markings, we equivalently consider it as a subset of markings or as a logic formula, so that we may equivalently write $m \models C$ or $m \in C$ to indicate that $C(m)$ is true.

Definition 2 (Coherent Net). *Given a Petri net N and a predicate C on markings, we say that N satisfies the coherency constraint C, or equivalently, that (N, C) is a coherent net, if and only if for all firing sequences $m \overset{\sigma}{\Rightarrow} m'$ with $m \in C$, we have*

$$\exists m'' \in C \,.\, m \overset{\sigma}{\Rightarrow} m'' \wedge m'' \overset{\epsilon}{\Rightarrow} m'$$

Intuitively, if we consider that all τ transitions are irreversible choices, then we can define a partial order on markings with $m < m'$ whenever $m \overset{\tau}{\rightarrow} m'$ holds. Then, markings satisfying the coherency constraint C must be minimal with respect to this partial order.

In this paper, we wish to prove automatically the soundness of a given reduction rule. A reduction rule basically consists of two nets with their coherency constraints, and a Presburger relation between markings.

Definition 3 (Parametric Reduction Rule). *A parametric reduction rule is written $(N_1, C_1) >_E (N_2, C_2)$, where (N_1, C_1) and (N_2, C_2) are both coherent nets, and C_1, C_2, and E are Presburger formulas whose free variables are in $P_1 \cup P_2$.*

A given reduction rule $(N_1, C_1) >_E (N_2, C_2)$ is a candidate, which we will analyze to prove its soundness: is it an E-abstraction equivalence?

Our analysis relies on a richer definition of E-abstraction, namely parametric E-abstraction (Definition 4, next), which includes the coherency constraints C_1, C_2. Parametric E-abstraction entails E-abstraction for each instance of its parameters (Theorem 2, below). Essentially, for any sequence $m_1 \overset{\sigma}{\Rightarrow} m_1'$ with $m_1 \in C_1$, there exists a marking m_2' such that $m_1' \equiv_E m_2'$; and for every marking $m_2 \in C_2$ compatible with m_1, i.e., $m_1 \equiv_E m_2$, all markings m_2' compatible with m_1' (i.e., $m_1' \equiv_E m_2'$) can be reached from m_2 by the same observable sequence σ. To ease the presentation, we define the notation

$$m_1 \langle C_1 E C_2 \rangle m_2 \triangleq m_1 \models C_1 \wedge m_1 \equiv_E m_2 \wedge m_2 \models C_2 \tag{2}$$

Definition 4 (Parametric E-abstraction). *Assume $(N_1, C_1) >_E (N_2, C_2)$ is a parametric reduction rule. We say that (N_2, C_2) is a parametric E-abstraction of (N_1, C_1), denoted $(N_1, C_1) \preceq_E (N_2, C_2)$ if and only if:*

(S1) *For all markings m_1 satisfying C_1 there exists a marking m_2 such that $m_1 \langle C_1 E C_2 \rangle m_2$.*

(S2) *For all firing sequences $m_1 \overset{\epsilon}{\Rightarrow} m_1'$ and all markings m_2, we have $m_1 \equiv_E m_2$ implies $m_1' \equiv_E m_2$.*

(S3) *For all firing sequences $m_1 \overset{\sigma}{\Rightarrow} m_1'$ and all marking pairs m_2, m_2', if $m_1 \langle C_1 E C_2 \rangle m_2$ and $m_1' \equiv_E m_2'$ then we have $m_2 \overset{\sigma}{\Rightarrow} m_2'$.*

We say that (N_1, C_1) and (N_2, C_2) are in parametric E-equivalence, denoted $(N_1, C_1) \approx_E (N_2, C_2)$, when we have both $(N_1, C_1) \preceq_E (N_2, C_2)$ and $(N_2, C_2) \preceq_E (N_1, C_1)$.

Condition (S1) corresponds to the solvability of the Presburger formula E with respect to the marking predicates C_1 and C_2. Condition (S2) ensures that silent transitions of N_1 are abstracted away by the formula E, and are therefore invisible to N_2. Condition (S3) follows closely condition (A2) of the standard E-abstraction equivalence.

Note that equivalence $\approx$ is not a bisimulation, in the same way that $\equiv$ from Definition 1. It is defined only for observable sequences starting from states satisfying the coherency constraint C_1 of N_1 or C_2 of N_2, and so this relation is usually not true on every pair of equivalent markings $m_1 \equiv_E m_2$.

Instantiation Law. Parametric E-abstraction implies E-abstraction for every instance pair satisfying the coherency constraints C_1, C_2.

Theorem 2 (Parametric E-abstraction Instantiation). *Assume $(N_1, C_1) \preceq_E (N_2, C_2)$ is a parametric E-abstraction. Then for every pair of markings m_1, m_2, $m_1 \langle C_1 E C_2 \rangle m_2$ implies $(N_1, m_1) \sqsubseteq_E (N_2, m_2)$.*

Proof. Consider $(N_1, C_1) \preceq_E (N_2, C_2)$, a parametric E-abstraction, and m_1, m_2 such that $m_1 \langle C_1 E C_2 \rangle m_2$ holds. By definition of $m_1 \langle C_1 E C_2 \rangle m_2$, see Eq. (2), condition (A1) of Definition 1 is immediately satisfied. We show (A2) by considering an observable sequence $(N_1, m_1) \overset{\sigma}{\Rightarrow} (N_1, m_1')$. Since m_1 satisfies the coherency constraint C_1, we get from Definition 2 a marking $m_1'' \in C_1$ such that $m_1 \overset{\sigma}{\Rightarrow} m_1'' \overset{\epsilon}{\Rightarrow} m_1'$ holds. By applying (S1) to m_1'', we get a marking m_2' such that $m_1'' \langle C_1 E C_2 \rangle m_2'$ holds, which implies $m_1'' \equiv_E m_2'$. Then, by applying (S2) to $m_1'' \overset{\epsilon}{\Rightarrow} m_1'$, we obtain the expected result $m_1' \equiv_E m_2'$. Finally, for all markings m_2' such that $m_1' \equiv_E m_2'$, we conclude $m_2 \overset{\sigma}{\Rightarrow} m_2'$ from (S3). Condition (A2) is proved, hence $(N_1, m_1) \sqsubseteq_E (N_2, m_2)$ holds. $\square$

4 Automated Proof Procedure

Our automated proof procedure receives a candidate reduction rule (Definition 3) as input, and has three possible outcomes: (i) the candidate is proven sound, congratulations you have established a new parametric E-abstraction equivalence; (ii) the candidate is proven unsound, try to understand why and fix it; or (iii) we cannot conclude, because part of our procedure relies on a semi-algorithm

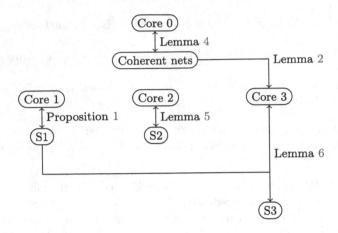

Fig. 3. Detailed dependency relations.

(see Sect. 5) for expressing the set of reachable markings of a flat subnet as a linear constraint.

Given the candidate reduction rule, the procedure generates SMT queries, which we call *core requirements* (defined in Sect. 4.2) that are solvable if and only if the candidate is a parametric E-abstraction (Theorems 3 and 4, Sect. 4.3). We express these constraints into Presburger predicates, so it is enough to use solvers for the theory of formulas on Linear Integer Arithmetic, what is known as LIA in SMT-LIB [9]. We illustrate the results given in this section using a diagram (Fig. 3) that describe the dependency relations between conditions (S1), (S2), (S3) and their encoding as core requirements.

4.1 Presburger Encoding of Petri Net Semantics

We start by defining a few formulas that ease the subsequent expression of core requirements. This will help with the most delicate point of our encoding, which relies on how to encode sequences of transitions. Note that the coherency constraints of reduction rules are already defined as such.

In the following, we use $\boldsymbol{x}$ for the vector of variables $(x_1, \ldots, x_n)$, corresponding to the places $p_1, \ldots, p_n$ of P, and $F(\boldsymbol{x})$ for a formula whose variables are included in $\boldsymbol{x}$. We say that a mapping m of $\mathbb{N}^P$ is a *model* of F, denoted $m \models F$, if the ground formula $F(m) = F(m(p_1), \ldots, m(p_n))$ is true. Hence, we can also interpret F as a predicate over markings. Finally, we define the semantics of F as the set $\llbracket F \rrbracket = \{m \in \mathbb{N}^P \mid m \models F\}$. As usual, we say that a predicate F is *valid*, denoted $\models F$, when all its interpretations are true ($\llbracket F \rrbracket = \mathbb{N}^P$). In order to keep track of fired transitions in our encoding, and without any loss of generality we assume that our alphabet of labels Σ is a subset of the natural numbers ($\Sigma \subset \mathbb{N}^*$), except 0 that is reserved for τ.

We define next a few Presburger formulas that express properties on markings of a net N. For instance, Equation (3) below defines the predicate ENBL_t, for a given transition t, which corresponds exactly to the markings that enable t. We also define a linear predicate $\mathrm{T}(\boldsymbol{x}, \boldsymbol{x}', a)$ that describes the relation between the markings before $(\boldsymbol{x})$ and after $(\boldsymbol{x}')$ firing a transition with label a. With this convention, formula $\mathrm{T}(m, m', a)$ holds if and only if $m \xrightarrow{t} m'$ holds for some transition t such that $l(t) = a$ (which implies $a \neq 0$).

$$\mathrm{ENBL}_t(\boldsymbol{x}) \triangleq \bigwedge_{i \in 1..n} (x_i \geqslant \mathrm{Pre}(t, p_i)) \tag{3}$$

$$\Delta_t(\boldsymbol{x}, \boldsymbol{x}') \triangleq \bigwedge_{i \in 1..n} (x_i' = x_i + \mathrm{Post}(t, p_i) - \mathrm{Pre}(t, p_i)) \tag{4}$$

$$\mathrm{T}(\boldsymbol{x}, \boldsymbol{x}', a) \triangleq \bigvee_{t \in T} (\mathrm{ENBL}_t(\boldsymbol{x}) \wedge \Delta_t(\boldsymbol{x}, \boldsymbol{x}') \wedge a = l(t)) \tag{5}$$

We admit the following, for all markings m, m' and label a:

$$\models T(m, m', a) \iff \exists t . m \xrightarrow{t} m' \wedge l(t) = a \tag{6}$$

In order to define the core requirements, we additionally require a predicate $\tau_C^*(\boldsymbol{x}, \boldsymbol{x}')$ encoding the markings reachable by firing any sequence of silent transitions from a state satisfying the coherency constraint C. And so, the following constraint must hold:

$$\models m \in C \implies (\tau_C^*(m, m') \iff m \xRightarrow{\epsilon} m') \tag{7}$$

Since $m \xRightarrow{\epsilon} m'$ may fire an arbitrary number of silent transitions τ, the predicate τ_C is not guaranteed to be expressible as a Presburger formula in the general case. Yet, in Sect. 5, we characterize the Petri nets for which τ_C can be expressed in Presburger logic, which include all the polyhedral reductions that we meet in practice (we explain why).

Thanks to this predicate, we define the formula $\acute{T}_C(\boldsymbol{x}, \boldsymbol{x}', a)$ encoding the reachable markings from a marking satisfying the coherency constraint C, by firing any number of silent transitions, followed by a transition labeled with a. Then, we define $\hat{T}$ which extends $\acute{T}$ with any number of silent transitions after a and also allows for only silent transitions (no transition a).

$$\acute{T}_C(\boldsymbol{x}, \boldsymbol{x}', a) \triangleq \exists \boldsymbol{x}'' . \tau_C^*(\boldsymbol{x}, \boldsymbol{x}'') \wedge T(\boldsymbol{x}'', \boldsymbol{x}', a) \tag{8}$$

$$\hat{T}_C(\boldsymbol{x}, \boldsymbol{x}', a) \triangleq \left(\exists \boldsymbol{x_1} . \acute{T}_C(\boldsymbol{x}, \boldsymbol{x_1}, a) \wedge C(\boldsymbol{x_1}) \wedge \tau_C^*(\boldsymbol{x_1}, \boldsymbol{x}') \right) \tag{9}$$

$$\vee (a = 0 \wedge \tau_C^*(\boldsymbol{x}, \boldsymbol{x}')) \tag{10}$$

Lemma 1. *For any markings m, m' and label a such that $m \in C$, we have $\models \acute{T}_C(m, m', a)$ if and only if $m \xRightarrow{a} m'$ holds.*

Proof. We show both directions separately.

– Assume $m \xRightarrow{a} m'$. By definition, this implies that there exists m'' and a transition t such that $l(t) = a$ and $m \xRightarrow{\epsilon} m'' \xrightarrow{t} m'$. Therefore, $\tau_C^*(m, m'')$ is valid by (7), and $T(m'', m', a)$ is valid by (6), hence the expected result $\models \acute{T}_C(m, m', a)$.

- Conversely, assume $\acute{T}_C(m, m', a)$ is valid. Then, by (8) there exists a marking m'' such that both $\tau_C^*(m, m'')$ and $T(m'', m', a)$ are valid. From (7), we get $m \overset{\epsilon}{\Rightarrow} m''$, and (6) implies $\exists t \, . \, m'' \overset{t}{\rightarrow} m' \wedge l(t) = a$. Thus, $m \overset{\epsilon}{\Rightarrow} m'' \overset{t}{\rightarrow} m'$, that is the expected result $m \overset{a}{\Rightarrow} m'$. □

Lemma 2. *Given a coherent net* (N, C), *for any markings* m, m' *such that* $m \in C$ *and* $a \in \Sigma \cup \{0\}$, *we have* $\models \hat{T}_C(m, m', a)$ *if and only if either* $m \overset{\epsilon}{\Rightarrow} m'$ *and* $a = 0$, *or* $m \overset{a}{\Rightarrow} m'$.

Proof. We show both directions separately.

- Assume $m \overset{\epsilon}{\Rightarrow} m'$ and $a = 0$, then $\tau_C^*(m, m')$ is valid by (7), hence the expected result $\models \hat{T}_C(m, m', a)$ from (10).
- Assume $m \overset{a}{\Rightarrow} m'$. From Definition 2 (coherent net), there exists $m'' \in C$ such that $m \overset{a}{\Rightarrow} m'' \overset{\epsilon}{\Rightarrow} m'$. Then, we get $\models \acute{T}_C(m, m'', a)$ from Lemma 1, and $\models \tau_C^*(m'', m')$ from (7). Consequently, $\hat{T}_C(m, m', a)$ is valid from (9).
- Conversely, assume $\hat{T}_C(m, m', a)$ holds by (10), then $a = 0$ and $\models \tau_C^*(m, m')$, which implies $m \overset{\epsilon}{\Rightarrow} m'$ by (7). This is the expected result.
- Finally, assume $\hat{T}_C(m, m', a)$ holds by (9), then there exists a marking $m'' \in C$ such that $\models \acute{T}_C(m, m'', a)$ and $\models \tau_C^*(m'', m')$. This implies $m \overset{a}{\Rightarrow} m'' \overset{\epsilon}{\Rightarrow} m'$ from Lemma 1 and (7). This implies the expected result $m \overset{a}{\Rightarrow} m'$. □

Finally, we denote $\tilde{E}(\boldsymbol{x}, \boldsymbol{y})$ the formula obtained from E where free variables are substituted as follows: place names in N_1 are replaced with variables in $\boldsymbol{x}$, and place names in N_2 are replaced with variables in $\boldsymbol{y}$ (making sure that bound variables of E are renamed to avoid interference). When the same place occurs in both nets, say $p_i^1 = p_j^2$, we also add the equality constraint $(x_i = y_j)$ to $\tilde{E}$ in order to preserve this equality constraint.

4.2   Core Requirements: Parametric E-abstraction Encoding

In order to check conditions (S1)–(S3) of parametric E-abstraction (Definition 4), we define a set of Presburger formulas, called *core requirements*, to be verified using an external SMT solver ((Core 1) to (Core 3)). You will find an illustration of these requirements in Figs. 4–7. The satisfaction of these requirements entail the parametric E-abstraction relation. We have deliberately stressed the notations to prove that (N_2, C_2) is a parametric E-abstraction of (N_1, C_1). Of course, each constraint must be checked in both directions to obtain the equivalence. Also, to not overload the notations, we assume that the transition relations are clear in the context if they belong to N_1 or N_2.

Verifying that a Net is Coherent. The first step consists in verifying that both nets N_1 and N_2 satisfy their coherency constraints C_1 and C_2 (the coherency constraint is depicted in Fig. 4). We recall Definition 2:

Definition (Coherent Net). *For all firing sequence* $m \overset{\sigma}{\Rightarrow} m'$ *with* $m \in C$, *there exists a marking* m'' *satisfying* C *such that* $m \overset{\sigma}{\Rightarrow} m''$ *and* $m'' \overset{\epsilon}{\Rightarrow} m'$.

We encode a simpler relation, below, with sequences σ of size 1. This relies on the following result:

Lemma 3. (N, C) is coherent if and only if for all firing sequence $m \overset{a\rangle}{\Longrightarrow} m'$ with $m \in C$ and $a \in \Sigma$, we have $\exists m'' \in C$. $m \overset{a\rangle}{\Longrightarrow} m'' \wedge m'' \overset{\epsilon}{\Rightarrow} m'$.

We deliberately consider a firing sequence $m \overset{a\rangle}{\Longrightarrow} m'$ (and not $m \overset{a}{\Rightarrow} m'$), since the encoding relies only on $\acute{T}_C$ (that is, $\overset{a\rangle}{\Longrightarrow}$), not on $\hat{T}_C$ (that is, $\overset{a}{\Rightarrow}$).

Proof. The "only if" part is immediate, as a particular case of Definition 2 and noting that $m \overset{a\rangle}{\Longrightarrow} m'$ implies $m \overset{a}{\Rightarrow} m'$. Conversely, assume the property stated in the lemma is true. Then, we show by induction on the size of σ, that Definition 2 holds for any σ. Note that the base case $\sigma = \epsilon$ always holds, for any net, by taking $m'' = m$. Now, consider a non-empty sequence $\sigma = \sigma'.a$ and $m \overset{\sigma'.a}{\Longrightarrow} m'$ with $m \in C$. By definition, there exists m_1 and m_2 such that $m \overset{\sigma'}{\Longrightarrow} m_1 \overset{a\rangle}{\Longrightarrow} m_2 \overset{\epsilon}{\Rightarrow} m'$. By induction hypothesis, on $m \overset{\sigma'}{\Longrightarrow} m_1$, there exists $m_3 \in C$ such that $m \overset{\sigma'}{\Longrightarrow} m_3 \overset{\epsilon}{\Rightarrow} m_1$. Therefore, we have $m \overset{\sigma'}{\Longrightarrow} m_3 \overset{\epsilon}{\Rightarrow} m_1 \overset{a\rangle}{\Longrightarrow} m_2 \overset{\epsilon}{\Rightarrow} m'$, which can simply be written $m \overset{\sigma'}{\Longrightarrow} m_3 \overset{a\rangle}{\Longrightarrow} m_2 \overset{\epsilon}{\Rightarrow} m'$. Using the property stated in the lemma on $m_3 \overset{a\rangle}{\Longrightarrow} m_2$, we get a marking $m_4 \in C$ such that $m_3 \overset{a\rangle}{\Longrightarrow} m_4 \overset{\epsilon}{\Rightarrow} m_2$. Hence, $m \overset{\sigma'}{\Longrightarrow} m_3 \overset{a\rangle}{\Longrightarrow} m_4 \overset{\epsilon}{\Rightarrow} m_2 \overset{\epsilon}{\Rightarrow} m'$ holds, which can be simplified as $m \overset{\sigma'.a}{\Longrightarrow} m_4 \overset{\epsilon}{\Rightarrow} m'$. This is the expected result. $\qquad \square$

Therefore, we can encode Definition 2 using the following formula:

$$\forall \boldsymbol{p}, \boldsymbol{p}', a \, . \, C(\boldsymbol{p}) \wedge \acute{T}_C(\boldsymbol{p}, \boldsymbol{p}', a)$$
$$\implies \exists \boldsymbol{p}'' \, . \, C(\boldsymbol{p}'') \wedge \acute{T}_C(\boldsymbol{p}, \boldsymbol{p}'', a) \wedge \tau_C^*(\boldsymbol{p}'', \boldsymbol{p}') \qquad \text{(Core 0)}$$

Lemma 4. *Given a Petri net N, the constraint (Core 0) is valid if and only if the net satisfies the coherency constraint C.*

Proof. Constraint (Core 0) is an immediate translation of the property stated in Lemma 3. $\qquad \square$

Given a net N, a constraint C expressed as a Presburger formula, and a formula τ_C^* that captures $\overset{\epsilon}{\Rightarrow}$ transitions (as obtained in Sect. 5), we are now able to check automatically that a net (N, C) is coherent. Thus, from now on, we assume that the considered nets (N_1, C_1) and (N_2, C_2) are indeed coherent.

Coherent Solvability. The first requirement of the parametric E-abstraction relates to the solvability of formula E with regard to the coherency constraint C_1, and is encoded by (Core 1). This requirement ensures that every marking of N_1 satisfying C_1 can be associated to at least one marking of N_2 satisfying C_2. Let us recall (S1), taken from Definition 4:

Definition (S1). *For all markings m_1 satisfying C_1 there exists a marking m_2 such that $m_1 \langle C_1 E C_2 \rangle m_2$.*

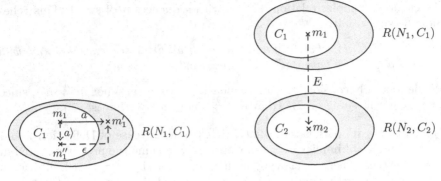

Fig. 4. Illustration of (Core 0).   **Fig. 5.** Illustration of (Core 1).

Condition (S1) is depicted in Fig. 5. We propose to encode it by the following Presburger formula:

$$\forall \boldsymbol{x} \, . \, C_1(\boldsymbol{x}) \implies \exists \boldsymbol{y} \, . \, \tilde{E}(\boldsymbol{x}, \boldsymbol{y}) \wedge C_2(\boldsymbol{y}) \qquad \text{(Core 1)}$$

Since the encoding is immediate, we admit this proposition:

Proposition 1. *The constraint (Core 1) is valid if and only if (S1) holds.*

Silent Constraints. So far, we have focused on the specific case of coherent nets, which refers to intermediate coherent markings. Another notable feature of parametric E-abstractions is the ability to fire any number of silent transitions without altering the solutions of E. In other words, if two markings, m_1 and m_2, are solutions of E, then firing any silent sequence from m_1 (or m_2) will always lead to a solution of $E \wedge m_2$ (or $E \wedge m_1$). This means that silent transitions must be invisible to the other net.

Let us recall (S2), taken from Definition 4:

Definition (S2). *For all firing sequences $m_1 \overset{\epsilon}{\Rightarrow} m_1'$ and all markings m_2, we have $m_1 \equiv_E m_2$ implies $m_1' \equiv_E m_2$.*

It actually suffices to show the result for each silent transition $t \in T_1$ taken separately:

Lemma 5. *Condition (S2) holds if and only if, for all markings m_1, m_2 such that $m_1 \equiv_E m_2$, and for all $t_1 \in T_1$ such that $l_1(t_1) = \tau$, we have $m_1 \overset{t_1}{\longrightarrow} m_1' \implies m_1' \equiv_E m_2$.*

Proof. The "only if" way is only a particular case of (S2) with a single silent transition t_1. For the "if" way, (S2) is shown from the given property by transitivity.   □

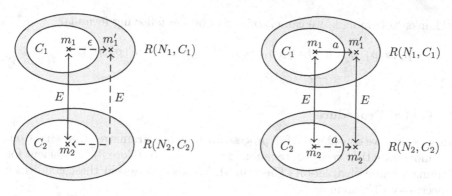

Fig. 6. Illustration of (Core 2).   **Fig. 7.** Illustration of (Core 3).

Thanks to this result, we encode (S2) by the following core requirement:

$$\forall p_1, p_2, p_1' \; . \; \tilde{E}(p_1, p_2) \wedge \tau(p_1, p_1') \implies \tilde{E}(p_1', p_2) \qquad \text{(Core 2)}$$

where $\tau(x, x')$ is defined as $\tau(x, x') \triangleq \bigvee_{t \in T | l(t) = \tau} (\text{ENBL}_t(x) \wedge \Delta_t(x, x'))$
(Fig. 6)

Reachability. Let us recall the definition of (S3), taken from Definition 4:

Definition (S3). *For all firing sequences $m_1 \overset{\sigma}{\Rightarrow} m_1'$ and all marking pairs m_2, m_2', if $m_1 \langle C_1 E C_2 \rangle m_2$ and $m_1' \equiv_E m_2'$ then we have $m_2 \overset{\sigma}{\Rightarrow} m_2'$.*

Condition (S3) mentions sequences σ of arbitrary length. We encode it with a formula dealing only with sequences of length at most 1, thanks to the following result:

Lemma 6. *Given a parametric reduction rule $(N_1, C_1) >_E (N_2, C_2)$ which satisfies condition (S1), then condition (S3) holds if and only if for all firing sequence $m_1 \overset{\sigma}{\Rightarrow} m_1'$ with $\sigma = \epsilon$ or $\sigma = a$ with $a \in \Sigma$, and all markings m_2, m_2', we have $m_1 \langle C_1 E C_2 \rangle m_2 \wedge m_1' \equiv_E m_2' \implies m_2 \overset{\sigma}{\Rightarrow} m_2'$.*

Proof. The given property is necessary as a particular case of (S3) taking $\sigma = a$ or $\sigma = \epsilon$. Conversely, assume the given property holds. We show by induction on the size of σ that (S3) holds for any sequence σ. The base cases $\sigma = a$ and $\sigma = \epsilon$ are ensured by hypothesis. Now, consider a non-empty sequence $\sigma = \sigma'.a$, and $m_1 \overset{\sigma}{\Rightarrow} m_1'$ (i), as well as markings m_2, m_2' such that $m_1 \langle C_1 E C_2 \rangle m_2$ and $m_1' \equiv_E m_2'$ holds. We have to show $m_2 \overset{\sigma}{\Rightarrow} m_2'$. From (i), we have $m_1 \overset{\sigma'.a}{\Rightarrow} m_1'$, that is, there exists a marking u_1 such that $m_1 \overset{\sigma'}{\Rightarrow} u_1 \overset{a}{\Rightarrow} m_1'$ (ii). By Definition 2, there exists $u_1' \in C_1$ such that $m_1 \overset{\sigma'}{\Rightarrow} u_1' \overset{\epsilon}{\Rightarrow} u_1$ (iii). Also, by condition (S1), there exists a marking u_2' of N_2 such that $u_1' \langle C_1 E C_2 \rangle u_2'$, which implies $u_1' \equiv_E u_2'$ (iv). Hence, by induction hypothesis on $m_1 \overset{\sigma'}{\Rightarrow} u_1'$, we have $m_2 \overset{\sigma'}{\Rightarrow} u_2'$ (α) From (iii) and (ii), we get $u_1' \overset{a}{\Rightarrow} m_1'$ (v). Applying the property of the lemma on (iv) and (v), we get $u_2' \overset{a}{\Rightarrow} m_2'$ (β). Combining (α) and (β) leads to $m_2 \overset{\sigma'.a}{\Rightarrow} m_2'$, that is the expected result $m_2 \overset{\sigma}{\Rightarrow} m_2'$. $\qquad\square$

Thanks to Lemma 6, we can encode (S3) by the following formula:

$$\forall \boldsymbol{p_1}, \boldsymbol{p_2}, a, \boldsymbol{p_1'}, \boldsymbol{p_2'} . \ \langle C_1 E C_2 \rangle(\boldsymbol{p_1}, \boldsymbol{p_2}) \wedge \hat{T}_{C_1}(\boldsymbol{p_1}, \boldsymbol{p_1'}) \wedge \tilde{E}(\boldsymbol{p_1'}, \boldsymbol{p_2'}) \qquad \text{(Core 3)}$$
$$\implies \hat{T}_{C_2}(\boldsymbol{p_2}, \boldsymbol{p_2'})$$

4.3 Global Procedure

In this section, we consider the full process for proving parametric E-abstraction. We demonstrate that verifying requirements (Core 0) to (Core 3) is sufficient for obtaining a sound abstraction (Theorem 3). We also prove that these conditions are necessary (Theorem 4).

Theorem 3 (Soundness). *Given two nets N_1, N_2 and constraints C_1, C_2 expressed as Presburger formulas, if core requirement (Core 0) holds for both (N_1, C_1) and (N_2, C_2), and if core requirements (Core 1), (Core 2), and (Core 3) are valid, then the rule is a parametric E-abstraction: $(N_1, C_1) \preceq_E (N_2, C_2)$.*

Proof. If (Core 0) holds for (N_1, C_1), then (N_1, C_1) is a coherent net by Lemma 4. Similarly for (N_2, C_2). Hence, $(N_1, C_1) >_E (N_2, C_2)$ is a parametric reduction rule. By Proposition 1, and since (Core 1) is valid, we get (S1) from Definition 4. Similarly, by Lemma 5, and since (Core 2) is valid, we get (S2). Finally, (S3) holds by Lemma 6 since (Core 3) is valid and since (S1) is known to hold. (S1), (S2), (S3) entail $(N_1, C_1) \preceq_E (N_2, C_2)$ by Definition 4.   □

The converse also holds:

Theorem 4 (Completeness). *Given a parametric E-abstraction $(N_1, C_1) \preceq_E (N_2, C_2)$, then core requirements (Core 1), (Core 2), and (Core 3) are valid, and (Core 0) holds for both (N_1, C_1) and (N_2, C_2).*

Proof. By hypothesis, conditions (S1), (S2) and (S3) hold and (N_1, C_1) and (N_2, C_2) are coherent nets. Then, Lemma 4 implies that (Core 0) holds for both nets. Besides, Proposition 1 and Lemmas 5 and 6 ensure that (Core 1), (Core 2), and (Core 3) are valid.   □

Consequently, checking E-abstraction equivalence, i.e., $(N_1, C_1) \cong_E (N_2, C_2)$, amounts to check that SMT formulas (Core 0)-(Core 3) are valid on both nets.

Our approach relies on our ability to express (arbitrarily long) sequences $m \overset{\epsilon}{\Rightarrow} m'$ thanks to a formula $\tau_C^*(\boldsymbol{x}, \boldsymbol{x'})$. This is addressed in the next section.

5 Silent Transition Relation Acceleration

The previous results, including Theorems 3 and 4, rely on our ability to express the reachability set of silent transitions as a Presburger predicate, denoted τ_C^*. Finding a finite formula τ_C^* that captures an infinite state-space is not granted, since τ-sequences may be of arbitrary length. However, we now show that, since

τ transitions must be abstracted away by E in order to define a valid parametric E-equivalence (condition (S2)), and since E is itself a Presburger formula, this implies that τ_C^* corresponds to the reachability set of a *flattable* subnet [24], which is expressible as a Presburger formula too.

We define the *silent reachability set* of a net N from a coherent constraint C as $R_\tau(N,C) \triangleq \{m' \mid m \models C \wedge m \overset{\epsilon}{\Rightarrow} m'\}$. We now want to find a predicate $\tau_C^*(\boldsymbol{x}, \boldsymbol{x}')$ that satisfies the relation:

$$R_\tau(N,C) = \{m' \mid m' \models \exists \boldsymbol{x} \ . \ C(\boldsymbol{x}) \wedge \tau_C^*(\boldsymbol{x}, \boldsymbol{x}')\} \tag{7}$$

In order to express the formula τ_C^*, we first use the tool FAST [7], designed for the analysis of infinite systems, and that permits to compute the reachability set of a given Vector Addition System with States (VASS). Note that a Petri net can be transformed to an equivalent VASS with the same reachability set, so the formal presentation of VASS can be skipped. The algorithm implemented in FAST is a semi-procedure, for which we have some termination guarantees whenever the net is flattable [8], i.e. its corresponding VASS can be unfolded into a VASS without nested cycles, called a flat VASS. Equivalently, a net N is flattable for some coherent constraint C if its language is flat, that is, there exists some finite sequence $\varrho_1 \ldots \varrho_k \in T^*$ such that for every initial marking $m \models C$ and reachable marking m' there is a sequence $\varrho \in \varrho_1^* \ldots \varrho_k^*$ such that $m \overset{\varrho}{\Rightarrow} m'$. In short, all reachable markings can be reached by simple sequences, belonging to the language: $\varrho_1^* \ldots \varrho_k^*$. Last but not least, the authors stated in Theorem 5 from [24] that a net is flattable if and only if its reachability set is Presburger-definable:

Theorem 5 ([24]). *For every VASS V, for every Presburger set C_{in} of configurations, the reachability set $\mathrm{ReachV}(C_{in})$ is Presburger if, and only if, V is flattable from C_{in}.*

As a consequence, FAST's algorithm terminates when its input is Presburger-definable. We show in Theorem 6 that given a parametric E-abstraction equivalence $(N_1, C_1) \cong_E (N_2, C_2)$, the silent reachability sets for both nets N_1 and N_2 with their coherency constraints C_1 and C_2 are indeed Presburger-definable – we can even provide the expected formulas. Yet, our computation is complete only if the candidate reduction rule is a parametric E-abstraction equivalence (then, we are able to compute the τ_C^* relation), otherwise FAST, and therefore our procedure too, may not terminate.

Theorem 6. *Given a parametric E-abstraction equivalence $(N_1, C_1) \cong_E (N_2, C_2)$, the silent reachability set $R_\tau(N_1, C_1)$ is Presburger-definable.*

Proof. We prove only the result for (N_1, C_1), the proof for (N_2, C_2) is similar since $\cong$ is a symmetric relation. We first propose an expression that computes $R_\tau(N_1, m_1)$ for any marking m_1 satisfying C_1. Consider an initial marking m_1 in C_1. From condition (S1) (solvability of E), there exists a compatible marking m_2 satisfying C_2, meaning $m_1 \langle C_1 E C_2 \rangle m_2$ holds. Now, take a silent sequence

$m_1 \xrightarrow{\epsilon} m_1'$. From condition (S2) (silent stability), we have $m_1' \equiv_E m_2$. Hence, $R_\tau(N_1, m_1) \subseteq \{m_1' \mid \exists m_2 . \tilde{E}(m_1, m_2) \wedge \tilde{E}(m_1', m_2)\}$. Conversely, we show that all m_1' solution of $\tilde{E}(m_1', m_2)$ are reachable from m_1. Take m_1' such that $m_1' \equiv_E m_2$. Since we have $m_2 \xrightarrow{\epsilon} m_2$, by condition (S3) we must have $m_1 \xrightarrow{\epsilon} m_1'$. And finally we obtain $R_\tau(N_1, m_1) = \{m_1' \mid m_1' \models \exists \boldsymbol{p_1}, \boldsymbol{p_2} . \underline{m_1}(\boldsymbol{p1}) \wedge \tilde{E}(\boldsymbol{p_1}, \boldsymbol{p_2}) \wedge \tilde{E}(\boldsymbol{p_1'}, \boldsymbol{p_2})\}$.

We can generalize this reachability set for all coherent markings satisfying C_1. We first recall its definition, $R_\tau(N_1, C_1) = \{m_1' \mid \exists m_1 . m_1 \models C_1 \wedge m_1 \xrightarrow{\epsilon} m_1'\}$. From condition (S1), we can rewrite this set as $\{m_1' \mid \exists m_1, m_2 . m_1 \langle C_1 E C_2 \rangle m_2 \wedge m_1 \xrightarrow{\epsilon} m_1'\}$ without losing any marking. Finally, thanks to the previous result we get $R_\tau(N_1, C_1) = \{m_1' \mid m_1' \models P\}$ with $P = \exists \boldsymbol{p_1}, \boldsymbol{p_2} . \langle C_1 E C_2 \rangle (\boldsymbol{p_1}, \boldsymbol{p_2}) \wedge \tilde{E}(\boldsymbol{p_1'}, \boldsymbol{p_2})$ a Presburger formula. Because of the E-abstraction equivalence, (S1) holds in both directions, which gives $\forall \boldsymbol{p_2} . C_2(\boldsymbol{p_2}) \implies \exists \boldsymbol{p_1} . \tilde{E}(\boldsymbol{p_1}, \boldsymbol{p_2}) \wedge C_1(\boldsymbol{p_1})$. Hence, P can be simplified into $\exists \boldsymbol{p_2} . C_2(\boldsymbol{p_2}) \wedge \tilde{E}(\boldsymbol{p_1'}, \boldsymbol{p_2})$.

Note that this expression of $R_\tau(N, C)$ relies on the fact that the equivalence $(N_1, C_1) \cong_E (N_2, C_2)$ already holds. Thus, we cannot conclude that a candidate rule is an E-abstraction equivalence by using this formula at once, without the extra validation of FAST.   □

Verifying FAST Results. We have shown that FAST terminates in case of a correct parametric E-abstraction. We now show that it is possible to check that the predicates $\tau_{C_1}^*$ and $\tau_{C_2}^*$, computed from the result of FAST (see Theorem 6) are indeed correct.

Assume τ_C^* is, according to FAST, equivalent to the language $\varrho_1^* \ldots \varrho_n^*$ with $\varrho_i \in T^*$. We encode this language with the following Presburger predicate (similar to the one presented in [4]), which uses the formulas $H(\sigma^{k_i})$ and $\Delta(\sigma^{k_i})$ defined later:

$$\tau_C^*(\boldsymbol{p^1}, \boldsymbol{p^{n+1}}) \triangleq \exists k_1 \ldots k_n, \boldsymbol{p^2} \ldots \boldsymbol{p^{n-1}} . \bigwedge_{i \in 1..n} \left((\boldsymbol{p^i} \geqslant H(\sigma^{k_i})) \wedge \Delta(\sigma^{k_i})(\boldsymbol{p^i}, \boldsymbol{p^{i+1}}) \right) \tag{11}$$

This definition introduces acceleration variables k_i, encoding the number of times we fire the sequence ϱ_i. The hurdle and delta of the sequence of transitions ϱ_i^k, which depends on k, are written $H(\sigma^{k_i})$ and $\Delta(\sigma^{k_i})$, respectively. Their formulas are given in equations (14) and (15) below. Let us explain how we obtain them.

First, we define the notion of hurdle $H(\varrho)$ and delta $\Delta(\varrho)$ of an arbitrary sequence ϱ, such that $m \xrightarrow{\varrho} m'$ holds if and only if (1) $m \geqslant H(\varrho)$ (the sequence ϱ is fireable), and (2) $m' = m + \Delta(\varrho)$. This is an extension of the hurdle and delta of a single transition t, already used in formulas (3) and (4). The definition of H and Δ is inductive:

$$H(\epsilon) = \boldsymbol{0}, \ H(t) = Pre(t) \ \text{ and } \ H(\varrho_1.\varrho_2) = \max(H(\varrho_1), H(\varrho_2) - \Delta(\varrho_1)) \tag{12}$$

$$\Delta(\epsilon) = \boldsymbol{0}, \ \Delta(t) = Post(t) - Pre(t) \ \text{ and } \ \Delta(\varrho_1.\varrho_2) = \Delta(\varrho_1) + \Delta(\varrho_2) \tag{13}$$

where max is the component-wise max operator. The careful reader will check by herself that the definitions of $H(\varrho_1.\varrho_2)$ and $\Delta(\varrho_1.\varrho_2)$ do not depend on the way the sequence $\varrho_1.\varrho_2$ is split.

From these, we are able to characterize a necessary and sufficient condition for firing the sequence ϱ^k, meaning firing the same sequence k times. Given $\Delta(\varrho)$, a place p with a negative displacement (say $-d$) means that d tokens are consumed each time we fire ϱ. Hence, we should budget d tokens in p for each new iteration, and this suffices to enable the $k-1$ more iterations following the first transition ϱ. Therefore, we have $m \overset{\varrho^k}{\Longrightarrow} m'$ if and only if (1) $m \models m \geqslant \mathbb{1}_{>0}(k) \times (H(\varrho) + (k-1) \times \max(\mathbf{0}, -\Delta(\varrho)))$, with $\mathbb{1}_{>0}(k) = 1$ if and only if $k > 0$, and 0 otherwise, and (2) $m' = m + k \times \Delta(\varrho)$. Concerning the token displacement of this sequence ϱ^k, it is k times the one of the non-accelerated sequence ϱ. Equivalently, if we denote by m^+ the "positive" part of a mapping m, such that $m^+(p) = 0$ when $m(p) \leqslant 0$ and $m^+(p) = m(p)$ when $m(p) > 0$, we get:

$$H(\varrho^k) = \mathbb{1}_{>0}(k) \times (H(\varrho) + (k-1) \times (-\Delta(\varrho))^+) \tag{14}$$

$$\Delta(\varrho^k) = k \times \Delta(\varrho) \tag{15}$$

Finally, given a parametric rule $(N_1, C_1) >_E (N_2, C_2)$ we can now check that the reachability expression $\tau^*_{C_1}$ provided by FAST, and encoded as explained above, corresponds to the solutions of $\exists p_2 \,.\, \tilde{E}(p_1, p_2)$ using the following additional SMT query:

$$\forall p_1, p'_1 \,.\, C_1(p_1) \implies (\exists p_2 \,.\, \tilde{E}(p_1, p_2) \wedge \tilde{E}(p'_1, p_2) \iff \tau^*_{C_1}(p_1, p'_1)) \tag{16}$$

(and similarly for $\tau^*_{C_2}$).

Once the equivalence (16) above has been validated by a solver, it is in practice way more efficient to use the formula $(\exists p_2 \,.\, \tilde{E}(p_1, p_2) \wedge \tilde{E}(p'_1, p_2))$ inside the core requirements, rather than the formula $\tau^*_{C_1}(p_1, p'_1)$ given by FAST, since the latter introduces many new acceleration variables.

6   Generalizing Equivalence Rules

Before looking at our implementation, we discuss some results related with the *genericity* and *generalisability* of our abstraction rules. We consider several "dimensions" in which a rule can be generalized. A first dimension is related with the parametricity of the initial marking, which is taken into account by our use of a parametric equivalence, $\approx$ instead of $\equiv$, see Theorem 2. Next, we show that we can infer an infinite number of equivalences from a single abstraction rule using compositionality, transitivity, and structural modifications involving labels. Therefore, each abstraction law can be interpreted as a schema for several equivalence rules.

Definition 5 (Transition Operations). *Given a Petri net $N = (P, T, \mathrm{Pre}, \mathrm{Post})$ and its labeling function $l : T \to \Sigma \cup \{\tau\}$, we define two operations: T^-, for removing, and T^+, for duplicating transitions. Let a and b be labels in Σ.*

– *$T^-(a)$ is a net $(P, T', \mathrm{Pre}', \mathrm{Post}')$, where $T' \triangleq T \setminus l^{-1}(a)$, and Pre' (resp. Post') is the projection of Pre (resp. Post) to the domain T'.*

- $T^+(a, b)$ *is a net* $(P, T', \text{Pre}', \text{Post}')$, *where* T' *is a subset of* $T \times \{0, 1\}$ *defined by* $T' \triangleq T \times \{0\} \cup l^{-1}(a) \times \{1\}$. *Additionally, we define* $\text{Pre}'(t, i) \triangleq \text{Pre}(t)$ *and* $\text{Post}'(t, i) \triangleq \text{Post}(t)$ *for all* $t \in T$ *and* $i \in \{0, 1\}$. *Finally, the labeling function* l' *is defined with* $l'(t, 0) \triangleq l(t)$ *and* $l'(t, 1) = b$ *for all* $t \in T$.

The operation $T^-(a)$ removes transitions labeled by a, while $T^+(a, b)$ duplicates all transitions labeled by a and labels the copies with b. We illustrated T^+ in the nets of rule (MAGIC), in Fig. 2, where the "dashed" transition c' can be interpreted has the result of applying operation $T^+(c, c')$. Note that these operations only involve labeled transitions. Silent transitions are kept untouched—up-to some injection.

Theorem 7 (Preservation by Transition Operations). *Assume we have a parametric E-abstraction equivalence* $(N_1, C_1) \cong_E (N_2, C_2)$, a *and* b *are labels in* Σ. *Then,*

- $T_i^-(a)$ *and* $T_i^+(a, b)$ *satisfy the coherency constraint* C_i, *for* $i = 1, 2$.
- $(T_1^-(a), C_1) \cong_E (T_2^-(a), C_2)$.
- $(T_1^+(a, b), C_1) \cong_E (T_2^+(a, b), C_2)$.

where T_i^-, T_i^+ *is (respectively) the operation* T^-, T^+ *on* N_i.

Finally, we recall a previous result from [1,2] (Theorem 8), which states that equivalence rules can be combined together using synchronous composition, relabeling, and chaining. Note that, in order to avoid inconsistencies that could emerge if we inadvertently reuse the same variable in different reduction equations (variable escaping its scope), we require that conditions can be safely composed: the equivalence statements $(N_1, m_1) \equiv_E (N_2, m_2)$ and $(N_2, m_2) \equiv_{E'} (N_3, m_3)$ are *compatible* if and only if $P_1 \cap P_3 = P_2 \cap P_3$. We also rely on classical operations for relabeling a net, and for synchronous product, $N_1 \parallel N_2$, which are defined in [2] for instance.

Theorem 8 (E-equivalence is a Congruence [1,2]). *Assume we have two compatible equivalence statements* $(N_1, m_1) \equiv_E (N_2, m_2)$ *and* $(N_2, m_2) \equiv_{E'} (N_3, m_3)$, *and that* M *is a Petri net such that* $N_1 \parallel M$ *and* $N_2 \parallel M$ *are defined, then*

- $(N_1, m_1) \parallel (M, m) \equiv_E (N_2, m_2) \parallel (M, m)$.
- $(N_1, m_1) \equiv_{E, E'} (N_3, m_3)$.
- $(N_1[a/b], m_1) \equiv_E (N_2[a/b], m_2)$ *for any* $a \in \Sigma$ *and* $b \in \Sigma \cup \{\tau\}$.

7 Validation and Conclusion

We have implemented our automated procedure in a new tool called `Reductron`. The tool is open-source, under the GPLv3 license, and is freely available on GitHub [23]. The repository contains a subdirectory, `rules`, that provides examples of equivalence rules that can be checked using our approach. Each test contains two Petri nets, one for N_1 (called `initial.net`) and another for N_2 (called

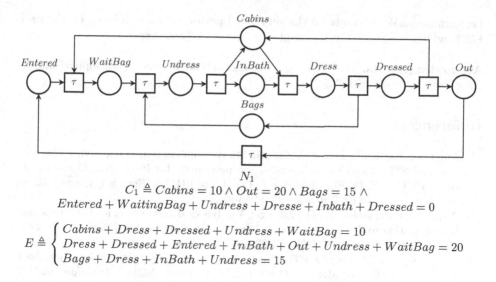

$$N_1$$
$$C_1 \triangleq Cabins = 10 \wedge Out = 20 \wedge Bags = 15 \wedge$$
$$Entered + WaitingBag + Undress + Dresse + Inbath + Dressed = 0$$

$$E \triangleq \begin{cases} Cabins + Dress + Dressed + Undress + WaitBag = 10 \\ Dress + Dressed + Entered + InBath + Out + Undress + WaitBag = 20 \\ Bags + Dress + InBath + Undress = 15 \end{cases}$$

Fig. 8. A Petri net modeling users in a swimming pool, see e.g. [10].

reduced.net), defined using the syntax of Tina. These nets also include declarations for constraints, C_1 and C_2, and for the equation system E. Our list contains examples of laws that are implemented in Tedd and SMPT, such as rule (CONCAT) depicted in Fig. 1, but also some examples of unsound equivalences rules. For instance, we provide example (FAKE_CONCAT), which corresponds to the example of Fig. 1 with transition d added.

An interesting feature of Reductron, when a rule is unsound, is to return which core requirement failed. For instance, with (FAKE_CONCAT), we learn that (N_1, C_1) is not coherent because of d (we cannot reach a coherent marking after firing d using only silent transitions). We can also detect many cases in which there is an error in the specification of either C or E.

We performed some experimentation using z3 [14] (version 4.8) as our target SMT solver, and FAST (version 2.1). All the examples given in our repository can be solved in a few seconds. Although we focus on the automatic verification of abstraction laws, we have also tested our tool on moderate-sized nets, such as the swimming pool example given in Fig. 8. In this context, we use the fact that an equivalence of the form $(N, C) \approx_E (\emptyset, \text{True})$, between N and a net containing an empty set of places, entails that the reachability set of (N, C) must be equal to the solution set of E. In this case, also, results are almost immediate.

These very good results depend largely on the continuous improvements made by SMT solvers. Indeed, we generate very large LIA formulas, with sometimes hundreds of quantified variables, and a moderate amount of quantifier alternation (formulas of the form $\forall \exists \forall$). For instance, experiments performed with older versions of z3 (such as 4.4.1, October 2015) exhibit significantly degraded performances. We also rely on the very good performances exhibited by the tool FAST, which is essential in the implementation of Reductron.

Acknowledgements. We would like to thanks Jérôme Leroux for his support during our experimentation with FAST.

References

1. Amat, N., Berthomieu, B., Dal Zilio, S.: On the combination of polyhedral abstraction and SMT-based model checking for petri nets. In: Buchs, D., Carmona, J. (eds.) PETRI NETS 2021. LNCS, vol. 12734, pp. 164–185. Springer, Cham (2021). https://doi.org/10.1007/978-3-030-76983-3_9
2. Amat, N., Berthomieu, B., Dal Zilio, S.: A polyhedral abstraction for Petri nets and its application to SMT-based model checking. Fundamenta Informaticae **187**(2–4) (2022). https://doi.org/10.3233/FI-222134
3. Amat, N., Dal Zilio, S.: SMPT: A testbed for reachabilty methods in generalized Petri nets. In: Formal Methods (FM). LNCS, Springer (2023). https://doi.org/10.1007/978-3-031-27481-7_25
4. Amat, N., Zilio, S.D., Hujsa, T.: Property Directed Reachability for Generalized Petri Nets. In: TACAS 2022. LNCS, vol. 13243, pp. 505–523. Springer, Cham (2022). https://doi.org/10.1007/978-3-030-99524-9_28
5. Amat, N., Dal Zilio, S., Le Botlan, D.: Accelerating the computation of dead and concurrent places using reductions. In: Laarman, A., Sokolova, A. (eds.) SPIN 2021. LNCS, vol. 12864, pp. 45–62. Springer, Cham (2021). https://doi.org/10.1007/978-3-030-84629-9_3
6. Amat, N., Dal Zilio, S., Le Botlan, D.: Leveraging polyhedral reductions for solving Petri net reachability problems. Int. J. Softw. Tools Technol. Transfer (2022). https://doi.org/10.1007/s10009-022-00694-8
7. Bardin, S., Finkel, A., Leroux, J., Petrucci, L.: FAST: fast acceleration of symbolic transition systems. In: Hunt, W.A., Somenzi, F. (eds.) CAV 2003. LNCS, vol. 2725, pp. 118–121. Springer, Heidelberg (2003). https://doi.org/10.1007/978-3-540-45069-6_12
8. Bardin, S., Finkel, A., Leroux, J., Petrucci, L.: FAST: acceleration from theory to practice. Int. J. Softw. Tools Technol. Transf. **10**(5) (2008). https://doi.org/10.1007/s10009-008-0064-3
9. Barrett, C., Fontaine, P., Tinelli, C.: The SMT-LIB Standard: Version 2.6. Tech. rep., Department of Computer Science, The University of Iowa (2017). http://www.smt-lib.org/
10. Bérard, B., Fribourg, L.: Reachability analysis of (timed) petri nets using real arithmetic. In: Baeten, J.C.M., Mauw, S. (eds.) CONCUR 1999. LNCS, vol. 1664, pp. 178–193. Springer, Heidelberg (1999). https://doi.org/10.1007/3-540-48320-9_14
11. Berthelot, G.: Transformations and decompositions of nets. In: Brauer, W., Reisig, W., Rozenberg, G. (eds.) ACPN 1986. LNCS, vol. 254, pp. 359–376. Springer, Heidelberg (1987). https://doi.org/10.1007/978-3-540-47919-2_13
12. Berthomieu, B., Le Botlan, D., Dal Zilio, S.: Petri net reductions for counting markings. In: Gallardo, M.M., Merino, P. (eds.) SPIN 2018. LNCS, vol. 10869, pp. 65–84. Springer, Cham (2018). https://doi.org/10.1007/978-3-319-94111-0_4

13. Berthomieu, B., Le Botlan, D., Dal Zilio, S.: Counting Petri net markings from reduction equations. Int. J. Softw. Tools Technol. Transfer **22**(2), 163–181 (2019). https://doi.org/10.1007/s10009-019-00519-1
14. de Moura, L., Bjørner, N.: Z3: an efficient SMT solver. In: Ramakrishnan, C.R., Rehof, J. (eds.) TACAS 2008. LNCS, vol. 4963, pp. 337–340. Springer, Heidelberg (2008). https://doi.org/10.1007/978-3-540-78800-3_24
15. Esparza, J.: Decidability and complexity of Petri net problems — an introduction. In: Reisig, W., Rozenberg, G. (eds.) ACPN 1996. LNCS, vol. 1491, pp. 374–428. Springer, Heidelberg (1998). https://doi.org/10.1007/3-540-65306-6_20
16. Esparza, J., Nielsen, M.: Decidability issues for Petri nets. BRICS Report Series 1(8) (1994)
17. Hack, M.H.T.: Decidability questions for Petri Nets. Ph.D. thesis, Massachusetts Institute of Technology (1976)
18. Hirshfeld, Y.: Petri nets and the equivalence problem. In: Börger, E., Gurevich, Y., Meinke, K. (eds.) CSL 1993. LNCS, vol. 832, pp. 165–174. Springer, Heidelberg (1994). https://doi.org/10.1007/BFb0049331
19. Hujsa, T., Berthomieu, B., Dal Zilio, S., Le Botlan, D.: Checking marking reachability with the state equation in Petri net subclasses. CoRR abs/2006.05600 (2020)
20. Hujsa, T., Berthomieu, B., Dal Zilio, S., Le Botlan, D.: On the Petri nets with a single shared place and beyond. CoRR abs/2005.04818 (2020)
21. LAAS-CNRS: SMPT (2020). https://github.com/nicolasAmat/SMPT/
22. LAAS-CNRS: Tina Toolbox (2020). http://projects.laas.fr/tina
23. LAAS-CNRS: Reductron (2023). https://github.com/nicolasAmat/Reductron/
24. Leroux, J.: Presburger vector addition systems. In: 2013 28th Annual ACM/IEEE Symposium on Logic in Computer Science (2013). https://doi.org/10.1109/LICS.2013.7
25. Thierry-Mieg, Y., Poitrenaud, D., Hamez, A., Kordon, F.: Hierarchical Hierarchical Set Decision Diagrams and regular models. In: Kowalewski, S., Philippou, A. (eds.) TACAS 2009. LNCS, vol. 5505. Springer, Heidelberg (2009). https://doi.org/10.1007/978-3-642-00768-2

Experimenting with Stubborn Sets on Petri Nets

Sami Evangelista[✉] [ID]

LIPN, CNRS UMR 7030, Université Paris 13, Sorbonne Paris Cité,
99, av. J.-B. Clément, 93430 Villetaneuse, France
sami.evangelista@lipn.univ-paris13.fr

Abstract. The implementation of model checking algorithms on real life systems usually suffers from the well known state explosion problem. Partial order reduction addresses this issue in the context of asynchronous systems. We review in this article algorithms developed by the Petri nets community and contribute with simple heuristics and variations of these. We also report on a large set of experiments performed on the models of a Model Checking Contest hosted by the Petri Nets conference since 2011. Our study targets the verification of deadlock freeness and liveness properties.

1 Introduction

System verification based on an exhaustive simulation suffers from the well known state explosion problem: the system state space often grows exponentially with respect to the system structure, making it hard if not impossible to apply it to real life systems. One major source of this problem lies in the concurrent execution of system components that often leads to a blowup of possible interleavings.

When dealing with asynchronous systems, it is often the case that the execution order of system transitions is irrelevant because their occurences can be swapped without consequences on the observed system. This observation has led to the development of some partial order reduction algorithms [10, 15, 18] that exploit this independence relation between transitions. Although they differ in their implementation of this general principle, they rely on a selective search within the state space: when considering a system state only a subset of allowed transitions are considered to pursue the exploration while the execution of other allowed transitions is postponed to a future state. Such a subset is called stubborn [18], persistent [10] or ample [15] in the literature.

Ignoring some transitions has the consequence of ruling out some system states and building a reduced state space that is more suitable for verification purposes. The filtering mechanism must however fulfill some conditions for the reduced state space to be of any use. Hence, several variations of the method have been designed depending on the property being investigated.

We focus in this article on deadlock freeness and liveness properties for which we review several algorithms. For deadlock freeness we consider Petri nets tailored algorithms while for liveness properties the algorithms usually operate on the underlying reduced state space and are thus language independent.

The contribution of this article is twofold. First, we introduce several simple heuristics and optimisations for existing algorithms. Second, in order to evaluate some of these

L. Gomes and R. Lorenz (Eds.): PETRI NETS 2023, LNCS 13929, pp. 346–365, 2023.
https://doi.org/10.1007/978-3-031-33620-1_19

algorithms, in particular the benefits of our contributions, we report on a large series of experiments performed on the Petri net models of the Model Checking Contest [] which resulted in approximately 150,000 runs.

The rest of this paper is organised as follows. Background on Petri nets and partial order reduction is given in Sect. 2. Section 3 recalls the elements of the stubborn set theory for deadlock detection, reviews some algorithms developed for that purpose, and presents our experimental evaluation of these. Likewise, in Sect. 4 we review partial order reduction algorithms for liveness verification and present experimental observations on these. Section 5 concludes our work and introduces some perspectives.

2 Background

This section introduces notations and definitions used in the remainder of the paper.

Definition 1. *A Petri net is a tuple* (P,T,W), *where P is a set of* places *; T is a set of* transitions *such that* $T \cap P = \emptyset$ *; and* $W : (P \times T) \cup (T \times P) \to \mathbb{N}$ *is a* weighting function.

From now on, we assume a Petri net $N = (P,T,W)$. For any $n \in P \cup T$, $^{\bullet}n$ and $n^{\bullet}$ respectively denote the sets $\{o \in P \cup T \mid W(o,n) > 0\}$ and $\{o \in P \cup T \mid W(n,o) > 0\}$.

Definition 2. *The set* $\mathcal{M} = \{m \in P \to \mathbb{N}\}$ *is the set of* markings *of N. Let* $m \in \mathcal{M}$ *and* $t \in T$. *If* $W(p,t) \geq m(p), \forall p \in P$ *then t is* firable *at m* $(m[t\rangle$ *for short). The* firing *of t at m leads to* $m' \in \mathcal{M}$ $(m[t\rangle m'$ *for short) defined by* $m'(p) = m(p) - W(p,t) + W(t,p), \forall p \in P$. *The set* $en(m) = \{t \in T \mid m[t\rangle\}$ *is the set of* firable *(or enabled) transitions at m. A* deadlock *is a marking m such that* $en(m) = \emptyset$.

The firing rule is extended to sequences of transitions (i.e., elements of T^*). Let $m \in \mathcal{M}$ and $\sigma \in T^*$. σ is *firable* at m, $(m[\sigma\rangle$ for short) if $\sigma = \varepsilon$ or if $\sigma = t.\sigma'$, $m[t\rangle m'$ and $m'[\sigma'\rangle$ where $t.\sigma'$ is the concatenation of $t \in T$ and $\sigma' \in S^*$; and ε the empty sequence.

Definition 3. *The* state space *of* (N,m_0) *(with* $m_0 \in \mathcal{M}$ *an* initial marking) *is a couple* (R,A) *such that R and A are the smallest sets respecting:* $m_0 \in R$ *and if* $m \in R$ *and* $m[t\rangle m'$ *for some* $t \in T$ *then* $m' \in R$ *and* $(m,t,m') \in A$.

Stubborn, ample, or persistent sets reductions rely on the use of a reduction function that filters transitions to be used to generate the successors of a marking, leading to the construction of a reduced state space.

Definition 4. *A* reduction function *is a mapping f from* $\mathcal{M}$ *to* 2^T. *The reduced state space of* (N,m_0) *with respect to f is a couple* (R_f, A_f) *such that* R_f *and* A_f *are the smallest sets respecting:* $m_0 \in R_f$ *and if* $m \in R_f$ *and* $m[t\rangle m'$ *for some* $t \in f(m)$ *then* $m' \in R_f$ *and* $(m,t,m') \in A_f$. *If* $m[t\rangle$ *and* $t \in f(m)$ *then we note* $m[t\rangle_f$. *Likewise,* $m[t\rangle_f m'$ *denotes that* $m[t\rangle_f$ *and* $m[t\rangle m'$.

If a marking m is such that $en(m) \cap f(m) \subset en(m)$, then it is said to be *reduced*. Otherwise, it is said to be *fully expanded*. An exhaustive state space construction algorithm can be modified to build a reduced state space, simply by considering $en(m) \cap f(m)$ instead of $en(m)$ when processing a marking m.

Obviously, a reduction function must respect some conditions for the reduced state space to be of any use. We review in subsequent sections sufficient conditions to preserve deadlocks and liveness properties.

3 Stubborn Sets for Deadlock State Detection

We recall in this section the theoretical background of the stubborn set theory for deadlock detection. We then review different algorithms that can be used in that context before presenting our experimental results.

3.1 Stubborn Set Theory for Deadlock Detection

Dynamic stubbornness is a key concept in the stubborn sets theory. Whatever the property being investigated, it is used as the starting point to define reduction functions.

Definition 5. *Let $m \in \mathcal{M}$. $S \subseteq T$ is dynamically stubborn at m [20] if conditions* **D1** *and* **D2** *hold, where:*

D1 $\forall \sigma \in (T \setminus S)^*, t \in S: m[\sigma.t\rangle \Rightarrow m[t.\sigma\rangle$
D2 *if* $en(m) \neq \emptyset$ *then* $\exists k \in S \mid \forall \sigma \in (T \setminus S)^*: m[\sigma\rangle \Rightarrow m[\sigma.k\rangle$

A transition k in condition D2 is called a *key transition*. If all transitions of S are key transitions, then S is *strongly dynamically stubborn* at m. A reduction function producing dynamically stubborn sets preserves all deadlocks [19] in the reduced state space. Such a reduction function is also characterised as *dynamically stubborn*.

The two conditions of Definition 5 rely on a notion of dependency as defined below.

Definition 6. *A dependency relation $\mathcal{D}$ is a symmetric and reflexive relation over $T \times T$ such that $(t,u) \notin \mathcal{D}$ implies that for all $m \in R$: $m[t\rangle m' \wedge m[u\rangle \Rightarrow m'[u\rangle$.*

Given a dependency relation $\mathcal{D}$, we will note $\mathcal{D}(t)$ the set $\{t' \in T \mid (t,t') \in \mathcal{D}\}$.

Generally speaking, it is also required for $\mathcal{D}$ that t and u commute (that their execution order is irrelevant) but we have left out this condition, as it is obviously superfluous in the case of Petri nets.

We will use the following proposition to serve as a basis for the implementation of the stubborn set computation algorithms that we will experiment with.

Proposition 1. *Let $\mathcal{D}$ be a dependency relation, $m \in \mathcal{M}$ and $S \subseteq T$ be such that:*

1. *if $en(m) \neq \emptyset$ then $S \cap en(m) \neq \emptyset$;*
2. *if $t \in S \cap en(m)$ then $\mathcal{D}(t) \subseteq S$;*
3. *if $t \in S \setminus en(m)$ then $\exists p \in P \mid m(p) < W(p,t)$ and $\{t \in T \mid W(t,p) > W(p,t)\} \subseteq S$.*

Then S is strongly dynamically stubborn at m.

According to Item 1 a non deadlock marking may not have an empty stubborn set. If an enabled transition is stubborn then so are all its dependent transitions (Item 2). Last, Item 3 states that if a disabled transition is stubborn then there is a place that disables its firing and such that all transitions that could increase its marking are also stubborn. It is easy to prove that the firing of any transition outside S cannot alter the firability of transitions of S and conversely. Hence, stubborn sets respecting conditions of Proposition 1 are indeed strongly dynamically stubborn sets. Note however, that conditions could be relaxed to ensure dynamic stubborness, see [20].

Proposition 1 is parametrized by $\mathcal{D}$. We define below two such dependency relations.

Definition 7. *The exact dependency relation $\mathcal{D}_e$ is such that $(t,u) \in \mathcal{D}_e$ if and only if $\exists m \in R$ such that $m[t\rangle m' \wedge m[u\rangle \wedge \neg m'[u\rangle$. The static dependency relation $\mathcal{D}_s$ is such that $(t,u) \in \mathcal{D}_s$ if and only if $\exists p \in P$ such that $min(W(t,p),W(u,p)) < min(W(p,t),W(p,u))$.*

It is straightforward to show that $\mathcal{D}_e$ and $\mathcal{D}_s$ are dependency relations (as defined by Definition 6) and that $\mathcal{D}_e \subseteq \mathcal{D}_s$. Relation $\mathcal{D}_e$ is the smallest dependency relation as its definition is based on the state space. It is however useless in practice since our goal is precisely to avoid the construction of this state space. Nevertheless, we will use it in our experiments (as first done in [9]) for comparison purposes in situations where the full state space can be computed with available resources. Relation $\mathcal{D}_s$ (from [19,20]) might be

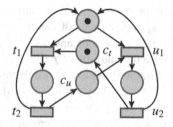

Fig. 1. $\mathcal{D}_e \subset \mathcal{D}_s$ for this net

larger and may thus have a smaller reduction power but it has the advantage to only rely on the structure of the net and can therefore be the basis of a practical implementation of the stubborn set reduction.

It is noted in [9] that Definition 5 actually considers for a marking m only its possible futures rather than the full state space. It is thus possible to refine $\mathcal{D}_e$ and define a context-dependent relation (see [9], Sec. 3.1, page 44). We have not considered such a possibility and leave it for future experiments.

The net depicted on Fig. 1 shows a simple example net for which $\mathcal{D}_e$ and $\mathcal{D}_s$ differ. Places c_t and c_u force an alternation between transitions $t_1.t_2$ and transitions $u_1.u_2$. So there actually is no conflict between t_1 and u_1 whereas the structure of the net tells us a different story. Therefore, we have $\mathcal{D}_e = \emptyset$ and $\mathcal{D}_s = \{(t_1, u_1), (u_1, t_1)\}$.

3.2 Stubborn Set Algorithms for Deadlock Detection

We chose to introduce and experiment with three algorithms: the closure algorithm, the deletion algorithm and a combination of these two. The first one has been chosen for its simplicity and because one of our contributions is to introduce a simple optimisation to this one that occurs to be quite helpful for some models. The second algorithm has been chosen for its ability to produce minimal sets (with respect to inclusion and according to the conditions of Proposition 1) despite its quadratic complexity.

The Closure Algorithm This first algorithm (see Algorithm 1) is a straightforward implementation of Proposition 1. It initiates the stubborn set (S) construction by picking an enabled transition (condition 1). Based on conditions 2 and 3, it then inserts new transitions in S until all transitions of S have been treated.

When examining a disabled transition t, Algorithm 1 chooses a place that disables its firing. Such a place is called a *scapegoat* place in the literature because it is considered as responsible of t being

---

Algorithm 1.clo, the closure algorithm

1: $S := \{ \textbf{pick from } en(m) \}$; $Q := S$
2: **while** $Q \neq \emptyset$ **do**
3:    $t := \textbf{pick from } Q$; $Q := Q \setminus \{t\}$
4:    **if** $t \in en(m)$ **then**
5:       $U := \mathcal{D}(t)$
6:    **else**
7:       $C := \{p \in P \mid m(p) < W(p,t)\}$
8:       $s := \textbf{pick from } C$
9:       $U := \{t \in T \mid W(t,s) > W(s,t)\}$
10:    $Q := Q \cup (U \setminus S)$; $S := S \cup U$
11: **return** S

---

disabled. The choice of this scapegoat largely impacts the construction of the stubborn set. To limit as much as possible the choices of undesirable scapegoats (in the sense that it may produce unnecessary large sets), we introduce a modification of this algorithm that exploits the past of the construction (see Algorithm 2). A counter I is associated with each place. For any $p \in P$, $I[p]$ is initialised with the number of transitions that increase the marking of p (lines 1–2). During the construction, each time a transition t is put in the stubborn set, we decrement the counter of any place p of which the marking is increased by t (lines 25–26). When the counter reaches 0 for some $p \in P$, then we know that p cannot gain any token without a transition of S occurring first. Consequently, all output transitions of p that are disabled by p can be put in S (lines 27–30). However, a transition u put in S by this way does not need to be further considered by the algorithm (this is the purpose of the *enqueue* parameter of procedure *new_stub*) since we know that p is already a valid scapegoat place for t: all transitions that may increase its marking are already stubborn.

Algorithm 2.$clo^\star$, an optimised closure algorithm

1: **for** p in P **do**	16: **procedure** $init\_stub()$ **is**
2: $I[p] := \lvert\{t \in T \mid W(t,p) > W(p,t)\}\rvert$	17: $t_0 := \textbf{pick from } en(m)$
3: $init\_stub()$	18: $S := \emptyset$
4: **while** $Q \neq \emptyset$ **do**	19: $Q := \emptyset$
5: $t := \textbf{pick from } Q$	20: $new\_stub(t_0, \textbf{true})$
6: $Q := Q \setminus \{t\}$	21: **procedure** $new\_stub(t, enqueue)$ **is**
7: **if** $t \in en(m)$ **then**	22: $S := S \cup \{t\}$
8: $U := \mathcal{D}(t)$	23: **if** $enqueue$ **then**
9: **else**	24: $Q := Q \cup \{t\}$
10: $C := \{p \in P \mid m(p) < W(p,t)\}$	25: **for** p in $\{p \in P \mid W(t,p) > W(p,t)\}$ **do**
11: $s := \textbf{pick from } C$	26: $I[p] := I[p] - 1$
12: $U := \{t \in T \mid W(t,s) > W(s,t)\}$	27: **if** $I[p] = 0$ **then**
13: **for** u in $U \setminus S$ **do**	28: **for** u in $p^\bullet \setminus S$ **do**
14: $new\_stub(u, \textbf{true})$	29: **if** $W(p,u) > m(p)$ **then**
15: **return** S	30: $new\_stub(u, \textbf{false})$

We illustrate the principle of our modification with the help of Fig. 2. Let us first see how the basic algorithm proceeds. Suppose that the algorithm is instantiated with $\mathcal{D}_s = \{(t,u),(u,t)\}$ and that the stubborn set construction is initiated with t. Since $u \in \mathcal{D}_s(t)$, u must also be put in the stubborn set. When processing u, the algorithm has to choose among two scapegoats: r and s. Choosing r causes the insertion of v while choosing s does not cause any new transition to be put in the stubborn set and halts the construction. Hence, the algorithm may produce either $\{t,u,v\}$ or either $\{t,u\}$ depending on the scapegoat choice.

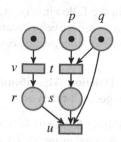

Fig. 2. An example net illustrating Algorithm 2

With our modification, if the construction is initiated with t, the insertion of t in the stubborn sets causes $I[s]$ to reach 0. This causes the insertion of u to S without u being put in Q, and immediately stops the construction. Hence, with the same starting transition, our modified algorithm can only produce $\{t,u\}$ as a resulting set.

Note that, as illustrated by our example, our modification does not improve on the basic closure algorithm as any set produced by the former can also be produced by the latter. Our modification must therefore be thought as a way to equip the basic algorithm with a mechanism that can avoid the choice of inappropriate scapegoat places.

The nondeterminism of the closure algorithm stems from the two choices done for the transition picked to initiate the construction of the set at l. 1 of Algorithm 1; and for the scapegoat place picked at l. 8 of Algorithm 1.

We considered 4 strategies to choose the starting transition t to compute a stubborn set S. They rely on a bijective mapping $ord : P \cup T \rightarrow \{1,\ldots,|P \cup T|\}$ and two heuristics h_t^e and h_t^f where $h_t^e(t)$ and $h_t^f(t)$ are respectively the number of enabled transitions and the number of forward transitions (i.e., enabled transitions for which the firing leads to an undiscovered marking) in S if t is chosen to initiate S. These 4 strategies are:

- rnd_t — Pick t randomly.
- fst_t — Pick t st. $ord(t)$ is minimal.
- min_t^e — Pick t st. $(h_t^e(t), ord(t))$ is minimal.
- min_t^f — Pick t st. $(h_t^f(t), h_t^e(t), ord(t))$ is minimal.

Note that, using strategies min_t^e, min_t^f we trade the linear complexity of the algorithm for a quadratic complexity since the closure algorithm is now invoked on every possible starting transition.

For the choice of a scapegoat place s we considered 8 strategies. They rely on 3 heuristics h_s^t, h_s^e and h_s^f where $h_s^t(s)$, $h_s^e(s)$ and $h_s^f(s)$ are respectively the number of transitions, the number of enabled transitions and the number of forward transitions inserted in S if s is chosen as a scapegoat. These 8 strategies are:

- rnd_s — Pick s randomly.
- fst_s — Pick s st. $ord(s)$ is minimal.
- min_s^t and max_s^t — Pick s st. $(h_s^t(s), ord(s))$ is minimal (maximal).
- min_s^e and max_s^e — Pick s st. $(h_s^e(s), h_s^t(s), ord(s))$ is minimal (maximal).
- min_s^f and max_s^f — Pick s st. $(h_s^f(s), h_s^e(s), h_s^t(s), ord(s))$ is minimal (maximal).

In the following, $clo(t,s)$ and $clo^\star(t,s)$ denote algorithms Algorithm 1 and Algorithm 2 respectively, instantiated with strategies t and s for choosing the starting transition and scapegoat places respectively.

The Deletion Algorithm. This second algorithm avoids the necessity of choosing scapegoats. It relies on the construction of a graph capturing transition dependencies.

Definition 8. *Let $\mathcal{D}$ be a dependency relation, and $m \in \mathcal{M}$. A dependency graph for m is a directed graph (V,E) with $V = P \cup T$ and $E = E_1 \cup E_2 \cup E_3$ where:*

$$E_1 = (en(m) \times T) \cap \mathcal{D}$$
$$E_2 = \{(t,p) \in (T \setminus en(m)) \times P \mid m(p) < W(p,t)\}$$
$$E_3 = \{(p,t) \in P \times T \mid W(t,p) > W(p,t)\}$$

A dependency graph is nothing more than a reformulation of Proposition 1 as a graph structure. The deletion algorithm iteratively tries to delete enabled transitions from this graph. When a node is deleted then so are its immediate predecessors that are places or enabled transitions. Disabled transitions are deleted when they do not have any successor remaining in V. If, after a deletion step, the graph does not contain an enabled transition anymore, then the deletion is undone and the algorithm tries to delete another transition. Termination occurs when no transition can be further deleted. Set $V \cap T$ is then a valid stubborn set. Indeed, after a successful deletion, edges of E_1 ensure that, for any $t \in en(m) \cap V$, all its dependent transitions are still in V; and edges of E_2 ensure that, for any $t \in (T \setminus en(m)) \cap V$, there is at least one $p \in P \cap V$ such that $m(p) < W(p,t)$, and, due to E_3, all transitions that increase the marking of p are also in V.

The only source of nondeterminism of the deletion algorithm is in the choice of the transition t to delete at each step. We considered 6 strategies to make that choice. They rely on two heuristics h_d^e and h_d^f where $h_d^e(t)$ and $h_d^f(t)$ are respectively the number of enabled transitions and the number of forward transitons deleted by the algorithm if t is picked as the transition to be deleted from the graph. These 6 strategies are:

- rnd_t — Pick t randomly.
- fst_t — Pick t st. $ord(t)$ is minimal.
- min_t^e and max_t^e — Pick t st. $(h_d^e(t), ord(t))$ is minimal (maximal).
- min_t^f and max_t^f — Pick t st. $(h_d^f(t), h_d^e(t), ord(t))$ is minimal (maximal).

For the four last strategies, the algorithm has to simulate the deletion of all enabled transitions remaining in the graph before picking one. This however does not impact the algorithm complexity since in the worst case every transition has to be checked anyway.

In the following, $del(t)$ denotes the deletion algorithm instantiated with strategy t for choosing the transition to be deleted at each iteration.

The Clodel Algorithm. It is also possible to chain both algorithms: a stubborn set is first computed using the closure algorithm; then the deletion algorithm is used on the resulting set to try to further eliminate transitions. We call this combination the clodel algorithm. Its principle has been given by Valmari and Hansen [21] (Sect. 7, page 58).

The clodel algorithm can be instanciated as the closure algorithm is. We could also consider in the instanciation the strategy followed by the deletion algorithm to pick transitions to delete but, to avoid a blowup of experimented configurations, we only considered the fst_t strategy.

In the following, $clodel(t,s)$ denotes the stubborn set construction algorithm that first invokes $clo^\star(t,s)$ and then tries to reduce it with $del(fst_t)$.

3.3 Experimentation Context

We have implemented the algorithms introduced in the previous section in the Helena [6] tool and we have performed experiments on models of the MCC model database. We also experimented with the Prod [23] tool that implements the deletion algorithm and the incremental algorithm based on strongly connected components and that is parametrised as is the closure algorithm. In the following, $inc(t,s)$ denotes the incremental algorithm instanciated with strategies t and s for choosing starting transitions and scapegoat places respectively.

All our experimental data are available on the following web page:

https://www.lipn.univ-paris13.fr/~evangelista/recherche/por-xp

Input Models. The MCC model database[1] comprises 128 Petri net models ranging from simple ones used for educational purposes to complex models corresponding to real life systems. Most of these are obtained from parametrised higher level descriptions (e.g., colored Petri nets) and can be instanciated. Although we have experimented with instances of 130 models (all models of the MCC database as well as two models of our own) we have voluntarily left out some of these. Several reasons can explain this: inability of partial order reduction to reduce the state space, timeout in the state space exploration, timeout in the model compilation, ... The reader may find on the aforementioned web page the details on our selection process. As a result, our report deals with 76 models. For each of these we considered two of its instances, or a single one for non parametrized models. This resulted in 140 instances.

Algorithmic Configurations. With Helena, we have experimented with all algorithm instances considered in Sect. 3.2: 32 instances for algorithms clo, $clodel$ and $clo^\star$ and 6 instances for algorithm del. Moreover, we have experimented with the $\mathcal{D}_s$ and, when available, the $\mathcal{D}_e$ dependency relations. Since the computation of $\mathcal{D}_e$ required to first perform a full state space exploration to store the state space on disk (as done in [9]) we could not experiment with $\mathcal{D}_e$ on instances for which this operation was not feasible.

With Prod, we experimented with the 8 algorithmic configurations it provides: $del(d)$, $\forall d \in \{fst_t, rnd_t\}$, and $inc(t,s)$, $\forall (t,s) \in \{fst_t, rnd_t, min_t^e\} \times \{fst_s, rnd_s\}$.

Randomness and Static Node Ordering. All algorithm instances rely either on a random selection of nodes, or either on a static ordering of nodes computed prior to the exploration (even strategies based on, e.g., a minimisation process, rely on a static ordering

---

[1] See https://mcc.lip6.fr/models.php for the list of models. All models from 2011 to 2022 (included) have been considered.

when several sets of minimal size are available). It is therefore relevant to explore to which extent these mechanisms alter the reduced state space size. Thus, for each model instance, we randomly shuffled the net description 5 times to generate as many different static orderings of nodes and launched each algorithm instance with these 5 settings.

Considering all parameters, our experiments resulted in 114780 runs. We checked that all runs on the same model instance produced the same number of deadlock states.

3.4 Experimental Observations

We start by general observations before presenting a sample of our results. We only consider for now, the static dependency relation $\mathcal{D}_s$. The comparison with the exact relation $\mathcal{D}_e$ will be addressed later in this section.

First, strategies based on a random selection process perform generally worse. For the deletion algorithm, the rnd_t strategy outperformed others for three instances only (airplane(20), airplane(50) and erk(100)). Likewise, for the three variants of the closure algorithm, selecting the starting transition randomly was the best strategy for only one instance (erk(100)). On all other instances it performed (sometimes significantly) worse. The same remark applies to the choice of the scapegoat place. It is somehow surprising that, all things being equal, choosing the first node according to some static ordering is generally preferable to choosing the node randomly. We conjecture that, unlike random strategies, choosing the first node leads to compute similar stubborn sets when processing similar markings (i.e., whose marking differs on a small number of places) which is probably preferable.

Algorithm $clo^\star$ outperforms clo on several non trivial instances while we did not find out any instance for which the converse holds. Moreover, the scapegoat choice strategy has a lesser impact with algorithm $clo^\star$, which is not surprising considering that the goal of $clo^\star$ is precisely to restrict the number of scapegoat candidates. Nevertheless, when both perform comparably, clo can be significantly faster than $clo^\star$.

For the clo and $clo^\star$ algorithms strategy fst_t (and rnd_t as said above) for choosing the starting transition is largely outperformed by strategies based on a minimisation process. We only found out one model instance for which always choosing the first candidate transition to build the stubborn set produced a smaller state space (qcertifprotocol(6)). Nevertheless when a run based on that strategy could terminate with a number of markings in the same order of magnitude as those based on strategies min_t^e or min_t^f, it was usually much faster due to its linear complexity.

For the closure algorithm and its variants, minimisation based strategies (min_s^t, min_s^e and min_s^f) are clearly preferable for the choice of the scapegoat. They exhibit similar performances. As one could expect maximisation based strategies (max_s^t, max_s^e and max_s^f) perform the worse. As noted above, the strategy used to pick a scapegoat place has clearly a lesser impact with algorithm $clo^\star$.

Identifying forward transitions and using this information generally has a small impact. Moreover, this identification has a non-negligible cost as it requires to execute all enabled transitions and check for the existence of successors in the state space. Hence, algorithmic instances relying on that process are generally slower by approximately 20% compared to strategies that only require to count enabled transitions. On four model instances (those of models di usion2d and neighborgrid) they significantly outperformed other algorithmic instances, even guaranteeing the success of the run in two cases.

For the deletion algorithm, strategies max_t^e and max_t^f are clearly the best. On only 13 instances (over 140) did none of these two perform the best (compared to the four other strategies). Moreover, when this occured, the differences observed were negligible whereas strategies max_t^e and max_t^f often significantly outperformed their rivals, sometimes making the run successful.

Sadly, *clodel* does not bring an improvement with respect to *clo**. For the models for which *clodel* built smaller state spaces (e.g., aslink, lamport, shieldrvt) the gain was very negligible in terms of reduction (typically less than 10% w.r.t. *clo**) and it often led to an important increase of the search time (remember that *clodel* first invokes *clo** then tries to reduce the stubborn set with *del*). This seems to indicate that *clo** already often produces stubborn sets that are minimal (w.r.t. inclusion).

The way nodes are ordered can have a large impact on the reduction. In a few pathological situations, we observed that an unfortunate ordering could lead to a state explosion. However, this observation seems more valid for "toy" examples, although there still are real life models (tagged as industrial on the MCC webpage) for which significant differences could be observed according to the ordering (e.g., gpufp or shieldrvt) whatever the algorithm used.

Algorithms *clo** (using strategies min_t^e or min_t^f) and *del* (using strategies max_t^e or max_t^f) have, on the average, comparable performances regarding both the reduction power and the search time. Nevertheless, significant differences can be observed when using both algorithms on the same instance.

We conclude our observations with a comparison of relations $\mathcal{D}_s$ and $\mathcal{D}_e$. We could compute the exact relation $\mathcal{D}_e$ for 73 model instances (over 140). For most of these 73 instances the use of $\mathcal{D}_e$ was useless or of very little help (with an additional reduction typically less than 5%). Table 1 gives, for the 12 instances for which relation $\mathcal{D}_e$ performed the best (w.r.t. $\mathcal{D}_s$) the minimal numbers of states in the reduced state space over all runs using the static (column $min(\mathcal{D}_s)$) and exact relations (column $min(\mathcal{D}_e)$). Table is sorted according to the ratio $\frac{min(\mathcal{D}_e)}{min(\mathcal{D}_s)}$. This observation is somehow disappointing as it seems to indicate that there is not much thing that can be expected from refining the dependency relation. Data reported in [9] (see Table 2, p. 49) exhibit better performances of relations based on the analysis of the full state space. We believe the difference with respect to our results can be explained by the DVE modelling language used in [9]. DVE processes synchronise through shared variables or rendez-vous and it is hard, in contrast to Petri nets, to perform a precise static analysis of such models which can in turn explain why semantic based relations space can fill that gap. Moreover algorithms in [9] are parametrized by two relations: the dependency and precedence relations while we only considered the first one here.

3.5   Experimental Results Sample

To back up our observations, we present in this section a sample of our experimental results. We only consider here the static dependency relation $\mathcal{D}_s$. For comparison purposes, we computed a *state score* (or more simply score) defined for an algorithm $alg \in \mathcal{A}$ ($\mathcal{A}$ being the set of all algorithmic instances) and a model instance *inst* as:

$$score(alg, inst) = \sum_{i \in \{1,...,5\}} 20 \cdot \frac{S_{\min}(inst)}{S(alg, inst, i)} \tag{1}$$

where $S(alg, inst, i)$ is the number of states in the reduced state space built by algorithm alg on model instance *inst* during run $i \in \{1,...,5\}$ if the run terminated within our time limit (30 min.), or ∞ otherwise; and $S_{\min}(inst) = \min_{alg \in \mathcal{A}, i \in \{1,...,5\}} S(alg, inst, i)$. Thus a score ranges from 0 if the algorithm did not terminate on the instance for any of the 5 runs to 100 if the algorithm performed the best on all its 5 runs.

Table 2 provides scores for 15 non trivial model instances as well as the average over the 140 model instances we experimented with. The bottom row gives the number of successful runs of an algorithmic instance over all model instances, which is at most 700 (5 runs × 140 model instances). On the basis of our previous observations and to lighten the table, we voluntarily ruled out several algorithmic configurations. We provide next to each model name, the minimal number of visited states over all algorithms ($S_{\min}$).

In general, Prod's implementation of algorithm *del* performs better than Helena's as evidenced by a comparison of columns $del(fst_t)$ of the two tools. We conjecture that this may be due to a finer implementation of the dependency graph (see, e.g., [22], Def. 3.5, page 136) that permits the computation of smaller sets. Instance smhome(8) is an interesting case from that perspective as Prod, using $del(fst_t)$, significantly outperforms all its competitors. Nevertheless, using the max_t^e and max_t^f strategies, Helena's deletion algorithm usually performs better than Prod's. It could be worthwhile experimenting with these two strategies on a refined dependency graph as computed by Prod.

Instance ibmb2s565s3960 is one the few for which the simplest algorithm (*clo* with strategy fst_t) is competitive with other algorithmic instances. Moreover it naturally significantly outperforms these regarding the execution time due to its linear complexity.

Instances aslink(1,a), deploy(3,a), or lamport(4) illustrate that $clo^\star$ can significantly outperform *clo* reducing further the state space by a factor of approximately 2.

Instance aslink(1,a) illustrates the impact of the scapegoat choice strategy with algorithm *clo* and its lesser importance with algorithm $clo^\star$. With the former, using the same starting transition choice strategy, strategy fst_s performs clearly worse that min_s^e and min_s^f while this observation is less valid when using algorithm $clo^\star$.

Table 1. Comparison of relations and $\mathcal{D}_s$ and $\mathcal{D}_e$

Model instance	$min(\mathcal{D}_s)$	$min(\mathcal{D}_e)$	Model instance	$min(\mathcal{D}_s)$	$min(\mathcal{D}_e)$
hexagonalgrid(1,2,6)	111,684	901	anderson(5)	219,420	104,406
triangulargrid(1,50,0)	81,198	764	anderson(4)	12,519	6,753
triangulargrid(1,20,0)	13,563	288	safebus(3)	3,052	2,784
hexagonalgrid(1,1,0)	6,708	196	egfr(20,1,0)	162	159
robot(5)	196	10	shieldsppp(1,a)	5,453	5,423
mapk(8)	3,483	619	deploy(4,a)	571,200	568,234

Table 2. Scores (according to Eq. (1)) of selected algorithmic instances on 15 model instances and average scores over all experimented model instances

	clo						clo*						del			Prod del	Prod inc	
fst_s	min^*_s	min^*_s	fst_s	min^*_s	min^*_s	fst_s	min^*_s	min^*_s	fst_s	min^*_s	min^*_s	fst_s	min^*_s	min^*_s	fst_s	min^*_s	min^*_s	fst_s	max^*_s	max^*_s	fst_s	fst_s	min^*_s
Model instance aslink(1,a), $S_{min} = 960{,}868$ states																							
1.6	14.1	14.8	7.8	46.8	50.2	6.8	46.9	51.3	18.4	36.5	32.3	79.2	93.8	93.3	80.3	94.6	94.9	74.0	97.9	99.0	80.9	5.2	5.3
Model instance deploy(3,a), $S_{min} = 28{,}510$ states																							
16.9	16.8	17.3	32.5	34.8	38.8	32.9	35.4	39.9	24.3	30.7	25.7	58.3	55.9	56.5	60.9	57.4	57.6	77.6	87.0	98.1	76.0	25.9	26.2
Model instance des(5,a), $S_{min} = 1{,}752{,}989$ states																							
0.6	1.9	2.7	51.2	63.7	68.2	55.1	70.0	75.2	9.9	9.9	6.2	80.2	79.1	79.4	85.1	83.8	84.4	72.0	78.8	84.7	66.9	22.3	45.3
Model instance exbar(4,b), $S_{min} = 2{,}017{,}473$ states																							
0.6	0.6	0.0	70.1	82.1	84.0	70.9	83.2	85.2	0.8	0.8	0.0	85.2	86.7	86.8	86.1	87.9	87.9	60.7	85.2	87.0	82.2	59.2	70.5
Model instance gpufp(08,a), $S_{min} = 47{,}410$ states																							
25.9	25.9	25.9	82.2	82.2	82.2	70.9	70.2	70.3	26.0	25.9	25.9	82.2	82.2	82.2	70.2	70.3	70.3	82.0	96.2	90.5	83.6	71.4	84.4
Model instance ibmb2s565s3960, $S_{min} = 1{,}500{,}964$ states																							
63.7	63.7	63.7	68.1	68.1	68.1	67.2	67.2	67.2	63.7	63.7	63.7	68.1	68.1	68.1	67.2	67.2	67.2	54.9	66.4	65.4	79.2	54.3	67.1
Model instance lamport(4), $S_{min} = 206{,}527$ states																							
15.4	17.3	19.5	23.7	26.6	31.2	23.7	27.4	32.9	27.6	31.1	31.2	76.2	85.2	85.8	78.2	88.4	88.3	81.4	97.0	99.4	88.9	23.1	23.1
Model instance peterson(3), $S_{min} = 102{,}371$ states																							
11.2	14.2	14.5	82.9	89.1	90.0	82.9	89.2	90.0	13.1	14.1	14.3	87.7	90.5	91.4	88.7	91.2	91.5	94.9	99.6	100.0	96.7	39.5	39.5
Model instance raft(3), $S_{min} = 7{,}262{,}240$ states																							
27.6	28.2	17.9	65.1	74.9	76.2	65.1	76.5	76.5	28.2	28.4	17.9	73.7	75.5	76.4	73.9	75.7	76.6	73.5	76.4	76.6	90.9	70.4	72.8
Model instance satmem(1000,32), $S_{min} = 1{,}880{,}803$ states																							
36.8	36.8	36.8	50.2	50.2	50.2	50.2	50.2	50.2	36.8	36.8	36.8	50.2	50.2	50.2	50.2	50.2	50.2	50.1	50.1	50.1	68.9	50.2	50.2
Model instance shieldsrv(3,a), $S_{min} = 386{,}893$ states																							
32.6	36.3	40.7	50.5	68.8	72.4	50.7	73.2	77.0	38.4	44.7	45.1	76.8	84.1	84.0	88.9	97.2	97.0	72.7	86.4	95.7	82.5	38.0	40.3
Model instance smhome(8), $S_{min} = 14{,}434$ states																							
0.0	0.0	0.0	19.8	18.3	18.3	20.4	18.8	18.8	0.0	0.0	0.0	18.3	18.3	18.3	18.8	18.8	18.8	7.5	18.4	18.6	76.0	13.0	19.0
Model instance stigcomm(2,b), $S_{min} = 165{,}682$ states																							
0.5	0.6	0.6	34.1	34.1	34.2	34.1	34.1	34.2	2.2	1.3	1.8	53.1	38.3	79.7	53.1	38.4	79.8	40.3	95.3	92.2	40.8	28.2	34.2
Model instance stigelec(4,b), $S_{min} = 133{,}892$ states																							
0.0	0.0	0.0	12.5	12.7	14.0	12.5	12.8	14.1	0.5	0.5	0.5	19.5	17.2	26.0	19.5	17.2	26.2	31.2	83.9	63.7	54.4	9.4	12.5
Model instance tcp(5), $S_{min} = 601{,}458$ states																							
29.2	30.1	29.9	90.3	92.4	92.5	93.9	97.4	97.4	29.6	30.2	30.0	93.1	93.4	93.4	97.7	98.2	98.2	60.2	87.8	91.9	69.5	42.1	87.3
Average scores and total number of successful runs over 140 instances																							
22.1	23.4	23.7	59.8	67.2	68.4	62.8	70.3	72.0	24.6	25.5	25.3	70.8	72.9	73.8	74.4	76.4	77.4	51.7	72.1	75.1	61.5	39.2	59.6
605	613	594	684	688	688	681	685	691	622	625	598	698	698	698	700	699	700	681	687	685	675	663	691

4 Stubborn Sets for Liveness Verification

The principle of state space reduction based on stubborn set reduction is somehow to reorder transitions in such a way that only stubborn transitions are considered to generate the successors of a marking while the firing of non stubborn transitions is postponed to a future marking. However, such a marking may never occur due to the so called *ignoring problem*. To illustrate this situation, let us assume a net having a transition t disconnected from the rest of the net (i.e., $\forall p \in P, W(p,t) = W(t,p) = 0$). Then, since $\mathcal{D}_s(t) = \emptyset$ and $m_0[t\rangle m_0$, a dynamically stubborn reduction function may build a reduced state with a single self loop marking. In other words, t hides the dynamics of the rest of the net. The reduced state space is therefore of very little use besides the one of proving that the system does not halt.

As in the previous section, we first recall in this section the theoretical background of the stubborn set theory for liveness verification. We then review different algorithms that can be used in that context and introduce two variations of previous algorithms before presenting our experimental results. Algorithms considered here are not specific to Petri nets. Therefore we will often use here more generic terms, such as *state* instead of marking, or *action* instead of transition.

4.1   Stubborn Set Theory for Liveness Verification

The above example has shown that dynamic stubborness is not sufficient for the verification of several properties, including liveness properties, because of the ignoring problem. Formally, transition ignoring occurs when a transition is enabled for some state of a cycle but never executed along that cycle. Otherwise, the reduced state space fullfils the *strong cycle proviso* defined below:

Definition 9. *Let f be a reduction function. The reduced state space has the* strong cycle proviso *property if, for any $m_1[t_1\rangle_f m_2[t_2\rangle_f \ldots m_n[t_n\rangle_f m_1$, the following holds for any $t \in T$: $(\exists i \in \{1,\ldots,n\}$ such that $m_i[t\rangle) \Rightarrow (\exists j \in \{1,\ldots,n\}$ such that $m_j[t\rangle_f)$.*

To verify linear time temporal logic properties, an addition condition linked to the visibility of transitions is required [15, 18] but this is out of the scope of our study.

A sufficient condition for a reduction function to ensure the strong cycle proviso is that along any cycle of the reduced state space, there is at least one fully expanded state:

Proposition 2. *Let f be a reduction function. If, for any $m_1[t_1\rangle_f m_2[t_2\rangle_f \ldots m_n[t_n\rangle_f m_1$, there is $i \in \{1,\ldots n\}$ such that $en(m_i) \subseteq f(m_i)$ then the reduced state space has the strong cycle proviso property.*

Hereafter, we refer to the condition of Proposition 2 as the *weak cycle proviso* or more simply *cycle proviso*. Checking that each cycle contains a fully expanded state is easier and can be done on-the-fly, i.e., during the construction of the reduced state space. Hence, while it is a stronger condition that may bring less reduction, this proposition serves as a basis for all the algorithms we review in the following.

4.2   Stubborn Set Algorithms for Liveness Verification

State of the Art. Most algorithms operate on-the-fly: they address the cycle proviso as they generate the state space. Therefore they are tightly linked to a specific search order.

For DFS, a sufficient condition to ensure the cycle proviso is to forbid a cycle-closing edge (i.e., an edge of which the destination is in the DFS stack) outgoing from a reduced state [15]. An alternate implementation for DFS has also been introduced in [7]. For BFS, a dual sufficient condition is that a reduced state only has successors in the BFS queue [2]. This principle has been generalised in [3] to any search order.

Several optimisations and variations for DFS (including Tarjan algorithm) have been proposed in [5] that lead to an improvement over [7, 15] in practice. A lesson that can be drawned from the experimentation is that the full expansion of the destination state of a cycle-closing edge should be preferred in practice (rather than the full expansion of the source state, as done in [7, 15]). Indeed, the destination state needs to be fully expanded when leaving the stack and at that moment, the algorithm may have discovered that the full expansion is no more required (e.g., if all its successors have been fully expanded). This can save useless full state expansions.

The algorithm of [1] alternates expansion phases, during which states are expanded without taking care of the cycle proviso, with topological sorts (efficiently performed in a distributed way) of the resulting reduced state space to detect states to be fully expanded to prevent action ignoring. The full expansion of these states may then lead to new states used to initiate a new expansion phase. The algorithm stops when the topological sort does not produce any new states.

We finally mention the static algorithm of [12] and the two-phase algorithm of [14].

New Algorithms Ensuring the Cycle Proviso

A BFS Based on Destination State Revisit. The principle of the dst proviso of [5] can be combined with the proviso of [2] for BFS (see Algorithm 3). S denotes in the algorithm a set of safe destination states, in the sense that any reduced state may have successors in S without endangering the cycle condition. S consists of all fully expanded states (see line 6). An invalid cycle may be closed each time the algorithm discovers an edge from an unsafe state to another state that is neither safe, neither in the queue. In that case, the destination state s' of the edge is reinserted in the queue to be fully reexpanded (see lines 15–16). This is the purpose of the second component of items put in Q: if set to **true**, the state must be fully expanded. Otherwise it can be reduced using function *stub* that can be any dynamically stubborn function (see line 5).

This proviso is especially suited for distributed model checking based on state space partitioning (as done in [17]). In that context, whenever a process p generates a state s' it puts s' in the queue of the owner process p' of s'. Ownership is determined using typically the hash value of the state. If the generating process is not the owner process, a communication is needed. With the proviso of [2], a round trip between the two processes would be necessary for p' to notify p whether s' is in its queue (or unvisited) which makes it unusable in that context. With a proviso based on the full expansion of the destination state, p now delegates the responsability of checking the cycle proviso to p'. Hence, it does not require additional communications.

Algorithm 3. Algorithm BFS^{dst} ensuring the cycle proviso

1: $R := \{m_0\}$; $Q := \{(m_0, \textbf{false})\}$; $S := \emptyset$	9: **procedure** *expand_marking*(m, U) **is**
2: **while** $Q \neq \emptyset$ **do**	10: **for** $t \in U \cap en(m)$ **do**
3: $(m, fexp) :=$ **pick from** Q	11: **let** m' be such that $m[t\rangle m'$
4: $Q := Q \setminus \{(m, fexp)\}$	12: **if** $m' \notin R$ **then**
5: $U :=$ **if** $fexp$ **then** $en(m)$ **else** $stub(m)$	13: $R := R \cup \{m'\}$
6: **if** $U = en(m)$ **then** $S := S \cup \{m\}$	14: $Q := Q \cup \{(m', \textbf{false})\}$
7: *expand_marking*(m, U)	15: **else if** $\neg(m \in S \vee m' \in S \cup Q)$ **then**
8: **return** S	16: $Q := Q \cup \{(m', \textbf{true})\}$

New Offline Optimal Provisos. We propose two algorithms that perform optimally in the sense that they do not uselessly (fully) re-expand reduced states to verify the cycle proviso: if the algorithm fully re-expands a state, then it is because it is part of a cycle of reduced states. Such an algorithm will be characterised as *WCP-optimal* (Weak Cycle Proviso-optimal) hereafter. It is easy to find counter examples showing that all algorithms we previously reviewed are not WCP-optimal. Likewise, an algorithm is *SCP-optimal* (Strong Cycle Proviso-optimal) if it does not uselessly visit new transitions to verify the strong cycle proviso: if the algorithm forces the execution of some transition t at a state s, then it is necessarily because s is part of a cycle that ignores t.

The two algorithms have very little practical use as they operate offline and consume a significant additional amount of memory per state (the adjacency list) but can be used experimentally to evaluate how other algorithms perform. Both are a variation of the topological sort based algorithm [1] and rely on an alternation of an expansion phasis with a cycle proviso checking phasis.

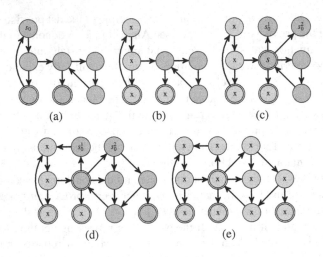

Fig. 3. Expansion and checking phases of our WCP-optimal algorithm. (a): after a 1st expansion step; (b): after a 1st checking step; (c): full expansion of s; (d): after a 2nd expansion step; (e): after a 2nd checking step.

Our WCP-optimal algorithm can be illustrated with the example depicted on Fig. 3. In a first step, the state space is generated starting from the initial state s_0 using a dynamically stubborn reduction function, (i.e., without taking care of transition ignoring). The reduced state space obtained after this first expansion step is depicted on Fig. 3(a) where states with double circles are fully expanded states. The second step, the checking step, consists of marking fully expanded states as *safe* meaning that these cannot be part of a cycle of reduced states. States of which all successors or all predecessors are safe are also marked as such and this procedure is repeated until no more state can be marked. This leads us to the configuration of Fig. 3(b) where safe states are green (and marked with a cross). The outcome of the checking step is to pick an unsafe state (s in our example) and fully expand it (see Fig. 3(c)). This may generate new states which are then used as initial states for a new expansion step (s_0^1 and s_0^2 in our example). After this one (see Fig. 3(d)) a new checking step is triggered and the algorithm may terminate if all states are safe (see Fig. 3(e)).

It is easy to see to that this algorithm is WCP-optimal. Indeed, if after an expansion step the reduced state space already has the weak cycle proviso property then all states will be marked as safe and the algorithm will immediately terminate whereas if it fully expands an unsafe state then it is because this state belongs to a cycle of reduced states.

For the strong proviso, this algorithm can be adapted by repeating the checking step for each transition. A state s is marked as safe for some transition t (denoted by t-safe hereafter) if either t is disabled at s, or either t has been executed at s (i.e., it is enabled and stubborn at s). The checking step then proceeds similarly: any state of which all the successors or all the predecessors are t-safe becomes t-safe. If some state is detected as not being t-safe after this step, the algorithm picks such a state and computes a stubborn set that includes t (both the closure and the deletion algorithms can be easily modified to compute stubborn sets including a specific transition). As in the WCP-optimal algorithm, new states that may be reached through this process are used as initial states for a new expansion step. The algorithm may terminate if, after a checking step, all states are marked as t-safe for each transition t.

4.3   Experimentation Context

We experimented with the following algorithms:

- DFS^{src}([7]): DFS + full expansion of source states of cycle closing edges
- DFS^{dst}([5]): DFS + full expansion of destination states of cycle closing edges [2]
- BFS^{src}([2]): BFS + full expansion of source states of backward edges
- BFS^{dst}: BFS + full expansion of destination states of backward edges (i.e., Algorithm 3)
- OPT^{wcp}: the WCP-optimal algorithm presented above

All these algorithms have been integrated in Helena.

Based on the outcome of our first experiment, we selected the following dynamically stubborn functions: $clo^\star(t,s)$, $\forall(t,s) \in \{min_t^f, min_t^e\} \times \{min_s^t, min_s^f, min_s^t\}$, and $del(t)$, $\forall t \in \{max_t^e, max_t^f, fst_t\}$. These were among the best strategies for algorithms Clo* and Del.

We also included to this second experiment a few additional model instances that were left out in our first experiment (for the reason that all dynamically stubborn function computed the exact same reductions for these) and removed some for which no run could terminate within our time limit (set, as in our first experiment, to 30 min.). Overall this second experiment was performed on 137 model instances of 75 models (over 130) resulting in 32400 runs.

4.4   Experimental Observations

As in the previous section, we start with general observations before presenting some selected results.

First, as noted elsewhere [2], BFS based provisos perform significantly worse than DFS based provisos. On only one model instance (MAPK(20)) did a BFS based proviso perform significantly better than its DFS analogous. BFS based provisos may however still be useful for specific contexts such as distributed model checking which do not allow a depth-first search order.

---

[2] We implemented the ColoredDest variant of [5] which makes use of state tagging mechanisms to avoid useless full expansions.

Our results confirm those of [5]: in DFS, the full expansion of destination states of cycle closing edges (rather that source states) is preferable, i.e., DFS^{dst} outperforms DFS^{src} in general. This also holds for BFS: BFS^{dst} performs better than BFS^{src}.

We observe that OPT^{wcp} does not bring any improvement with respect to DFS^{dst}. When both algorithms could terminate, they performed comparably — DFS^{dst} being even slightly better. This seems to indicate that DFS^{dst} is already close to optimal in the sense that it never uselessly reexpands states. Moreover, when it does, it "picks" better states (destinations of cycle closing edges) than OPT^{wcp}, that picks them randomly.

4.5 Experimental Results Sample

Table 3 provides scores of the 5 search algorithms for 15 selected model instances and 4 selected dynamically stubborn functions. Average scores and total number of successful runs (over 685) are also provided at the bottom of the table.

Instance deploy(3,a) illustrates that choosing stubborn sets reducing forward transitions may be unappropriate for liveness properties as it tends to trigger more state reexpansions. Indeed, we observe that, regardless of the search algorithm used, $del(max_t^e)$ and $clo^*(min_t^e, min_s^e)$ perform better than $del(max_t^f)$ and $clo^*(min_t^f, min_s^f)$ respectively.

A single run of the OPT^{wcp} algorithm on instance ibmb2s565s3960 coupled with reduction function $del(fst_t)$ (not shown on the table) could produce a reduced graph with an order of magnitude smaller — which explains the low scores reported in the table. We plan to further investigate the net structure and the conditions that made such a drastic reduction feasible.

Instance shieldtppp(1,b) is one of the few instances for which DFS^{src} competes favorably against DFS^{dst}.

As said above, algorithm OPT^{wcp} does not improve on DFS^{dst}. It performed slightly better on 5 of the 15 model instances selected: aslink(1,a), eisenbergmcguire(4), lamport(4), raft(2) and shieldtppp(1,b). Note however that the average score computed over all instances must be taken with care since, as witnessed by the last row, runs of algorithm OPT^{wcp} timed out more frequently. It is likely that with a higher time limit, the average scores of OPT^{wcp} and DFS^{dst} would have been very close.

Table 3. Scores (according to Eq. (1)) of selected algorithmic instances on 15 model instances and average scores over all experimented model instances

	BFS^{src} (from [2])				BFS^{dst} (from Alg. 3)				DFS^{src} (from [7])				DFS^{dst} (from [5])				OPT^{wkp} (from Fig. 3)			
	clo*		del		clo*		del		clo*		del		clo*		del		clo*		del	
	min_c^e min_s^e	min_c^f min_s^f	max_c^e	max_c^f	min_c^e min_s^e	min_c^f min_s^f	max_c^e	max_c^f	min_c^e min_s^e	min_c^f min_s^f	max_c^e	max_c^f	min_c^e min_s^e	min_c^f min_s^f	max_c^e	max_c^f	min_c^e min_s^e	min_c^f min_s^f	max_c^e	max_c^f
Model instance aslink(1,a), S_{min} = 975,971 states																				
0.0	0.0	0.0	0.0	51.0	40.5	43.2	39.7	93.4	94.5	97.5	98.9	92.7	93.9	96.2	97.5	93.0	94.5	97.2	98.5
Model instance deploy(3,a), S_{min} = 67,212 states																				
45.0	35.4	40.4	35.8	75.4	36.0	65.3	41.7	92.8	57.3	88.8	82.5	94.5	57.6	95.5	85.8	87.0	50.4	81.1	72.0
Model instance des(5,a), S_{min} = 1,900,083 states																				
0.0	0.0	0.0	0.0	0.0	0.0	0.0	0.0	76.4	82.0	75.2	80.5	77.8	83.4	75.5	80.6	0.0	0.0	0.0	0.0
Model instance eisenbergmcguire(4), S_{min} = 282,317 states																				
20.1	19.0	20.7	19.7	65.5	50.5	71.2	70.6	89.9	91.7	99.3	99.6	90.2	92.7	99.6	99.8	90.4	93.0	99.7	100.0
Model instance exbar(6,a), S_{min} = 148,514 states																				
5.0	5.0	5.0	5.0	10.1	5.4	8.2	6.4	14.8	15.1	14.7	22.5	66.2	30.8	67.4	65.3	57.2	32.7	60.0	68.1
Model instance gpufp(08,b), S_{min} = 500,514 states																				
0.0	0.0	0.0	0.0	0.0	0.0	0.0	0.0	66.0	66.2	66.6	67.1	66.9	66.9	68.2	68.2	60.7	60.7	62.1	62.1
Model instance ibmb2s565s3960, S_{min} = 1,567,329 states																				
0.0	0.0	0.0	0.0	0.0	0.0	0.0	0.0	5.7	8.4	14.8	17.5	6.5	9.1	15.6	16.5	0.0	0.0	0.0	0.0
Model instance lamport(4), S_{min} = 208,895 states																				
12.9	12.6	12.7	12.5	56.9	42.7	73.0	65.6	73.1	74.8	87.0	89.5	83.3	85.9	95.9	98.3	85.1	87.5	97.0	99.5
Model instance peterson(3), S_{min} = 125,748 states																				
6.6	5.3	6.6	5.8	54.5	28.5	58.6	53.0	87.3	87.5	93.8	94.1	92.5	92.6	99.7	100.0	84.8	84.5	90.8	91.1
Model instance raft(2), S_{min} = 3,116 states																				
60.9	59.6	50.8	59.1	82.9	68.7	80.5	80.4	93.3	78.1	92.2	94.4	98.7	79.6	97.8	98.7	98.8	80.0	99.3	99.7
Model instance shieldsrv(3,b), S_{min} = 1,832,300 states																				
0.0	0.0	0.0	0.0	1.7	0.0	0.7	0.0	83.0	93.8	75.6	80.5	87.5	97.2	81.7	86.5	0.0	0.0	0.0	0.0
Model instance shieldtppp(1,b), S_{min} = 7,450 states																				
0.0	0.1	0.0	0.1	0.5	0.2	0.2	0.1	77.6	78.9	51.6	54.9	56.4	56.3	39.5	41.0	63.2	63.3	42.0	43.1
Model instance smhome(8), S_{min} = 155,000 states																				
0.0	0.0	0.0	0.0	0.0	0.0	0.0	0.0	19.3	42.9	0.0	2.0	98.5	82.1	10.5	12.6	97.5	81.7	0.0	0.0
Model instance stigcomm(2,b), S_{min} = 468,130 states																				
0.0	0.0	0.0	0.0	4.7	1.5	5.0	0.5	33.8	55.1	55.5	66.5	35.8	59.1	61.2	62.6	20.0	28.0	43.1	37.1
Model instance tcp(5), S_{min} = 787,898 states																				
26.7	26.9	26.5	26.6	42.7	30.5	34.7	34.6	37.6	36.1	36.2	36.3	95.2	98.0	88.3	91.3	0.0	0.0	0.0	0.0
Average scores and total number of successful runs over 137 instances																				
23.4	22.8	22.4	22.4	37.1	31.8	35.1	34.7	69.2	69.5	65.0	69.1	82.5	79.6	76.4	79.1	70.8	65.9	65.2	65.6
509	512	492	485	573	537	543	526	665	664	643	645	671	660	656	653	553	541	531	538

5 Conclusion

We have contributed with this paper with a number of simple algorithmic variants and heuristics for stubborn set construction algorithms. In the context of deadlock detection, we introduced for the closure algorithm an optimisation used to avoid the selection of unappropriate scapegoats and, for the deletion algorithm, we introduced very simple heuristics to choose the transition to delete. For liveness properties we introduce a BFS algorithm based on [2,5] and an offline algorithm that has the property to fully expand states only when absolutely needed.

A second contribution of our work is a large experimentation of these algorithms and variants on models of the MCC database resulting in approximately 150 000 runs using both Prod and Helena tools. These showed that our algorithmic contributions, despite their simplicity, can bring significant results on many instances.

We plan to pursue these experiments in several directions in order to identify room for improvements or to design new heuristics.

First, we would like to experiment with other forms of dependence and precedence relations (as done in [9]) based on the full state graph in order identify if there are still room for improvement in that perspective.

We also plan to implement, for the deletion algorithm, a finer dependency graph, as done by Prod, to study the impact of our heuristics (or others) on it.

Stubborn set construction algorithms may also exploit other informations than the net structures, such as place invariants, or unit decomposition [8] and we plan to investigate how these can be used.

There also exists some algorithms based on integer linear programming techniques [13]. Comparing them to algorithms discussed here is relevant.

For liveness properties, we have only considered the resolution of the ignoring problem, putting aside other conditions required for, e.g., LTL model checking, regarding the visibility of transitions. It seems worth experimenting with such conditions.

Considering the ignoring problem, our experiments are somehow disappointing in the sense that the DFS^{dst} algorithm seems hard to outperform. Still we have not experimented with algorithm OPT^{scp} and plan to do so in order to check whether new algorithms reasoning on the strong cycle proviso could be of practical use.

Last we have not considered safety properties in our study and we would like to experiment with an algorithm tailored to these, e.g., [11, 16], and compare them to general purpose algorithms, i.e., stubborn sets construction algorithm coupled with a cycle proviso and conditions on transition visibility.

Acknowledgments. We thank the organizers of the Model Checking Contest and all people that contribute to its model database for providing such a database. Experiments presented in this paper were carried out using the Grid'5000 [4] testbed, supported by a scientific interest group hosted by Inria and including CNRS, RENATER and several Universities as well as other organisations (see https://www.grid5000.fr). We thank Laure Petrucci for her comments on the first version of this paper.

References

1. Barnat, J., Brim, L., Rockai, P.: Parallel partial order reduction with topological sort proviso. In: SEFM'2010, pp. 222–231. IEEE Computer Society Press (2010)
2. Bošnački, D., Holzmann, G.J.: Improving spin's partial-order reduction for breadth-first search. In: Godefroid, P. (ed.) SPIN 2005. LNCS, vol. 3639, pp. 91–105. Springer, Heidelberg (2005). https://doi.org/10.1007/11537328_10
3. Bošnački, D., Leue, S., Lafuente, A.L.: Partial-order reduction for general state exploring algorithms. In: Valmari, A. (ed.) SPIN 2006. LNCS, vol. 3925, pp. 271–287. Springer, Heidelberg (2006). https://doi.org/10.1007/11691617_16
4. Cappello, F., et al.: Grid'5000: A large scale and highly reconfigurable grid experimental testbed. In SC'05: Proc. The 6th IEEE/ACM International Workshop on Grid Computing CD, pp. 99–106, Seattle, Washington, USA, November 2005. IEEE/ACM
5. Duret-Lutz, A., Kordon, F., Poitrenaud, D., Renault, E.: Heuristics for checking liveness properties with partial order reductions. In: Artho, C., Legay, A., Peled, D. (eds.) ATVA 2016. LNCS, vol. 9938, pp. 340–356. Springer, Cham (2016). https://doi.org/10.1007/978-3-319-46520-3_22

6. Evangelista, S.: High level petri nets analysis with Helena. In: Ciardo, G., Darondeau, P. (eds.) ICATPN 2005. LNCS, vol. 3536, pp. 455–464. Springer, Heidelberg (2005). https://doi.org/10.1007/11494744_26

7. Evangelista, S., Pajault, C.: Solving the ignoring problem for partial order reduction. Int. J. Softw. Tools Technol. Transfer (STTT) **12**(2) (2010)

8. Garavel, H.: Nested-unit Petri nets. J. Log. Algebraic Methods Program. **104**, 60–85 (2019)

9. Geldenhuys, J., Hansen, H., Valmari, A.: Exploring the scope for partial order reduction. In: Liu, Z., Ravn, A.P. (eds.) ATVA 2009. LNCS, vol. 5799, pp. 39–53. Springer, Heidelberg (2009). https://doi.org/10.1007/978-3-642-04761-9_4

10. Godefroid, P. (ed.): Partial-Order Methods for the Verification of Concurrent Systems. LNCS, vol. 1032, pp. 41–73. Springer, Heidelberg (1996). https://doi.org/10.1007/3-540-60761-7_31

11. Kristensen, L.M., Valmari, A.: Improved question-guided stubborn set methods for state properties. In: Nielsen, M., Simpson, D. (eds.) ICATPN 2000. LNCS, vol. 1825, pp. 282–302. Springer, Heidelberg (2000). https://doi.org/10.1007/3-540-44988-4_17

12. Kurshan, R., Levin, V., Minea, M., Peled, D., Yenigün, H.: Static partial order reduction. In: Steffen, B. (ed.) TACAS 1998. LNCS, vol. 1384, pp. 345–357. Springer, Heidelberg (1998). https://doi.org/10.1007/BFb0054182

13. Lehmann, A., Lohmann, N., Wolf, K.: Stubborn sets for simple linear time properties. In: Haddad, S., Pomello, L. (eds.) PETRI NETS 2012. LNCS, vol. 7347, pp. 228–247. Springer, Heidelberg (2012). https://doi.org/10.1007/978-3-642-31131-4_13

14. Nalumasu, R., Gopalakrishnan, G.: An efficient partial order reduction algorithm with an alternative proviso implementation. FMSD **20**(3), 231–247 (2002)

15. Peled, D.: All from one, one for all: on model checking using representatives. In: Courcoubetis, C. (ed.) CAV 1993. LNCS, vol. 697, pp. 409–423. Springer, Heidelberg (1993). https://doi.org/10.1007/3-540-56922-7_34

16. Schmidt, K.: Stubborn sets for standard properties. In: Donatelli, S., Kleijn, J. (eds.) ICATPN 1999. LNCS, vol. 1639, pp. 46–65. Springer, Heidelberg (1999). https://doi.org/10.1007/3-540-48745-X_4

17. Stern, U., Dill, D.L.: Parallelizing the MurΦ verifier. In: Grumberg, O. (ed.) CAV 1997. LNCS, vol. 1254, pp. 256–267. Springer, Heidelberg (1997). https://doi.org/10.1007/3-540-63166-6_26

18. Valmari, A.: A stubborn attack on state explosion. In: Clarke, E.M., Kurshan, R.P. (eds.) CAV 1990. LNCS, vol. 531, pp. 156–165. Springer, Heidelberg (1991). https://doi.org/10.1007/BFb0023729

19. Valmari, A.: Stubborn sets for reduced state space generation. In: Rozenberg, G. (ed.) ICATPN 1989. LNCS, vol. 483, pp. 491–515. Springer, Heidelberg (1991). https://doi.org/10.1007/3-540-53863-1_36

20. Valmari, A.: The state explosion problem. In: Reisig, W., Rozenberg, G. (eds.) ACPN 1996. LNCS, vol. 1491, pp. 429–528. Springer, Heidelberg (1998). https://doi.org/10.1007/3-540-65306-6_21

21. Valmari, A., Hansen, H.: Can stubborn sets be optimal? In: Lilius, J., Penczek, W. (eds.) PETRI NETS 2010. LNCS, vol. 6128, pp. 43–62. Springer, Heidelberg (2010). https://doi.org/10.1007/978-3-642-13675-7_5

22. Varpaaniemi, K.: Finding small stubborn sets automatically. In: Proceedings of the 11th International Symposium on Computer and Information Sciences, pp. 133–142 (1996)

23. Varpaaniemi, K., Heljanko, K., Lilius, J.: Prod 3.2 an advanced tool for efficient reachability analysis. In: Grumberg, O. (ed.) CAV 1997. LNCS, vol. 1254, pp. 472–475. Springer, Heidelberg (1997). https://doi.org/10.1007/3-540-63166-6_51

Timed Models

Symbolic Analysis and Parameter Synthesis for Time Petri Nets Using Maude and SMT Solving

Jaime Arias[1], Kyungmin Bae[2], Carlos Olarte[1(✉)],
Peter Csaba Ölveczky[3], Laure Petrucci[1], and Fredrik Rømming[4]

[1] LIPN, CNRS UMR 7030, Université Sorbonne Paris Nord, Villetaneuse, France
olarte@lipn.fr
[2] Pohang University of Science and Technology, Pohang, South Korea
[3] University of Oslo, Oslo, Norway
[4] University of Cambridge, Cambridge, UK

Abstract. In this paper we present a concrete and a symbolic rewriting logic semantics for parametric time Petri nets with inhibitor arcs (PITPNs). We show how this allows us to use Maude combined with SMT solving to provide sound and complete formal analyses for PITPNs. We develop a new general folding approach for symbolic reachability that terminates whenever the parametric state-class graph of the PITPN is finite. We explain how almost all formal analysis and parameter synthesis supported by the state-of-the-art PITPN tool Roméo can be done in Maude with SMT. In addition, we also support analysis and parameter synthesis from *parametric* initial markings, as well as full LTL model checking and analysis with user-defined execution strategies. Experiments on three benchmarks show that our methods outperform Roméo in some cases.

Keywords: parametric timed Petri nets · semantics · rewriting logic · Maude · SMT · parameter synthesis · symbolic reachability analysis

1 Introduction

System designers often do not know in advance the concrete values of key system parameters, and want to find those values that make the system behave as desired. *Parametric time Petri nets with inhibitor arcs* (PITPNs) [2,20,28,49] extend the popular time(d) Petri nets [22,30,51] to the setting where bounds on when transitions can fire are unknown or only partially known.

The formal analysis of PITPNs—including synthesizing the values of the parameters which make the system satisfy desired properties—is supported by the state-of-the-art tool Roméo [29], which has been applied to a number of applications, e.g., [3,19,46]. Roméo supports the analysis and parameter synthesis for reachability (is a certain marking reachable?), liveness (will a certain marking be reached in all behaviors?), time-bounded "until," and bounded response (will each P-marking be followed by a Q-marking within time Δ?), all from *concrete* initial markings. Roméo does not support a number of desired features, including:

L. Gomes and R. Lorenz (Eds.): PETRI NETS 2023, LNCS 13929, pp. 369–392, 2023.
https://doi.org/10.1007/978-3-031-33620-1_20

- Broader set of system properties, e.g., full (i.e., nested) temporal logic.
- Starting with *parametric* initial markings and synthesizing also the initial markings that make the system satisfy desired properties.
- Analysis with user-defined execution strategies. For example, what happens if I always choose to fire transition t instead of t' when they are both enabled?
- Providing a "testbed" for PITPNs in which different analysis methods can quickly be developed and evaluated. This is not supported by Roméo, which is a high-performance tool with dedicated algorithms implemented in C++.

PITPNs do not support many features needed for large distributed systems, such as user-defined data types and functions. Rewriting logic [31, 32]—supported by the Maude language and tool [18], and by Real-Time Maude [37, 43] for real-time systems—is an expressive logic for distributed and real-time systems. In rewriting logic, any computable data type can be specified as an (algebraic) equational specification, and the dynamic behaviors of a system are specified by rewriting rules over terms (representing states). Because of its expressiveness, Real-Time Maude has been successfully applied to a number of large and sophisticated real-time systems—including 50-page active networks and IETF protocols [27, 44], industrial cloud systems [13, 21], scheduling algorithms with unbounded queues [39], airplane turning algorithms [8], and so on—beyond the scope of most popular formalisms for real-time systems. Its expressiveness has also made Real-Time Maude a useful semantic framework and formal analysis backend for industrial modeling languages [1, 9, 36, 38].

This expressiveness comes at a price: most analysis problems are undecidable in general. Real-Time Maude uses explicit-state analysis where only *some* points in time are visited. All possible system behaviors are therefore *not* analyzed (for dense time domains), and hence the analysis is unsound in many cases [41].

This paper exploits the recent integration of SMT solving into Maude to address the first problem above (more features for PITPNs) and to take the second step towards addressing the second problem (developing sound and complete analysis methods for rewriting-logic-based real-time systems).

Maude combined with SMT solving, e.g., as implemented in the Maude-SE tool [53], allows us to perform *symbolic rewriting* of "states" $\phi \parallel t$, where the term t is a state pattern that contains variables, and ϕ is an SMT constraint restricting the possible values of those variables.

Section 3 defines a (non-executable) "concrete" rewriting logic semantics for (instantiated) PITPNs in "Real-Time Maude style" [42], and proves that this semantics is bisimilar to the one for PITPNs in [49]. Section 4 transforms this semantics into a Maude-with-SMT semantics for *parametric* PITPNs, and shows how to perform sound symbolic analysis of such nets using Maude-with-SMT. However, existing symbolic reachability analysis methods may fail to terminate even when the state class graph of the PITPN is finite (and hence Roméo analysis terminates). We therefore develop a new method for "folding" symbolic states, and show that reachability analysis with such folding terminates whenever the state class graph of the PITPN is finite.

In Sect. 5 we show how all analysis methods supported by Roméo—with one small exception: the time bounds in some temporal formulas cannot be parameters—also can be performed using Maude-with-SMT. In addition, we support analysis and parameter synthesis for *parametric* initial markings, model checking full temporal logic formulas, and analysis w.r.t. user-defined execution strategies. Our methods are implemented in Maude, using its meta-programming features. This makes it very easy to develop new analysis methods for PITPNs.

This work also constitutes the second step in our quest to develop sound and complete analysis methods for dense-time real-time systems in Real-Time Maude. We present both a Real-Time Maude-style semantics in Sect. 3 *and* the symbolic semantics in Sect. 4 to explore how we can transform Real-Time Maude models into Maude-with-SMT models for symbolic analysis. In our first step in this quest, we studied symbolic rewrite methods for the much simpler parametric timed automata [4]; see Sect. 7 for a comparison with that work.

In Sect. 6 we benchmark both Roméo and our Maude-with-SMT methods, and find that in some cases our high-level prototype outperforms Roméo.

The longer report [6] has proofs of all results in this paper and much more detail. All executable Maude files with analysis commands, tools for translating Roméo files into Maude, and data from the benchmarking are available at [5].

2   Preliminaries

Transition Systems. A *transition system* $\mathcal{A}$ is a triple $(A, a_0, \rightarrow_{\mathcal{A}})$, where A is a set of *states*, $a_0 \in A$ is the *initial state*, and $\rightarrow_{\mathcal{A}} \subseteq A \times A$ is a *transition relation*. We call $\mathcal{A}$ *finite* if the set of states reachable by $\rightarrow_{\mathcal{A}}$ from a_0 is finite. A relation $\sim \subseteq A \times B$ is a *bisimulation* [16] between $\mathcal{A}$ and $\mathcal{B} = (B, b_0, \rightarrow_{\mathcal{B}})$ iff: (i) $a_0 \sim b_0$; and (ii) for all a, b s.t. $a \sim b$: if $a \rightarrow_{\mathcal{A}} a'$ then there is a b' s.t. $b \rightarrow_{\mathcal{B}} b'$ and $a' \sim b'$, and, vice versa, if $b \rightarrow_{\mathcal{B}} b''$, then there is a a'' s.t. $a \rightarrow_{\mathcal{A}} a''$ and $a'' \sim b''$.

Parametric Time Petri Nets with Inhibitor Arcs (PITPN). $\mathbb{N}$, $\mathbb{Q}_+$, and $\mathbb{R}_+$ denote, resp., the natural numbers, the non-negative rational numbers, and the non-negative real numbers. We assume a finite set $\Lambda = \{\lambda_1, \dots, \lambda_l\}$ of *time parameters*. A *parameter valuation* π is a function $\pi : \Lambda \rightarrow \mathbb{R}_+$. A (linear) *inequality* over Λ is an expression $\sum_{1 \le i \le l} a_i \lambda_i \prec b$, where $\prec \in \{<, \le, =, \ge, >\}$ and $a_i, b \in \mathbb{R}$. A *constraint* is a conjunction of such inequalities. $\mathcal{L}(\Lambda)$ denotes the set of all constraints over Λ. A parameter valuation π *satisfies* a constraint $K \in \mathcal{L}(\Lambda)$, written $\pi \models K$, if the expression obtained by replacing each λ in K with $\pi(\lambda)$ evaluates to true. An interval I of $\mathbb{R}_+$ is a $\mathbb{Q}_+$-interval if its left endpoint ${}^{\uparrow}I$ belongs to $\mathbb{Q}_+$ and its right endpoint $I^{\uparrow}$ belongs to $\mathbb{Q}_+ \cup \{\infty\}$. We denote by $\mathcal{I}(\mathbb{Q}_+)$ the set of $\mathbb{Q}_+$-intervals. A parametric time interval is a function $I : \mathbb{Q}_+{}^{\Lambda} \rightarrow \mathcal{I}(\mathbb{Q}_+)$ that associates with each parameter valuation a $\mathbb{Q}_+$-interval. The set of parametric time intervals over Λ is denoted $\mathcal{I}(\Lambda)$.

Definition 1. *A parametric time Petri net with inhibitor arcs (PITPN) [49] is a tuple $\mathcal{N} = \langle P, T, \Lambda, {}^{\bullet}(.), (.)^{\bullet}, {}^{\circ}(.), M_0, J, K_0 \rangle$ where*

- $P = \{p_1, \dots, p_m\}$ *is a non-empty finite set (of* places),
- $T = \{t_1, \dots, t_n\}$ *is a non-empty finite set (of* transitions), *with* $P \cap T = \emptyset$,

- $\Lambda = \{\lambda_1, \ldots, \lambda_l\}$ *is a finite set of* parameters,
- $^\bullet(.) \in [T \to \mathbb{N}^P]$ *is the* backward incidence function,
- $(.)^\bullet \in [T \to \mathbb{N}^P]$ *is the* forward incidence function,
- $^\circ(.) \in [T \to \mathbb{N}^P]$ *is the* inhibition function,
- $M_0 \in \mathbb{N}^P$ *is the* initial marking,
- $J \in [T \to \mathcal{I}(\Lambda)]$ *assigns a parametric time interval to each transition, and*
- $K_0 \in \mathcal{L}(\Lambda)$ *is the* initial constraint *over* Λ.

If $\Lambda = \emptyset$ *then* $\mathcal{N}$ *is a (non-parametric) time Petri net with inhibitor arcs (ITPN).*

A *marking* of $\mathcal{N}$ is an element $M \in \mathbb{N}^P$, where $M(p)$ is the number of tokens in place p. $\pi(\mathcal{N})$ denotes the ITPN where each occurrence of λ_i in the PITPN $\mathcal{N}$ has been replaced by $\pi(\lambda_i)$ for a parameter valuation π.

The *concrete semantics* of a PITPN $\mathcal{N}$ is defined in terms of concrete ITPNs $\pi(\mathcal{N})$ where $\pi \models K_0$. A transition t is *enabled* in M if $M \geq {}^\bullet t$ (the number of tokens in M in each input place of t is greater than or equal to the value on the arc between this place and t). A transition t is *inhibited* if the place connected to one of its inhibitor arcs is marked with at least as many tokens as the weight of the inhibitor arc. A transition t is *active* if it is enabled and not inhibited. The sets of enabled and inhibited transitions in marking M are denoted *Enabled(M)* and *Inhibited(M)*, respectively. Transition t is *firable* if it has been (continuously) enabled for at least time ${}^\uparrow J(t)$, without counting the time it has been inhibited. Transition t is *newly enabled* by the firing of transition t_f in M if it is enabled in the resulting marking $M' = M - {}^\bullet t_f + t_f^\bullet$ but was not enabled in $M - {}^\bullet t_f$:

$$NewlyEnabled(t, M, t_f) = ({}^\bullet t \leq M - {}^\bullet t_f + t_f^\bullet) \wedge ((t = t_f) \vee \neg({}^\bullet t \leq M - {}^\bullet t_f)).$$

NewlyEnabled(M, t_f) denotes the transitions newly enabled by firing t_f in M.

The semantics of an ITPN is defined as a transition system with states (M, I), where M is a marking and I is a function mapping each transition enabled in M to a time interval, and two kinds of transitions: *time* transitions where time elapses, and *discrete* transitions when a transition in the net is fired.

Definition 2 (ITPN Semantics [49]). *The transition system for an ITPN* $\pi(\mathcal{N})$ *is* $\mathcal{S}_{\pi(\mathcal{N})} = (\mathcal{A}, a_0, \to)$, *where:* $\mathcal{A} = \mathbb{N}^P \times [T \to \mathcal{I}(\mathbb{Q})]$, $a_0 = (M_0, J)$ *and* $(M, I) \to (M', I')$ *if there exist* $\delta \in \mathbb{R}_+$, $t \in T$, *and state* (M'', I'') *such that* $(M, I) \xrightarrow{\delta} (M'', I'')$ *and* $(M'', I'') \xrightarrow{t} (M', I')$, *for the following relations:*

- *the* time transition relation, *defined* $\forall \delta \in \mathbb{R}_+$ *by:* $(M, I) \xrightarrow{\delta} (M, I')$ *iff* $\forall t \in T$:
$$\begin{cases} I'(t) = \begin{cases} I(t) \text{ if } t \in Enabled(M) \text{ and } t \in Inhibited(M) \\ {}^\uparrow I'(t) = \max(0, {}^\uparrow I(t) - \delta) \text{ and } I'(t)^\uparrow = I(t)^\uparrow - \delta \text{ otherwise} \end{cases} \\ M \geq {}^\bullet(t) \implies I'(t)^\uparrow \geq 0 \end{cases}$$

- *the* discrete transition relation, *defined* $\forall t_f \in T$ *by:* $(M, I) \xrightarrow{t_f} (M', I')$ *iff*
$$\begin{cases} t_f \in Enabled(M) \wedge t_f \notin Inhibited(M) \wedge M' = M - {}^\bullet t_f + t_f^\bullet \wedge {}^\uparrow I(t_f) = 0 \\ \forall t \in T, I'(t) = \begin{cases} J(t) \text{ if } NewlyEnabled(t, M, t_f) \\ I(t) \text{ otherwise} \end{cases} \end{cases}$$

The *symbolic* semantics of PITPNs is given in [2] as a transition system $(\mathbb{N}^P \times \mathcal{L}(\Lambda), (M_0, K_0), \Rightarrow)$ on *state classes*, i.e., pairs $c = (M, D)$ consisting of a marking M and a constraint D over Λ. The firing of a transition leads to a new marking as in the concrete semantics, and also captures the new constraints induced by the time that has passed for the transition to fire. See [2] for details.

Rewrite Theories. A *rewrite theory* [31] is a tuple $\mathcal{R} = (\Sigma, E, L, R)$ where

- the signature Σ declares sorts, a subsort partial order, and function symbols;
- E is a set of equations of the form $t = t'$ **if** ψ, where t and t' are Σ-terms of the same sort, and ψ is a conjunction of equations;
- L is a set of *labels*; and
- R is a set of rewrite rules of the form $l : q \longrightarrow r$ **if** ψ, where $l \in L$ is a label, q and r are Σ-terms of the same sort, and ψ is a conjunction of equations.

$T_{\Sigma,s}$ denotes the set of ground (i.e., variable-free) terms of sort s, and $T_{\Sigma}(X)_s$ the set of terms of sort s over a set of variables X. $T_{\Sigma}(X)$ and T_{Σ} denote all terms and ground terms, respectively. A substitution $\sigma : X \to T_{\Sigma}(X)$ maps each variable to a term of the same sort, and $t\sigma$ denotes the term obtained by simultaneously replacing each variable x in a term t with $\sigma(x)$.

A *one-step rewrite* $t \longrightarrow_{\mathcal{R}} t'$ holds if there is a rule $l : q \longrightarrow r$ **if** ψ, a subterm u of t, and a substitution σ such that $u = q\sigma$ (modulo equations), t' is the term obtained from t by replacing u with $r\sigma$, and $v\sigma = v'\sigma$ holds for each $v = v'$ in ψ. We denote by $\longrightarrow_{\mathcal{R}}^*$ the reflexive-transitive closure of $\longrightarrow_{\mathcal{R}}$. A rewrite theory $\mathcal{R}$ is *topmost* iff there is a sort *State* at the top of one of the connected components of the subsort partial order such that for each rule, both sides have the top sort *State*, and no operator has sort *State* or any of its subsorts as an argument sort.

Rewriting with SMT [47]. For a signature Σ and equations E, a *built-in theory* $\mathcal{E}_0$ is a first-order theory with a signature $\Sigma_0 \subseteq \Sigma$, where (1) each sort s in Σ_0 is minimal in Σ; (2) $s \notin \Sigma_0$ for each operator $f : s_1 \times \cdots \times s_n \to s$ in $\Sigma \setminus \Sigma_0$; and (3) f has no other subsort-overloaded typing in Σ_0. The satisfiability of a constraint in $\mathcal{E}_0$ is assumed decidable using the SMT theory $T_{\mathcal{E}_0}$.

A *constrained term* is a pair $\phi \parallel t$ of a constraint ϕ in $\mathcal{E}_0$ and a term t in $T_{\Sigma}(X_0)$ over variables $X_0 \subseteq X$ of the built-in sorts in $\mathcal{E}_0$ [11,47]. A constrained term $\phi \parallel t$ *symbolically* represents all instances of the pattern t such that ϕ holds: $\llbracket \phi \parallel t \rrbracket = \{t' \mid t' = t\sigma \text{ (modulo } E) \text{ and } T_{\mathcal{E}_0} \models \phi\sigma \text{ for ground } \sigma : X_0 \to T_{\Sigma_0}\}$.

Let $\mathcal{R}$ be a topmost theory such that for each rule $l : q \longrightarrow r$ **if** ψ, extra variables not occurring in the left-hand side q are in X_0, and ψ is a constraint in a built-in theory $\mathcal{E}_0$. A *one-step symbolic rewrite* $\phi \parallel t \leadsto_{\mathcal{R}} \phi' \parallel t'$ holds iff there exist a rule $l : q \longrightarrow r$ **if** ψ and a substitution $\sigma : X \to T_{\Sigma}(X_0)$ such that (1) $t = q\sigma$ and $t' = r\sigma$ (modulo equations), (2) $T_{\mathcal{E}_0} \models (\phi \wedge \psi\sigma) \Leftrightarrow \phi'$, and (3) ϕ' is $T_{\mathcal{E}_0}$-satisfiable. We denote by $\leadsto_{\mathcal{R}}^*$ the reflexive-transitive closure of $\leadsto_{\mathcal{R}}$.

A *symbolic rewrite* on constrained terms symbolically represents a (possibly infinite) set of system transitions. If $\phi_t \parallel t \leadsto^* \phi_u \parallel u$ is a symbolic rewrite, then there exists a "concrete" rewrite $t' \longrightarrow^* u'$ with $t' \in \llbracket \phi_t \parallel t \rrbracket$ and $u' \in \llbracket \phi_u \parallel u \rrbracket$. Conversely, for any concrete rewrite $t' \longrightarrow^* u'$ with $t' \in \llbracket \phi_t \parallel t \rrbracket$, there exists a symbolic rewrite $\phi_t \parallel t \leadsto^* \phi_u \parallel u$ with $u' \in \llbracket \phi_u \parallel u \rrbracket$.

Maude. Maude [18] is a language and tool supporting the specification and analysis of rewrite theories. We summarize its syntax below:

```
sorts S ... Sk .            --- Declaration of sorts S1,..., Sk
subsort S1 < S2 .           --- Subsort relation
vars X1 ... Xm : S .        --- Logical variables of sort S
op f : S1 ... Sn -> S .     --- Operator S1 x ... x Sn -> S
ceq t = t' if c .           --- Conditional equation
crl [l] : q => r if c .     --- Conditional rewrite rule
```

Maude provides a number of analysis methods, including computing the normal form of a term t (`red t`), simulation by rewriting (`rew t`), and rewriting following a given strategy (`srew t using str`). Basic strategies include $r[\sigma]$ (apply rule r once with the optional ground substitution σ) and `all` (apply any of the rules once). Compound strategies include concatenation (α ; β), α `or-else` β (execute β if α fails), normalization α! (execute α until it cannot be further applied), etc.

Maude also offers explicit-state reachability analysis from a ground term t (`search [n,m] t =>* t'` such that Φ) and model checking an LTL formula F (`red modelCheck(t, F)`). For symbolic reachability analysis, the command

`smt-search [n, m]: t =>* t'` such that Φ `--- n and m are optional`

symbolically searches for n states, reachable from $t \in T_\Sigma(X_0)$ within m steps, that match the pattern $t' \in T_\Sigma(X)$ and satisfy the constraint Φ in $\mathcal{E}_0$.

Maude provides built-in sorts `Boolean`, `Integer`, and `Real` for the SMT theories of booleans, integers, and reals. Rational constants of sort `Real` are written n/m (e.g., `0/1`). Maude-SE [53] extends Maude with additional functionality for rewriting modulo SMT and bindings with different SMT solvers.

3   A Rewriting Logic Semantics for ITPNs

This section presents a rewriting logic semantics for (non-parametric) ITPNs, using a (non-executable) rewrite theory $\mathcal{R}_0$. We provide a bisimulation relating the concrete semantics of a net $\mathcal{N}$ and a rewrite relation in $\mathcal{R}_0$, and discuss variants of $\mathcal{R}_0$ to avoid consecutive tick steps and to enable time-bounded analysis.

3.1   Formalizing ITPNs in Maude: The Theory $\mathcal{R}_0$

We fix $\mathcal{N}$ to be the ITPN $\langle P, T, \emptyset, {}^\bullet(.), (.)^\bullet, {}^\circ(.), M_0, J, true\rangle$, and show how ITPNs and markings of such nets can be represented as Maude terms.

The usual approach is to represent a transition t_i and a place p_j as a constant of sort `Label` and `Place`, respectively (e.g., `ops p1 p2 ... pm : -> Place [ctor]`). To use a single rewrite theory $\mathcal{R}_0$ to define the semantics of all ITPNs, we instead assume that places and transition (labels) can be represented as strings; i.e., there is an injective naming function $\eta : P \cup T \to$ `String` which we usually do not mention explicitly.[1]

---

[1] We do not show variable declarations, but follow the convention that variables are written in (all) capital letters.

```
protecting STRING .    protecting RAT .
sorts Label Place .    --- identifiers for transitions and places
subsorts String < Label Place .    --- we use strings for simplicity
sorts Time TimeInf .   --- time values
subsort Zero PosRat < Time   < TimeInf .
op inf :  -> TimeInf [ctor] .
eq T <= inf = true .
```

The sort `TimeInf` adds an "infinity" value `inf` to the sort `Time` of time values, which are the non-negative rational numbers (`PosRat`).

The "standard" way of formalizing Petri nets in rewriting logic [31, 48] represents, e.g., a marking with two tokens in place p and three tokens in place q as the Maude term $p\ p\ q\ q\ q$. This is crucial to support *concurrent* firings of transitions in a net. Since the semantics of PITPNs is an *interleaving* semantics, to enable rewriting-with-SMT-based analysis from *parametric* initial markings, we instead represent markings as maps from places to the number of tokens in that place, so that the above marking is represented by the Maude term $\eta(p)\ |\text{-}>2$; $\eta(q)\ |\text{-}>3$ of sort `Marking`. The Maude term $\eta(t):$ *pre* -> *post* inhibit *inhibit* in *interval* represents a transition $t \in T$, where *pre*, *post*, and *inhibit* are markings representing, respectively, ${}^\bullet(t), (t)^\bullet, {}^\circ(t)$; and *interval* represents the interval $J(t)$. A `Net` is represented as a ;-separated set of such transitions:

```
sort Marking . --- Markings
op empty :  -> Marking [ctor] .
op _|->_ : Place Nat -> Marking [ctor] .
op _;_ : Marking Marking -> Marking [ctor assoc comm id: empty] .
sort Interval .  --- Time intervals (upper bound can be infinite)
op '[_:_'] : Time TimeInf -> Interval [ctor] .
sorts Net Transition .   subsort Transition < Net .
op _':_-->_inhibit_in_ :
  Label Marking Marking Marking Interval -> Transition [ctor] .
op emptyNet :  -> Net [ctor] .
op _;_ : Net Net -> Net [ctor assoc comm id: emptyNet] .
```

Example 1. Assuming the obvious naming function η mapping A to `"A"`, and so on, the Maude term `net3(a)` represents the net in Fig. 1:

```
op net3 : Time -> Net .
eq net3(T) =
    "t1" : "p5" |-> 1 --> "p1" |-> 1 inhibit empty in [2 : 6] ;
    "t2" : "p1" |-> 1 --> "p2" |-> 1 ; "p5" |-> 1 inhibit empty in [2 : 4] ;
    "t3" : "p2" |-> 1 ; "p4" |-> 1 --> "p3" |-> 1 inhibit empty in [T : T] ;
    "t4" : "p3" |-> 1 --> "p4" |-> 1 inhibit empty in [0 : 0] .
```

It is very easy to define operations +, -, and <= on markings (see [6]); we can then check whether a transition is active in a marking:

```
op active : Marking Transition -> Bool .  --- Active transition
eq active(M, L : PRE --> POST inhibit INHIBIT in INTERVAL) =
    (PRE <= M) and not inhibited(M, INHIBIT) .
op inhibited : Marking Marking -> Bool .  --- Inhibited transition
eq inhibited(M, empty) = false .
eq inhibited((P |-> N2) ; M, (P |-> N) ; INHIBIT) =
    ((N > 0) and (N2 >= N)) or inhibited(M, INHIBIT) .
```

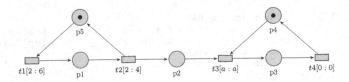

Fig. 1. A simple production-consumption system taken from [52]

Dynamics. We define the dynamics of ITPNs as a Maude "interpreter" for such nets. The definition of the semantics in [49] adjusts the "time intervals" of non-inhibited transitions when time elapses, but seems slightly "inconsistent": Time interval end-points should be non-negative, and only enabled transitions have intervals in the states; however, the definition of time and discrete transitions in [49] mentions $\forall t \in T, I'(t) = ...$ and $M \geq {}^\bullet(t) \implies I'(t)^\uparrow \geq 0$. Taking the definition of time and transition steps in [49] leads us to time intervals where the right end-points of disabled transitions could have arbitrarily large *negative* values. To have a simple and well-defined semantics, we use "clocks" instead of "decreasing intervals"; a clock denotes how long the corresponding transition has been enabled (but not inhibited). Our semantics is equivalent to the (natural interpretation of the) one in [49] in a way made precise in Theorem 1.

The sort `ClockValues` (see [6]) denotes sets of ;-separated terms $\eta(t) \rightarrow \tau$, where t is the (label of the) transition and τ represents the current value of t's "clock." The states in $\mathcal{R}_0$ are terms $m : clocks : net$ of sort `State`, where m represents the current marking, *clocks* the current values of the transition clocks, and *net* the representation of the Petri net:

```
sort State .    op _:_:_ : Marking ClockValues Net -> State [ctor] .
```

The following rewrite rule models the application of a transition L. Since `_;_` is associative and commutative, *any* transition L in the net can be applied:

```
crl [applyTransition] :
     M  :  (L -> T) ; CLOCKS  :
       (L : PRE --> POST inhibit INHIBIT in INTERVAL) ; NET
  => (M - PRE) + POST : L -> 0 ; updateClocks(CLOCKS, M - PRE, NET) :
       (L : PRE --> POST inhibit INHIBIT in INTERVAL) ; NET'
  if active(M, L : PRE --> POST inhibit INHIBIT in INTERVAL)
     and (T in INTERVAL) .
```

The transition L is active in the marking M and its clock value T is in the INTERVAL. After performing the transition, the marking is (M - PRE) + POST, the clock of L is reset and the other clocks are updated using the following function:

```
eq updateClocks((L' -> T') ; CLOCKS, INTERM-M,
                  (L' : PRE --> POST inhibit INHIBIT in INTERVAL) ; NET)
 = if PRE <= INTERM-M then (L' -> T') else (L' -> 0) fi ;
   updateClocks(CLOCKS, INTERM-M, NET) .
eq updateClocks(empty,  INTERM-M, NET) = empty .
```

The second rewrite rule in $\mathcal{R}_0$ specifies how time advances. Time can advance by *any* value T, as long as time does not advance beyond the time when an active transition must be taken. The clocks are updated according to the elapsed time T, except for those transitions that are disabled or inhibited:

```
crl [tick] : M : CLOCKS : NET => M : increaseClocks(M, CLOCKS, NET, T) : NET
    if  T <= mte(M, CLOCKS, NET) [nonexec] .
```

This rule is not executable (`[nonexec]`), since the variable `T`, which denotes how much time advances, only occurs in the right-hand side of the rule. `T` is therefore *not* assigned any value by the substitution matching the rule with the state being rewritten. This time advance `T` must be less or equal to the minimum of the upper bounds of the enabled transitions in the marking `M`:[2]

```
op mte : Marking ClockValues Net -> TimeInf .
eq mte(M, (L -> T) ; CLOCKS, (L : PRE --> POST ... in [T1 : inf]) ; NET)
 = mte(M, CLOCKS, NET) .
eq mte(M, (L -> T) ; CLOCKS, (L : PRE --> ... in [T1 : T2]) ; NET)
 = if active(M, L : PRE --> ...) then min(T2 - T, mte(M, CLOCKS, NET))
   else mte(M, CLOCKS, NET) fi .
eq mte(M, empty, NET) = inf .
```

The function `increaseClocks` increases the transitions clocks according to the elapsed time, except for those transitions that are disabled or inhibited:

```
op increaseClocks : Marking ClockValues Net Time -> ClockValues .
eq increaseClocks(M, (L -> T1) ; CLOCKS, (L : PRE --> ...) ; NET, T)
 = if active(M, L : PRE --> ...)
   then (L -> T1 + T) else (L -> T1) fi ; increaseClocks(M, CLOCKS,NET,T) .
eq increaseClocks(M, empty, NET, T) = empty .
```

The function $[\![\_]\!]_{\mathcal{R}_0}$ (see [6] for its formal definition) formalizes how markings and nets are represented as terms in rewriting logic.[3]

To show that $\mathcal{R}_0$ correctly simulates any ITPN $\mathcal{N}$, we provide a bisimulation relating behaviors from $a_0 = (M_0, J)$ in $\mathcal{N}$ with behaviors in $\mathcal{R}_0$ starting from initial state $[\![M_0]\!]_{\mathcal{R}_0} : \text{initClocks}([\![\mathcal{N}]\!]_{\mathcal{R}_0}) : [\![\mathcal{N}]\!]_{\mathcal{R}_0}$, where $\text{initClocks}(net)$ is a clock valuation which assigns the value 0 to each transition (label) $\eta(t)$ in net.

Since a transition in $\mathcal{N}$ consists of a delay followed by a discrete transition, we define a corresponding rewrite relation $\mapsto$ combining the `tick` and `applyTransition` rules, and prove (in [6]) the bisimulation for this relation. The following relation relates our clock-based states with the interval-based states:

Definition 3. *Let $\mathcal{N}$ be an ITPN and $\mathcal{S}_{\mathcal{N}} = (\mathcal{A}, a_0, \rightarrow)$ its concrete semantics. We define a relation $\approx \subseteq \mathcal{A} \times T_{\Sigma, State}$, relating states in the concrete semantics of $\mathcal{N}$ to states in $\mathcal{R}_0$, where for all states $(M, I) \in \mathcal{A}$, $(M, I) \approx m : clocks : net$ if and only if $m = [\![M]\!]_{\mathcal{R}_0}$ and $net = [\![\mathcal{N}]\!]_{\mathcal{R}_0}$ and for each transition $t \in T$,*

- *the value of $\eta(t)$ in clocks is 0 if t is not enabled in M;*
- *otherwise:*
  - *if $J(t)^{\uparrow} \neq \infty$ then the value of clock $\eta(t)$ in clocks is $J(t)^{\uparrow} - I(t)^{\uparrow}$;*
  - *otherwise, if $^{\uparrow}I(t) > 0$ then $\eta(t)$ has the value $^{\uparrow}J(t) - {}^{\uparrow}I(t)$ in clocks; otherwise, the value of $\eta(t)$ in clocks could be any value $\tau \geq {}^{\uparrow}J(t)$.*

---

[2] Parts of Maude specification will be replaced by '...' throughout the paper.
[3] $[\![\_]\!]_{\mathcal{R}_0}$ is parametrized by the naming function η, not shown explicitly here.

Theorem 1. $\approx$ *is a bisimulation between transition systems* $\mathcal{S}_\mathcal{N} = (\mathcal{A}, a_0, \rightarrow)$ *and* $(T_{\Sigma, state}, (\llbracket M_0 \rrbracket_{\mathcal{R}_0} : \text{initClocks}(\llbracket \mathcal{N} \rrbracket_{\mathcal{R}_0}) : \llbracket \mathcal{N} \rrbracket_{\mathcal{R}_0}), \mapsto)$.

3.2   Some Variations of $\mathcal{R}_0$

The theory $\mathcal{R}_1$ avoids consecutive applications of the tick rule by adding a new component—with value tickOk or tickNotOk—to the global state. The tick rule can only be applied when this component is tickOk. We add a new constructor `_:_:_:_` for these global states, a new sort TickState with values tickOk and tickNotOk, and add two rewrite rules:

```
sort TickState .        ops tickOk tickNotOk : -> TickState [ctor] .
op _:_:_:_ : TickState Marking ClockValues Net -> State [ctor] .
crl [applyTransition] :
  TS : M : ((L -> T) ; CLOCKS) : (L : PRE --> ...) ; NET) =>
  tickOk : ((M - PRE) + POST) :  ... if active(...) and (T in INTERVAL) .
crl [tick] : tickOk : M : ...  => tickNotOk : M : increaseClocks(...) ...
  if  T <= mte(M, CLOCKS, NET) [nonexec] .
```

We prove in [6] that $m : cs : net \longrightarrow^*_{\mathcal{R}_0} m' : cs' : net$ iff $\text{tickOk} : m : cs : net$ $\longrightarrow^*_{\mathcal{R}_1} \text{tickNotOk} : m' : cs' : net$. While reachability is preserved, a tick rule application in $\mathcal{R}_1$, where time does not advance far enough for a transition to be taken, could lead to a deadlock in $\mathcal{R}_1$ which cannot happen in $\mathcal{R}_0$.

The theory $\mathcal{R}_2$ adds a "global clock", denoting how much time has elapsed in the system, to answer questions such as whether a certain state can be reached in a certain time interval, and to enable time-bounded analysis where behaviors beyond the time bound are not explored. $\mathcal{R}_2$ adds the "global time," to the state:

```
op _:_:_:_@_ :  TickState Marking ClockValues Net Time -> State [ctor] .
```

The rewrite rules are modified as expected. For instance, the rule tick becomes:

```
crl [tick] :   tickOk    : M : CLOCKS : NET @ GT
         =>    tickNotOk : M : increaseClocks(..., T) : NET @ GT + T
if  T <= mte(M, CLOCKS, NET) [nonexec] .
```

where GT is a variable of sort Time. For a time bound Δ, we can add a conjunct GT + T <= Δ in the condition of this rule to stop executing beyond the time bound.

3.3   Explicit-state Analysis of ITPNs in Maude

The theories $\mathcal{R}_0$–$\mathcal{R}_2$ cannot be directly executed in Maude, since the tick rule introduces a new variable T in its right-hand side. Following the Real-Time Maude methodology, we can "sample" system execution at *some* time points, e.g., by changing the tick rule to increase time by *one time unit* in each application:

```
crl [tickOne] : M : CLOCKS : NET => M : increaseClocks(M, CLOCKS, NET, 1) : NET
              if  1 <= mte(M, CLOCKS, NET) .
```

Such time sampling analysis is in general not sound and complete, since it does not cover all possible system behaviors for dense time domains. Nevertheless, if all interval bounds are natural numbers, then "all behaviors" should be covered.

We can quickly experiment with different parameter values for our model, before applying the sound and complete symbolic methods developed in Sects. 4 and 5. Our report [6] describes a wealth of such analyses, including LTL model checking and time-bounded analysis. Here we just check whether the net in Fig. 1 is 1000-safe when $a = 5$, where the term init3 denotes the initial marking in Fig. 1. We define a function k-safe, where k-safe(n, m) holds iff the marking m does not have any place with more than n tokens:

```
op k-safe : Nat Marking -> Bool .
eq k-safe(N, empty) = true .
eq k-safe(N1, P |-> N2 ; M) = N2 <= N1 and k-safe(N1, M) .
```

We can then quickly check whether the net is 1000-safe when $a = 5$:

```
Maude> search [1] init3 : initClocks(net3(5)) : net3(5)  =>*
                 M : CLOCKS : NET such that not k-safe(1000, M) .
```

```
Solution 1 (state 83924)
M -->"p1" |-> 0 ; "p2" |-> 1001 ; "p3" |-> 0 ; "p4" |-> 1 ; "p5" |-> 1
```

The net is not 1000-safe: we reached a state with 1001 tokens in place p_2. Similar searches show that the net is 2-safe (but not 1-safe) if $a = 4$ and 1-safe if $a = 3$.

4 Parameters and Symbolic Executions

Standard explicit-state Maude analysis of $\mathcal{R}_0$–$\mathcal{R}_2$ cannot be used to analyze all behaviors of PITPNs for two reasons: **(1)** The rule tick introduces a new variable T in its right-hand side, reflecting that time can advance by *any* value T <= mte(...); and **(2)** analyzing nets with *uninitialized* parameters is impossible with explicit-state Maude analysis of concrete states. (For example, the condition T in INTERVAL in rule applyTransition does not evaluate to true if INTERVAL is not a *concrete* interval, and hence the rule will never be applied.) Maude-SE analysis of *symbolic* states with SMT variables can solve both issues, by symbolically representing the time advances T and the uninitialized parameters.

This section defines a rewrite theory $\mathcal{R}_1^{\mathbf{S}}$ that faithfully models PITPNs and that can be symbolically executed using Maude-SE. We prove that (concrete) executions in $\mathcal{R}_1$ are captured by (symbolic) executions in $\mathcal{R}_1^{\mathbf{S}}$, and vice versa. We also show that standard folding techniques [33] in rewriting modulo SMT are not sufficient for collapsing equivalent symbolic states in $\mathcal{R}_1^{\mathbf{S}}$. We therefore propose a new folding technique that guarantees termination of the reachability analyses of $\mathcal{R}_1^{\mathbf{S}}$ when the state-class graph of the encoded PITPN is finite.

4.1 The Symbolic Rewriting Logic Semantics

We define the "symbolic" semantics of PITPNs using the rewrite theory $\mathcal{R}_1^{\mathbf{S}}$, which is the symbolic counterpart of $\mathcal{R}_1$, instead of basing it on $\mathcal{R}_0$, since a symbolic "tick" step represents all possible tick steps from a symbolic state. We therefore do not introduce deadlocks not possible in the corresponding PITPN.

$\mathcal{R}_1^S$ is obtained from $\mathcal{R}_1$ by replacing the sort Nat in markings and the sort PosRat for clock values with the corresponding SMT sorts Integer and Real. (The former is only needed to enable reasoning with *symbolic* initial states where the number of tokens in a location is unknown). Conditions in rules (e.g., M1 <= M2) are replaced with the corresponding SMT expressions of sort Boolean. The symbolic execution of $\mathcal{R}_1^S$ in Maude-SE will accumulate and check the satisfiability of the constraints needed for a parametric transition to happen.

We start by declaring the sort Time as follows:

```
sorts Time TimeInf .  subsort Real < Time < TimeInf .
op inf : -> TimeInf [ctor] .
```

where Real is the sort for SMT reals (constraints in rewrite rules guarantee that only non-negative numbers are considered). Intervals are defined as in $\mathcal{R}_0$. Since Real is a subsort of Time, a parametric interval $[a, b]$ in a PITPN can be represented in $\mathcal{R}_1^S$ as the term [a:Real : b:Real], where a and b are variables of sort Real. The definition and operations on markings, nets, and clock values are similar to those in Sect. 3.1, albeit with the appropriate SMT sorts.

The rewrite rules in $\mathcal{R}_1^S$ act on symbolic states that may contain SMT variables. Although these rules are similar to those in $\mathcal{R}_1$, their symbolic execution is completely different. Maude-SE defines a theory transformation to implement symbolic rewriting. In the resulting theory $\widehat{\mathcal{R}_1^S}$, when a rule is applied, the variables occurring in the right-hand side but not in the left-hand side are replaced by fresh variables. Moreover, rules in $\widehat{\mathcal{R}_1^S}$ act on constrained terms of the form $\phi \parallel t$, where t in this case is a term of sort State and ϕ is a satisfiable SMT boolean expression. The constraint ϕ is obtained by accumulating the conditions in rules, thereby restricting the possible values of the variables in t.

The tick rewrite rule in $\mathcal{R}_1^S$ is

```
crl [tick] :  tickOk    : M : CLOCKS     : NET
           => tickNotOk : M : increaseClocks(M, CLOCKS, NET, T) : NET
if (T >= 0/1 and mte(M, CLOCKS, NET, T)) = true .
```

The variable T is restricted to be a non-negative real number and to satisfy the following *predicate* mte, which gathers the constraints to ensure that time cannot advance beyond the point in time when an enabled transition *must* fire:

```
op mte : Marking ClockValues Net Real -> Boolean .
eq mte(M, empty, NET, T) = true .
eq mte(M, (L -> R1) ; CLOCKS, (L : PRE --> ... in [T1 : inf]) ; NET, T)
 = mte(M, CLOCKS, NET, T) .
eq mte(M, (L -> R1) ; CLOCKS, (L : PRE --> ... in [T1 : T2]) ; NET, T)
 = (active(M, L : ...) ? T <= T2 - R1 : true) and mte(M, CLOCKS, NET, T) .
```

This means that, for every transition L, if the upper bound of the interval in L is inf, no restriction on T is added. Otherwise, if L is active at marking M, the SMT ternary operator C ? E1 : E2 (checking C to choose either E1 or E2) further constrains T to be less than T2 - R1. The definition of increaseClocks also uses this SMT operator to represent the new values of the clocks:

```
eq increaseClocks(M, (L -> R1) ; CLOCKS, (L : PRE --> ... ) ; NET, T)
```

```
= (L -> (active(M, L : PRE ...) ? R1 + T : R1 )) ;
  increaseClocks(M, CLOCKS, NET, T) .
```

The rule for applying a transition is defined as follows:

```
crl [applyTransition] :
    TS : M : ((L -> T) ; CLOCKS) : (L : PRE --> ...) ; NET)
=> tickOk : ((M - PRE) + POST) :  updateClocks(...) :
    (L : PRE --> ... ; NET) if (active(...) and (T in INTERVAL)) = true .
```

When applied, this rule adds new constraints asserting that the transition L can be fired (predicates active and _in_) and updates the state of the clocks:

```
eq updateClocks((L' -> R1) ; CLOCKS, INTERM-M, (L' : PRE --> ...); NET)
 = (L -> PRE <= INTERM-M ? R1 : 0/1) ; updateClocks(...) .
```

Example 2. Let *net* and m_0 be the Maude terms representing, respectively, the PITPN and the initial marking shown in Fig. 1. The term *net* includes a variable a:Real representing the parameter a. The command

```
smt-search tickOk : m₀ : initClocks(net) : net  =>*  TICK : M : CLOCKS : NET
    such that (a:Real >= 0/1 and not k-safe(1, M)) = true .
```

checks whether it is possible to reach a non-1-safe marking. Maude positively answers this question, with resulting accumulated constraint telling us that such a state is reachable (with 2 tokens in p_2) if a:Real >= 4/1.

Terms of sort Marking in $\mathcal{R}_1^{\mathbf{S}}$ may contain expressions with parameters (i.e., variables) of sort Integer. Let Λ_m denote the set of such parameters and $\pi_m : \Lambda_m \to \mathbb{N}$ a valuation function for them. We use m_s to denote a mapping from places to Integer expressions including parameter variables. Similarly, $clocks_s$ denotes a mapping from transitions to Real expressions (including variables). We write $\pi_m(m_s)$ to denote the ground term where the parameters in markings are replaced by the corresponding values $\pi_m(\lambda_i)$. Similarly for $\pi(clocks_s)$. We use $[\![\mathcal{N}]\!]_{\mathcal{R}_1^{\mathbf{S}}}$ to denotes the above rewriting logic representation of nets in $\mathcal{R}_1^{\mathbf{S}}$.

Recall that $t \in [\![\phi \parallel t_s]\!]$ is a ground instance, with a suitable ground substitution σ, of the constrained term $\phi \parallel t_s$. By construction, in $\mathcal{R}_1^{\mathbf{S}}$, if for all $t \in [\![\phi \parallel t_s]\!]$ all markings (sort Integer), clocks and parameters (Real) are non-negative numbers, then this is also the case for all reachable states from $\phi \parallel t_s$. Hence, there is a one-to-one correspondence for ground terms in $\mathcal{R}_1^{\mathbf{S}}$ satisfying that condition with terms in $\mathcal{R}_1$. We use $t \approx \in [\![\phi \parallel t_s]\!]$ to denote that there exists a $\mathcal{R}_1^{\mathbf{S}}$-term $t' \in [\![\phi \parallel t_s]\!]$ and t is its corresponding term in $\mathcal{R}_1$. Note that the ground substitution σ $(t' = t_s\sigma)$ determines a parameter (π) and a marking (π_m) valuation consistent with the constraint ϕ $(\mathcal{T}_{\mathcal{E}_0} \models \phi\sigma)$.

The following theorem states that the symbolic semantics matches all the behaviors resulting from a concrete execution of $\mathcal{R}_1$ with arbitrary parameter valuations π and π_m. Furthermore, for all symbolic executions with parameters, there exists a corresponding concrete execution where the parameters are instantiated with values consistent with the resulting accumulated constraint.

Theorem 2 (Soundness and Completeness). *Let $\mathcal{N}$ be a PITPN and ϕ be the constraint $\bigwedge_{J(t), t \in T}(0 \leq {}^{\uparrow}J(t) \leq J(t)^{\uparrow}) \wedge \bigwedge_{\lambda_i \in \Lambda_m}(0 \leq \lambda_i)$. (1) If $\phi \parallel t_s \leadsto_{\mathcal{R}_1^{\mathbf{S}}}^*$*

$\phi' \parallel t'_s$ then, there exist t' and $t \approx \in [\![\phi \parallel t_s]\!]$ (and the corresponding valuations π and π_m) such that $t \longrightarrow^*_{\mathcal{R}_1} t'$ and $t' \approx \in [\![\phi' \parallel t'_s]\!]$.

(2) If $t \longrightarrow^*_{\mathcal{R}_1} t'$ with $t \approx \in [\![\phi \parallel t_s]\!]$, then there exists $\phi' \parallel t'_s$ such that $t' \approx \in [\![\phi' \parallel t'_s]\!]$ and $\phi \parallel t_s \leadsto^*_{\mathcal{R}^S_1} \phi' \parallel t'_s$.

The symbolic counterpart $\mathcal{R}^S_2$ of the theory $\mathcal{R}_2$ can be defined similarly.

4.2 A New Folding Method for Symbolic Reachability

Reachability analysis should terminate for both positive and negative queries for nets with finite parametric state-class graphs. However, this is not the case in analysis with $\mathcal{R}^S_1$: the symbolic state space generated by smt-search is infinite even for such nets. The problem is that smt-search stops exploring from a symbolic state only if it has already visited the *same* state. Moreover, due to the fresh variables created in $\mathcal{R}^S_1$, symbolic states representing the same set of concrete states are not the same, even though they are *logically* equivalent. For instance, if we use smt-search to try to show that the PITPN in Fig. 1 is 1-safe if $0 \le a < 4$, such a command does not terminate. In fact, the command

```
smt-search tickOk : m_0 : 0-clock(net) : net  =>*  TICK : M : CLOCKS : NET
such that (a:Real >= 0/1 and a:Real < 4 and M <= m_0 and m_0 <= M) = true.
```

searching for reachable states where $M = m_0$ will produce infinitely many equivalent solutions, where the state of the system is represented by different (new) variables but subject to equivalent constraints.

The usual approach for collapsing equivalent symbolic states in rewriting modulo SMT is subsumption [33]. Essentially, we stop searching from a symbolic state if, during the search, we have already encountered another state that subsumes ("contains") it. Let $U = \phi_u \parallel t_u$ and $V = \phi_v \parallel t_v$ be constrained terms. Then $U \sqsubseteq V$ if there is a substitution σ such that $t_u = t_v\sigma$ and the implication $\phi_u \Rightarrow \phi_v\sigma$ holds. In that case, $[\![U]\!] \subseteq [\![V]\!]$ and U does not need to be further explored if V has already been encountered.

Reachability analysis with folding is sound [7] but not necessarily complete (since $[\![U]\!] \subseteq [\![V]\!]$ does not imply $U \sqsubseteq V$) [33]. In fact, if we take two solutions U and V from the above smt-search command and use the Maude's command match to find the needed substitution σ, the SMT solver determines that the formula $\neg(\phi_u \Rightarrow \phi_v\sigma)$ is satisfiable (and therefore $\phi_u \Rightarrow \phi_v\sigma$ is not valid). Hence, a procedure based on checking this implication will fail to determine that $U \sqsubseteq V$.

The satisfiability witnesses of $\neg(\phi_u \Rightarrow \phi_v\sigma)$ show that the values for markings and clocks in the *current* time instant are equally constrained in ϕ_u and ϕ_v (and hence, they represent the same set of concrete states). However, since the variables representing the *current* state are different, the implication is falsifiable.

In the following, we propose a subsumption relation that solves the aforementioned problems. Let $\phi \parallel t$ be a constrained term where t is a term of sort State. Consider an *abstraction of built-ins* (t°, σ°) for t [47], where t° is as t but it replaces the expression e_i in markings $(p_i \mapsto e_i)$ and clocks $(l_i \to e_i)$ with new fresh variables. The substitution σ° is defined accordingly in such a way that

$t = t^\circ \sigma^\circ$. Let $\Psi_{\sigma^\circ} = \bigwedge_{x \in dom(\sigma^\circ)} x = x\sigma^\circ$, where $dom(\sigma) = \{x \in X \mid \sigma(x) \neq x\}$. We use $(\phi \parallel t) \Downarrow_{\texttt{now}}$ to denote the constrained term $\phi \wedge \Psi_{\sigma^\circ} \parallel t^\circ$. Intuitively, $(\phi \parallel t) \Downarrow_{\texttt{now}}$ replaces the clock values and markings with fresh variables, and the boolean expression Ψ_{σ° constrains those variables to take the values of clocks and the marking in t. From [47] we can show that $[\![\phi \parallel t]\!] = [\![(\phi \parallel t) \Downarrow_{\texttt{now}}]\!]$.

Note that the only variables occurring in $(\phi \parallel t) \Downarrow_{\texttt{now}}$ are those for parameters (if any) and the fresh variables in $dom(\sigma^\circ)$ (representing the symbolic state of clocks and markings). For a constrained term $\phi \parallel t$, we use $\exists(\phi \parallel t)$ to denote the formula $(\exists X)\phi$ where $X = vars(\phi) \setminus vars(t)$.

Definition 4 (Relation $\preceq$). *Let $U = \phi_u \parallel t_u$ and $V = \phi_v \parallel t_v$ be constrained terms where t_u and t_v are terms of sort* State*. Moreover, let $U \Downarrow_{\texttt{now}} = \phi'_u \parallel t'_u$ and $V \Downarrow_{\texttt{now}} = \phi'_v \parallel t'_v$, where $vars(t'_u) \cap vars(t'_v) = \emptyset$. We define the relation $\preceq$ on constrained terms so that $U \preceq V$ whenever there exists a substitution σ such that $t'_u = t'_v \sigma$ and the formula $\exists(U \Downarrow_{\texttt{now}}) \Rightarrow \exists(V \Downarrow_{\texttt{now}})\sigma$ is valid.*

The formula $\exists(U \Downarrow_{\texttt{now}})$ hides the information about all the tick variables as well as the information about the clocks and markings in previous time instants. What we obtain is the information about the parameters, clocks and markings "now". Moreover, if t_u and t_v above are both `tickOk` states (or both `tickNotOk` states), and they represent two symbolic states of the same PITPN, then t'_u and t'_v always match (σ being the identity on the variables representing parameters and mapping the corresponding variables created in $V \Downarrow_{\texttt{now}}$ and $U \Downarrow_{\texttt{now}}$).

Theorem 3 (Soundness and Completeness) *Let U and V be constrained terms for two symbolic states of the same PITPN. Then, $[\![U]\!] \subseteq [\![V]\!]$ iff $U \preceq V$.*

We have implemented a new symbolic reachability analysis using the folding relation in Definition 4. Building on the theory transformation [47] implemented in Maude-SE, we transform $\mathcal{R}_1^S$ into a rewrite theory $\mathcal{R}_1^{fS}$ that rewrites terms of the form $S : \phi \parallel t$ where S is a set of constrained terms (the already visited states). Theory $\mathcal{R}_1^{fS}$ defines an operator $\texttt{subsumed}(\phi \parallel t , S)$ that reduces to true—by a call to the SMT solver Z3 for quantifier elimination and satisfiability checking—iff there exists $\phi' \parallel t' \in S$ s.t $\phi \parallel t \preceq \phi' \parallel t'$. Rules in $\mathcal{R}_1^S$ are systematically transformed to add a further constraint: the new state on the right-hand side of the rule is not subsumed by a state in the set S.

In $\mathcal{R}_1^{fS}$, for an initial constraint ϕ on the parameters, the Maude command $\texttt{search } [n,m] \texttt{ empty:} \phi \parallel t \texttt{ =>* } S : \phi' \parallel t' \texttt{ such that smtCheck}(\phi' \texttt{ and } \Phi)$ d answers the question whether it is possible to reach a symbolic state that matches t' and satisfies the condition Φ. In the following, we use $\texttt{init}(net, m_0, \phi)$ to denote the term $\texttt{empty :} \phi \parallel \texttt{tickOk :} m_0 \texttt{ :initClocks}(net) : net$.

Example 3. Consider the PITPN in Fig. 1. Let m_0 be the marking in the figure and $\phi = 0 \leq a < 4$. The command

```
search init(net, m₀, φ) =>* S : φ' ‖ ( TICK : M : CLOCKS : NET )
   such that smtCheck(φ' and not k-safe(1,M)) .
```

terminates returning No solution, showing that the net is 1-safe if $0 \leq a < 4$.

If the set of reachable state classes in the symbolic semantics of $\mathcal{N}$ in [2] is finite, then so is the set of reachable symbolic states with the new folding method:

Corollary 1. *For any PITPN $\mathcal{N}$ and state class (M, D), if the transition system $(\mathcal{C}, (M, D), \Rightarrow)$ is finite, then so is $\left(T_{\Sigma, State}, init(\mathcal{N}, M, D), \rightsquigarrow_{\mathcal{R}_1^{fs}} \right)$.*

The new folding relation is applicable to any rewrite theory $\mathcal{R}$ that satisfies the requirements for rewriting with SMT [47], briefly explained in Sect. 2.

5   Parameter Synthesis and Symbolic Model Checking

This section shows how Maude-SE can be used for solving parameter synthesis problems, model checking the non-nested timed temporal logic properties supported by Roméo (in addition to LTL model checking), reasoning with *parametric* initial states, and analyzing nets with user-defined execution strategies.

5.1   Parameter Synthesis

A *state predicate* is a boolean expression on markings (e.g., k-safe$(1, m)$) and clocks (e.g., $c_1 < c_2$). EF-*synthesis* (resp. *safety synthesis* (AG$\neg\phi$)) is the problem of computing parameter values π such that *some* (resp. *no*) run of $\pi(\mathcal{N})$ reaches a state satisfying a given state predicate ϕ.

search in $\mathcal{R}_1^{fs}$ provides semi-decision procedures for solving these parameter synthesis problems (which are undecidable in general). As illustrated below, the resulting constraint computed by search can be used to synthesize the parameter values that allow such behaviors. The safety synthesis problem AG$\neg\phi$ can be solved by finding all solutions for EFϕ and negating the resulting constraint.

Example 4. The following command solves the EF-synthesis problem of finding values for a in Fig. 2 such that the net is not 1-safe, where $\phi = 0 \leq a$:

```
search [1] init(net, m₀, φ) =>* S : PHI' || ( TICK : M : CLOCKS : NET )
   such that smtCheck(PHI' and not k-safe(1,M)) .
```

It returns one solution, and the resulting constraint ϕ', instantiating the pattern PHI', can be used to extract the parameter values as follows. Let X be the set of SMT variables in ϕ' *not* representing parameters. A call to the quantifier elimination procedure (qe) of the SMT solver Z3 on the formula $\exists X.\phi'$ reduces to a:Real >= 4/1, giving us the desired values for the parameter a.

To solve the safety synthesis problem AG$\neg\phi$, we have used Maude's meta-programming facilities [18] to implement a command safety-syn(net, m_0, ϕ_0, ϕ) where m_0 is a marking, ϕ_0 a constraint on the parameters and ϕ a constraint involving the variables M and CLOCKS as in the search command in Example 4. This command iteratively calls search to find a state reachable from m_0, with initial constraint ϕ_0, where ϕ does not hold. If such state is found, with accumulated constraint ϕ', the search command is invoked again with initial constraint $\phi_0 \wedge \neg\phi'$. This process stops when no more reachable states where ϕ does not hold are found, thus solving the $\mathsf{AG}\neg\phi$ synthesis problem.

Example 5. The following command synthesizes the values of the parameter a, so that $30 \leq a \leq 70$, that make the scheduling system in [6,50] 1-safe:

```
safety-syn(net, m₀, a:Real >= 30/1 and a:Real <= 70/1, k-safe(1,M)) .
```

The first counterexample found assumes that $a \leq 48$. If $a > 48$, search does not find any counterexample. This is the same answer that Roméo found.

Since we can have Integer variables in initial markings, we can use Maude-SE to synthesize the initial markings that, e.g., make the net k-safe or alive:

Example 6. Consider a *parametric* initial marking m_s for the net in Fig. 1, with parameters x_1, x_2, and x_3 denoting the number of tokens in places p_1, p_2, and p_3, respectively, and the initial constraint ϕ_0 stating that $a \geq 0$ and $0 \leq x_i \leq 1$. The execution of the command execution of the command safety-syn($net, m_s, \phi_0, \texttt{k-safe(1,M)}$) determines that the net is 1-safe when $x_1 = x_3 = 0$ and $0 \leq x_2 \leq 1$.

Analysis with Strategies. Maude's strategy facilities [17] allow us to analyze PITPNs whose executions follow some user-defined strategy:

Example 7. We execute the net in Fig. 1 with the following strategy t3-first: whenever transition t_3 and some other transition are enabled at the same time, then t_3 fires first. The following strategy definition (sd) specifies this strategy:

```
sd t3-first := ( applyTransition[L <- "t3"] or-else all )!
```

Running srew init($net, m_0, a \geq 0$) using t3-first in $\mathcal{R}_1^{fS}$ finds all symbolic states reachable with this strategy, and all of them are 1-safe. Therefore, all parameter values $a \geq 0$ guarantee the desired property with this execution strategy.

5.2 Analyzing Temporal Properties

This section shows how Maude-SE can be used to analyze the temporal properties supported by Roméo [29], albeit in a few cases without parametric bounds in the temporal formulas. Roméo can analyze the following temporal properties:

$$\mathbf{Q}\,\phi\,\mathsf{U}_J\,\psi \mid \mathbf{Q}\mathsf{F}_J\,\phi \mid \mathbf{Q}\mathsf{G}_J\,\phi \mid \phi \rightsquigarrow_{\leq b}\psi$$

where $\mathbf{Q} \in \{\exists, \forall\}$, ϕ and ψ are *state predicates* on *markings*, and J is a time interval $[a, b]$, where a and/or b can be parameters and b can be ∞. For example, $\forall\mathsf{F}_{[a,b]}\,\phi$ says that in *each* path from the initial state, a marking satisfying ϕ is reachable in some time in $[a, b]$. The bounded response $\phi \rightsquigarrow_{\leq b} \psi$ says that each ϕ-marking *must* be followed by a ψ-marking within time b.

Since queries include time bounds, we use $\mathcal{R}_2^{fS}$, and init(net, m_0, ϕ) will denote the term empty : $\phi \parallel$ tickOk : m_0 : initClocks(net) : net @ 0/1.

State predicates, including inequalities on markings and clocks, and also a test whether the global clock is in a given interval are defined as follows:

```
ops _>=_ _>_ _<_ _<=_ _==_ : Place Integer -> Prop .
ops _>=_ _>_ _<_ _<=_ _==_ : Clock Real -> Prop .
op in-time : Interval -> Prop .
eq S : C || (TICK : M ; (P |-> N1) : CLOCKS : NET) @ G-CLOCK  |=  P >= N1'
 = smtCheck(C and N1 >= N1' ) . --- similarly for >, <=, < and ==
eq S : C || (TICK : M : CLOCKS : NET) @ G-CLOCK  |=  in-time INTERVAL
 = smtCheck(C and (G-CLOCK in INTERVAL )) .
```

Atomic propositions (Prop) are evaluated (|=) on symbolic states represented as constrained terms $S : \phi \parallel t$. Since they may contain variables, a call to the SMT solver (smtCheck) is needed to determine whether ϕ entails the proposition.

Some of the temporal formulas supported by Roméo can be easily verified using the reachability commands presented in the previous section. The property $\exists F_{[a,b]} \, \psi$ can be verified using the command:

```
search [1] init(net, m0, φ) =>* S : PHI' || TICK : M : CLOCKS : NET @ G-CLOCK
    such that (STATE' |= ψ) and G-CLOCK in [a : b] .
```

where ϕ states that all parameters are non-negative numbers and $STATE'$ is the expression to the right of =>*. a and b can be variables representing parameters to be synthesized; and ψ can be an expression involving CLOCKS. For example,

```
search [1] init(net, m0, φ) =>*
 S' : PHI' || TICK : (M ; "p1" |-> P1) : (CLOCKS ; "t2" -> C2) : NET @ G-CLOCK
 such that (STATE' |= P1 > 1 /\ C2 < 2/1) and G-CLOCK in [a : b] .
```

checks whether it is possible to reach a marking, in some time in $[a, b]$, with more than one token in place p_1, when the value of the clock of transition t_2 is < 2.

The dual property $\forall G_{[a,b]} \, \phi$ can be checked by analyzing $\exists F_{[a,b]} \, \neg \phi$.

Example 8. Consider the PITPN in Example 5 with (interval) parameter $\phi = 30 \le a \le 70$. The property $\exists F_{[b,b]}(\neg 1\text{-}safe)$ can be verified with the following command, which determines that the parameter b satisfies $60 \le b \le 96$.

```
search [1] init(net, m0, φ) =>* S : PHI' || TICK : M : CLOCKS : NET @ G-CLOCK
    such that STATE' |= b:Real >= 0/1 and (G-CLOCK in [b:Real : b:Real])
                      and not (k-safe(1,M)) .
```

$\phi \rightsquigarrow_{\le b} \psi$ can be verified using a simple theory transformation on $\mathcal{R}_0^S$ followed by reachability analysis. The theory transformation adds a new "clock," which is either noClock or clock(τ), to the state. The latter represents the time (τ) since a ϕ-state was visited without having been followed by a ψ-state. The applyTransition rule is modified as follows: when the clock is noClock and the new marking satisfies $\phi \wedge \neg\psi$, this clock is set to clock(0), and when a ψ-marking is reached, the clock is set to noClock. The tick rule updates clock(T1) to clock(T1 + T) and leaves noClock unchanged. $\phi \rightsquigarrow_{\le b} \psi$ can be checked by searching for a "bad" state with "clock" clock(T) where $T > b$. See [6] for details.

Reachability analysis cannot be used to analyze the other properties supported by Roméo ($Q \, \phi U_J \, \psi$, and $\forall F_J \, \phi$ and its dual $\exists G_J \, \phi$). We combine Maude's explicit-state model checker and SMT solving to solve these (and other) queries.

The timed temporal operators can be defined on top of the (untimed) LTL temporal operators in Maude (<>, [] and U) :

```
ops <_>_ [_]_ : Interval Prop -> Formula .  --- F_J φ and G_J φ
op _U__ : Prop Interval Prop -> Formula .   --- φU_j ψ
eq < INTERVAL > PR1 = <> (PR1 /\ in-time INTERVAL) .
eq [ INTERVAL ] PR1 = ~ (< INTERVAL > (~ PR1)) .
eq PR1 U INTERVAL PR2 = PR1 U (PR2 /\ in-time INTERVAL) .
```

For this fragment of non-nested timed temporal logic formulas, universally quantified properties can be model checked directly by Maude; for $\exists \Phi$ it is enough to model check $\neg \Phi$: any counterexample to this is a witness for $\exists \Phi$, and vice versa.

6 Benchmarking

We have compared the performance of Maude-with-SMT analysis with that of Roméo on three case studies: the producer-consumer [52] system in Fig. 1, the scheduling system in [50], and the tutorial system taken from the Roméo website. We modified tutorial to produce two tokens in the loop-back, which leads to infinite behaviors. We compared the performance of solving the synthesis problem $EF(p > n)$ (place p holds more than n tokens), for different p and n, and of checking whether the net is 1-safe. In each experiment, Maude was executed with two different SMT solvers: Yices and Z3. The benchmarking data are available in the repository [5] and in the technical report [6].

The results show that using Maude with Yices is faster than using it with Z3. For negative queries, as expected, $\mathcal{R}_0^S$ and $\mathcal{R}_1^S$ time out (set to 10 minutes), while $\mathcal{R}_1^{fS}$ (which uses folding) completes the analysis before the timeout.

Maude-SE outperforms Roméo in some reachability queries, and sometimes our analysis terminates when Roméo does not, which may happen when the search order leads Roméo to explore an infinite branch with an unbounded marking.

7 Related Work

Tool Support for Parametric Time Petri Nets. We are not aware of any tool for analyzing parametric time(d) Petri nets other than Roméo [29].

Petri Nets in Rewriting Logic. Formalizing Petri nets algebraically [34] partly inspired rewriting logic. Different kinds of Petri nets are given a rewriting logic semantics in [48], and in [40] for timed nets. In contrast to our paper, these papers focus on the semantics of such nets, and do not consider execution and analysis (or inhibitor arcs or parameters). Capra [14,15], Padberg and Schultz [45], and Barbosa et al. [12] use Maude to formalize dynamically reconfigurable Petri nets and I/O Petri nets. In contrast to our work, these papers target untimed and non-parametric nets, and do not focus on formal analysis, but only show examples of standard (explicit-state) search and LTL model checking.

Symbolic Methods for Real-Time Systems in Maude. We develop symbolic analysis methods for parametric time automata (PTA) in [4]. The differences with the current paper include: PTAs are very simple structures compared to PITPNs (with inhibitor arcs, no bounds on the number of tokens in a state), so the semantics of PITPNs is more sophisticated than the one for PTAs, which does not use "structured" states, equations, or user-defined functions; defining a new rewrite theory for each PTA in [4] compared to having a single rewrite theory for all nets in this work; obtaining desired symbolic reachability properties using "standard" folding methods for PTAs compared to having to develop a new folding mechanism for PITPNs; analysis in [4] does not include temporal logic model checking; and so on. In addition, a number of real-time systems have been formally analyzed using rewriting with SMT, including PLC ST programs [26], virtually synchronous cyber-physical systems [23–25], and soft agents [35]. These papers differ from our work in that they use guarded terms [10, 11] for state-space reduction instead of folding, and do not consider parameter synthesis problems.

8 Concluding Remarks

We have provided a "concrete" rewriting logic semantics for PITPNs, and proved that this semantics is bisimilar to the semantics of such nets in [49]. We then *systematically transformed* this non-executable "Real-Time Maude-style" model into a "symbolic" rewrite model which is amenable to sound and complete symbolic analysis for dense-time systems using Maude combined with SMT solving.

We have shown how almost all analysis and parameter synthesis supported by the PITPN tool Roméo can be done using Maude-with-SMT. We have also shown how Maude-with-SMT can provide additional capabilities for PITPNs, including synthesizing initial markings (and not just firing bounds) from *parametric* initial markings so that desired properties are satisfied, full LTL model checking, and analysis with user-defined execution strategies. We developed a new "folding" method for symbolic states, so that symbolic reachability analysis using Maude-with-SMT terminates whenever the corresponding Roméo analysis terminates.

Our benchmarking shows that our symbolic methods using Maude combined with the SMT solver Yices in some cases outperforms Roméo, whereas Maude with Z3 is significantly slower.

This paper has not only provided new features for PITPNs. It has also shown that even a model like our Real-Time Maude-inspired PITPN interpreter—with functions, equations, and unbounded markings—can easily be turned into a symbolic rewrite theory for which Maude-with-SMT provides very useful sound and complete analyses even for dense-time systems.

Acknowledgments. We thank the anonymous reviewers for their insightful comments. Arias, Olarte, Ölveczky, Petrucci, and Rømming acknowledge support from CNRS INS2I project ESPRiTS and the PHC project Aurora AESIR. Bae was supported by the NRF grants funded by the Korea government (No. 2021R1A5A1021944 and No. 2022R1F1A1074550).

References

1. AlTurki, M., Dhurjati, D., Yu, D., Chander, A., Inamura, H.: Formal specification and analysis of timing properties in software systems. In: Chechik, M., Wirsing, M. (eds.) FASE 2009. LNCS, vol. 5503, pp. 262–277. Springer, Heidelberg (2009). https://doi.org/10.1007/978-3-642-00593-0_18
2. André, É., Pellegrino, G., Petrucci, L.: Precise robustness analysis of time Petri nets with inhibitor arcs. In: Braberman, V., Fribourg, L. (eds.) FORMATS 2013. LNCS, vol. 8053, pp. 1–15. Springer, Heidelberg (2013). https://doi.org/10.1007/978-3-642-40229-6_1
3. Andreychenko, A., Magnin, M., Inoue, K.: Analyzing resilience properties in oscillatory biological systems using parametric model checking. Biosystems **149**, 50–58 (2016)
4. Arias, J., Bae, K., Olarte, C., Ölveczky, P.C., Petrucci, L., Rømming, F.: Rewriting logic semantics and symbolic analysis for parametric timed automata. In: Proceedings of the 8th ACM SIGPLAN International Workshop on Formal Techniques for Safety-Critical Systems (FTSCS 2022), pp. 3–15. ACM (2022)
5. Arias, J., Bae, K., Olarte, C., Ölveczky, P.C., Petrucci, L., Rømming, F.: PITPN2Maude (2023). https://depot.lipn.univ-paris13.fr/arias/pitpn2maude
6. Arias, J., Bae, K., Olarte, C., Ölveczky, P.C., Petrucci, L., Rømming, F.: Symbolic analysis and parameter synthesis for time Petri nets using Maude and SMT solving (2023). https://doi.org/10.48550/ARXIV.2303.08929
7. Bae, K., Escobar, S., Meseguer, J.: Abstract logical model checking of infinite-state systems using narrowing. In: Rewriting Techniques and Applications (RTA 2013). LIPIcs, vol. 21, pp. 81–96. Schloss Dagstuhl - Leibniz-Zentrum für Informatik (2013)
8. Bae, K., Krisiloff, J., Meseguer, J., Ölveczky, P.C.: Designing and verifying distributed cyber-physical systems using Multirate PALS: an airplane turning control system case study. Sci. Comput. Program. **103**, 13–50 (2015). https://doi.org/10.1016/j.scico.2014.09.011
9. Bae, K., Ölveczky, P.C., Feng, T.H., Lee, E.A., Tripakis, S.: Verifying hierarchical Ptolemy II discrete-event models using Real-Time Maude. Sci. Comput. Program. **77**(12), 1235–1271 (2012)
10. Bae, K., Rocha, C.: Guarded terms for rewriting modulo SMT. In: Proença, J., Lumpe, M. (eds.) FACS 2017. LNCS, vol. 10487, pp. 78–97. Springer, Cham (2017). https://doi.org/10.1007/978-3-319-68034-7_5
11. Bae, K., Rocha, C.: Symbolic state space reduction with guarded terms for rewriting modulo SMT. Sci. Comput. Program. **178**, 20–42 (2019)
12. Barbosa, P., et al.: SysVeritas: a framework for verifying IOPT nets and execution semantics within embedded systems design. In: Camarinha-Matos, L.M. (ed.) DoCEIS 2011. IAICT, vol. 349, pp. 256–265. Springer, Heidelberg (2011). https://doi.org/10.1007/978-3-642-19170-1_28
13. Bobba, R., et al.: Survivability: design, formal modeling, and validation of cloud storage systems using Maude. In: Assured Cloud Computing, Chap. 2, pp. 10–48. Wiley (2018)
14. Capra, L.: Canonization of reconfigurable PT nets in Maude. In: Lin, A.W., Zetzsche, G., Potapov, I. (eds.) Reachability Problems. RP 2022. LNCS, vol. 13608, pp. 160–177. Springer, Cham (2022). https://doi.org/10.1007/978-3-031-19135-0_11

15. Capra, L.: Rewriting logic and Petri nets: a natural model for reconfigurable distributed systems. In: Bapi, R., Kulkarni, S., Mohalik, S., Peri, S. (eds.) ICDCIT 2022. LNCS, vol. 13145, pp. 140–156. Springer, Cham (2022). https://doi.org/10.1007/978-3-030-94876-4_9

16. Clarke, E.M., Grumberg, O., Peled, D.A.: Model Checking. MIT Press, Amsterdam/Cambridge (2001)

17. Clavel, M., et al.: Maude Manual (Version 3.2.1). SRI International (2022). http://maude.cs.illinois.edu

18. Clavel, M., et al.: All About Maude - A High-Performance Logical Framework. LNCS, vol. 4350. Springer, Heidelberg (2007). https://doi.org/10.1007/978-3-540-71999-1

19. Coullon, H., Jard, C., Lime, D.: Integrated model-checking for the design of safe and efficient distributed software commissioning. In: Ahrendt, W., Tapia Tarifa, S.L. (eds.) IFM 2019. LNCS, vol. 11918, pp. 120–137. Springer, Cham (2019). https://doi.org/10.1007/978-3-030-34968-4_7

20. Grabiec, B., Traonouez, L.-M., Jard, C., Lime, D., Roux, O.H.: Diagnosis using unfoldings of parametric time Petri nets. In: Chatterjee, K., Henzinger, T.A. (eds.) FORMATS 2010. LNCS, vol. 6246, pp. 137–151. Springer, Heidelberg (2010). https://doi.org/10.1007/978-3-642-15297-9_12

21. Grov, J., Ölveczky, P.C.: Formal modeling and analysis of Google's Megastore in Real-Time Maude. In: Iida, S., Meseguer, J., Ogata, K. (eds.) Specification, Algebra, and Software. LNCS, vol. 8373, pp. 494–519. Springer, Heidelberg (2014). https://doi.org/10.1007/978-3-642-54624-2_25

22. Jensen, K., Kristensen, L.M.: Coloured Petri Nets - Modelling and Validation of Concurrent Systems. Springer, Heidelberg (2009). https://doi.org/10.1007/b95112

23. Lee, J., Bae, K., Ölveczky, P.C.: An extension of HybridSynchAADL and its application to collaborating autonomous UAVs. In: Margaria, T., Steffen, B. (eds.) Leveraging Applications of Formal Methods, Verification and Validation. Adaptation and Learning (ISoLA 2022). LNCS, vol. 13703, pp. 47–64. Springer, Cham (2022). https://doi.org/10.1007/978-3-031-19759-8_4

24. Lee, J., Bae, K., Ölveczky, P.C., Kim, S., Kang, M.: Modeling and formal analysis of virtually synchronous cyber-physical systems in AADL. Int. J. Software Tools Technol. Transf. **24**(6), 911–948 (2022)

25. Lee, J., Kim, S., Bae, K., Ölveczky, P.C.: HYBRIDSYNCHAADL: modeling and formal analysis of virtually synchronous CPSs in AADL. In: Silva, A., Leino, K.R.M. (eds.) CAV 2021. LNCS, vol. 12759, pp. 491–504. Springer, Cham (2021). https://doi.org/10.1007/978-3-030-81685-8_23

26. Lee, J., Kim, S., Bae, K.: Bounded model checking of PLC ST programs using rewriting modulo SMT. In: Proceedings of the 8th ACM SIGPLAN International Workshop on Formal Techniques for Safety-Critical Systems (FTSCS 2022), pp. 56–67. ACM (2022)

27. Lien, E., Ölveczky, P.C.: Formal modeling and analysis of an IETF multicast protocol. In: Seventh IEEE International Conference on Software Engineering and Formal Methods (SEFM 2009), pp. 273–282. IEEE Computer Society (2009)

28. Lime, D., Roux, O.H., Seidner, C.: Cost problems for parametric time Petri nets. Fundam. Informaticae **183**(1-2), 97–123 (2021). https://doi.org/10.3233/FI-2021-2083

29. Lime, D., Roux, O.H., Seidner, C., Traonouez, L.-M.: Romeo: a parametric model-checker for Petri nets with stopwatches. In: Kowalewski, S., Philippou, A. (eds.) TACAS 2009. LNCS, vol. 5505, pp. 54–57. Springer, Heidelberg (2009). https://doi.org/10.1007/978-3-642-00768-2_6

30. Merlin, P.M.: A study of the recoverability of computing systems. Ph.D. thesis, University of California, Irvine, CA, USA (1974)
31. Meseguer, J.: Conditional rewriting logic as a unified model of concurrency. Theor. Comput. Sci. **96**(1), 73–155 (1992)
32. Meseguer, J.: Twenty years of rewriting logic. J. Log. Algebraic Methods Program. **81**(7–8), 721–781 (2012)
33. Meseguer, J.: Generalized rewrite theories, coherence completion, and symbolic methods. J. Log. Algebraic Methods Program. 110 (2020)
34. Meseguer, J., Montanari, U.: Petri nets are monoids. Inform. Comput. **88**(2), 105–155 (1990)
35. Nigam, V., Talcott, C.L.: Automating safety proofs about cyber-physical systems using rewriting modulo SMT. In: Bae, K. (ed.) Rewriting Logic and Its Applications (WRLA 2022). LNCS, vol. 13252, pp. 212–229. Springer, Cham (2022). https://doi.org/10.1007/978-3-031-12441-9_11
36. Ölveczky, P.C.: Semantics, simulation, and formal analysis of modeling languages for embedded systems in Real-Time Maude. In: Agha, G., Danvy, O., Meseguer, J. (eds.) Formal Modeling: Actors, Open Systems, Biological Systems. LNCS, vol. 7000, pp. 368–402. Springer, Heidelberg (2011). https://doi.org/10.1007/978-3-642-24933-4_19
37. Ölveczky, P.C.: Real-Time Maude and its applications. In: Escobar, S. (ed.) WRLA 2014. LNCS, vol. 8663, pp. 42–79. Springer, Cham (2014). https://doi.org/10.1007/978-3-319-12904-4_3
38. Ölveczky, P.C., Boronat, A., Meseguer, J.: Formal semantics and analysis of behavioral AADL models in Real-Time Maude. In: Hatcliff, J., Zucca, E. (eds.) FMOODS/FORTE -2010. LNCS, vol. 6117, pp. 47–62. Springer, Heidelberg (2010). https://doi.org/10.1007/978-3-642-13464-7_5
39. Ölveczky, P.C., Caccamo, M.: Formal simulation and analysis of the CASH scheduling algorithm in Real-Time Maude. In: Baresi, L., Heckel, R. (eds.) FASE 2006. LNCS, vol. 3922, pp. 357–372. Springer, Heidelberg (2006). https://doi.org/10.1007/11693017_26
40. Ölveczky, P.C., Meseguer, J.: Specification of real-time and hybrid systems in rewriting logic. Theor. Comput. Sci. **285**(2), 359–405 (2002)
41. Ölveczky, P.C., Meseguer, J.: Abstraction and completeness for Real-Time Maude. In: 6th International Workshop on Rewriting Logic and its Applications (WRLA 2006). Electronic Notes in Theoretical Computer Science, vol. 174, pp. 5–27. Elsevier (2006)
42. Ölveczky, P.C., Meseguer, J.: Semantics and pragmatics of Real-Time Maude. High. Order Symb. Comput. **20**(1–2), 161–196 (2007)
43. Ölveczky, P.C., Meseguer, J.: The Real-Time Maude tool. In: Ramakrishnan, C.R., Rehof, J. (eds.) TACAS 2008. LNCS, vol. 4963, pp. 332–336. Springer, Heidelberg (2008). https://doi.org/10.1007/978-3-540-78800-3_23
44. Ölveczky, P.C., Meseguer, J., Talcott, C.L.: Specification and analysis of the AER/NCA active network protocol suite in Real-Time Maude. Formal Methods Syst. Des. **29**(3), 253–293 (2006)
45. Padberg, J., Schulz, A.: Model checking reconfigurable Petri nets with Maude. In: Echahed, R., Minas, M. (eds.) ICGT 2016. LNCS, vol. 9761, pp. 54–70. Springer, Cham (2016). https://doi.org/10.1007/978-3-319-40530-8_4
46. Parquier, B., et al.: Applying parametric model-checking techniques for reusing real-time critical systems. In: Artho, C., Ölveczky, P.C. (eds.) FTSCS 2016. CCIS, vol. 694, pp. 129–144. Springer, Cham (2017). https://doi.org/10.1007/978-3-319-53946-1_8

47. Rocha, C., Meseguer, J., Muñoz, C.A.: Rewriting modulo SMT and open system analysis. J. Log. Algebraic Methods Program. **86**(1), 269–297 (2017)
48. Stehr, M.-O., Meseguer, J., Ölveczky, P.C.: Rewriting logic as a unifying framework for Petri nets. In: Ehrig, H., Padberg, J., Juhás, G., Rozenberg, G. (eds.) Unifying Petri Nets. LNCS, vol. 2128, pp. 250–303. Springer, Heidelberg (2001). https://doi.org/10.1007/3-540-45541-8_9
49. Traonouez, L.-M., Lime, D., Roux, O.H.: Parametric model-checking of time Petri nets with stopwatches using the state-class graph. In: Cassez, F., Jard, C. (eds.) FORMATS 2008. LNCS, vol. 5215, pp. 280–294. Springer, Heidelberg (2008). https://doi.org/10.1007/978-3-540-85778-5_20
50. Traonouez, L., Lime, D., Roux, O.H.: Parametric model-checking of stopwatch Petri nets. J. Univers. Comput. Sci. **15**(17), 3273–3304 (2009)
51. Vernadat, F., Berthomieu, B.: State space abstractions for time Petri nets. In: Son, S.H., Lee, I., Leung, J.Y. (eds.) Handbook of Real-Time and Embedded Systems. Chapman and Hall/CRC (2007)
52. Wang, J.: Time Petri nets. In: Timed Petri Nets: Theory and Application, pp. 63–123. Springer, Cham (1998)
53. Yu, G., Bae, K.: Maude-SE: a tight integration of Maude and SMT solvers. In: Preliminary Proceedings of WRLA@ETAPS, pp. 220–232 (2020)

A State Class Based Controller Synthesis Approach for Time Petri Nets

Loriane Leclercq[✉][iD], Didier Lime[iD], and Olivier H. Roux[iD]

École Centrale de Nantes, CNRS, LS2N, Nantes, France
{Loriane.Leclercq,Didier.Lime,Olivier-h.Roux}@ec-nantes.fr

Abstract. We propose a new algorithm for reachability controller synthesis with time Petri nets (TPN). We consider an unusual semantics of time Petri nets in which the firing date of a transition is chosen in its static firing interval when it becomes enabled. This semantics is motivated i) *by a practical concern:* it aims at approaching the implementation of the controller on a real-time target; ii) *by a theoretical concern:* it ensures that in the classical state class graph [6], every state in each state class is an actual reachable state from the TPN, which is not the case with the usual interval-based semantics. We define a new kind of two-player timed game over the state class graph and we show how to efficiently and symbolically compute the winning states using state classes. The approach is implemented in the tool Roméo [23]. We illustrate it on various examples including a case-study from [2].

Keywords: Time Petri nets · state classes · timed games · controller synthesis

1 Introduction

Reactive systems allow multiple components to work and interact together and with the environment. In order to ensure the correctness of such systems, we can use controllers to restrict their behavior. Unlike the model-checking problem in which systems are mainly represented as standalone (open-loop systems), the control framework models the interaction between a controller and its environment by using controllable and non-controllable actions. The problem is to design this controller (or a strategy for the controller) that ensures a given specification is valid whatever the environment does (closed-loop system).

The theory of control was first defined over discrete event systems in [25] and then extended to various models. The idea is to model the system and the properties we are interested in, and to implement a controller that makes sure the model behaves correctly w.r.t the properties. Some of the basic properties are reachability and safety. Given a state of the system, the *reachability problem* consists in deciding if this good state is reachable in every execution. Dually the

This work has been partially funded by ANR project ProMiS ANR-19-CE25-0015.

L. Gomes and R. Lorenz (Eds.): PETRI NETS 2023, LNCS 13929, pp. 393–414, 2023.
https://doi.org/10.1007/978-3-031-33620-1_21

safety problem consists in deciding if we can stay in good states forever for every execution. These control problems usually use controllable events, for example transitions in our case, that will be handled by the controller.

Timed Games and Time Petri Nets for Control. Games on graphs [26] are a classical and successful framework for controller synthesis in reactive systems. Many such systems however have strong timing requirements and must therefore be modelled using timed formalisms such as timed automata [1]. This leads to the notion of timed games [24], that have been much studied theoretically, and for which efficient clock based algorithms using the so-called zones [15] have been devised [14] and implemented, e. g., in the state-of-the art tool Uppaal-Tiga [3]. The clock-zone based algorithm faces a termination problem that can be solved with an extrapolation/approximation operation. This operation makes things more complex by adding states that are actually not reachable but, in general, do not interfere with the properties we are interested in. In [14], for instance, the authors just assume the clocks to be bounded to avoid dealing with it. And in some cases the added states do interfere with the property of interest [12,13].

Another classical formalism for timed models is time Petri nets (TPNs) [6]. It is possible to apply to TPNs a semantics very close to the one of timed automata by making clocks appear that measure the duration for which transitions have been enabled and then a state-space computation can be done in a manner similar to timed automata [18]. It thus no surprise that the algorithm of [14] can be lifted to TPNs [17]. It is indeed implemented in the tool Roméo [23], and even extended to account for timing parameters [22].

The classical semantics for TPNs is not the « explicit clocks » one of [18] however, but rather the interval-based semantics of [6]. The latter semantics leads naturally to a different kind of abstraction called state classes, which has some advantages: in particular it does not require the use of an extrapolation/approximation operation to ensure termination.

The control problem for time Petri nets has been studied in [2,19] with forward approaches computing the reachable states over a modified state class graph (SCG) in order to synthesize new constraints to reach winning states. In [2], no constraints are back propagated but the time constraints of a transition that remains enabled contain all its past and have the size of the path during which it remains enabled. In [19], the new constraints are back-propagated to previous classes, until the events when transitions were first newly enabled. Only rectangular constraints are propagated without splitting the state classes, making it impossible to synthesize a controller with a state where a controllable action should be done in disjoint intervals.

Controller Implementation. Implementing a timed controller on a hardware target such as a microcontroller is not a trivial operation. If a controllable action is to be performed after waiting for a duration within a time interval $[a, b]$, it is necessary to first choose a duration d within this interval and then to go into a non-active wait which is usually achieved by using a timer of duration d that will trigger an interrupt. The program associated with this interruption then executes the controllable action. If an action of the environment occurs in the meantime that requires a change of the d duration, then the controller must have the ability to change the timer value. This is implicitly considered in most of the works on timed controller synthesis for time Petri nets and timed automata whose computation is based on a backward clock-based algorithm [14,17,24]. This implicit re-evaluation of waiting durations makes the implementation of the controller difficult (unless active waiting, i. e. polling, is used, which is not acceptable in a real-time context) because it is not known a priori which actions should cause the re-evaluation of durations.

Contribution. We propose an unusual semantics of time Petri nets in which the firing date of a transition is chosen in its static firing interval when it becomes enabled. This semantics is motivated by a practical concern: it aims at approaching the implementation of the controller on a real-time target. The choice of the timer value must be made as soon as the controllable transition is enabled. If this value is to be re-evaluated then the Petri net must model it explicitly.

It is also motivated by a theoretical concern: there is a tight correspondence between this semantics and the construction of the state class graph. It ensures that in the classical state class graph [6], every state in each state class is an actual reachable state from the TPN, which is not the case with the usual interval-based semantics. This semantics was already used in [8], motivated by the use of dynamic firing dates, that can be chosen again at every firing event.

We then leverage the state class abstraction to solve timed games. We define a new kind of timed games based on TPNs and we show how to efficiently and symbolically compute the winning states using state classes. Our method computes backward winning states on the SCG using predecessor operators to split the state classes. The approach, implemented in the tool Roméo [23] is applied on two examples including a case-study from [2].

The rest of this article is organized as follows: Sect. 2 introduces our new semantics nets and provides the necessary basic definitions of time Petri nets, state classes and the two-player game over this graph, Sect. 3 describes the computations of winning states for the controller leading to the strategy. Section 4 applies our approach to two case studies. We conclude in Sect. 5.

2   Definitions

2.1   Preliminaries

We denote the set of natural numbers (including 0) by $\mathbb{N}$ and the set of real numbers by $\mathbb{R}$. We note $\mathbb{R}_{\geq 0}$ the set of non-negative real numbers. For $n \in \mathbb{N}$, we let $[\![0, n]\!]$ denote the set $\{i \in \mathbb{N} \mid i \leq n\}$. For a finite set X, we denote its size by $|X|$.

Given a set X, we denote by $\mathcal{I}(X)$, the set of non-empty, non necessarily bounded, real intervals that have their finite end-points in X. We say that an interval I is non-negative if $I \subseteq \mathbb{R}_{\geq 0}$.

Given sets V and X, a V-valuation (or simply valuation when V is clear from the context) of X is a mapping from X to V. We denote by V^X the set of V-valuations of X. When X is finite, given an arbitrary fixed order on X, we often equivalently consider V-valuations as vectors of $V^{|X|}$.

2.2   Time Petri Nets

A time Petri net is a Petri net with time intervals associated with each transition. We propose a slightly different semantics than the one commonly used, in which firing dates are decided at the moment transitions are newly enabled. We consider that input tokens are consumed before the firing of a transition and produced after, so transitions using one of these input tokens have their firing date chosen again.

Definition 1 (Time Petri net). *A* time Petri net *(TPN) is a tuple* $\mathcal{N} = (P, T, F, I_s)$ *where:*

- *P is a finite non-empty set of* places,
- *T is a finite set of* transitions *such that* $T \cap P = \emptyset$,
- *F* : $(P \times T) \cup (T \times P)$ *is the* flow function,
- *I$_s$* : $T \to \mathcal{I}(\mathbb{N})$ *is the* static firing interval *function,*

We assume T contains at least one transition t_{init} *and P contains at least a place* p_0 *such that* $(p_0, t_{\text{init}}) \in F$, $I_s(t_{\text{init}}) = [0, 0]$, *for all* $p \in P \backslash \{p_0\}$, $(p, t_{\text{init}}) \notin F$ *and for all* $t \in T$, $(p_0, t) \notin F$. *We also assume that for all* $t \in T$ *there exists* $p \in P$ *such that* $(p, t) \in F$.

Places of a Petri net can contain *tokens*. A *marking* is then usually an $\mathbb{N}$-valuation of P giving the number of tokens in each place.

Remark 1. For the sake of simplicity, we consider only *safe* nets, i.e., nets in which there is always at most 1 token in each place and where all arcs have weight 1. All subsequent developments can be generalised without any difficulty to more complex discrete dynamics provided the net remains *bounded*, i.e., there is a constant K such that all places never contain more than K tokens. Boundedness is an appropriate restriction since the control problem is undecidable for unbounded TPN. The proof of [21] extends directly to our semantics.

We therefore define a marking as the set of the places of P containing a token. We say those places are *marked*.

Usually we define an initial marking for the net. Without loss of generality, we consider here that all places are initially empty except p_0 which is marked. An immediate transition t_{init} sets the initial marking by firing. By construction, while it has not fired, no other transition can fire.

Given a transition t, we define the sets of its input places $\mathsf{Pre}(t) = \{p \mid (p, t) \in F\}$ and of its output places $\mathsf{Post}(t) = \{p \mid (t, p) \in F\}$.

Definition 2 (Enabled and persistent transitions). *A transition t is said to be* enabled *by marking m if all its input places are marked:* $\mathsf{Pre}(t) \subseteq m$. *A transition t is said to be* persistent *by firing transition t' from marking m if it is not fired and still enabled when removing tokens from the input places of t':* $t \neq t'$ *and* $\mathsf{Pre}(t) \subseteq m \setminus \mathsf{Pre}(t')$. *We say that t is* newly enabled *by firing t' from m, if t is enabled before and after the firing of t' but not persistent. We denote by* $\mathsf{en}(m)$, $\mathsf{pers}(m,t)$ *and* $\mathsf{newen}(m,t)$ *respectively the sets of enabled, persistent and newly enabled transitions.*

Remark 2. The definition of newly enabled transitions uses a reset policy in which every transition that is disabled by a token taken by $\mathsf{Pre}(t)$ is considered newly enabled even if it is enabled again after putting back tokens from $\mathsf{Post}(t)$. And in particular the fired transition itself is always considered newly enabled. This is the usual memory policy called intermediate semantics, see [5] for details and comparison with other semantics

Definition 3 (States and semantics of a TPN). *A state of a TPN is a pair $s = (m, \theta)$ with $m \subseteq P$ a marking and $\theta : T \to \mathbb{R}_{\geq 0} \cup \{\bot\}$ a function that associates a firing date with every transition t enabled at marking m ($t \in \mathsf{en}(m)$) and $\bot$ to all other transitions. For any valuation on transitions θ, we denote by $\mathsf{tr}(\theta)$ the set of transitions t such that $\theta(t) \neq \bot$. We will use θ_i to denote $\theta(t_i)$.*

The semantics of a TPN is a Timed Transition System $(S, s_0, \Sigma, \to)$ with:

- *S the set of all possible states,*
- *an initial state $s_0 = (\{p_0\}, \theta_0) \in S$ with $\theta_0(t_{\mathsf{init}}) = 0$, and $\forall t \neq t_{\mathsf{init}}, \theta_0(t) = \bot$,*
- *a labelling alphabet Σ divided between two types of letters: $t_f \in T$ and $d \in \mathbb{R}_{\geq 0}$,*
- *the transition relation between states $\to \subseteq S \times \Sigma \times S$ and, $(s, a, s') \in \to$, denoted by $s \xrightarrow{a} s'$:*
  - *either $(m, \theta) \xrightarrow{t_f} (m', \theta')$ for $t_f \in T$ when:*
    1. *$t_f \in \mathsf{en}(m)$ and $\theta_f = 0$*
    2. *$m' = (m \setminus \mathsf{Pre}(t_f)) \cup \mathsf{Post}(t_f)$*
    3. *$\forall t_k \in T, \theta'_k \in I_s(t_k)$ if $t_k \in \mathsf{newen}(m, t_f)$, $\theta'_k = \theta_k$ if $t_k \in \mathsf{pers}(m, t_f)$, and $\theta'_k = \bot$ otherwise*
  - *or $(m, \theta) \xrightarrow{d} (m, \theta')$ when: $d \in \mathbb{R}_{\geq 0} \setminus \{0\}$, $\forall t_k \notin \mathsf{en}(m), \theta_k = \bot$, and $\forall t_k \in \mathsf{en}(m), \theta_k - d \geq 0$ and $\theta'_k = \theta_k - d$.*

Remark 3. Had we not assumed a unique transition t_{init} enabled, with a time interval reduced to $[0, 0]$, we would have in general an infinity of initial states corresponding to all possible choices of function θ_0 with values in the static firing intervals of enabled transitions, which is not a problem but is a small inconvenience. Otherwise for the two-player game construction to follow, we would have needed a first half turn to reach a correct state before even starting.

A *run* in the semantics of a TPN is a possibly infinite sequence $s_0 a_0 s_1 a_1 s_2 a_2 \cdots$ such for all i, $s_i \xrightarrow{a_i} s_{i+1}$. We denote by $\mathsf{seq}(\rho)$ the subsequence of ρ containing exactly the transitions $a_0 a_1 a_2 \cdots$.

In this semantics the choice of firing date occurs directly when the transition is newly enabled whereas in the classical semantics of [6] this choice is postponed to the moment the transition fires. This is already used in [20] but with probabilistic choices instead of non-deterministic one.

We have chosen this semantics because it will allow us to more precisely relate states and state classes as defined in the next section. Such a close relation has never been achieved with the semantics of [6], leading to further refinements into so-called atomic state classes [7].

2.3   State Classes

The number of states from a TPN is not finite in general because of the density of the static intervals. There are several finite representations abstracting the state space of a TPN using various methods and one of them is the state class graph. One of its benefits is to be finite as long as the TPN is bounded, i.e. the number of tokens in each place is bounded (by 1 in the case of safe nets).

Definition 4 (State class). *Let $\sigma = t_1...t_n$ be a sequence of transitions. The state class K_σ is the set of all states obtained by firing σ in order, with all possible delays before each fired transition. Clearly, all states in K_σ share the same marking m, and so we write $K_\sigma = (m, D)$ where D, called the firing domain, is the union of all possible firing date functions for those states.*

The firing domain D is a set of valuations of transitions. With an arbitrary order on transitions, and ignoring $\perp$ values, such a valuation can be seen as a point in $\mathbb{R}_{\geq 0}^{|\mathsf{en}(m)|}$. We will therefore consider such sets of valuations as subsets of $\mathbb{R}_{\geq 0}^{|\mathsf{en}(m)|}$. And as we will see, firing domains are actually a special kind of convex polyhedra in that space.

As a direct consequence of Definition 4, we have the following lemma:

Lemma 1. *Let $K_\sigma = (m, D)$. Let $s = (m, \theta)$; then $\theta \in D$ if and only if there exists a run ρ from the initial state s_0 to s, such that $\mathsf{seq}(\rho) = \sigma$ and either σ is empty or ρ ends with a transition firing.*

Remark 4. With the usual semantics of [6], only the *if* part holds [7], because the timing part of states in that semantics assigns intervals to enabled transition and an arbitrary interval taken from D does not necessarily correspond to a reachable state. A state can contain an interval overlapping two adjacent intervals grouped in a class, but that is not reachable.

We can naturally now extend the notions of enabled, persistent, and newly enabled transitions to state classes: $\mathsf{en}((m, D)) = \mathsf{en}(m)$, $\mathsf{newen}((m, D), t) = \mathsf{newen}(m, t)$, and $\mathsf{pers}((m, D), t) = \mathsf{pers}(m, t)$.

We have the following lemma:

Lemma 2. *Let $K_\sigma = (m, D)$ and $K_{\sigma.t_f} = (m', D')$, with $t_f \in \text{en}(m)$. We have:*

$$\theta' \in D' \text{ iff } \exists \theta \in D \text{ s. t. } \begin{cases} \forall i \in \text{en}(m), \theta_i - \theta_f \geq 0 \\ \forall i \in \text{pers}(m, t_f), \theta_i' = \theta_i - \theta_f \\ \forall i \in \text{newen}(m, t_f), \theta_i' \in I_s(i) \end{cases}$$

Proof. By Lemma 1, for all states $s' = (m', \theta') \in K_{\sigma.t_f}$ there exists a run ρ' that goes from the initial state s_0 to s' such that $\text{seq}(\rho') = \sigma.t_f$. Also ρ ends with the firing of t_f.

Let $s = (m, \theta)$ and $s'' = (m, \theta'')$ be the states in ρ' such that $\exists d.s \xrightarrow{d} s'' \xrightarrow{t_f} s'$. Possibly, we have $s = s''$. Let ρ be the prefix of ρ' ending in s, then $\text{seq}(\rho) = \sigma$ and ρ does not end with a delay, so $s \in K_\sigma$ by Lemma 1. We thus have $\theta \in D$ and Definition 4 directly implies the three expected conditions because, from top to bottom, t_f is firable, we must delay until θ_f is 0, and the firing dates for newly enabled transitions are chosen in their static firing intervals.   □

From Lemma 2, D' is not empty if and only if there exists θ in D such that for all $i \in \text{en}(m)$, $\theta_i \geq \theta_f$. In that case we say that t_f is *firable* from (m, D).

Algorithm 1 then follows straightforwardly from Lemma 1 to compute $K' = (m', D')$ from $K = (m, D)$ by firing firable transition t_f.

Algorithm 1. Successor (m', D') of (m, D) by firing firable transition t_f

---
1: $m' \leftarrow (m \setminus \text{Pre}(t_f)) \cup \text{Post}(t_f)$
2: $D' \leftarrow D \wedge \bigwedge_{i \neq f, i \in \text{en}(m)} \theta_f \leq \theta_i$
3: for all $i \in \text{en}(m \setminus \text{Pre}(t_f)), i \neq f$, add variable θ_i' to D', constrained by $\theta_i' = \theta_i - \theta_f$
4: eliminate (by existential projection) variables θ_i for all i from D'
5: for all $i \in \text{newen}(m, t_f)$, add variable θ_i'' to D', constrained by $\theta_i'' \in I_s(i)$

The state class associated with the empty sequence ϵ contains the set of initial states, here reduced to a singleton: $K_\epsilon = (m_0, \{\theta_0\})$.

Algorithm 1 corresponds to the classical state class computation from [6]. The initial class is also what we would obtain with that construction. It is well-known that those state classes can be represented and computed using a special kind of convex polyhedra encoded in the efficient data structure called *difference bound matrix* (DBM) [6,16]. An efficient way to directly compute successor classes is given in [10,11].

Definition 5. *Starting from K_ϵ, we can construct an infinite directed tree (labeled by fired transitions) by inductively computing successors by firable transitions. The* State class graph *(SCG) $\mathcal{G}$ is the graph obtained by quotienting this tree with the equality relation on state classes (same marking, and same firing domain).*

2.4 Two-player Game on the State Class Graph

Since we are interested in controller synthesis, from now on, the set of transitions is partitioned between two sets T_c and T_u which contain respectively controllable and uncontrollable transitions. The controllable transitions are controlled by a controller, in the sense that it can choose their firing dates, and the order of firing but uncontrollable transitions can be fired in between.

For short, we define $\text{newen}_u(m, t) = T_u \cap \text{newen}(m, t)$, $\text{newen}_c(m, t) = T_c \cap \text{newen}(m, t)$, $\text{en}_u(m) = T_u \cap \text{en}(m)$ and $\text{en}_c(m) = T_c \cap \text{en}(m)$. We also extend these notations to state classes as before.

We now define a game over the TPN $\mathcal{N}$ that simulates the behavior of controllable and uncontrollable transitions in order to decide whether a set of states is always reachable by choosing the right controllable firing dates or not. And if this is the case, a strategy for the controller will be constructed.

A round in the game is in three steps:

1. the controller chooses a firable transition $t_c \in T_c$ that he wants to fire first;
2. the environment chooses either to fire a firable transition $t_u \in T_u$ or to let the controller fire t_c;
3. both choose independently the firing dates of their newly enabled transitions.

Definition 6. *Let $\mathcal{N} = (P, T, F, I_s)$, a time Petri net with $T = T_c \cup T_u$ and $T_c \cap T_u = \emptyset$ and $(S, s_0, \Sigma, \rightarrow)$, its semantics, an* arena *is a tuple $\mathcal{A} = (S, \rightarrow, Pl, (\text{Movt}_i)_{i \in Pl}, (\text{Movf}_i)_{i \in Pl}, \text{Trans})$ with:*

- *$Pl = (Pl_u, Pl_c)$ the two players of the game: the environment (Pl_u) and the controller (Pl_c). The controller plays over controllable transitions, whereas the environment plays over uncontrollable transitions.*
- *$\text{Movt}_u : S \times T_c \rightarrow 2^T$ and $\text{Movt}_c : S \rightarrow 2^T$ rule the choices of transitions:*

$$\text{Movt}_c(m, \theta) = \{t_i \mid t_i \in \text{en}_c(m) \wedge \theta_i = \min_{t_k \in \text{en}(m)} \theta_k\}$$

$$\text{Movt}_u((m, \theta), t_c) = \{t_i \mid t_i \in \text{en}_u(m) \wedge \theta_i = \min_{t_k \in \text{en}(m)} \theta_k\} \cup \{t_c\}$$

- *$\text{Movf}_u : S \times T \rightarrow 2^{\mathbb{R}_{\geq 0}^{T_u}}$ and $\text{Movf}_c : S \times T \rightarrow 2^{\mathbb{R}_{\geq 0}^{T_c}}$ rule the choices of firing dates:*

$$\text{Movf}_c((m, \theta), t_i) = \left\{ \theta^c \in \mathbb{R}_{\geq 0}^{T_c} \;\middle|\; \begin{array}{l} \theta_k^c \in I_s(t_k) \text{ if } t_k \in \text{newen}(m, t_i) \\ \theta_k^c = \theta_k - \theta_i \text{ if } t_k \in \text{pers}(m, t_i) \\ \theta_k^c = \bot \text{ otherwise} \end{array} \right\}$$

$$\text{Movf}_u((m, \theta), t_i) = \left\{ \theta^u \in \mathbb{R}_{\geq 0}^{T_u} \;\middle|\; \begin{array}{l} \theta_k^u \in I_s(t_k) \text{ if } t_k \in \text{newen}(m, t_i) \\ \theta_k^u = \theta_k - \theta_i \text{ if } t_k \in \text{pers}(m, t_i) \\ \theta_k^u = \bot \text{ otherwise} \end{array} \right\}$$

– *finally,* $\mathsf{Trans} : S \times T \times T \times \mathbb{R}_{\geq 0}^{T_c}, \mathbb{R}_{\geq 0}^{T_u} \rightarrow S$ *combines all the choices of the players and gives the resulting state:*

$$\mathsf{Trans}(s, t_c, t_u, \theta^c, \theta^u) = ((m \setminus \mathsf{Pre}(t_u)) \cup \mathsf{Post}(t_u), \theta^c \cup \theta^u)$$

when $t_c \in \mathsf{Movt}_c(s)$, $t_u \in \mathsf{Movt}_u(s, t_c)$, $\theta(t_u) = \min_k(\theta(t_k))$, $t_u \in T_u \vee t_u = t_c$, $\theta^c \in \mathsf{Movf}_c(s, t_u)$ *and* $\theta^u \in \mathsf{Movf}_u(s, t_u)$.
Note that $\theta^u \cup \theta^c$ *is a disjoint union and is in* $\mathbb{R}_{\geq 0}^T$ *because* $\mathbb{R}_{\geq 0}^{T_u}$ *and* $\mathbb{R}_{\geq 0}^{T_c}$ *are disjoints and their union is* $\mathbb{R}_{\geq 0}^T$.

A reachability game $\mathcal{R} = (\mathcal{A}, \mathsf{Goal})$ consists of an arena and a set $\mathsf{Goal} \subseteq S$ of goal states. The objective of Pl_c, the controller, is to reach a state in Goal and the objective of Pl_u, the environment, is to avoid these states.

Definition 7. *A* play *in an arena is a finite or infinite word* $s_0 s_1 ... s_n$ *over the alphabet* S *such that*

$$s_0 \xrightarrow{t_1} s_1 \xrightarrow{t_2} s_2 .... \xrightarrow{t_n} s_n$$

with $\forall i, \exists \theta_i^u, \theta_i^c, t_{ci}, t_{ui}.\mathsf{Trans}(s_i, t_{ci}, t_{ui}, \theta_i^c, \theta_i^u) = s_{i+1}$ *and* $t_{ci} \in \mathsf{Movt}_c(s_i)$, $t_{ui} \in \mathsf{Movt}_u(s_i, t_{ci})$, $\theta_i^u \in \mathsf{Movf}_u(s_i, t_{ui})$ *and* $\theta_i^c \in \mathsf{Movf}_c(s_i, t_{ui})$.

Definition 8. *A* strategy *for the environment* Pl_u *(resp. the controller* Pl_c*) is a function* $\sigma_u : S \times T_c \rightarrow T \times \mathbb{R}_{\geq 0}^{T_u}$ *(resp.* $\sigma_c : S \rightarrow T_c \times \mathbb{R}_{\geq 0}^{T_c}$*).*[1]

Definition 9. *A play* $s_0 s_1 ...$ conforms *to strategy* σ_c *(resp.* σ_u*) if at each position* i *(except the last in case of a finite play):* $\exists t_{ci}, t_{ui}, \theta_i^c, \theta_i^u$ *s. t.* $\sigma_c(s_i) = (t_{ci}, \theta_i^c)$ *(resp.* $\sigma_u(s_i, t_{ci}) = (t_{ui}, \theta_i^u)$*) and* $Trans(s_i, t_{ci}, t_{ui}, \theta^c, \theta^u) = s_{i+1}$.

Definition 10. *A* maximal play *is a play that is either infinite or finite and such that from the last state no new marking is reachable.*

Definition 11. *A maximal play is* winning *for the controller if there is a position* n *such that* $s_n \in \mathsf{Goal}$. *Otherwise, the play is winning for the environment.*
   A strategy is winning *for a player if and only if all plays conforming to this strategy are winning for that player.*

Remark 5. This game is a *two-player determined concurrent game* and we always have either Pl_c wins or Pl_u wins, but not both.

3   Computing the Winning States

The construction of a strategy for the controller is based on the state class graph $\mathcal{G}$. To construct such a winning strategy over this graph, we will use a backward process to recursively compute the controllable predecessors of the target states until a fixed point is reached.

---

[1] Usualy, strategies are defined with the whole trace as memory but we will see in Subsect. 3.3 that by construction we only need memoryless strategies.

A TPN $\mathcal{N}$ and its state class graph $\mathcal{G}$ are shown in Fig. 1a and 1b. We use black squares to depict controllable transitions and white ones for uncontrollable transitions. In $\mathcal{G}$, dashed arrows are used for uncontrollable transitions. Goal states are those with a token in place p_5. To reach such a state from states in $ClassG_{toa}$, we must fire transition c before b. This is the condition that we will have to propagate during the backward process.

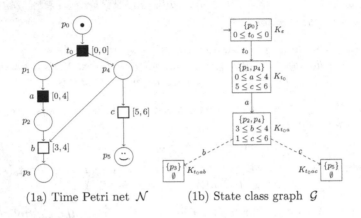

(1a) Time Petri net $\mathcal{N}$ (1b) State class graph $\mathcal{G}$

Fig. 1. (a) Time Petri net $\mathcal{N}$. (b) State class graph $\mathcal{G}$

Definition 12. *Let $C \xrightarrow{t_f} C'$ be a transition in $\mathcal{G}$ and let B be a subset of the class C'.*

We define the set of predecessors $\mathsf{Pred}_{C\xrightarrow{t_f}C'}(B)$ *of $B \subseteq C'$ in C by transition t_f:* $\mathsf{Pred}_{C\xrightarrow{t_f}C'}(B) = \{s \in C \mid \exists s'.s \xrightarrow{t_f} s' \in B\}.$

We further define two sets $\mathsf{cPred}_{C\xrightarrow{t_f}C'}(B)$ and $\mathsf{uPred}_{C\xrightarrow{t_f}C'}(B)$ for the *controllable* and *uncontrollable predecessors* of a subset B of C' in class C and by firing transition t_f. The controllable predecessors correspond to states from which the controller can force to reach B and the uncontrollable predecessors correspond to states from which the controller can not force to avoid B.

Definition 13. *Without loss of generality we suppose $\{t_1,...,t_n\} = \mathsf{newen_c}(C,t_f)$ and $\{t_{n+1},...,t_{n+k}\} = \mathsf{newen_u}(C,t_f)$. We define:*

$$\mathsf{cPred}_{C\xrightarrow{t_f}C'}(B) = \left\{ (m,\theta) \in C \;\middle|\; \begin{array}{l} \forall t_i \in \mathsf{newen_c}(C,t_f), \exists \theta'_i \in I_s(t_i) \text{ s. t.} \\ \forall t_{n+j} \in \mathsf{newen_u}(C,t_f), \forall \theta'_{n+j} \in I_s(t_{n+j}), \\ s \xrightarrow{t_f} s' = (m',\theta') \in B \\ \text{where } \forall i \in [\![1,n+k]\!], \theta'(t_i) = \theta'_i \\ \text{and } \forall i \in [\![1,l]\!], \theta'(t_{n+k+i}) = \theta(t_{n+k+i}) - \theta(t_f) \end{array} \right\}$$

And:

$$\mathsf{uPred}_{C \xrightarrow{t_f} C'}(B) = \left\{ (m,\theta) \in C \middle| \begin{array}{l} \forall t_i \in \mathsf{newen}_c(C,t_f), \forall \theta_i' \in I_s(t_i) \text{ s. t.} \\ \forall t_{n+j} \in \mathsf{newen}_u(C,t_f), \exists \theta_{n+j}' \in I_s(t_{n+j}), \\ s \xrightarrow{t_f} s' = (m',\theta') \in B \\ where \; \forall i \in [\![1,n+k]\!], \theta'(t_i) = \theta_i' \\ and \; \forall i \in [\![1,l]\!], \theta'(t_{n+k+i}) = \theta(t_{n+k+i}) - \theta(t_f) \end{array} \right\}$$

In order to symbolically compute these sets of states, we will need a few operators on sets of valuations.

In the following, D and D' are sets of valuations and we denote valuations by the sequences of their non-$\perp$ values.

We first define the classical *existential projection*:

Definition 14. *For any set of valuations D s. t. $\forall \theta \in D, \mathsf{tr}(\theta) = \{t_1,\ldots,t_{n+k}\}$,*

$$\pi^{\exists}_{\{t_1,\ldots,t_n\}}(D) = \{(\theta_1 \ldots \theta_n) \mid \exists \theta_{n+1},\ldots,\theta_{n+k}, (\theta_1 \ldots \theta_{n+k}) \in D\}$$

We also define a less usual *universal projection* of D' inside D:

Definition 15. *For any two sets of valuations D and D' such that $D' \subseteq D$ and $\forall \theta \in D, \mathsf{tr}(\theta) = \{t_1,\ldots,t_{n+k}\}$,*

$$\pi^{\forall}_{\{t_1,\ldots,t_n\}}(D,D') = \left\{ (\theta_1 \ldots \theta_n) \middle| \begin{array}{l} \exists \theta_{n+1},\ldots,\theta_{n+k}, (\theta_1 \ldots \theta_{n+k}) \in D \\ \wedge \; \forall \theta_{n+1},\ldots,\theta_{n+k}, (\theta_1 \ldots \theta_{n+k}) \in D \\ \qquad\qquad \implies (\theta_1 \ldots \theta_{n+k}) \in D' \end{array} \right\}$$

We also need an *extension operation*:

Definition 16. *For any set of valuations D s. t. $\forall \theta \in D, \mathsf{tr}(\theta) = \{t_1,\ldots,t_n\}$,*

$$\pi^{-1}_{\{t_1,\ldots,t_{n+k}\}}(D) = \{(\theta_1 \ldots \theta_{n+k}) \mid (\theta_1 \ldots \theta_n) \in D \text{ and } \forall i, \theta_{n+i} \geq 0\}$$

Finally, we define a *backward in time* operator:

Definition 17. *For any set of valuations D s. t. $\forall \theta \in D, \mathsf{tr}(\theta) = \{t_1,\ldots,t_n\}$ and for $t_f \neq t_i$ for all $i \in [\![1,n]\!]$,*

$$D + t_f = \{(\theta_1' \ldots \theta_n' \theta_f') \mid (\theta_1 \ldots \theta_n) \in D, \theta_f' \geq 0 \text{ and } \forall i, \theta_i' = \theta_i + \theta_f'\}$$

Remark 6. The universal projection is parameterized by two sets of valuations unlike the existential projection. The reason for this choice is that we only want states that after extension are correct regarding the semantics of the TPN. And since in our construction we will use this projection with $\{t_{n+1},\ldots t_{n+k}\} \in \mathsf{newen}(A,t)$ we can easily justify this choice because choosing a firing date θ_{n+i} outside of $I_s(t_{n+i})$ is not relevant for a newly enabled transition. So it is natural to restrict projections inside a set of valuations to those that can possibly be extended in some correct and reachable states, namely states that are part of a class in the SCG.

Note that for now the extension operation gives us this kind of irrelevant valuations, therefore we will need to intersect them with the domain of a state class from the graph beforehand.

The universal projection is expressible with set complements and existential projections only, as stated in the following proposition. We denote by $\overline{D}$ the complement of D, i.e., $\overline{D} = \{s \mid s \notin D\}$.

Proposition 1. *Let* $\tau = \{t_1, ..., t_n\}$ $\forall \tau \subseteq T, \pi_\tau^\forall(D, D') = \pi_\tau^\exists(D) \cap \overline{\pi_\tau^\exists(\overline{D'} \cap D)}$.

Due to the lack of space we ommit the proof.

Example 1. A graphical way to see the intuition behind the universal projection in two dimensions is given in Fig. 2. We use for this example the TPN in Fig. 1a, with D' being the part of the domain of the state class K_σ for the firing sequence $\sigma = t_0.a$, that allows to put a token in p_5.

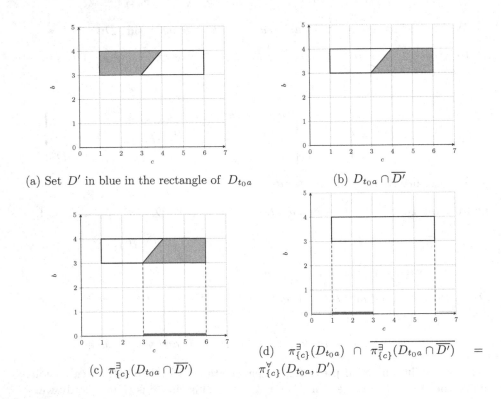

(a) Set D' in blue in the rectangle of $D_{t_0 a}$

(b) $D_{t_0 a} \cap \overline{D'}$

(c) $\pi_{\{c\}}^\exists(D_{t_0 a} \cap \overline{D'})$

(d) $\pi_{\{c\}}^\exists(D_{t_0 a}) \cap \overline{\pi_{\{c\}}^\exists(D_{t_0 a} \cap \overline{D'})} = \pi_{\{c\}}^\forall(D_{t_0 a}, D')$

Fig. 2. Example of universal projection

3.1 Symbolic Computation for Pred ()

We now have the necessary operations on sets of valuations to be able to symbolically compute the predecessors of a subset of a class. In the following we always assume that $B \subseteq C'$ and in particular we thus have $\mathsf{en}(C') = \mathsf{en}(B)$.

We will use the projection operators as building blocks for the $\mathsf{cPred}()$ and $\mathsf{uPred}()$ predecessor operators.

Proposition 2 ($\mathsf{cPred}_{C \xrightarrow{t_f} C'}(B)$ computation). *Let $C = (m, D)$ and $C' = (m', D')$. Consider $B = (m', D'') \subseteq C'$, and let $\mathsf{cPred}_{C \xrightarrow{t_f} C'}(B) = (m, D_p)$. Then:*

$$D_p = D \cap \pi^{-1}_{\mathsf{en}(C)} \left(\pi^{\exists}_{\mathsf{pers}(C,t_f)} \left(\pi^{\forall}_{\mathsf{newen_c}(C,t_f) \cup \mathsf{pers}(C,t_f)}(D', D'') \right) + t_f \right)$$

Proof. As in the definition of $\mathsf{cPred}()$, we have chosen without loss of generality that $\mathsf{newen_c}(C, t_f) = \{t_1, ..., t_n\}$, $\mathsf{newen_u}(C, t_f) = \{t_{n+1}, ..., t_{n+k}\}$ and $\mathsf{pers}(C, t_f) = \{t_{n+k+1}, ..., t_{n+k+l}\}$.

$\rightarrow$: First suppose $s = (m, \theta) \in \mathsf{cPred}_{C \xrightarrow{t_f} C'}(B)$. By Definition 13, we have $s \in C$ and $\forall t_i \in \mathsf{newen_c}(C, t_f), \exists \theta'_i \in I_s(t_i)$ s. t. $\forall t_{n+j} \in \mathsf{newen_u}(C, t_f), \forall \theta'_{n+j} \in I_s(t_{n+j}), s \xrightarrow{t_f} s' = (m', \theta') \in B$ where $\forall i \in [\![1, n+k]\!], \theta'(t_i) = \theta'_i$ and $\forall i \in [\![1, l]\!], \theta'_{n+k+i} = \theta_{n+k+i} - \theta_f$.

Since $B \subseteq C'$, we also have $s' \in C'$. Let θ^1 be the valuation such that $\forall i \in [\![1, n]\!], \theta^1_i = \theta'_i$ and $\forall i \in [\![1, l]\!], \theta^1_{n+k+i} = \theta_{n+k+i} - \theta_f$ and $\forall i \in [\![1, k]\!], \theta^1_{n+i} = \bot$, i.e., we have removed uncontrollable newly enabled transitions from θ''. Then from Definition 15, $\theta^1 \in \pi^{\forall}_{\mathsf{newen_c}(C,t_f) \cup \mathsf{pers}(C,t_f)}(D', D'')$ because the only way to assign values to newly enabled transitions within D' is to take them in their static firing interval, and because those intervals are all non-empty. Further define θ^2 from θ^1 by removing controllable newly enabled transitions: $\forall t_i \in \mathsf{newen_c}(C, t_f), \theta^2_i = \bot$ and $\forall t_i \notin \mathsf{newen_c}(C, t_f), \theta^2_i = \theta^1_i$. Then by construction, $\theta^2 \in \pi^{\exists}_{\mathsf{pers}(C,t_f)}(\{\theta^1\})$, and hence $\theta^2 \in \pi^{\exists}_{\mathsf{pers}(C,t_f)}(\pi^{\forall}_{\mathsf{newen_c}(C,t_f) \cup \mathsf{pers}(C,t_f)}(D', D''))$. In θ^2, exactly all persistent transitions have a value different from $\bot$, so if we add θ_f to all of those, we obtain a new valuation θ^3, with $\theta^3 = \theta^2 = (\theta_i - \theta_f) + \theta_f = \theta_i$ for all persistent transition t_i, and $\theta^3_i = \bot$ for all other transitions. By construction, $\theta^3 \in \pi^{\exists}_{\mathsf{pers}(C,t_f)}(\pi^{\forall}_{\mathsf{newen_c}(C,t_f) \cup \mathsf{pers}(C,t_f)}(D', D'')) + t_f$. Finally, we can extend θ^3 with values for the transitions in $\mathsf{en}(m) \setminus \mathsf{pers}(C, t_f)$ in the following manner: let θ^4 defined by $\theta^4_i = \theta^3_i$ for all persistent transitions t_i, $\theta^4_i = \theta_i$ for all $t_i \in \mathsf{en}(m) \setminus \mathsf{pers}(C, t_f)$ and $\theta^4_i = \bot$ for all other transitions. Then clearly, $\theta^4 \in \pi^{-1}_{\mathsf{en}(C)} \left(\pi^{\exists}_{\mathsf{pers}(C,t_f)} \left(\pi^{\forall}_{\mathsf{newen_c}(C,t_f) \cup \mathsf{pers}(C,t_f)}(D', D'') \right) + t_f \right)$. But since θ^4 has exactly all the same values for transitions as θ, we have the expected result.

$\Leftarrow$: consider $\theta \in D \cap \pi^{-1}_{\mathsf{en}(C)} \left(\pi^{\exists}_{\mathsf{pers}(C,t_f)} \left(\pi^{\forall}_{\mathsf{newen_c}(C,t_f) \cup \mathsf{pers}(C,t_f)}(D', D'') \right) + t_f \right)$. Let $s = (m, \theta)$. By definition of π^{-1}, there exists a θ^1 in $\pi^{\exists}_{\mathsf{pers}(C,t_f)}(\pi^{\forall}_{\mathsf{newen_c}(C,t_f) \cup \mathsf{pers}(C,t_f)}(D', D'')) +$

t_f such that for all persistent transitions t_i, $\theta_i^1 = \theta_i$, and $\theta_f \neq \bot$, and for all other transitions t_i, $\theta_i^1 = \bot$. By definition of the extension operator, there exists a valuation $\theta^2 \in \pi^{\exists}_{\text{pers}(C,t_f)}(\pi^{\forall}_{\text{newen}_c(C,t_f)}(D', D''))$, such that for all persistent $\cup \text{pers}(C,t_f)$
transitions t_i, $\theta_i^2 + \theta_f = \theta_i$, and for all other transitions t_i, $\theta_i^2 = \bot$.

By definition of $\pi^{\exists}$, there exists a valuation $\theta^3 \in \pi^{\forall}_{\text{newen}_c(C,t_f)}(D', D'')$ such $\cup \text{pers}(C,t_f)$
that for all persistent transitions t_i, we still have $\theta_i^3 = \theta_i^2 = \theta_i - \theta_f$, and for all newly enabled controllable transitions t_i we have $\theta_i^3 \neq \bot$.

By definition of $\pi^{\forall}$, there exists a valuation $\theta^4 \in D''$ such that for all persistent transitions t_i, we still have $\theta_i^4 = \theta_i - \theta_f$, and for all newly enabled (controllable and uncontrollable) transitions t_i we have $\theta_i^4 \neq \bot$. In addition, we know that for any other valuation $\theta^5 \in D'$ that equals θ^4 on all but the newly enabled uncontrollable transitions, we also have $\theta^5 \in D''$.

By construction, we have $\text{tr}(\theta^4) = \text{en}(m')$ (persistent plus all newly enabled transitions) and since $\theta^4 \in D'' \subseteq D'$, we have for all newly enabled transitions t_i, $\theta_i^4 \in I_s(t_i)$, and $s = (m, \theta) \xrightarrow{t_f} (m', \theta^4)$ and we have the same properties for all θ^5 as defined above. This, with $\theta^5 \in D''$ implies that $s = (m, \theta) \in \text{cPred}_{C \xrightarrow{t_f} C'}(B)$. $\qquad\square$

Proposition 3 is similar to Proposition 2 and its proof follows the same steps so we omit it.

Proposition 3 (uPred$_{C \xrightarrow{t_f} C'}(B)$ **computation**). *Let* $C = (m, D)$ *and* $C' = (m', D')$. *Consider* $B = (m', D'') \subseteq C'$, *and let* uPred$_{C \xrightarrow{t_f} C'}(B) = (m, D_p)$. *Then:*

$$D_p = D \cap \pi^{-1}_{\text{en}(C)}\left(\pi^{\forall}_{\text{pers}(C,t_f)}\left(\pi^{\exists}_{\substack{\text{newen}_u(C,t_f)\\ \cup \text{pers}(C,t_f)}}(D'), \pi^{\exists}_{\substack{\text{newen}_u(C,t_f)\\ \cup \text{pers}(C,t_f)}}(D'')\right) + t_f\right)$$

3.2 Predecessor Computations with DBMs

First recall that a DBM is a matrix in which coefficient $(d_{ij}, \prec)$ in row i and column j encodes a *diagonal* constraint $\theta_i - \theta_j \prec d_{ij}$, with $\prec \in \{\leq, <\}$. Variable θ_0 is assumed to always be equal to 0 so this also encodes *rectangular* constraints of the form $\theta_i \prec d_{i0}$ and $-\theta_i \prec d_{0i}$. DBMs can be put in a canonical form so that the DBM for a given set of valuations is unique [4].

The formulas we have given for cPred() and uPred() can be implemented with DBM operations. Indeed, existential projection and intersection on DBMs are classical operations and can be performed efficiently [4].

Universal projection is more complex. Most importantly, we need to complement a DBM. This can be done easily by creating, for each (non-redundant) constraint $\theta_i - \theta_j \prec d_{ij}$ of the DBM, a new DBM with only negated constraint $\theta_j - \theta_i \prec' -d_{ij}$, with $\prec'$ being strict if $\prec$ was weak and vice-versa. Then we take the union of all those DBMs. The result is therefore not a single DBM but a *finite* union of those. The rest of the operations is classical. This is kind of similar to subtraction between DBMs that are involved in computing the controllable predecessors for timed automata [3].

The extension operator just consists in resizing the DBM and initializing the new variables so they are not constrained.

Finally, the backward in time operator is more tricky: we need to add a new variable θ'_f (the delay) and do changes of variables for all other variables θ_i as follows: $\theta'_i = \theta_i + \theta'_f$. Then we existentially project out the θ_i variables. This is actually easier than it sounds because, assuming the DBM is in canonical form, diagonal constraints $\theta_i - \theta_j \prec d_{ij}$ are left unchanged by the transformation, the θ'_f cancelling each other, while rectangular constraints $\theta_i \prec d_{i0}$ or $-\theta_i \prec d_{0i}$ just become diagonal constraints: $\theta'_i - \theta'_f \prec d_{i0}$ or $\theta'_f - \theta'_i \prec d_{0i}$ respectively.

All these operations are straightforwardly extended to finite unions of DBMs, though at a price in terms of computation cost.

3.3 Winning States

The aim of this part is to define the set Win of winning states for the controller. We will start by defining inductively Win_n for strategies in less than n steps and we then show that it admits a fixpoint that corresponds to the full set of winning states.

Definition 18. *We start by defining the following sets of states that we will need in order to construct* Win_{k+1} *using* Win_k:

$$\text{uGood}_k(C) = \bigcup_{\substack{(C \xrightarrow{t_f} C') \in \mathcal{G}, \\ t_f \in \text{en}_u(C)}} \left(\text{cPred}_{C \xrightarrow{t_f} C'}(\text{Win}_k \cap C') \right)$$

$$\text{cGood}_k(C) = \bigcup_{\substack{(C \xrightarrow{t_f} C') \in \mathcal{G}, \\ t_f \in \text{en}_c(C)}} \left(\text{cPred}_{C \xrightarrow{t_f} C'}(\text{Win}_k \cap C') \right)$$

$$\text{uBad}_k(C) = \bigcup_{\substack{(C \xrightarrow{t_f} C') \in \mathcal{G}, \\ t_f \in \text{en}_u(C)}} \left(\text{uPred}_{C \xrightarrow{t_f} C'}(\overline{\text{Win}_k} \cap C') \right)$$

$$\text{cBad}_k(C) = \bigcup_{\substack{(C \xrightarrow{t_f} C') \in \mathcal{G}, \\ t_f \in \text{en}_c(C)}} \left(\text{uPred}_{C \xrightarrow{t_f} C'}(\overline{\text{Win}_k} \cap C') \right)$$

Intuitively, a state is in $\text{cGood}_k(C)$ if there is a controllable transition that can be fired, for which when arriving in C' we can choose a firing date for newly enabled controllable transitions such that no matter what firing date the environment chooses for its newly enabled uncontrollable transitions, we end up in Win_k. The set $\text{uGood}_k(C)$ is the same except the transition that is fired is uncontrollable.

Conversely, a state is in $\mathsf{uBad}_k(C)$ if there is an uncontrollable transition that can be fired, for which when arriving in C', no matter what firing dates for newly enabled controllable transitions we choose, we cannot be sure to end up in Win_k. The set $\mathsf{cBad}_k(C)$ is the same except the transition that is fired is controllable.

From those sets of states, we can inductively define the set Win_n that contains exactly the states from which the controller has a winning strategy in at most n steps:

$$\mathsf{Win}_0 = \mathsf{Goal}$$

$$\mathsf{Win}_{k+1} = \mathsf{Win}_k \cup \bigcup_{C \in \mathcal{G}} \left(\left[(\mathsf{uGood}_k(C) \setminus \mathsf{cBad}_k(C)) \cup \mathsf{cGood}_k(C) \right] \setminus \mathsf{uBad}_k(C) \right)$$

Lemma 3. *For all state s of $\mathcal{N}$, $s \in \mathsf{Win}_n$ if and only if from s the controller has a strategy to reach Goal in at most n steps.*

Proof. Proof by induction on n the number of steps.

The base case, $n = 0$, is straightforward, so we focus on the induction step.

Suppose that for some k we have $\forall s', s' \in \mathsf{Win}_k$ if and only if the controller has a strategy from s' to reach Goal in at most k steps. Let s be a state of the TPN.

$\Rightarrow$: Let $s \in \mathsf{Win}_{k+1}$. The case $s \in \mathsf{Win}_k$ is trivially true by the induction hypothesis. Assume therefore that s is in some class C and either $s \in \mathsf{uGood}_k(C) \setminus (\mathsf{cBad}_k(C) \cup \mathsf{uBad}_k(C))$ or $s \in \mathsf{cGood}_k(C) \setminus \mathsf{uBad}_k(C)$.

- In the first case, since $s \in \mathsf{uGood}_k(C)$, then Definition 13 ensures that the controller can choose firing dates θ^c to force that all its successors by an uncontrollable transition t_u are in Win_k. We also have that $s \notin (\mathsf{cBad}_k(C) \cup \mathsf{uBad}_k(C))$, hence Definition 13 gives us that no other controllable or uncontrollable transition that could lead to $\overline{\mathsf{Win}_k}$ can be fired before one of the above-mentioned favorable uncontrollable transitions.
- In the second case, s is in $\mathsf{cGood}_k(C)$ and not in $\mathsf{uBad}_k(C)$. The same arguments as before allows us to say that the controller has a way to choose firing dates to force that all its successors by a controllable transition are in Win_k and that no unfavorable uncontrollable transition can be fired before. Note that there is no need to guard against unfavorable controllable transition firing because we are in the case where the transition choice is made by the controller.

Clearly in both cases from s the controller has a strategy to choose t_c and θ^c such that $\mathsf{Trans}(s, t_c, t_u, \theta^c, \theta^u) \in \mathsf{Win}_k, \forall t_u, \theta^u$. And so it has a strategy to reach some $s' \in \mathsf{Win}_k$ in a single step. The induction hypothesis allows us to conclude that from any state in Win_{k+1}, the controller has a strategy to reach Goal in at most $k + 1$ steps.

$\Leftarrow$: If there is a strategy to reach Goal in (strictly) less than $k + 1$ steps from a state s, then there is a strategy in at most k steps and by the induction hypothesis, $s \in \mathsf{Win}_k \subseteq \mathsf{Win}_{k+1}$.

So we focus on the case in which the controller has a strategy to reach Goal in exactly $k + 1$ steps from a state s. To begin with, the controller can choose t_c and θ^c such that from all states $s' \in \mathsf{Trans}(s, t_c, t_u, \theta^c, \theta^u)$ it can still force to reach Goal in at most k steps, for all permitted choices of t_u and θ^u. These states s' are each in $C' \cap \mathsf{Win}_k$ for some class C' and not in $C'' \cap \overline{\mathsf{Win}_k}$ for any class C''. There is two main cases: either the environment lets the controller play t_c or it chooses to play some other $t_u \in T_u$.

- If t_u has been played, then the choices of firing dates in θ^c were such that whatever the environment chooses for θ^u, the successor of s by firing of t_u with these firing dates for newly enabled transitions is in Win_k. So using Definition 13 we get that $s \in \mathsf{uGood}_k(C)$. And the environment had no way to let another transition (uncontrollable or not) fire that would have led to $\overline{\mathsf{Win}_k}$ and would thus have been unfavorable to the controller. Definition 13 ensures that $s \notin (\mathsf{uBad}_k(C) \cup \mathsf{cBad}_k(C))$.
- If t_c has been played, then the environment had no way to have a disadvantageous uncontrollable transition fire first, then using Definition 13, $s \notin \mathsf{uBad}_k(C)$. And since the resulting state s' is in Win_k regardless of the environment choices, it follows from Definition 13 that $s \in \mathsf{cGood}_k(C)$.

Bringing all these sets together we have that all such states s are in:

$$\left(\left[(\mathsf{uGood}_k(C) \setminus \mathsf{cBad}_k(C)) \cup \mathsf{cGood}_k(C) \right] \setminus \mathsf{uBad}_k(C) \right)$$

This leads us to conclude that $s \in \mathsf{Win}_{k+1}$.   □

Proposition 4. *For all $\mathcal{N}$ such that $\mathcal{G}$ is finite (i.e. $\mathcal{N}$ is bounded as proved in [9]), $\exists n, \mathsf{Win}_n = \mathsf{Win}_{n+l}, \forall l > 0$*

Proof. Let $b \in \mathbb{N}$ and $\mathcal{M}$ a DBM in canonical form. Let us call a b-DBM a DBM in which all finite coefficients are smaller or equal to b in absolute value. We have shown in Subsect. 3.2, that all $\mathsf{uPred}()$ and $\mathsf{cPred}()$ computations done on DBMs give finite unions of DBMs. And furthermore, these operations preserve b-DBMs. It is well-known for the intersection because each coefficient of the result is the minimum of the corresponding coefficients in the operands [4]. The other operations are immediate using the constructions described above.

Now, let $b_{\max}$ be the greatest of the finite coefficients in the DBMs representing the domains of all classes in the state class graph. Since that graph is finite, this maximum is well-defined. Then all those DBMs in the state class graph are $b_{\max}$-DBMs.

It follows that all $\mathsf{uGood}_k(C)$ and its three variants, which are computed from them, are finite unions of $b_{\max}$-DBMs, and so are then all the Win_k's. Clearly, there is a finite number of $b_{\max}$-DBMs because DBM coefficients are non-negative integers. By enforcing that a given union does not contain twice the same DBM, we also make sure that there are only a finite number of different finite unions of $b_{\max}$-DBMs.

Since Win_k is clearly non-decreasing with k, we can then conclude that there must be an n such that $\text{Win}_{n+1} = \text{Win}_n$ and, by a simple induction, that all subsequent Win_{n+l}, for $l \geq 0$, are also equal to Win_n.   □

We can now define the set of winning states for the controller: $\text{Win} = \text{Win}_n$ for the smallest n such that we have reached a fixpoint in the construction of Win_n (i.e. $\text{Win}_n = \text{Win}_{n+1}$).

Using this set of winning states, the controller has a winning strategy if and only if the initial state in is Win. A strategy for the controller is then to choose firing dates and transitions to fire in order to stay in the set Win. Therefore, if the current state s is in some Win_{i+1} for $i \geq 0$, the controller will choose a successor of s that is in $\text{Win}_i \setminus \text{Win}_{i+1}$ in order to avoid infinite loops. As long as $s \notin \text{Goal}$, it is always possible by construction of Win. To make this choice deterministic, we could assume states are ordered, e.g. in lexicographic order, and that the controller will always choose the smaller one first. Successors by a controllable transition t_c will be given priority (in this order) because the controller has to propose a transition first. Then the new valuation θ^c will be chosen depending of the transition t_u selected by the environment (t_u might be the transition t_c). The current state is the only information used to make the choices. Hence the strategy is memoryless since no information from the previous turns are needed.

4   Case Studies

In the two following examples, for the sake of readability, we omit the immediate initialization transition and start the net directly in the initial marking.

Note that for these two examples, the classical clock-based method of [14, 17, 24], does not provide a winning strategy since the firing of an uncontrollable transition is needed to reach the goal state. A solution in this case is to add a controllable transition with firing interval $[b, b]$ in parallel of an uncontrollable transition whose firing interval is $[a, b]$.

4.1   Supply Chain

We consider the model of Fig. 3 of two production lines starting respectively in p_1 and p_4 associated with a sorting and assembly cell. The two lines start by bringing products to $p_2 + p_3$ and p_5, respectively with transitions t_1 and t_6. The products in p_5 are either discharged through transition t_7 or assembled with the products in p_2 or p_3. The products in p_5 assembled with the products in p_3 are unloaded through transition t_4. The products in p_3 which are not assembled with p_5 are supplied through transition t_3 to another line W_3. The products in p_5 assembled with the products in p_2 are supplied through transition t_5 to another line W_2. The products in p_2 which are not assembled with p_5 are supplied to another line W_1 through transition t_2.

Transition t_2 is the only controllable transition. We wish to synthesize a controller that will enforce the products reaching places W_1, W_2 or W_3, depending on the case we consider.

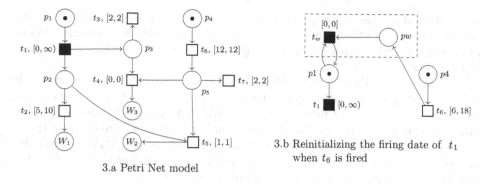

3.a Petri Net model

3.b Reinitializing the firing date of t_1 when t_6 is fired

Fig. 3. Production lines

Our approach is implemented in the tool Roméo [23]. We ask for a controller to reach one of the goal states chosen successively among W_1, W_2 and W_3, and we obtain three winning strategies that consists in initializing the firing date of t_1 in the initial state. The results are as follows:

– If the goal is W_1, initialize t_1 such that: $\theta_1 \in [0,3)$ or $\theta_1 \in (10,+\infty)$
– If the goal is W_2, initialize t_1 such that: $\theta_1 \in (0,3)$
– If the goal is W_3, initialize t_1 such that: $\theta_1 \in (10,12)$ or $\theta_1 \in (12,14)$

If we extend the firing interval of t_6 to $[6,18]$ then there is no strategy for obtaining a token in W_3 with our method and our semantics because the choice of the firing date of t_1 has to be re-evaluated depending on what the environment does.

Indeed, we are looking for controllers that can be implemented with classical real-time methods and in particular with timer-triggered interrupts. We therefore need to explicitly specify which action of the environment should cause the controller to re-evaluate the firing dates of its transitions. This can be done easily with a widget that allows to disable and then re-enable a given controllable transition (here t_1) when a given transition of the environment (here t_6) is fired as shown in Fig. 3b.

We then obtain a winning strategy to reach a marking with a token in W_3 as follows:

In the initial state, initialize t_1 such that: $t_1 \in (16,+\infty)$

After the firing of t_6 (and then of t_w), initialize t_1 such that: $t_1 \in [0,2)$.

4.2   AGV

We now consider the TPN proposed in [2] that models a materials handling system with two Automated Guided Vehicle (AGV) systems and a workstation. Places p_1 and p_{10} are associated with the AGVs starting positions, and the other places in each AGV subnet correspond to the presence of the AGV in a section. Transitions t_1 and t_9 represent the start commands of the respective missions, t_7

is the start of a cycle of the workstation, and the other transitions of each AGV subnet (except t_5) correspond to the movement of the vehicle from one section to another. Finally, transition t_5 is the unloading of a part into the workstation input buffer.

Transitions t_1, t_7 and t_9 are controllable actions (commands can be activated at any time), t_6 is controllable since the speed of the AGV in this section can be set so that the time spent in the section is within the interval $[30, 40]$, while the other transitions are uncontrollable, and their static intervals are given in [2].

Places p_3 and p_{12} represent a shared zone between the two AGVs where only one vehicle at a time can stay. We then add a transition *bad* that remove the tokens in p_3 and p_{12} when two vehicles are in this zone.

The goal of the control problem proposed in [2] is to first reach a state with a marking $\{p_5, p_7, p_{10}\}$ in the time interval $[30, 65]$ and then to reach the goal state with $\{p_1, p_8, p_{13}\}$ or $\{p_6, p_9, p_{13}\}$ in the time interval $[90, 135]$. To express this goal we can use an observer as defined in [27] such that there is a token in a place WIN iff the goal is achieved within the constraints.

We then ask for a controller to reach a state with a token in the place WIN and we obtain the following winning strategy:

- In the initial state, initialize t_1, t_7 and t_9 such that: $t_1 \in [0, 5)$, $t_7 \in (45, 55)$, $t_9 \in (50, 55)$, $45 < t_7 - t_1$, $50 < t_9 - t_1$ and $0 \le t_9 - t_7 < 5$
- when the marking is $p_1\ p_8\ p_{12}$, arriving with t_6, initialize t_1 such that: $t1 \in (10, +\infty)$ $t_8 \in [35, 55]$, $t_{11} \in [10, 40]$, $t_8 - t_1 < 35$, $t_{11} - t_1 < 0$, $-35 \le t_{11} - t_8 < 0$

5 Conclusion

We have defined a new kind of two-player reachability timed games over the state class graph of time Petri nets. This allows to synthesize a controller that chooses, as soon as a new transition is enabled, the date on which this transition will be fired. The interest of this type of controller is that it can be implemented in real-time context with interrupts triggered by timers whose durations are fixed as soon as the associated actions are planned.

In our future work, we will study how well this semantics fits to the problem of partial observation. Moreover we plan to study the controller synthesis problem for safety and for ω-regular properties. We will also consider the question of joint timing parameters and controller synthesis.

References

1. Alur, R., Dill, D.: A theory of timed automata. Theoret. Comput. Sci. **126**(2), 183–235 (1994)
2. Basile, F., Cordone, R., Piroddi, L.: Supervisory control of timed discrete-event systems with logical and temporal specifications. IEEE Trans. Autom. Control **67**(6), 2800–2815 (2022). https://doi.org/10.1109/TAC.2021.3093618

3. Behrmann, G., Cougnard, A., David, A., Fleury, E., Larsen, K.G., Lime, D.: UPPAAL-TIGA: time for playing games! In: 19th International Conference on Computer Aided Verification (CAV 2007). Lecture Notes in Computer Science, vol. 4590, pp. 121–125. Springer, Berlin (2007). https://doi.org/10.1007/978-3-540-73368-3_14

4. Bengtsson, J., Yi, W.: Timed automata: semantics, algorithms and tools. In: Desel, J., Reisig, W., Rozenberg, G. (eds.) ACPN 2003. LNCS, vol. 3098, pp. 87–124. Springer, Heidelberg (2004). https://doi.org/10.1007/978-3-540-27755-2_3

5. Bérard, B., Cassez, F., Haddad, S., Lime, D., Roux, O.H.: Comparison of different semantics for Time Petri Nets. In: Peled, D.A., Tsay, Y.-K. (eds.) ATVA 2005. LNCS, vol. 3707, pp. 293–307. Springer, Heidelberg (2005). https://doi.org/10.1007/11562948_23

6. Berthomieu, B., Diaz, M.: Modeling and verification of time dependent systems using Time Petri Nets. IEEE Trans. Soft. Eng. **17**(3), 259–273 (1991)

7. Berthomieu, B., Vernadat, F.: State class constructions for branching analysis of Time Petri Nets. In: Garavel, H., Hatcliff, J. (eds.) TACAS 2003. LNCS, vol. 2619, pp. 442–457. Springer, Heidelberg (2003). https://doi.org/10.1007/3-540-36577-X_33

8. Berthomieu, B., Dal Zilio, S., Fronc, Ł, Vernadat, F.: Time Petri Nets with dynamic firing dates: semantics and applications. In: Legay, A., Bozga, M. (eds.) FORMATS 2014. LNCS, vol. 8711, pp. 85–99. Springer, Cham (2014). https://doi.org/10.1007/978-3-319-10512-3_7

9. Berthomieu, B., Menasche, M.: An enumerative approach for analyzing Time Petri Nets. In: Proceedings IFIP, pp. 41–46. Elsevier Science Publishers (1983)

10. Boucheneb, H., Mullins, J.: Analyse des réseaux temporels?: Calcul des classes en $O(n^2)$ et des temps de chemin en $O(m \times n)$. TSI. Technique et science informatiques **22**(4), 435–459 (2003)

11. Bourdil, P.A., Berthomieu, B., Dal Zilio, S., Vernadat, F.: Symmetry reduction for Time Petri Net state classes. Sci. Comput. Program. **132**, 209–225 (2016)

12. Bouyer, P.: Untameable timed automata! In: Alt, H., Habib, M. (eds.) STACS 2003. LNCS, vol. 2607, pp. 620–631. Springer, Heidelberg (2003). https://doi.org/10.1007/3-540-36494-3_54

13. Bouyer, P., Colange, M., Markey, N.: Symbolic optimal reachability in weighted timed automata. In: Chaudhuri, S., Farzan, A. (eds.) CAV 2016. LNCS, vol. 9779, pp. 513–530. Springer, Cham (2016). https://doi.org/10.1007/978-3-319-41528-4_28

14. Cassez, F., David, A., Fleury, E., Larsen, K.G., Lime, D.: Efficient on-the-fly algorithms for the analysis of timed games. In: Abadi, M., de Alfaro, L. (eds.) CONCUR 2005. LNCS, vol. 3653, pp. 66–80. Springer, Heidelberg (2005). https://doi.org/10.1007/11539452_9

15. Daws, C., Tripakis, S.: Model checking of real-time reachability properties using abstractions. In: Steffen, B. (ed.) TACAS 1998. LNCS, vol. 1384, pp. 313–329. Springer, Heidelberg (1998). https://doi.org/10.1007/BFb0054180

16. Dill, D.L.: Timing assumptions and verification of finite-state concurrent systems. In: Sifakis, J. (ed.) CAV 1989. LNCS, vol. 407, pp. 197–212. Springer, Heidelberg (1990). https://doi.org/10.1007/3-540-52148-8_17

17. Gardey, G., Roux, O.F., Roux, O.H.: Safety control synthesis for Time Petri Nets. In: 8th International Workshop on Discrete Event Systems (WODES 2006), pp. 222–228. IEEE Computer Society Press, Ann Arbor, USA, July 2006

18. Gardey, G., Roux, O.H., Roux, O.F.: State space computation and analysis of time Petri nets. Theory and Practice of Logic Programming (TPLP). Special Issue Specif. Anal. Verif. React. Syst. **6**(3), 301–320 (2006)

19. Heidari, P., Boucheneb, H.: Maximally permissive controller synthesis for time petri nets. Int. J. Control **86** (2013). https://doi.org/10.1080/00207179.2012.743038

20. Horváth, A., Paolieri, M., Ridi, L., Vicario, E.: Transient analysis of non-markovian models using stochastic state classes. Perform. Eval. **69**(7), 315–335 (2012). https://doi.org/10.1016/j.peva.2011.11.002, https://www.sciencedirect. com/science/article/pii/S0166531611001520, selected papers from QEST 2010

21. Jones, N.D., Landweber, L.H., Edmund Lien, Y.: Complexity of some problems in petri nets. Theoret. Comput. Sci. **4**(3), 277–299 (1977). https://doi.org/10. 1016/0304-3975(77)90014-7, https://www.sciencedirect.com/science/article/pii/ 0304397577900147

22. Jovanović, A., Lime, D., Roux, O.H.: Control of real-time systems with integer parameters. IEEE Trans. Autom. Control **67**(1), 75–88 (2022). https://doi.org/10. 1109/TAC.2020.3046578

23. Lime, D., Roux, O.H., Seidner, C., Traonouez, L.-M.: Romeo: a parametric model-checker for Petri Nets with stopwatches. In: Kowalewski, S., Philippou, A. (eds.) TACAS 2009. LNCS, vol. 5505, pp. 54–57. Springer, Heidelberg (2009). https:// doi.org/10.1007/978-3-642-00768-2_6

24. Maler, O., Pnueli, A., Sifakis, J.: On the synthesis of discrete controllers for timed systems. In: Mayr, E.W., Puech, C. (eds.) STACS 1995. LNCS, vol. 900, pp. 229–242. Springer, Heidelberg (1995). https://doi.org/10.1007/3-540-59042-0_76

25. Ramadge, P.J., Wonham, W.M.: Supervisory control of a class of discrete event processes. SIAM J. Control Optim. **25**(1), 206–230 (1987). https://doi.org/10.1137/ 0325013

26. Thomas, W.: On the synthesis of strategies in infinite games. In: Mayr, E.W., Puech, C. (eds.) STACS 1995. LNCS, vol. 900, pp. 1–13. Springer, Heidelberg (1995). https://doi.org/10.1007/3-540-59042-0_57

27. Toussaint, J., Simonot-Lion, F., Thomesse, J.P.: Time constraint verifications methods based on time Petri nets. In: IEEE, Future Trends in Distributed Computing Systems (FTDCS 1997), pp. 262–267 (1997)

Model Transformation

Transforming Dynamic Condition Response Graphs to Safe Petri Nets

Vlad Paul Cosma[1,2]([✉]), Thomas T. Hildebrandt[2], and Tijs Slaats[2]

[1] KMD, Ballerup, Denmark
vco@kmd.dk
[2] Computer Science Department, Copenhagen University, Copenhagen, Denmark
{vco,hilde,slaats}@di.ku.dk

Abstract. We present a transformation of the Dynamic Condition Response (DCR) graph constraint based process specification language to safe Petri Nets with inhibitor and read arcs, generalized with an acceptance criteria enabling the specification of the union of regular and ω-regular languages. We prove that the DCR graph and the resulting Petri Net are bisimilar and that the bisimulation respects the acceptance criterium. The transformation enables the capturing of regular and omega-regular process requirements from texts and event logs using existing tools for DCR requirements mapping and process mining. A representation of DCR Graphs as Petri Nets advances the understanding of the relationship between the two models and enables improved analysis and model checking capabilities for DCR graph specifications through mature Petri net tools. We provide a python script implementing the transformation from the DCR XML export format to the PNML exchange format extended with arc types. In the implementation, all read arcs are replaced by a pair of standard input and output arcs. This directly enables the simulation and analysis of the resulting Petri Nets in tools such as TAPAAL, but means that the acceptance criterium for infinite runs is not preserved.

Keywords: Petri Nets · DCR graphs · Bisimilarity

1 Introduction

Whereas process control-flow is traditionally captured using imperative notations such as Business Process Modelling Notation (BPMN), process requirements for information systems are typically presented as declarative rules, describing the constraints (i.e. provisions and obligations) for the execution of individual tasks in a process. For instance, a requirement for an e-shop application may specify that payment information must be provided before a payment can be made, and that a payment can be made and is required to eventually happen, if an order has been made. The requirements are typically translated to imperative code when the system is implemented.

In this paper we consider the transformation from process requirements presented in the declarative Dynamic Condition Response (DCR) graphs notation

© The Author(s), under exclusive license to Springer Nature Switzerland AG 2023
L. Gomes and R. Lorenz (Eds.): PETRI NETS 2023, LNCS 13929, pp. 417–439, 2023.
https://doi.org/10.1007/978-3-031-33620-1_22

to processes expressed in a variant of the well-known process notation of Petri Nets [39]. The DCR graphs notation was introduced in [12, 25] as a formal specification language for distributed workflows and further developed in a range of papers, e.g. adding time, sub processes and data (see e.g. [13, 14, 27, 32]).

In its core form, DCR graphs is a graph-based notation with a single kind of node and a few basic relations. Nodes of the graph denote actions (or events) of the process and four kinds of directed edges between nodes denoting constraints and effects between actions, as will be explained below in our e-shop running example. The core DCR graph notation can express all regular and ω-regular languages [8] and in particular liveness properties, e.g. that some action must eventually happen (not to be confused with the standard notion of live Petri Nets). The fact that DCR graphs can express all ω-regular languages makes the notation more expressive than the classical declarative process language of Linear-time Temporal Logic (LTL) [28] that can only express the star-free omega-regular languages. The DCR graph notation is also different from LTL in that it has an operational execution semantics, similarly to Petri Nets expressed as a marking on the nodes of the graph.

The declarative nature and operational semantics makes DCR graphs similar to the model of Petri Nets, yet there are still notable differences. Firstly, DCR graphs abstracts from the notion of places, which is prominent for Petri Nets. Secondly, DCR graphs can directly express infinitary languages and liveness properties. This makes DCR graphs closer to traditional declarative notations such as LTL and textual representations of rules. Indeed, the highlighter tool [20] supports the mapping back and forth between textual requirement specifications and DCR graphs. On the other hand, Petri Nets with their notion of tokens, branching and loops are closer to imperative notations such as BPMN processes. Moreover, while DCR graphs are supported by design and specification tools and process engines used by industry[1], there are still no powerful model checking tools as it is the case for Petri Nets, such as the TAPAAL tool [6].

Thus, the motivation for providing the transformation of DCR graphs to safe Petri Nets with inhibitor and read arcs is threefold, as illustrated in Fig. 1: Firstly, we provide a path for transforming declarative requirements supported by industrial design tools to Petri Nets, which are closer to imperative process models such as BPMN. Secondly, the transformation enables the use of Petri Nets verification and analysis tools, notably the TAPAAL tool [6], for DCR graph specifications. Finally, the transformation allows us to use the DisCoveR miner [3] to mine Petri nets via an intermediate DCR graph representation. Hereby we get the high accuracy of DisCoveR in an imperative model and maintain a higher degree of concurrency in the model than is usually the case for block-structured approaches. This was already demonstrated with an early (unsound) translation of DCR graphs to Petri Nets, which managed to win the prize for best imperative miner in the 2021 Process Discovery Contest[2].

---

[1] Available freely for academic use at DCRSolutions.net.

[2] https://icpmconference.org/2021/process-discovery-contest/.

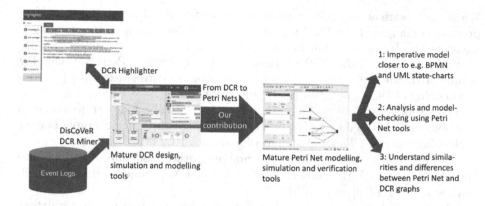

Fig. 1. Motivation for contributions of the paper

The paper is structured as follows. After the related work in Sect. 2, we give the definitions of core DCR graphs and Petri Nets with inhibitor arcs, read arcs and pending places in Sect. 3. We then proceed in Sect. 4 to provide the transformation of DCR graphs to Petri Nets, which is done by induction in the number of relations of the DCR graph. We also provide a sketch proof of the bisimilarity between the safe Petri Net and the DCR graph. Next we show how we reduce the size of the mapping in Sect. 5. We exemplify the mapping with a simple e-shop process. As usual we conclude and discuss future work in Sect. 6.

2 Related Work

Several notations for declarative process modelling have been developed. In addition to DCR graphs, the Declare [1] and Guard-Stage-Milestone (GSM) notations have also seen broad use in the business process management research community.

Declare provides a set of templates for modelling business constraints that are formalised as LTL formulae (parameterized by activities). A Declare model is the conjunction of a set of instantiated formulae. Given the limited expressiveness of the templates, a mapping from DCR graphs to Declare is not possible. Declare has been formalized in other languages such as coloured automata [21] and SCIFF [22,23]. Mappings from Declare to Petri Nets and R/I-nets were provided respectively in [30] and [7], however proofs of correctness are missing from each of these.

The GSM notation [19] takes a declarative data-centric approach to modelling processes, where stages of activities in the process are connected through guards that need to be satisfied for their activation and milestones that represent their acceptance criteria. A mapping has been proposed from Petri Nets to GSM [29], in particular with a focus on representing the output of process discovery algorithms (which usually produce Petri Nets) as GSM models.

We are not aware of any direct mappings in the opposite direction. Similarly [10] provides a mapping from DCR graphs to GSM models, an opposite mapping is mentioned as future work but has not yet materialised.

In [14] a subset of the DCR relations and their equivalent Petri Net mapping is presented, without inhibitor arcs without proof of correctness. [26] provides an encoding of DCR graphs as Büchi automata.

Petri Nets are widely used, and therefore there are also many translations to notations outside the declarative process modelling sphere, for example Ladder Logic Diagrams [35], Timed Automata [5] and mCRL2 [31].

Similarly much work has gone into mapping other modelling notations into Petri Nets, such as UML activity diagrams [34], UML sequence diagrams [38], UML state charts [18], and BPMN [9,31].

The work in [22] presents logic-based approaches which formalize regulatory models by relying on the deontic notions of obligations and permissions.

Different classes of ω-language Petri Nets have been introduced in [36] and their complexity has been studied in [11]. The definition of acceptance criteria for infinite words in [36] is based on markings being visited infinitely often, similar to the acceptance criteria of Büchi-automata. This differs from the acceptance criteria introduced in the present paper, which is based on pending places, for which tokens cannot rest infinitely without being consumed by a transition being fired.

3   Preliminaries

In this section we provide the running example and the formal definitions of Dynamic Condition Response graphs and safe Petri Nets with inhibitor and read arcs and pending places.

3.1   Running Example

We consider as running example a simple e-shop application that has the following specification:

(i) If an order is added, a payment for the order must eventually be made.
(ii) Payment information (eg. credit card number) must be provided before a payment can be executed.
(iii) The payment information can be edited any number of times.
(iv) A new order cannot be added before a subsequent payment has been made and payment can only be made if an order has been added and is not yet paid.

We can identify three actions in the system: Edit (or initially provide) Payment Information, Add Order and Make Payment. Below we will see how to model processes that fulfil these requirements as respectively DCR graphs and safe Petri Nets with inhibitor and read arcs and pending places.

3.2 Dynamic Condition Response Graphs

We give a formal definition of core Dynamic Condition Response (DCR) graphs as attributed directed graphs.[3] For a set A we write $\mathcal{P}(A)$ for the set of all subsets of A, i.e. the powerset of A and $\mathcal{P}_{ne}(A)$ for the set of all non-empty subsets of A.

Definition 1. *A DCR graph G is given by a tuple $(E, M, R, @, L, l)$ where*

(i) E is a finite set of events
(ii) $M = (Ex, Re, In) \in \mathcal{P}(E) \times \mathcal{P}(E) \times \mathcal{P}(E)$ is the marking
(iii) $R \subseteq E \times E$ is the set of relations *between events*
(iv) $@ : R \to \mathcal{P}_{ne}(\{\bullet\!\!-, \bullet\!\!\to, \to\!\!+, \to\!\%\})$ is the relation type assignment
(v) L is the set of action labels
(vi) $l : E \to L$ is the labelling *function assigning an action label to each event*

   The marking $M = (Ex, Re, In)$ describes the state of an event e in the following way. If e has been **executed** at least once then $e \in Ex$. If e is **pending** (i.e. it must eventually be executed) then $e \in Re$. If e is **included** (i.e. it is currently relevant) then $e \in In$.

   Assume a relation $r = (e, e') \in R$ from event e to e'. If $\bullet\!\!- \in @r$ we say r is a constraining relation. If $@r \cap \{\bullet\!\!\to, \to\!\!+, \to\!\%\} \neq \emptyset$ we say that r is an effect relation. Note that r can be both a constraining and an effect relation at the same time. We write $e \bullet\!\!- e'$ (or $e' \to\!\!\bullet e$) if $\bullet\!\!- \in @r$ and say there is a *condition* from e' to e. We write $e \bullet\!\!\to e'$ if $\bullet\!\!\to \in @r$ and say there is a response from e to e'. We write $e \to\!\!+ e'$ if $\to\!\!+ \in @r$ and say there is an include from e to e'. Finally, we write $e \to\!\% e'$ if $\to\!\% \in @r$ and say there is an exclude from e to e'.

   The behaviour of a DCR graph is given by a labelled transition system, where the states are markings and the transitions are the execution of a labelled event. Hereto comes a definition of when a finite or infinite execution sequence is accepting or not. We first define when events are enabled, i.e. can be executed.

Definition 2 (Event enabling). *Let $(E, M, R, @, L, l)$ be a DCR graph. An event $e \in E$ is* enabled *for the marking $M = (Ex, Re, In)$, writing* enabled(M, e) *if and only if:*

(i) $e \in In$
(ii) $\forall e' \in In.\ e' \to\!\!\bullet e \implies e' \in Ex$

   The conditions for event enabling state that for an event e to be enabled, (i) it must be included. (ii) Whenever e has a condition relation from an included event e', then this e' was executed at least once.

   We now define the effect of executing an event e for a given marking M.

---

[3] The presentation deviates slightly from the original definition given in [12] to facilitate the definition of the mapping to Petri Nets, but defines the same graph structures.

Definition 3. *Let G be a DCR graph with marking $M = (Ex, Re, In)$. The effect of executing an enabled event e is the marking* $\text{effect}_G(M, e) = (Ex', Re', In')$ *where*

$$
\begin{aligned}
Ex' &= Ex \cup \{e\} \\
Re' &= (Re \setminus \{e\}) \cup \{e' \mid e \bullet\!\!\rightarrow e'\} \\
In' &= (In \setminus \{e' \mid e \rightarrow\% e'\}) \\
&\quad \cup \{e' \mid e \rightarrow\!\!+ e'\}
\end{aligned}
$$

We are now ready to define the labelled transition semantics for DCR graphs.

Definition 4. *Let $G = (E, M, R, @, L, l)$ be a DCR graph. Define a labelled transition relation between markings by $M \xrightarrow{e}_G \text{effect}_G(M, e)$ if $\text{enabled}(M, e)$, where $e \in E$. Write $M \Rightarrow M'$ for $\exists e \in E.M \xrightarrow{e}_G M'$ and write $\Rightarrow^*$ for the reflexive and transitive closure of $\Rightarrow$. Define $\mathbb{M}_G = \{M' \mid M \Rightarrow^* M'\}$, i.e. the set of all reachable markings from the initial marking M of G. The labelled transition system for G is then defined as $[[G]] = (\mathbb{M}_G, M, \rightarrow_G \subset (\mathbb{M}_G \times E \times \mathbb{M}_G), L, l)$.*

Finally, we define when a finite or infinite execution sequence of a DCR graph is *accepting*. Intuitively, it is required that any included and pending event e in some intermediate state must eventually be executed or no longer included or pending in a later state. If one limits attention to finite execution sequences, the acceptance criteria is that no pending event is included in the final state.

Definition 5. *Let $G = (E, M_0, R, @, L, l)$ be a DCR graph. A finite or infinite sequence of transitions $M_0 \xrightarrow{e_0}_G M_1 \xrightarrow{e_1}_G \dots$ in $[[G]]$ with $M_i = (Ex_i, Re_i, In_i)$, is accepting if $e \in Re_i \cap In_i$ implies $\exists j \geq i.(e_j = e \vee e \notin Re_j \cap In_j)$.*

A DCR graph modelling our running example is shown in Fig 2. Events are depicted as boxes containing the action label of the event and relations as arrows. A relation with multiple types is depicted as multiple arrows between the same two events, one arrow for each type. Events that are included in the initial marking are drawn as boxes with a solid border, events that are excluded in the initial marking are drawn as boxes with a dashed border. Consequently, the events labelled EditPaymentInfo and AddOrder are initially included and the event labelled MakePayment is excluded in the initial marking of the graph.

The first requirement, "If an order is made, a payment for the order must eventually be made" is modelled by a response relation ($\bullet\!\!\rightarrow$ in blue) and an include relation ($\rightarrow\!\!+$ in green) from the event labelled AddOrder to the event labelled MakePayment. (The include relation is needed because of the interplay with the fourth requirement described below).

The second requirement, "Payment information (eg. credit card number) must be provided before a payment can be executed" is modelled by a condition relation ($\rightarrow\!\bullet$ or $\bullet\!\!-$ in orange) from the event labelled EditPaymentInfo to the event labelled MakePayment.

The third requirement, "The payment information may be provided at any time and any number of times." is modelled by having no condition relations

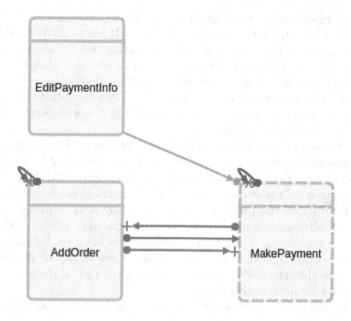

Fig. 2. DCR graph specification for the e-shop process

pointing to the event labelled EditPaymentInfo and making sure that it is included in the initial marking and never excluded.

The forth requirement is in two parts. The first part, "a new order cannot be made before a subsequent payment has been made" is modelled by an exclude relation (—% in red) from AddOrder to itself and an include relation from Make-Payment to AddOrder. The effect is that when AddOrder is executed, it excludes itself and is thus no longer available, except if MakePayment is executed, which will include AddOrder again. The second part, "payment can only be made if an order has been made and is not yet paid" is similarly modelled by an exclusion relation from MakePayment to itself and an inclusion relation from MakePayment to AddOrder.

3.3 Petri Nets with Inhibitor Arcs, Read Arcs and Pending Places

There are numerous variants of Petri Nets with different expressive power. As described in the introduction, we use safe Petri Nets with inhibitor and read arcs and a notion of both finite and infinite acceptance criteria. Inhibitor arcs (also called negative contextual arcs) are special arcs between places and transitions specifying the constraint that the transition is only enabled if all places related to it by inhibitor arcs are empty. In general, the addition of inhibitor arcs makes the model of Petri Nets Turing complete [2]. However, with the additional requirement of safeness, which means that places can hold at most one token (also known as the property of all the net places being 1-bounded), the notation is restricted to finite state models.

Read arcs (also called test, activator or positive contextual arcs) [4] specify the constraint that a transition is only enabled if all places related to it by read arcs have a token. A key difference between having a read arc and a pair of input and output arcs between a transition and a place, is that read arcs are not consuming the token. This means that two transitions with read arcs to the same place can occur concurrently [24]. However, if two transitions are connected to the same place by a read arc and a standard input arc respectively, the two transitions will still be in conflict.

The acceptance criteria we introduce is inspired by DCR graphs and allows us to conveniently express the union of regular and ω-regular languages, without needing to refer to explicit markings. The acceptance criteria is defined by indicating a subset of the states to be so-called *pending* places, and then define a finite or infinite execution sequence to be accepting if any token on a pending place is eventually subsequently consumed (but possibly placed back) by the execution of a transition. If one limits attention to finite execution sequences, the acceptance criteria is that all pending places are empty at the end of the execution. Note that the use of read arcs allows us to test, if there is a token on a pending place without consuming it.

We define Petri Nets with inhibitor arcs, read arcs and pending places as follows.

Definition 6. *A Petri Net with inhibitor and read arcs and pending places (PNirp) is a tuple $N = (P, T, A, Inhib, Read, Act, \lambda, Pe)$, where*

   (i) P is a finite set of places,
   (ii) T is a finite set of transitions s.t. $P \cap T = \emptyset$,
   (iii) $A = IA \sqcup OA$ is a finite set of input and output arcs, where:
     (a) $IA \subseteq P \times T$ is a finite set of input arcs,
     (b) $OA \subseteq T \times P$ is a finite set of output arcs,
   (iv) Inhib: $IA \longrightarrow \{true, false\}$ is a function defining inhibitor arcs,
   (v) Read: $IA \longrightarrow \{true, false\}$ is a function defining read arcs,
   (vi) Act is a set of labels (actions),
   (vii) $\lambda : T \to Act$ is a labelling function,
   (viii) $Pe \subseteq P$ is the set of pending places,

and the constraint that if $Inhib((p, t))$ then $\neg Read((p, t))$ and if $Read((p, t))$ then $\neg Inhib((p, t)) \wedge (t, p) \notin OA$. That is, an input arc cannot be both a read arc and an inhibitor arc. And if there is a read arc from place p to transition t, then there cannot be an output arc from transition t to p.

We only consider 1-bounded places in the present paper, which means that markings can be defined as simply a subset of places (the places containing a token).

Definition 7. (safe Marking). *Let $N = (P, T, A, Inhib, Read, Act, \lambda, Pe)$ be a PNirp. A safe marking M on N is a subset $M \subseteq P$ of places. We say there is a token x at a place $p \in P$, written $x \in M(p)$, if $p \in M$. The set of all markings over N is denoted by $M(N)$.*

We say that a Petri Net is safe if the execution of transitions preserves the safeness of markings. In this paper we will work only with safe Petri Nets, in particular we prove that the mapping from DCR graphs to Petri Nets provided in the next section always yields a safe Petri Net.

Assuming the Petri Net to be safe simplifies the definition of enabledness of transitions defined as follows.

Definition 8. (Enabledness). *Let* $N = (P, T, A, Inhib, Read, Act, \lambda, Pe)$ *be a PNirp. We say that a transition* $t \in T$ *is enabled in a marking* M, *if*

(i) *for* $t \in T$ *we have* $\{p \in P \mid (p, t) \in IA \wedge \neg Inhib((p, t))\} \subseteq M$, *i.e. for all input arcs except the inhibitor arcs there is a token in the input place,*

(ii) *for* $t \in T$ *we have* $\{p \in P \mid (p, t) \in IA \wedge Inhib((p, t))\} \cap M = \emptyset$, *i.e. for all inhibitor arcs there is* not *a token in the input place,*

We abuse notation and, just as for DCR graphs, let enabled(M, t) *denote that the transition* t *is enabled in marking* M.

Next we formalise the effect of executing (or firing) a transition. Again it is simplified by the assumption of safeness and we use the same notation as for DCR graphs to denote the result of firing a transition.

Definition 9. (Firing rule). *Let* $N = (P, T, A, Inhib, Read, Act, \lambda, Pe)$ *be a PNirp,* M *a marking on* N *and* $t \in T$ *a transition. If* enabled(M, t) *with* $Input(t) = \{p \in P \mid (p, t) \in IA \wedge \neg Inhib((p, t)) \wedge \neg Read((p, t))\}$ *and* $Output(t) = \{p \in P \mid (t, p) \in OA\}$ *then* t *can fire, i.e. be executed, and produce a marking* effect$_G(M, t) = (M \setminus Input) \cup Output$.

For convenience in the construction, we include the marking M in the $PNirp$ tuple and we use $N = (P, M, T, A, Inhib, Read, Act, \lambda, Pe)$ to refer to a safe marked $PNirp$ with marking $M \subseteq P$.

Similarly to how DCR graphs define labelled transition systems, the firing rule defines a labelled transition system for a $PNirp$ with markings as states and the labelled Petri Net transitions as labels.

Definition 10. *Let* $N = (P, M, T, A, Inhib, Read, Act, \lambda, Pe)$ *be a PNirp with safe marking* M. *Define a labelled transition relation between markings by* $M \xrightarrow{e}_N$ effect$_G(M, t)$ *if* enabled(M, t), *where* $t \in T$. *Write* $M \Rightarrow M'$ *for* $\exists t \in T . M \xrightarrow{t}_N M'$ *and write* $\Rightarrow^*$ *for the reflexive and transitive closure of* $\Rightarrow$. *Define* $\mathbb{M}_N = \{M' \mid M \Rightarrow^* M'\}$, *i.e. the set of all reachable markings from the initial marking* M *of* N. *The labelled transition system for* N *is then defined as* $[[N]] = (\mathbb{M}_N, M, \to_N \subset (\mathbb{M}_N \times T \times \mathbb{M}_N), Act, \lambda)$.

Finally, we define when a finite or infinite execution sequence of a $PNirp$ is accepting.

Definition 11. *Let* $N = (P, M, T, A, Inhib, Read, Act, \lambda, Pe)$ *be a PNirp with safe marking* M. *A finite or infinite sequence of transitions* $M_0 \xrightarrow{t_0}_N M_1 \xrightarrow{t_1}_N \ldots$ *in* $[[N]]$ *is accepting if* $p \in M_i \cap Pe$ *implies* $\exists j \geq i . (p, t_j) \in IA$.

Figure 3 shows the safe Petri Net resulting from the implemented optimized transformation of the running example DCR graph. The place pending_included_MakePayment is the only pending place. The arcs between the transition pend_MakePayment and the place executed_EditPaymentInfo are in fact a read arc in the transformation, but represented as a pair of standard input and output arcs in the implementation so we can simulate it in the TAPAAL tool.

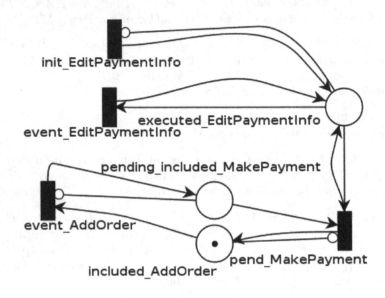

Fig. 3. E-shop Petri Net resulting from the transformation implementation.

It is worth noting that in the Petri Net we need two transitions labelled with the action EditPaymentInfo, namely a transition init_EditPaymentInfo mapping the initial execution (or the initial entry of the payment information) and a transition event_EditPaymentInfo mapping subsequent executions.

4   Mapping DCR Graphs to Petri Nets

In this section we provide the mapping from DCR graphs to marked safe Petri Nets with inhibitor and read arcs and pending places and prove that the DCR graph and the Petri Net have bisimilar transition semantics. The mapping has been implemented as a python script, which can be found at: https://github. com/paul-cvp/dcr-to-tapn.git, where we also provide some results of our mapping. We support the standard PNML [18] exchange format extended with arc types [16]. A key difference in the code mapping is that read arcs are automatically translated to a pair of input and output arcs. This was a design choice in order to maintain compatibility with a greater number of Petri Net verification tools and the difference only has consequences for the degree of concurrency and the acceptance criteria for infinite runs. In what follows we use the notation — (a straight line) to refer to read arcs and use the notation <—> to refer to a pair of input and output arcs.

The core part of the mapping is given as a function $DP : DCR \rightarrow PNirp$, defined inductively in the number of relations of the DCR graph. The Petri Net $DP(G)$ will have the events of G as labels (actions), since there will in general be more than one transition representing each event of G. To get the same observable behaviour as the DCR graph G, we then subsequently just need to compose the labelling function of $DP(G)$ with the event labelling function of G. Due to the rich structure of DCR graph markings, the basic inductive mapping in general produces a number of unused places and transitions. These can subsequently be removed by searching for unreachable transitions and places and merging places with the same arcs.

As part of the mapping $DP : DCR \rightarrow PNirp$, we also define for every $G \in DCR$ a mapping $DPM_G : \mathbb{M}_G \rightarrow \mathbb{M}_{DP(G)}$, i.e. from markings of G to the markings of $DP(G)$. For a DCR graph $G = (E, M, R, @, L, l)$ and $DP(G) = (P_{DP(G)}, M_{DP(G)}, T_{DP(G)}, A_{DP(G)}, Inhib_{DP(G)}, Read_{DP(G)}, Act_{DP(G)}, \lambda_{DP(G)}, Pe_{DP(G)})$ we then have $M_{DP(G)} = DPM_G(M)$ and $Act_{DP(G)} = E$.

The two mappings are defined so we get the following precise semantic correspondence between the two process models. (Note we write $\exists!$ to mean "there exists a unique").

Theorem 1. (Bisimilarity) *For $G \in DCR$ we have that the relation $Sim_G = \{(M, DPM_G(M)) \mid M \in \mathbb{M}_G\}$ is a bisimulation relation between $[[G]]$ and $[[DP(G)]]$, in the sense that $(M_0, DPM_G(M_0)) \in Sim_G$, where M_0 is the initial marking of G and for all $(M, DPM_G(M)) \in Sim$, we have*

(i) $M \xrightarrow{e} M'$ implies $\exists!t \in T_{DP(G)}.DPM_G(M) \xrightarrow{t} DPM_G(M')$ and $\lambda(t) = e$,

(ii) $DPM_G(M) \xrightarrow{t} M'$ and $\lambda(t) = e$ implies $M \xrightarrow{e} M''$ and $DPM_G(M'') = M'$.

That is, for every enabled event in a marking M of the DCR graph we have a unique enabled transition in the corresponding marking $DPM_G(M)$ of the Petri Net which is labelled by the event e and firing the transition changes the marking of the Petri Net to the marking corresponding to the DCR marking resulting from executing e - and vice versa. We will see below, that in addition the bisimulation also pairs accepting runs.

We now proceed to define the mapping function and outline the proof of Theorem 1 along the way. For each event $e \in E$ of the DCR graph G, there will be four places in $DP(G)$, which we will write as P_e^{Ex}, P_e^{In}, P_e^{Re}, and P_e^{Rex}. The first two places represent respectively if the event e has been executed and if it has been included. The last two places record the pending response state of the event e by a token in P_e^{Re} if and only if the event e is pending and included, and a token in P_e^{Rex} if and only if the event e is pending and excluded. The places P_e^{Re} will constitute the set $Pe_{DP(G)}$ of pending places.

Definition 12. (Places mapping) *Let $G = (E, M, R, @, L, l) \in DCR$. Define the corresponding Petri Net places of $DP(G)$ as $P_{DP(G)} = \{P_e^\gamma | e \in E, \gamma \in$*

$\{Ex, In, Re, Rex\}\}$. *Define the corresponding pending places of* $DP(G)$ *as* $Pe_{DP(G)} = \{P_e^{Re}|e \in E\}$.

Definition 13. *(Markings mapping)* *Let* $G = (E, M_0, R, @, L, l) \in DCR$ *and* $\mathbb{M}_G$ *be the reachable markings of* G *and* $\mathbb{M}_{DP(G)} = \mathcal{P}(P_{DP(G)})$, *i.e. all safe markings of the places* $P_{DP(G)}$ *defined above. Define* $DPM_G : \mathbb{M}_G \to \mathbb{M}_{DP(G)}$ *as follows. For* $M = (Ex, Re, In) \in \mathbb{M}_G$ *define* $DPM_G(M)$ *such that for any event* $e \in E$,

(i) $P_e^{Ex} \in DPM_G(M) \iff e \in Ex$
(ii) $P_e^{In} \in DPM_G(M) \iff e \in In$
(iii) $P_e^{Re} \in DPM_G(M) \iff e \in Re \wedge e \in In$
(iv) $P_e^{Rex} \in DPM_G(M) \iff e \in Re \wedge e \notin In$

The events and relations of a DCR graph are represented by respectively transitions and arcs in the Petri Net. Each event of the DCR graph will be represented by several transitions in the Petri Net. Indeed, the number of transitions representing each event depends on the number of relations in the DCR graph.

We define the corresponding Petri Net transitions $T_{PD(G)}$, arcs $A_{PD(G)}$ and labelling function $\lambda_{PD(G)}$ by induction in the number $k = |R|$ of relations. We will at the same time argue for the proof of Theorem 1, since it also follows from the inductive construction.

In the base case, $k = 0$ i.e. a DCR graph $G = G_0 = (E, M, \emptyset, @, L, l)$ without any relations, each event will be represented by a (sub) Petri Net as shown in Fig. 4 (assuming a marking M, where the event is included, not executed and not pending, i.e. $e \in In, e \notin Ex \cup Re$) which is completely independent of the similar sub Petri Nets representing the other events. We have four transitions for each event $e \in E$ of the DCR graph and arcs as shown in Fig. 4. We use the labelling function of the Petri Net to label all the transitions with the event e and thereby record that the transitions represent this event in the DCR graph. That is, we define the mapping formally as follows.

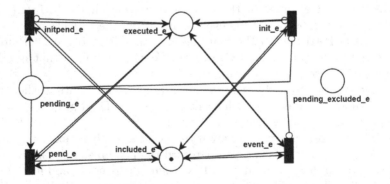

Fig. 4. Base case: Petri Net for a single included DCR event, which is not yet executed nor pending.

Definition 14. (Base case: Mapping a DCR graph with no relation)
Let $G = (E, M, \emptyset, @, L, l)$ be a DCR Graph with no relations. Then $P_{DP(G)} = \{p_e^\delta | e \in E, \delta \in \{In, Ex, Re, Rex\}\}$, $T_{DP(G)} = \{t_e^\delta | e \in E, \delta \in \{event, init, pend, initpend\}\}$ and $\lambda_{DP(G)}(t_e^\delta) = e$. The set of arcs $A_{DP(G)} = IA_{DP(G)} \cup OA_{DP(G)}$ and $Inhib_{DP(G)} : IA_{DP(G)} \to \{true, false\}$ are defined by Table 1. Each row in the table corresponds to one of the four transitions, and each column to one of the four places. Each entry is an arc in which the left arrow is an input arc and the right arrow is an output arc. That is, the input and output arc pair $<\!\!-\!\!>$ in the entry of column p_e^{Ex} and row t_e^{event} in the table means that we have $(p_e^{Ex}, t^{event}) \in IA_{DP(G)}$ and $(t_e^{event}, p_e^{Ex}) \in OA_{DP(G)}$. A table entry of $o\!-$ in the entry of column p_e^{Re} and row t_e^{event} means we add an arc $(p_e^{Re}, t^{event}) \in IA_{DP(G)}$, which is an inhibitor arc, i.e. $Inhib((p_e^{Re}, t_e^{event})) = true$, and we add no arc in $OA_{DP(G)}$. The read arc $-$ between column p_e^{In} and row t_e^{event} is mapped as $(p_e^{In}, t^{event}) \in IA_{DP(G)}$ and $Read((p_e^{In}, t_e^{event})) = true$.

Table 1. Arc patterns for an event in the base case

	p_e^{In}	p_e^{Ex}	p_e^{Re}	p_e^{Rex}
t_e^{event}	$-$	$<\!\!-\!\!>$	$o\!-$	
t_e^{init}	$-$	$o\!\!-\!\!>$	$o\!-$	
$t_e^{initpend}$	$-$	$o\!\!-\!\!>$	$<\!\!-$	
t_e^{pend}	$-$	$<\!\!-\!\!>$	$<\!\!-$	

Note that we have no arcs in any directions connected to the place p_e^{Rex}. This means that this place is redundant, unless more relations are added to the DCR graph, which will give rise to more transitions in the Petri Net. We will reuse the same table notation style for arc pattern mappings throughout the paper. Figure 4 is a visual representation of Table 1.

Proof sketch of Theorem 1: base case (1). It follows by a trivial inspection of the event execution cases and the different initial markings that we have the bisimulation property in Theorem 1 for the base case. If an event e is initially included and not executed nor pending, then we have the marking in Fig. 4. Observe that only the transition labelled *init_e* can fire, which will read the token at the place *included_e* and put a token at the place *executed_e*. This corresponds to the execution semantics of DCR graphs. Subsequently, only the transition labelled *event_e* can fire and firing the transition will read the tokens at the places *included_e* and *executed_e*. If the event e is initially pending, included and not executed, it will fire first the transition *initpend_e* after which only the transition *event_e* can fire. If the event e is initially pending, included and executed, it will fire first the transition *pend_e* after which only the transition *event_e* that can fire. Finally, if the event is not included in the initial DCR marking, there will be no token in the *inlcuded_e* place and consequently no transition can fire. □

Now consider the induction step. Let $G = (E, M, R, @, L, l)$ be a DCR Graph with $R = \{r_1, \ldots, r_k, r_{k+1}\}$ and assume we have defined the mapping and proven Theorem 1 for $G_k = (E, M, \{r_1, \ldots, r_k\}, @, L, l)$.

We proceed by cases of the type $@r_{k+1}$ of the relation $r_{k+1} = (e, e')$. Effect relations change the marking of e' when e fires, and thus refines the transitions for e and adds arcs connected to the places recording the marking for e'. Dually, constraining relations requires a refinement of the transitions for e', adding arcs connected to the places recording the marking for e.

We first consider the cases where the relation $r_{k+1} = (e, e')$ is a single effect, i.e. $@r_{k+1} \in \{\{\rightarrow+\}, \{\rightarrow\%\}, \{\bullet\rightarrow\}\}$ and $e \neq e'$.

For $@r_{k+1} = \{\rightarrow+\}$ we replace each transition t_e^δ representing the event e with three new transitions $t_e^{0,\delta}$, $t_e^{1,\delta}$ and $t_e^{2,\delta}$, which in addition to the arcs connected to t_e^δ also get the new arcs shown in Table 2a. More formally we say that we apply the Definition 15 below for relation $@r_{k+1} = \{\rightarrow+\}$ and the arc pattern table Table 2a.

Definition 15. *(Inductive step: Mapping a DCR relation)*
Let $G = (E, M, R, @, L, l)$ be a DCR Graph, $APT(@r)$ be the arc pattern table for any given relation $r \in @R$ and $|APT(@r)_r|$ be the number of rows (transition copies) in the arc pattern table. Then $T_{DP(G_{k+1})} = T_{DP(G_k)} \backslash \{t_e^\delta \mid t_e^\delta \in T_{DP(G_k)}$ and $\lambda_{DP(G_k)}(t_e^\delta) = e\} \cup \{t_e^{i,\delta} \mid i \in \{0, 1, .., |APT(@r)_r| - 1\}\}$ and let $(p, t_e^{i,\delta}) \in IA_{DP(G_{k+1})}$ for $i \in \{0, 1.., |APT(@r)_r| - 1\}$ if and only if $(p, t_e^\delta) \in IA_{DP(G_k)}$ and let $(t_e^{i,\delta}, p) \in OA_{DP(G_{k+1})}$ for $i \in \{0, 1.., |APT(@r)_r| - 1\}$ if and only if $(t_e^\delta, p) \in OA_{DP(G_k)}$ or $(t_e^{i,\delta}, p)$ is one of the arcs in $APT(@r)$. Finally, for $t \in A_{DP(G_k)}$ such that $\lambda_{DP(G_k)} \neq e$, let $(p, t) \in IA_{DP(G_{k+1})}$ and $(t, p) \in OA_{DP(G_{k+1})}$ if and only if $(p, t) \in IA_{DP(G_k)}$ or $(t, p) \in OA_{DP(G_k)}$.

Table 2. Arc patterns for $r_{k+1} = (e, e')$ effect relations

$\rightarrow+$	$p_{e'}^{In}$	$p_{e'}^{Re}$	$p_{e'}^{Rex}$
$t_e^{0,\delta}$			
$t_e^{1,\delta}$	o—>	—>	<—
$t_e^{2,\delta}$	o—>		o—

(a) Arc patterns for $\rightarrow+$

$\rightarrow\%$	$p_{e'}^{In}$	$p_{e'}^{Re}$	$p_{e'}^{Rex}$
$t_e^{0,\delta}$	o—		
$t_e^{1,\delta}$	<—	o—	
$t_e^{2,\delta}$	<—	<—	—>

(b) Arc patterns for $\rightarrow\%$

$\bullet\rightarrow$	$p_{e'}^{In}$	$p_{e'}^{Re}$	$p_{e'}^{Rex}$
$t_e^{0,\delta}$	—	o—>	
$t_e^{1,\delta}$	—	—	
$t_e^{2,\delta}$	o—		o—>
$t_e^{3,\delta}$	o—		—

(c) Arc patterns for $\bullet\rightarrow$

The cases for $@r_{k+1} = \{\rightarrow\%\}$ and $@r_{k+1} = \{\bullet\rightarrow\}$ follow the same approach, by applying Definition 15. Observe that we need four copies for each existing transition for e in the case of the response relation. The case for response is also illustrated graphically in Fig. 5.

Example 1. (Mapping the response relation) Fig. 5 shows how the response relation $e \bullet\rightarrow e'$ is mapped by replacing each existing transition t_e^δ (t_delta_e) representing e by four new copies, connected to the places representing the marking of the event e'.

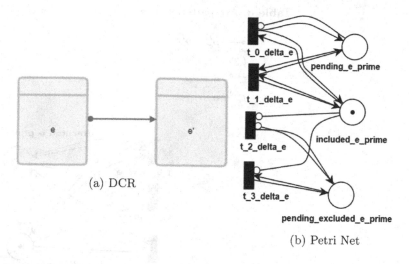

(a) DCR

(b) Petri Net

Fig. 5. Mapping (a) DCR response relation to a Petri Net notation (b)

Proof sketch of Theorem 1: *single effect mapping(2)*. First note that adding an effect relation from e to e' to the DCR graph only changes the output transitions representing e in $DP(G_k)$. Here we need three transitions, covering the different possibilities of the marking of e'. For the $e \rightarrow\!\!+ e'$, transition $t_e^{0,\delta}$ handles the case, where e' is already included, transition $t_e^{1,\delta}$ handles the case, where e' is not included, but pending and $t_e^{2,\delta}$ handles the case, where e' is not included and not pending. We follow a similar reasoning for $e \rightarrow\!\!+ e', e \rightarrow\!\!\% e'$ and $e \bullet\!\!\rightarrow e'$. □

We now consider the constraining relation consisting of a single condition, i.e. $@r_{k+1} = \{\bullet\!\!-\}$ and $e' r_{k+1} e$ and $e \neq e'$. For the condition relation we replace all existing transitions of e' with 3 new copies, again keeping the old arcs and adding new arcs to the places of e according to Table 3. This is also illustrated graphically in Fig. 6. Again we apply Definition 15.

Proof sketch of Theorem 1: *single constraint mapping(3)*. The transition copy $t_e^{0,\delta}$ handles the case where e is included and already executed. The transition $t_e^{1,\delta}$ handles the case where e is excluded and not already executed. Finally, the transition $t_e^{2,\delta}$ handles the case where e is excluded and already executed. □

Example 2. (Mapping a condition relation). Figure 6 shows how a condition relation is mapped between the transitions representing the DCR event e' and the places representing the execution and inclusion marking for the DCR event e.

Now we proceed to describe the cases of relations where the events e and e' are identical, and thereafter the cases of multiple relations between the same two events.

Table 3. Arc patterns for ●—

●—	$p_{e'}^{In}$	$p_{e'}^{Ex}$
$t_e^{0,\delta}$	—	—
$t_e^{1,\delta}$	o—	o—
$t_e^{2,\delta}$	o—	—

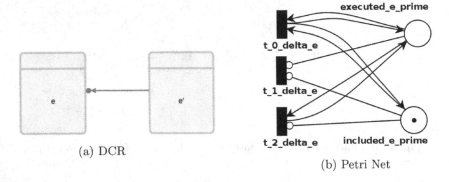

(a) DCR

(b) Petri Net

Fig. 6. Mapping (a) DCR condition relation to a Petri Net notation (b)

Case *e = e' and a single relation:* For single relations where the source and target is the same, we do not get the same multiplication of transitions:

(i) $@r = \{\rightarrow +\}$: Do nothing, since it can only take effect if e is already included.

(ii) $@r = \{-\%\}$: Remove all output arcs from transitions t_e^δ (i.e. transitions with label e) to the place p_e^{In}, because the event can only be executed if there is a token at p_e^{In}, and that token should not be put back.

(iii) $@r = \{\bullet\rightarrow\}$: Add output arcs from all transitions t_e^δ to p_e^{Re} and replace all read arcs from p_e^{Re} to transitions t_e^δ with standard input arcs.

(iv) $@r = \{\bullet-\}$: Remove the transitions $\{t_e^{init}, t_e^{initpend}\}$ and all their associated arcs.

We now consider the cases with multiple relations between the same two events, i.e. $|@r_{k+1}| > 1$. When we have both an include and exclude relation, i.e. $\{\rightarrow +, -\%\} \subseteq @r_{k+1}$ we only apply the include relation, as the DCR graph semantics stipulate that first the exclusion takes place and then the inclusion.

Case *e = e' and* $@r_{k+1} = \{\bullet\rightarrow, -\%\}$: For all transitions t such that $\lambda_{DP(G_k)}(t) = e$, add output arcs from t to p_e^{Rex} and remove all output arcs from t to p_e^{In}. Define $\lambda_{DP(G_{k+1})} = \lambda_{DP(G_k)}$.

The remaining cases look at multiple relations $|@r_{k+1}| > 1$ between different events $e \neq e'$. We again follow the same reasoning as in part 2 and 3 of the proof, i.e. for a given $e\, r_{k+1}\, e'$ use its arc pattern table to make copies of existing transitions and their arcs and create new arc mappings to the existing

Table 4. Effect and constraint pair arc patterns

•← ∧ →┼	$p_{e'}^{In}$	$p_{e'}^{Ex}$	$p_{e'}^{Re}$	$p_{e'}^{Rex}$
$t_e^{0,\delta}$	—	—		
$t_e^{1,\delta}$	o—>	o—	—>	<—
$t_e^{2,\delta}$	o—>	o—		o—
$t_e^{3,\delta}$	o—>	—	—>	<—
$t_e^{4,\delta}$	o—>	—		o—

(a) Arc patterns for $e' \bullet\!\!\leftarrow e \wedge e \rightarrow\!\!\!+ e'$

•← ∧ ⇸%	$p_{e'}^{In}$	$p_{e'}^{Ex}$	$p_{e'}^{Re}$	$p_{e'}^{Rex}$
$t_e^{0,\delta}$	<—	—	o—	
$t_e^{1,\delta}$	<—	—	<—	—>
$t_e^{2,\delta}$	o—	o—		
$t_e^{3,\delta}$	o—	—		

(b) Arc patterns for $e' \bullet\!\!\leftarrow e \wedge e \rightarrow\!\!\!\% e'$

•← ∧ •→	$p_{e'}^{In}$	$p_{e'}^{Ex}$	$p_{e'}^{Re}$	$p_{e'}^{Rex}$
$t_e^{0,\delta}$	—	—	o—>	
$t_e^{1,\delta}$	—	—	—	
$t_e^{2,\delta}$	o—	o—		o—>
$t_e^{3,\delta}$	o—	o—		—
$t_e^{4,\delta}$	o—	—		o—>
$t_e^{5,\delta}$	o—	—		—

(c) Arc patterns for $e' \bullet\!\!\leftarrow e \wedge e \bullet\!\!\rightarrow e'$

places. Formally we apply Definition 15 for each relation and arc pattern table mentioned.

Case $e \neq e'$ *and* r_{k+1} *is both constraining and effect:* When an effect constraint pair exists, i.e. $\bullet\!\!\leftarrow\, \in @r_{k+1}$ and $@r_{k+1} \cap \{\bullet\!\!\rightarrow, \rightarrow\!\!\%, \rightarrow\!\!\!+\} \neq \emptyset$, their mapping produces arcs that both check the necessary places and also change their marking. Given $e\, r_{k+1}\, e'$, we consider the different cases as follows:

  (i) $@r_{k+1} = \{\bullet\!\!\leftarrow, \rightarrow\!\!\!+\}$: The arc pattern is shown in Table 4a.
 (ii) $@r_{k+1} = \{\bullet\!\!\leftarrow, \rightarrow\!\!\%\}$: The arc pattern is shown in Table 4b.
(iii) $@r_{k+1} = \{\bullet\!\!\leftarrow, \bullet\!\!\rightarrow\}$: The arc pattern is shown in Table 4c.

Note that brown arcs show the changes done to the arc patterns for the condition relation from Table 3.

Table 5. Two effect relations arc patterns

•→ ∧ →┼	$p_{e'}^{In}$	$p_{e'}^{Re}$	$p_{e'}^{Rex}$
$t_e^{0,\delta}$	—	o—>	
$t_e^{1,\delta}$	—	—	
$t_e^{2,\delta}$	o—>	o—>	o—
$t_e^{3,\delta}$	o—>	—>	<—

(a) Arc pattern for $e \bullet\!\!\rightarrow e' \wedge e \rightarrow\!\!\!+ e'$

•→ ∧ ⇸%	$p_{e'}^{In}$	$p_{e'}^{Re}$	$p_{e'}^{Rex}$
$t_e^{0,\delta}$	<—	o—	—>
$t_e^{1,\delta}$	<—	<—	—>
$t_e^{2,\delta}$	o—		o—>
$t_e^{3,\delta}$	o—		—

(b) Arc pattern for $e \bullet\!\!\rightarrow e' \wedge e \rightarrow\!\!\%\, e'$

Case $e \neq e'$, $e\, r_{k+1}\, e'$ *and* r_{k+1} *is composed of 2 effect relations:*

(i) $@r_{k+1} = \{-\!\%, \rightarrow\!+\}$: Is equivalent to only mapping the include relation.
(ii) $@r_{k+1} = \{\bullet\!\rightarrow, \rightarrow\!+\}$: The arc pattern is shown in Table 5a.
(iii) $@r_{k+1} = \{\bullet\!\rightarrow, -\!\%\}$: The arc pattern is shown in Table 5b.
(iv) $@r_{k+1} = \{\bullet\!\rightarrow, \rightarrow\!+, -\!\%\}$: Equivalent to the case $@r_{k+1} = \{\bullet\!\rightarrow, \rightarrow\!+\}$.

Note that brown arcs show the changes done to the arc patterns for the response relation from Table 2c.

Table 6. Two effect relations and a condition relation arc patterns

•— ∧ •→ ∧ →+	p_e^{In}	$p_{e'}^{Ex}$	$p_{e'}^{Re}$	$p_{e'}^{Rex}$
$t_e^{0,\delta}$	—	—	o—>	
$t_e^{1,\delta}$	—	—	—	
$t_e^{2,\delta}$	o—>	o—	—>	<—
$t_e^{3,\delta}$	o—>	o—	o—>	o—
$t_e^{4,\delta}$	o—>	—	—>	<—
$t_e^{5,\delta}$	o—>	—	o—>	o—

(a) Arc pattern for $e' \bullet\!\!- e \wedge e \bullet\!\!\rightarrow e' \wedge e \rightarrow\!+ e'$

•— ∧ —% ∧ •→	$p_{e'}^{In}$	$p_{e'}^{Ex}$	$p_{e'}^{Re}$	$p_{e'}^{Rex}$
$t_e^{0,\delta}$	<—	—	o—	—>
$t_e^{1,\delta}$	<—	—	<—	—>
$t_e^{2,\delta}$	o—	o—		o—>
$t_e^{3,\delta}$	o—	o—		—
$t_e^{4,\delta}$	o—	—		o—>
$t_e^{5,\delta}$	o—	—		—

(b) Arc pattern for $e' \bullet\!\!- e \wedge e \bullet\!\!\rightarrow e' \wedge e -\!\% e'$

Case $e \neq e'$, $e\, r_{k+1}\, e'$ *and* r_{k+1} *is a composed of 2 effect relations and a condition relation:*

(i) $@r_{k+1} = \{\bullet\!\!-, \rightarrow\!+, -\!\%\}$: Equivalent to the case $@r_{k+1} = \{\bullet\!\!-, \rightarrow\!+\}$.
(ii) $@r_{k+1} = \{\bullet\!\!-, \rightarrow\!+, \bullet\!\rightarrow\}$: The arc pattern is shown in Table 6a.
(iii) $@r_{k+1} = \{\bullet\!\!-, \bullet\!\rightarrow, -\!\%\}$: The arc pattern is shown in Table 6b.

Note that brown arcs show the changes done to the arc patterns for the condition response relation mapping from Table 4c.

This completes the inductive definition of the mapping from DCR graphs to safe Petri Nets with inhibitor arcs, read arcs and pending places as there are no other exceptional cases.

Proof sketch of Theorem 1: exhaustive mapping of exceptional cases(4). We follow the same reasoning as part 2 and 3 of the proof i.e. we take each case sub-point and detail all the possible changes in marking of the DCR Graph in order to show that there is a transition and arc pattern that handles this in the mapped *PNirp*.   □

Parts 1 to 4 of the proof of Theorem 1 show how the strong bisimilarity property is preserved by each induction step in the definition. We believe the reader should be convinced of how the entire Petri Net is constructed by following the inductive transformation.

What remains to show is that the accepting runs in the two models are the same. This follows easily from the correspondence of markings in Definition 13, which is maintained by the bisimulation relation.

Proposition 1. *For a DCR Graph DCR graph $G = G_0 = (E, M, R, @, L, l)$ it holds that an execution sequence $M \rightarrow M_1 \rightarrow M_2 \rightarrow ..$ is accepting if and only if $DPM_G(M) \rightarrow DPM_G(M_1) \rightarrow DPM_G(M_2) \rightarrow ..$ is accepting.*

5   Pruning and Reachability Analysis

We report the steps needed to reduce the size of the mapped $PNirp$. This is achieved in two ways: pruning away transitions and places based on the DCR Graph relations and marking; and based on a reachability analysis of the $PNirp$.

5.1   Pruning Based on the DCR Graph

Creating Places. Given a DCR Graph we follow these intuitions when creating places. Only create an:

(i) Included Place for events that are included and may become excluded (have an exclusion relation towards them) and events that are not included and may become included (have an inclusion relation towards them);

(ii) Executed Place for events that have a condition from them;

(iii) Pending Place for events that have a response relation to them and events that are initially pending;

(iv) Pending Excluded Place for events that need both an Included Place and a Pending Place.

Creating Event Transitions. Given a DCR Graph and a $PNirp$ we follow these intuitions when creating transitions. Only create:

(i) init labelled transitions for events that need an Executed Place;

(ii) pend labelled transitions for events that need a Pending Place.

The pruning is done during the inductive construction of the Petri Net.[4]

5.2   Petri Net Reachability Analysis

Pruning Based on the Reachability Graph. Our mapping creates dead transitions because it preemptively creates arc patterns for both the marked and unmarked state of a place. Then at each induction step we expand the set of dead transitions, either because we need to copy the dead transition or if we create new transitions and map arcs from a place the dead transition should have an effect on.

Reachability analysis is done on the Petri Net reachability graph which is a labelled transition system where the states are the set of places and the transitions represent the set of transitions fired to move from one state to another. The optimization on the $PNirp$ is done by removing all places and transitions that are not part of the reachability graph.

---

[4] We direct the reader to the Appendix in our repository https://github.com/paul-cvp/dcr-to-tapn/blob/master/appendix/Appendix.pdf to see the simplified arc pattern tables.

Merging Places. Finally it is possible to merge places that label the same state in the reachability graph of the Petri Net. The merging also requires us to update the set of pending places accordingly. Notice that in the e-shop example AddOrder and MakePayment have the relation the $r@ = \{\bullet\rightarrow, \rightarrow\!\!+\}$ and the initial marking of AddOrder is not included and not pending. Therefore we merged the pending and included places of MakePayment. Notice that this would not have been possible if the event AddOrder was initially included.

Definition 16. *(Equivalent places) Let $p, p' \in P$. We say that $p \equiv p'$ if the set of input and output arcs is equal and also the arc type. We define a new place p'' with the merged ids of p and p' and copy their input and output arcs and also the arc type. (Updating Pe) If $(p \in Pe \vee p' \in Pe) \wedge p \equiv p' \iff p'' \in Pe$.*

5.3 Space Analysis on the Running Example

The unoptimized Petri Net of our e-shop has 12 places, 56 transitions and 488 arcs. The DCR analysis pruned one has 5 places, 10 transitions and 49 arcs. Doing just the Petri Net reachability analysis yields 7 places, 6 transitions and 42 arcs. The full optimization, as show in Fig. 3 has 3 places, 4 transitions and 12 arcs.

6 Conclusion and Future Work

We presented a transformation from the Dynamic Condition Response (DCR) graph constraint based process specification language to safe Petri Nets with inhibitor arcs and read arcs, generalized with an acceptance crietria for the modelling of ω-regular liveness properties. We outlined the proof for strong bisimilarity between the transition system for the DCR graph and the transition system for the resulting Petri Net, also preserving the acceptance criteria of finite and infinite executions.

We believe the work in the present paper provides a plethora of research avenues, which we aim to explore in future work. Concretely, we plan to extend the transformation from the core DCR relations to include features of later versions, in particular to cover Timed DCR graphs [14], thereby providing a complete mapping from Timed DCR graphs to safe Timed Arc Petri Nets and also extend the strong bisimulation correspondence to support this case. We plan to evaluate the space complexity of the mapping and the complexity of the resulting models by using the DisCoveR [3] miner to mine DCR Graphs from well-known, real-life, public event logs and map these to their Petri Net counter parts. We also aim to improve the optimization step by using DCR Event-Reachability [17] and by detecting handmade rules such as in [37].

As seen by our running example, the mapping nicely captures concurrency between independent events. This could potentially also be combined with a mapping from Petri Nets to BPMN [9], to provide an output following an ISO standard process notation. Finally we aim to integrate the existing mapping from safe Timed Arc Petri Nets to Timed Automata [33] to provide a link from DCR Graphs to Timed Automata.

References

1. van der Aalst, W.M.P., Pesic, M.: DecSerFlow: towards a truly declarative service flow language. In: Bravetti, M., Nunez, M., Zavattaro, G. (eds.) Proceedings of Web Services and Formal Methods (WS-FM 2006), vol. 4184, pp. 1–23 (2006)
2. Agerwala, T.: A complete model for representing the coordination of asynchronous processes. Hopkins Computer Research Report 32 (1974)
3. Back, C.O., Slaats, T., Hildebrandt, T.T., Marquard, M.: Discover: accurate and efficient discovery of declarative process models. Int. J. Soft. Tools Technol. Transfer **24**, 563–587 (2022)
4. Baldan, P., Busi, N., Corradini, A., Michele Pinna, G.: Functional concurrent semantics for Petri nets with read and inhibitor arcs. In: Palamidessi, C. (ed.) CONCUR 2000. LNCS, vol. 1877, pp. 442–457. Springer, Heidelberg (2000). https://doi.org/10.1007/3-540-44618-4_32
5. Byg, J., Jørgensen, K.Y., Srba, J.: An efficient translation of timed-arc Petri nets to networks of timed automata. In: Breitman, K., Cavalcanti, A. (eds.) ICFEM 2009. LNCS, vol. 5885, pp. 698–716. Springer, Heidelberg (2009). https://doi.org/10.1007/978-3-642-10373-5_36
6. Byg, J., Jørgensen, K.Y., Srba, J.: TAPAAL: editor, simulator and verifier of timed-arc Petri nets. In: Liu, Z., Ravn, A.P. (eds.) ATVA 2009. LNCS, vol. 5799, pp. 84–89. Springer, Heidelberg (2009). https://doi.org/10.1007/978-3-642-04761-9_7
7. De Smedt, J., Vanden Broucke, S., De Weerdt, J., Vanthienen, J.: A full r/i-net construct lexicon for declare constraints. Available at SSRN 2572869 (2015)
8. Debois, S., Hildebrandt, T.T., Slaats, T.: Replication, refinement & reachability: complexity in dynamic condition-response graphs. Acta Informatica **55**(6), 489–520 (2018)
9. Dijkman, R.M., Dumas, M., Ouyang, C.: Formal semantics and analysis of BPMN process models using petri nets. Queensland University of Technology, Tech. Rep, pp. 1–30 (2007)
10. Eshuis, R., Debois, S., Slaats, T., Hildebrandt, T.: Deriving consistent GSM schemas from DCR graphs. In: Sheng, Q.Z., Stroulia, E., Tata, S., Bhiri, S. (eds.) ICSOC 2016. LNCS, vol. 9936, pp. 467–482. Springer, Cham (2016). https://doi.org/10.1007/978-3-319-46295-0_29
11. Finkel, O.: On the high complexity of Petri nets ω-languages. In: Janicki, R., Sidorova, N., Chatain, T. (eds.) PETRI NETS 2020. LNCS, vol. 12152, pp. 69–88. Springer, Cham (2020). https://doi.org/10.1007/978-3-030-51831-8_4
12. Hildebrandt, T.T., Mukkamala, R.R.: Declarative event-based workflow as distributed dynamic condition response graphs. In: PLACES, pp. 59–73 (2010)
13. Hildebrandt, T.T., Normann, H., Marquard, M., Debois, S., Slaats, T.: Decision modelling in timed dynamic condition response graphs with data. In: Marrella, A., Weber, B. (eds.) BPM 2021. LNBIP, vol. 436, pp. 362–374. Springer, Cham (2022). https://doi.org/10.1007/978-3-030-94343-1_28
14. Hildebrandt, T., Mukkamala, R.R., Slaats, T., Zanitti, F.: Contracts for cross-organizational workflows as timed dynamic condition response graphs. J. Logic Algebraic Programm. **82**(5), 164–185 (2013). ISSN 1567-8326. https://doi.org/10.1016/j.jlap.2013.05.005. https://www.sciencedirect.com/science/article/pii/S1567832613000283. Formal Languages and Analysis of Contract-Oriented Software (FLACOS2011)

15. Hillah, L.M., Kordon, F., Petrucci, L., Trèves, N.: PNML framework: an extendable reference implementation of the petri net markup language. In: Lilius, J., Penczek, W. (eds.) PETRI NETS 2010. LNCS, vol. 6128, pp. 318–327. Springer, Heidelberg (2010). https://doi.org/10.1007/978-3-642-13675-7_20

16. Hillah, L.-M., Kordon, F., Lakos, C., Petrucci, L.: Extending PNML scope: a framework to combine Petri nets types. In: Jensen, K., van der Aalst, W.M., Ajmone Marsan, M., Franceschinis, G., Kleijn, J., Kristensen, L.M. (eds.) Transactions on Petri Nets and Other Models of Concurrency VI. LNCS, vol. 7400, pp. 46–70. Springer, Heidelberg (2012). https://doi.org/10.1007/978-3-642-35179-2_3

17. Høgnason, T., Debois, S.: DCR event-reachability via genetic algorithms. In: Daniel, F., Sheng, Q.Z., Motahari, H. (eds.) BPM 2018. LNBIP, vol. 342, pp. 301–312. Springer, Cham (2019). https://doi.org/10.1007/978-3-030-11641-5_24

18. Hu, Z., Shatz, S.M.: Mapping UML diagrams to a petri net notation for system simulation. In SEKE, pp. 213–219. CiteSeer (2004)

19. Hull, R., et al.: Introducing the guard-stage-milestone approach for specifying business entity lifecycles. In: Bravetti, M., Bultan, T. (eds.) WS-FM 2010. LNCS, vol. 6551, pp. 1–24. Springer, Heidelberg (2011). https://doi.org/10.1007/978-3-642-19589-1_1

20. López, H.A., Debois, S., Hildebrandt, T.T., Marquard, M.: The process highlighter: from texts to declarative processes and back. In: Proceedings of the Dissertation Award and Demonstration, Industrial Track at BPM 2018, vol. 2196 (2018). https://CEUR-WS.org/VOL-2196/

21. Maggi, F.M., Montali, M., Westergaard, M., van der Aalst, W.M.P.: Monitoring business constraints with linear temporal logic: an approach based on colored automata. In: Rinderle-Ma, S., Toumani, F., Wolf, K. (eds.) BPM 2011. LNCS, vol. 6896, pp. 132–147. Springer, Heidelberg (2011). https://doi.org/10.1007/978-3-642-23059-2_13

22. Montali, M.: Specification and Verification of Declarative Open Interaction Models. LNBIP, vol. 56. Springer, Heidelberg (2010). https://doi.org/10.1007/978-3-642-14538-4

23. Montali, M., Pesic, M., van der Aalst, W.M.P., Chesani, F., Mello, P., Storari, S.: Declarative specification and verification of service choreographies. ACM Trans. Web, 4(1), 1658376 (2010). ISSN 1559–1131. https://doi.org/10.1145/1658373.1658376. https://doi.org/10.1145/1658373.1658376

24. Montanari, U., Rossi, F.: Contextual nets. Acta Informatica 32, 545–596 (1995)

25. Mukkamala, R.R.: A formal model for declarative workflows: dynamic condition response graphs, Ph. D. thesis, IT University of Copenhagen (2012)

26. Mukkamala, R.R., Hildebrandt, T.T.: From dynamic condition response structures to büchi automata. In: 2010 4th IEEE International Symposium on Theoretical Aspects of Software Engineering, pp. 187–190, 2010. https://doi.org/10.1109/TASE.2010.22

27. Normann, H., Debois, S., Slaats, T., Hildebrandt, T.T.: Zoom and enhance: action refinement via subprocesses in timed declarative processes. In: Polyvyanyy, A., Wynn, M.T., Van Looy, A., Reichert, M. (eds.) BPM 2021. LNCS, vol. 12875, pp. 161–178. Springer, Cham (2021). https://doi.org/10.1007/978-3-030-85469-0_12

28. Pnueli, A.: The temporal logic of programs. In 18th Annual Symposium on Foundations of Computer Science (SFCS1977), pp. 46–57 (1977). https://doi.org/10.1109/SFCS.1977.32

29. Popova, V., Dumas, M.: From petri nets to guard-stage-milestone models. In: La Rosa, M., Soffer, P. (eds.) BPM 2012. LNBIP, vol. 132, pp. 340–351. Springer, Heidelberg (2013). https://doi.org/10.1007/978-3-642-36285-9_38

30. Prescher, J., Di Ciccio, C., Mendling, J.: From declarative processes to imperative models. SIMPDA **1293**, 162–173 (2014)
31. Raedts, I., et al.: Transformation of BPMN models for behaviour analysis. MSVVEIS **2007**, 126–137 (2007)
32. Slaats, T.: Flexible process notations for cross-organizational case management systems, Ph. D. thesis, IT University of Copenhagen (2015)
33. Srba, J.: Timed-arc Petri nets vs. networks of timed automata. In: Ciardo, G., Darondeau, P. (eds.) ICATPN 2005. LNCS, vol. 3536, pp. 385–402. Springer, Heidelberg (2005). https://doi.org/10.1007/11494744_22
34. Staines, T.S.: Intuitive mapping of UML 2 activity diagrams into fundamental modeling concept petri net diagrams and colored petri nets. In: 15th Annual IEEE International Conference and Workshop on the Engineering of Computer Based Systems (ECBS 2008), pp. 191–200 (2008). https://doi.org/10.1109/ECBS.2008.12
35. Thapa, D., Dangol, S., Wang, G.-N.: Transformation from petri nets model to programmable logic controller using one-to-one mapping technique. In: International Conference on Computational Intelligence for Modelling, Control and Automation and International Conference on Intelligent Agents, Web Technologies and Internet Commerce (CIMCA-IAWTIC'06), vol. 2, pp. 228–233 (2005). https://doi.org/10.1109/CIMCA.2005.1631473
36. Valk, R.: Infinite behaviour of Petri nets. Theoret. Comput. Sci. **25**(3), 311–341 (1983)
37. Verbeek, H.M.W., Wynn, M.T., van der Aalst, W.M., ter Hofstede, A.H.M.: Reduction rules for reset/inhibitor nets. J. Comput. Syst. Sci. **76**(2), 125–143 (2010)
38. Yang, N., Yu, H., Sun, H., Qian, Z.: Modeling UML sequence diagrams using extended Petri nets. In: 2010 International Conference on Information Science and Applications, pp. 1–8 (2010). https://doi.org/10.1109/ICISA.2010.5480384
39. Zaitsev, D.A.: Toward the minimal universal petri net. IEEE Trans. Syst. Man Cybern. Syst. **44**(1), 47–58 (2014). https://doi.org/10.1109/TSMC.2012.2237549

Enriching Heraklit Modules by Agent Interaction Diagrams

Daniel Moldt, Marcel Hansson, Lukas Seifert, Karl Ihlenfeldt,
Laif-Oke Clasen[✉], Kjell Ehlers, and Matthias Feldmann

Department of Informatics, Faculty of Mathematics, Informatics and
Natural Sciences, University of Hamburg, Hamburg, Germany
laif-oke.clasen@uni-hamburg.de
http://www.paose.de

Abstract. The modeling of systems in informatics has always been a challenge and the difficulty increases with the system's scale and complexity.
Since there is no direct way to turn complex systems into executable code, various modeling techniques are used to cover different perspectives of a system with models. These models must then be turned into code correctly and consistently. But how to create, structure, and compose the various models throughout the development process?
As a formal basis, Reisig proposes net modules, which inherit an associative calculus for composition. Practical modeling is addressed by the work of Fettke and Reisig with the HERAKLIT approach, which adopts especially the net modules as a basis. Based on this, we combined the HERAKLIT approach with our PAOSE approach and its multi-agent system elements.
As our main result we present HERAKLIT Interaction Diagrams which we obtain by enriching HERAKLIT modules by Agent Interaction Diagrams. We connect the concepts of reference nets and agents to HERAKLIT modules, and thereby construct HERAKLIT AGENTS.

Keywords: Petri Nets · Reference Nets · Modeling · HERAKLIT AGENTS · HERAKLIT Modules · Modularization · Associative Composition · Agent-oriented Petri Nets

1 Introduction

Informatics is always somehow involved in modeling. In [48] Thalheim emphasizes the importance of models and claims that on the level of Entity-Relationship (ER) models, information systems and conceptual database modeling have already been completely investigated. We are following his suggestion by investigating another branch of modeling: In this contribution, we are focussing on domain-specific language sets (DSL), which can directly be transformed into high-level Petri nets. Due to operational semantics, we follow a more process-oriented view than Thalheim. In application modeling, we currently focus primarily on software engineering, business informatics and distributed systems development.

Supported by participants of our teaching project classes and many student theses.

L. Gomes and R. Lorenz (Eds.): PETRI NETS 2023, LNCS 13929, pp. 440–463, 2023.
https://doi.org/10.1007/978-3-031-33620-1_23

In the context of software engineering Allen and Garlan made a similar proposal (see [2]): They suggested to define architectural connectors as explicit semantic entities, using a collection of protocols that characterize each of the participant roles in an interaction and how these roles interact.

During the last years, Reisig has developed an associative composition approach for Petri nets (see [43,44]) referring to [2]. He has extended his proposal of net modules from [42] in a relevant modeling way. While it narrows the system structures, it thereby provides a well-structured way to arrange system model components.

Reisig offers the associative calculus for the composition of systems, which is applicable in a wide range of system modeling and development. Working together with Fettke, he developed the HERAKLIT approach[1] (see http://www.heraklit.org) to address business informatics models. While their approach combines the Petri net composition calculus of Reisig without further restrictions to the business modeling area, we propose a more rigorous way and restrict the kinds of models even further.

Business models are nowadays often referred to as systems-of-systems or ultra-large-scale systems [23]. A modeling technique that matches the modeling requirements can be found in nets as tokens in net systems. The concept of nets-within-nets as defined by Valk in [50] can be found in several variants. As our modeling, execution and validation tool we use RENEW, which supports (Java) reference nets as defined by Kummer in [29]. Besides the true concurrency semantics of Petri nets, reference nets inherit the concept of synchronous channels and nets-within-nets paradigm.

Concerning application modeling, an adequate modeling paradigm is the use of agents or multi-agent systems (MAS) (see e.g. [24] for a standard definition). Agents and MAS extend the object-oriented concepts as they can incorporate social concepts in their models. Dynamic behavior, relationships etc. are enriched by the more abstract concept of interactions (e.g. speech acts), ontologies, belief-desire-intention (BDI) architectures, intrinsic goals, mobility, etc. In our PAOSE-approach (see [6,35]), we use reference nets as the basis of the units/agents that compose the system (multi-agent system). Reference nets offer the possibility to arbitrarily nested nets, by treating nets as a kind to objects in object-oriented software development. They can even have self-references.

In the context of this approach, we developed, based on the speech act theory of Austin [3] and Searle [47], our way of generating protocols for the communication of agents. As our modeling technique, we use AGENT INTERACTION PROTOCOL DIAGRAMS (AIP) (see [11]), which can be seen as a specialized kind of sequence diagram from UML. In [30] we showed how to verify such models for the interaction of agents with workflow concepts [52].

In this contribution, we combine our former research results and the new HERAKLIT approach of Fettke and Reisig [18,19]. This allows us to pick up the original idea of Allen and Garlan [2], as it directly supports our way of

---

[1] In the following, we use HERAKLIT in capital letters to refer to the work of Fettke and Reisig. For our terms we use HERAKLIT AGENTS or PAOSE in lower case.

building PAOSE applications based on AIPs. Reisig's associative calculus is the central background of the HERAKLIT modules. This calculus is a perfect fit for our AIPs. In this contribution, we therefore combine the PAOSE approach and the HERAKLIT approach to provide HERAKLIT AGENTS. With HERAKLIT AGENTS, our adapted AIPs, called HERAKLIT INTERACTION DIAGRAMS (HIDs), can be used to model the behavior of HERAKLIT modules in a similar way as for MULAN agents in PAOSE. While the use of HIDs limits the interaction modeling of HERAKLIT, we have the advantage of systematic construction of system models for which we have tool support.

In the following Sect. 2, we describe the context of our PAOSE-approach to clarify which concepts are used for HERAKLIT AGENTS, and how these concepts are applied. We introduce Fettke and Reisig's HERAKLIT modules in Sect. 3. Examples for modeling systems with the PAOSE approach can be found in Sect. 4. In Sect. 5, we describe how we conceptually extend MULAN agents to HERAKLIT AGENTS and how they can be generated. In Sect. 6 we explain the relationship of the HERAKLIT and PAOSE approaches before we conclude in Sect. 7.

2   PAOSE Background

The context of our PETRI NET-BASED, AGENT- AND ORGANIZATION-ORIENTED SOFTWARE ENGINEERING approach (PAOSE, [6]) with its manyfold starting points is explained briefly in the following.

As a basic concept we use agent-oriented Petri nets (see [4,32,38]). The key idea is to encode classes and objects via a specific structure of Petri nets. Kummer integrated this idea with the concepts of nets-within-nets [51] and synchronous channels [12] to invent the reference net formalism [29]. Helpful explanations can be found in [29], the RENEW handbook (as RENEW is the major tool to build models of reference nets) and several publications, e.g. see [6,8,9,37]. In this context the MULAN framework [28,45] with its main concept of the MULAN agent was developed. These agent-oriented Petri nets allow for a mapping of social metaphors of humans and social systems to our Petri net models.

2.1   Multi-agent System Modeling Paradigm

PAOSE is an approach for developing multi-agent systems. Cabac describes the modeling techniques and the MULAN foundations used in the PAOSE approach, which have been developed over the years, in [6].

Figure 1 shows a corresponding overview of the techniques and artifacts of PAOSE. [6,9] present details of the techniques and tools. In this contribution, there is a particular focus on AGENT INTERACTION PROTOCOL DIAGRAMS (AIPs). Therefore, these are considered in the separate Subsect. 2.3.

According to the PAOSE approach, a multi-agent system can be divided into three dimensions. The dimensions are:

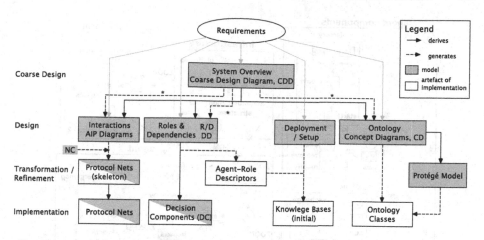

* From the system overview diagram a folder structure for the developed agent system is already generated.
This includes source for: basic ontology, basic R/D Diagram, basic AIP Diagrams, build files and start/setup scripts.

Fig. 1. Techniques and Artifacts of the PAOSE approach [6, p. 134]

1. **Structure**: Roles describe the structural aspects of a system. At runtime, agents adopt one or more roles. A role can be adopted by any number of agents. The agent-based modeling approach is an extension of object-oriented software development.
2. **Behavior**: Interactions describe the behavior. An interaction always has exactly one trigger, that starts a process within the system. A trigger can have parameters as further input to the system. Triggers are defined according to the size of the system, the application context and the capabilities of the participants in such a way that concrete solutions can be developed for them. One or more roles participate in an interaction. The participation of the roles take place in an order specified by the interaction modeled as an AIP.
3. **Ontology**: The ontology determines the common terminology of the agents, e.g. common language concepts. The ontology is determined application-wise for the system to be created.

The PAOSE approach applies some constraints on the net structures of individual net instances to the now well-established concepts for high-level Petri nets (via the underlying net template from which net instances are generated at runtime). This follows the idea of agent-oriented architectures as indicated by FIPA (see http://fipa.org/) for agent-oriented software engineering (AOSE) (see [24] for a consolidating definition). In [4,38], the net structure was restricted to model asynchronous communication of objects and agents. Synchronous communication was added as a structural gluing concept e.g. in [20,21,31,32]. Encapsulation of internal components and separation of concern was the driving modeling force, grounded by the true concurrency semantics of reference nets. A process

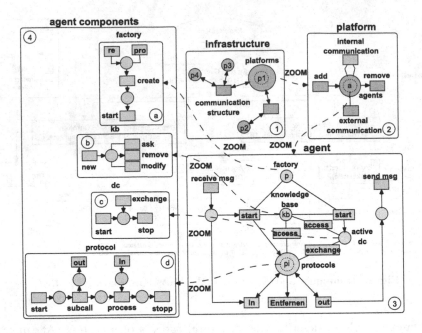

Fig. 2. The MULAN Petri net model. [6]

and agent integrating perspective (see e.g. [1,30,53,54]) is complementing the behavioral and structural modeling.

2.2   The MULAN Framework

For the structure of PAOSE models in general, the MULAN framework [45] is an important contribution and has been improved in several ways over the years. Figure 2 shows the general structure of the MULAN architecture. MULAN agents use a nested net structure (dashed arrows in Fig. 2 indicate that a reference net token points to the net instance). The use of references allows all net instances to be held as first-order objects and thus have references to them. An important advantage is that this allows for self-references. This implicitly opens up all Self-X concepts to our agent modeling systems. The MULAN framework consists of some key elements that allow each net to be used in several different environments, since references can be used arbitrarily often:

1. **Infrastructure:** The infrastructure supports the communication structure between platforms.
2. **Agent Platform:** An agent platform holds multiple agents and enables the communication between them. An agent platform can in itself be considered to be a special kind of agent.
3. **Agent:** The main part of MULAN is formed by the agents. They are the acting parts of the system.

4. **Agent Components:** An agent has a basic, specific structure. The tokens on the places are nets. These nets are synchronized via synchronous channels and the transitions of the agent structure. An agent can send and receive messages over synchronous channels to and from other agents via the platform (orange transitions in Fig. 2).

The net tokens used in the agent structure are (see left part of Fig. 2):

a) **Factory:** The factory is used to start instances of protocol nets or decision components via the **start** transition.

b) **Knowledge Base:** The knowledge base (short **KB**) contains the information an agent has. It is a net which has Strings of key-value pairs as tokens. Persistent data is kept within the KB, while all protocol nets and decision components cover transient data. A general perspective is that the KB usually covers the persistent states of an agent. The interface of the KB is accessed via the **access** transitions.

c) **Decision Components:** All decisions the agents take, based on their knowledge, are done in the decision component (short **DC**). DCs are arbitrary net structures that interact with the protocol nets and knowledge base via the **access** transitions. Calculations and computations are mainly done by decision component nets. Therefore DCs cover the functionality of an agent.

d) **Protocol:** The protocol nets represent the interaction between agents. Nets of these interactions in which an agent is participating are put in the place **protocols**. Each agent that participates in a trigger-specific communication adopts a part of the protocol net. Each protocol net part of an agent covers a specific part of the communication between agents defined for a trigger by an AIP.

The main structure of communication between agents is generated from AIPs (see Sect. 2.3 for details). Manually added inscriptions then represent the implementation part of protocol nets to make use of the agents' internal functions and states. Protocol nets are usually structured like branching processes or workflow nets with a finite number of repetitions. Instances executed by an agent are created by the factory when the agent receives a message for which it has not yet started a specific protocol net instance. Based on the reference net formalism, multiple instances of a single protocol net can be instantiated at the same time. The knowledge base contains the information about which protocol net to instantiate for which message as a key-value pair.

2.3   Agent Interaction Diagrams/Protocols

Sequence diagrams of UML model interactions between participants and show the sequence of the exchanged messages, where a message is a method call or signal. Agent interaction diagrams/protocols (AIPs), as described in [6], extend them with agent-oriented aspects, defined by the Foundation for Intelligent Physical Agents (FIPA, see http://fipa.org/) in the Agent Unified Modeling Language

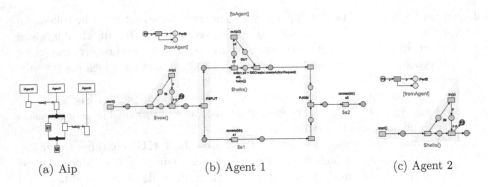

(a) Aip   (b) Agent 1   (c) Agent 2

Fig. 3. AIP transformed into Petri nets

(AUML) [22]. An AIP describes a conversation between roles of agents. To represent an agent's behavior for a given signal or message and its parameters, more than one partial order of the actions (causal nets) might have to be modeled. Therefore, control flow elements are introduced in AIPs as an extension of sequence diagrams of UML, through which alternatives and concurrency can be expressed. These elements can be categorized as *split* and *join* operators. *AND* splits and joins allow concurrency, *XOR* splits and joins allow explicit alternate path handling. *OR* operators split and join paths, but usually the semantics is not properly defined (see [5]). In AIPs operators split the lifeline horizontally and messages vertically. *OR* operators are not used in AIPs. Compared to UML not all kinds of messages can be drawn in RENEW (e.g.: lost and found messages are missing) as this usually requires a kind of global property of a system model. Sequence diagrams are extended for implementation purposes by annotations for KB and DC accesses.

The semantics of AIPs can be expressed by Petri nets, which enables both formal and operational semantics [10]. They can be transformed into Petri nets, also referred to as protocol nets, with a pattern based approach as they are mapped to reference net components as described in [5,6]. Protocol nets consist of the basic tasks of agents [22] that occur in agent protocols [10]. Basic tasks include, in particular, sending / receiving messages and the split / join operators. AIPs are better suited for our approach than extended UML sequence diagrams as they are, beside the two operators, extended with the KB and DC annotations to support an explicit design of the internal interfaces of a role already for interaction modeling. In addition speech acts [3,47] are supported by detailed message types with special semantics[2].

Within PAOSE, RENEW provides plugins that can model AIPs and convert them to protocol nets. Figure 3 a) shows a simple AIP modeled in RENEW, also with annotated lifelines and their conversion to Petri nets, to give an example of what protocol net generation looks like. Figure 3 b) and c) show the two roles involved for which the internal control flow structure is generated, here as a

---

[2] http://fipa.org/.

reference net. The initiating agent (Agent0) is not modeled because it generates the trigger of the process and is considered to be external to the process. Agent0 can be any kind of entity / unit inside the system or outside. In fact, it can be a time event or an agent-internal *decision* of any agent in the entire system. The $a1 and $a2 annotations indicate agent internal accesses to the KB of the agent. $hello represents an example for an internal calculation / action of the agent, performed as a necessary agent action during the conversation.

In addition to reference nets, AIPs can also be converted into other forms to serve other purposes such as P/T-nets for formal analyses [39]. A transformation into HERAKLIT modules with consistent semantics also allows for the execution of the AIPs and adds the aspect of composition as interaction. This is being worked on in RENEW and will be explained in Sect. 5.

3   HERAKLIT Background

In this section, we introduce the HERAKLIT modeling technique of Fettke and Reisig [14, 18] by explaining the concept of HERAKLIT modules and their composition. Whilst Fettke and Reisig describe HERAKLIT in three dimensions (architecture, static and dynamic), we are mainly interested in the architectural side of HERAKLIT. The architecture of HERAKLIT is described by Fettke and Reisig with three key components: modules, composition and refinements. Each of them will be presented here.

3.1   HERAKLIT Module

Modeling systems in terms of Petri nets has many advantages, especially for distributed systems, but the expressiveness of Petri nets means that a model can quickly become unwieldy as the size of the modeled system increases. Furthermore, changing small parts of the model can become quite complicated, as the entire model may need to be adjusted. Therefore, modularization plays an essential role in modeling large systems. It allows different parts of the overall system to be modeled as their own Petri nets or *submodules*. This reduces complexity and increases comprehensibility. These modules can then be combined to form a complete model of the system. Modularization can also aid in the validation of a system, since validating several small nets is generally less costly than validating their combination. Validation of such a submodule can be done via its interface, separate from any implementation. This further reduces the cost of validation. However, this requires a composition operation that preserves the properties of the combined parts.

Fettke and Reisig [18] have defined such a composition operation with their HERAKLIT modeling technique. They introduce HERAKLIT modules as graphs with interfaces. For the scope of this paper, however, we will only use Petri nets with interfaces and refer to them as *net module*.

Following [18], a net module, therefore, is a Petri net with a left and right interface. Each interface is a (possibly empty) subset of the union of a net's place and transition sets. Fettke and Reisig do not require the interfaces to be

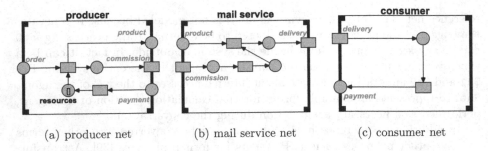

(a) producer net   (b) mail service net   (c) consumer net

Fig. 4. Three net modules, producer, mail service and consumer (Color figure online)

disjoint. In the general modeling we follow their semantics. For special context this can be restricted. Therefore, interfaces are not necessarily disjoint. For a given net module N, we will denote $^*N$ to refer to N's left interface, and N^* to refer to N's right interface. We will refer to all elements which are not part of an interface as *inner* elements of a net module. Elements in an interface subset have a label. Multiple elements can have the same label. Inner elements may have a label, but are not required to.

Definition 1: *A* net module *is a Petri net* $N = (P, T, F)$, *with* P *a set of places*, T *a set of transitions, and* $F \subseteq (T \times P) \cup (P \times T)$ *a set of arcs. A net module* N *has two interfaces: A left interface* $^*N \subseteq P \cup T$ *and a right interface* $N^* \subseteq P \cup T$. *Every element* $e \in {^*N} \cup N^*$ *has a label.*

As shown in Fig. 4, net modules are represented by a box. Transitions are represented by rectangles, places as ellipses and tokens as [], as can be seen in the *resource* place in the *producer* net. On the left and right side of the box, the interfaces of the net are highlighted by thick black lines forming a bracket. The elements that constitute the interface of a net are graphically placed on the brackets and their labels are highlighted in blue.

3.2 Composition

Two net modules can be *composed* to yield a new net module by combining one net module's right interface with the other net module's left interface. The following rules, analogous to those in [41], apply when composing two net modules $N_1 \cdot N_2$:

I Elements $a \in N_1^*$ and $b \in {^*N_2}$ are merged, if their label and type are the same (meaning: both are places or both are transitions, and the labels are the same). b is removed and all edges connected to it are connected to a instead, preserving the structure of N_2.

II Unmatched elements in unmatched elements in $^*N_2$ become part of $^*(N_1 \cdot N_2)$ and N_1^* become part of $(N_1 \cdot N_2)^*$.

III $^*N_1$ becomes a subset of $^*(N_1 \cdot N_2)$ and N_2^* becomes a subset of $(N_1 \cdot N_2)^*$.

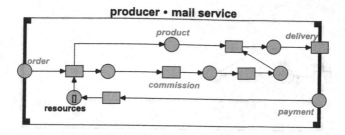

Fig. 5. The net module resulting from composing *producer* · *mail service*

When composing *producer* · *mail service*, the places labeled *product* and the transitions labeled *commission* are merged together and become inner elements of the newly composed module. The result can be seen in Fig. 5.

The composition operation presented so far is not well-defined and allows for some ambiguity: For any element of N_1^*, there might be multiple matching elements in $^*N_2$. Following rule I, it is not clear which of the matching elements in $^*N_2$ should be chosen. To eliminate this ambiguity and obtain a deterministic composition operation, Reisig defined in [41], as the basis for [18], to order all elements of an interface that have the same type and label by an *index function*. If there is more than one matching candidate element, the element with the lowest index is selected. Graphically, the order can be understood from top to bottom, with the top element having the lowest index. In the following, it is assumed that the elements of the interface are ordered and composition is used as a deterministic operation.

As shown by Reisig in [41], theorem 1, the deterministic version of the composition operation is associative. When composing *producer* · *mail service* · *consumer*, the order in which the composition operator is applied does not matter:
(*producer* · *mail service*) · *consumer* = *producer* · (*mail service* · *consumer*)
Fig. 5 shows the resulting net module of the composition (*producer* · *mail service*). If the *consumer* net module is combined with it, the net module in Fig. 7 results. Figure 6 shows the result of applying the composition operator to the *mail service* and *consumer* net modules first. Composing the *producer* net module with this result, Fig. 7 results as well.

However, the composition operation is not commutative. This can be seen clearly when comparing Fig. 7 with Fig. 8.

Refinement. The refinement and abstraction of modules are a core part of HERAKLIT's architectural view. Every module can be seen as the result of composing smaller modules, thus these smaller modules are refinements of the bigger module's parts. Applying this view recursively, we come to the net structures defining the behavior of the module. Inversely, abstraction allows us to view a module in a broader sense, separate from its concrete behavior. Multiple composed modules can, therefore, be seen as one single module.

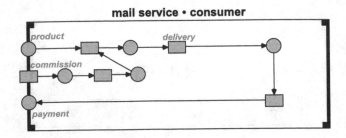

Fig. 6. The net module resulting from composing *mail service · consumer*

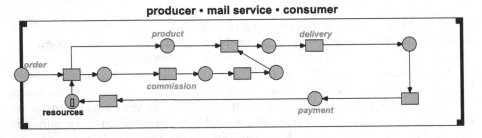

Fig. 7. The net module resulting from composing *producer · mail service · consumer*

4 Modularized Modeling Examples

After taking a brief look into Fettke and Reisig's approach to modeling with HERAKLIT, we will explore a classic example and RENEW's plugins using the PAOSE approach. By this we motivate our modeling motivation and the relation to multi-agent systems and software development, especially our RENEW modularization.

4.1 Modeling of Organizational Units

HERAKLIT concepts have already been taken up in the context of business modeling in various case studies (see e.g. [15–17]). HERAKLIT addresses in a constructive way the modeling of organizational units as can be found in normal organizations. Based on Petri nets and the idea of partial order semantics, causal nets are the backbone of the approach. Causal nets and scenarios, as used in HERAKLIT, are defined in [42]. Examples of this can be found in [18, Case studies], where Fettke and Reisig model different scenarios using causal nets and HERAKLIT modules. Ideas to extend the agent-oriented approach of PAOSE [6,9,36] can be found in [55]. Wester-Ebbinghaus describes how to model organizational units based on reference nets and interpret multi-agent systems as organizational units to reach a higher level of abstraction in system models.

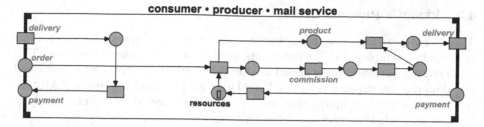

Fig. 8. The net module resulting from composing *consumer* • *producer* • *mail service*

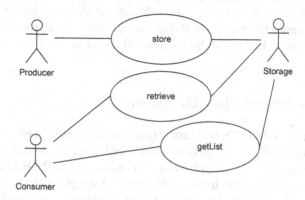

Fig. 9. Coarse Design Diagram (see [6, p. 264])

4.2 Software Engineering: Producer/Storage/Consumer Example

We'd like to broaden the context by introducing an example from the field of software engineering: The Producer/Storage/Consumer example (PSC) that Cabac proposes in [6, pp. 263–280] is a simple extended version of the classic Producer/Consumer problem. It consists of three roles (Producer, Storage, Consumer) and their associated interactions ("store", "getList", "retrieve"). An overview of these relationships is given in Fig. 9.

Using the PAOSE approach (see Sect. 2), Cabac utilized agent interaction protocol diagrams for his model in [6], thereby explicating the interfaces. Development of the example was conducted using RENEW and MULAN, which allowed for an agent-oriented implementation based on nets and Java code. Architectural components of the transformed PSC in our Java prototype reflect the static and the behavioral perspective in code. Roles cover states and functionality of the participating units. Interaction between roles is covered mainly by Java interfaces, supporting, like motivated in [2], a separation of interfaces and implementations. Details of that architecture are beyond the scope of this paper.

4.3 RENEW's Plugins

The agent approach of modeling systems is not only a theoretical concept, but has also been employed in the development of RENEW to some extend (see [37]).

RENEW's architecture is based on modularized plugins and was developed based on the concurrency theory of Petri [40] with the agent approach of MULAN in mind. RENEW's plugins, therefore, can be interpreted as agents (see [7]). RENEW as a whole can be seen as a multi-agent system [9].

Modeling techniques, as described in Fig. 1 are used. We currently focus on how to follow the original idea of Reisig (see [41]) and use his formal basis for our AIPs. AIPs, as used in PAOSE, directly inherit the associative composition, enabling us to apply the results of Reisig [41] to our Java modules resp. plugins. The here introduced HIDs shall be further enhanced for this kind of application area.

5 Heraklit Interaction Diagrams

HERAKLIT INTERACTION DIAGRAMS (HIDs) are restricted AIPs (see Sect. 2.3), that support the modeling of interactions of HERAKLIT modules. The restrictions result from our PAOSE approach. We associate a HERAKLIT module with an agent and derive the internal structure of the generated net modules for execution purposes. In the following first part we describe net modules, that are already mentioned in Sect. 3, with reference net semantics as the implementation of HERAKLIT modules. In the second part of this chapter we describe how HERAKLIT modules and their interfaces are generated from HIDs.

5.1 Extending Net Modules with Reference Nets

Using simple net modules to model the individual parts of a system has many advantages, but composing these parts can still result in large, complex, and difficult-to-understand nets. This can be solved by colored Petri nets, as Fettke and Reisig [18, 41] showed. Therefore, we use Petri nets in the form of higher level reference nets. Due to Kummer's definition of the high-level Petri net formalism of reference nets [29] condition/event nets (C/E nets, see [46, 49]) can also be seen as a basis. Beside the normal inscription, like in e.g. Coloured Petri Nets (see [25, 26]), reference nets provide synchronous channels. Synchronous channels go back to the work of the Weltlinie of Minkowski [34] picked up by Petri in his concurrency theory. Several authors have used the concept of synchronization. Jessen and Valk used it in [27] and Christensen and Hansen extended the definition of Coloured Petri Nets in [12].

In reference nets, synchronous channels are one of the core concepts. Reference nets do not have a page concept as in the original Coloured Petri Net definition of Kurt Jensen [25]. However, as reference nets have been implemented in Java, net templates can be seen as net classes and net instances as net objects. References to objects are created at instantiation time and can be used as tokens. While in Java methods of objects are being used to communicate between objects, we use synchronous channels to establish communication between net objects in reference nets.

Synchronous channels as introduced by [12] and used in the reference net formalism of Kummer [29] allow to bind an arbitrary large number of transitions to be synchronized via one or more synchronous channels. The underlying mechanisms are elaborated unification algorithms. These are used to realize several formalisms especially for reference nets and P/T-nets. Synchronous channels are defined by an *uplink* and a *downlink*. When a transition inscribed with a downlink fires, it binds itself to a transition inscribed with its channel's uplink and causes this transition to fire as well.

In reference nets, the tokens can be references to other nets and can, therefore, be used as such. The combination of tokens to other net instances and synchronous channels on transitions support dynamic net structures in a way similar to object based systems. As synchronous channels attached to transitions can be seen as method calls, values can be passed by parameters to and from the called transition. The uplink can be interpreted as a method's signature, and a downlink as a method call. The number of defined parameters is part of an uplink's definition. All unbound variables need to be bound for the synchronization. This mechanism allows us to execute arbitrarily complex algorithms in a single step being distributed over any number of net instances and their transitions with the pairwise bound synchronous channels.

We can combine the concepts of net modules and reference nets by extending our previous net module semantics for colored nets, and defining the behavior for merging transitions inscribed with synchronous channels. As shown by Reisig in [41], Sect. 4.3, the coloring of a net does not affect the composition or its properties. For colored nets N_1, N_2, we can sum up the tokens of to-be-merged places at their joint place in $N_1 \cdot N_2$ and thereby preserve associativity. For places $p_1 \in N_1^*, p_2 \in {}^*N_2$, transitions $t_1 \in N_1^*, t_2 \in {}^*N_2$, and inscribed arcs between p_1, t_1 and p_2, t_2, we can add the weights of the arcs together when merging p_1, p_2 and t_1, t_2, retaining associativity. Overall, we follow the same rules as in [18].

We use uplinks to model an agent's reception of a message or signal, and downlinks to model an agent sending a message or signal. Both of these actions represent an agent's interaction with its environment. Since we want the inner workings of an agent to be clearly distinguishable from its interaction with its environment. Only inter agent interaction is used for the composition, synchronous channels used for intra net communication are retained.

Furthermore, the interactions in which an agent can participate in are determined by its interface. Communication between agents, therefore, is defined by their composition. This is transferred directly to HERAKLIT modules, which places the focus on the interaction of the modules.

Therefore, we have to amend rule I of the previous composition rules (see Sect. 3.2), as we want the communication pattern of agents to be determined by their respective interfaces and their composition:

I.T Transitions $t_1 \in N_1^*$ and $t_2 \in {}^*N_2$ are merged, iff their label is the same, and either

a) t_1 is inscribed with an uplink and t_2 has its corresponding downlink, or

b) t_2 is inscribed with an uplink and t_1 has its corresponding downlink, or

c) neither has a synchronous channel inscription.

t_2 is removed and all edges connected to it are instead connected to t_1, retaining N_2's structure. Both t_1's and t_2's inter module up- and downlinks are removed, if they exist. All variables that need to be set as parameters between the two transitions are bound consistently. We assume that the respective internal module variables are bound locally and annotated consistently at the now internal transition t_1. If the variables of the parameters are locally named differently ($t_1(x)$ vs. $t_2(y)$), the corresponding variables are renamed accordingly.

I.P Places $p_1 \in N_1^*$ and $p_2 \in {}^*N_2$ are merged, iff their labels are the same. p_2 is removed, and all edges connected to it are instead connected to p_1, retaining N_2's structure. Any token contained in p_2 is added to p_1.

As discussed in Sect. 3.2, we assume all interface elements to be ordered like in [18]. This yields unambiguous communication patterns between the HERAK-LIT modules, since the composition of HERAKLIT modules is well-defined by this.

5.2   Generating HERAKLIT Modules

Our current goal is to enable an automatic translation of AIPs, now called HIDs due to the HERAKLIT context, into net modules in RENEW. For each HER-AKLIT module in an HID a (reference) net module shall be generated. The modeling basis for this generation is the transfer of our agent modeling concepts to the HERAKLIT approach.

In Sect. 2.3 we described how Petri nets/reference nets are generated from AIPs.[3]

For MULAN agents, there is a clear separation of the interfaces, the internal states, and the functionality by additional parts of the agent model. By restricting HERAKLIT modules in a very similar way, we thereby decrease the modeling capabilities of HERAKLIT modules and require more modules in total. These additional modules, however, can easily be added, as MULAN/PAOSE models already define a reference architecture. An example of a net module generated from the same AIP as the net in Fig. 3, is shown in Fig. 10. Internal parts are omitted, but the $a1$ annotation shows how to access the internal parts of an agent (the KB in this example), which could be any internal implementation function of a module.

As one agent can be involved in multiple interactions for different triggers, different parts of the agent would be translated into different net modules. The

---

[3] RENEW can be used to generate net modules also based on net components, as described in [6]. The structure of the reference net components can be kept. Additional elements can be added to model inner parts of agents' behaviors. In AIPs these were DCs and KBs.

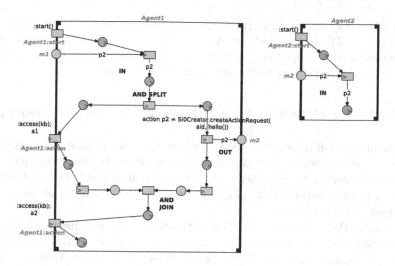

Fig. 10. AIP transformed into two net modules

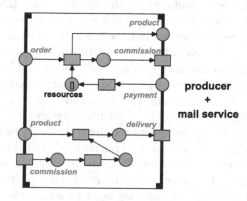

Fig. 11. The union over the *producer* and *mail service* nets

composition of the modules would not be sufficient, as the automatically generated modules could have unintentionally matching elements in their interfaces. We can, for example, interpret the *producer* and *mail service* modules from Fig. 4 as two interactions of one agent. Composing the two modules would result in Fig. 5, which does represent the wanted behavior. The agent shouldn't interact with himself, but instead offer exactly those elements in its interface, as the *producer* and *mail service* nets dictate. We can achieve this by performing a union over the two nets. The result is shown in Fig. 11.

HIDs specify the roles and interactions of HERAKLIT AGENTS without revealing the internal implementation. Following this modeling perspective, an HID does not only specify the interfaces, but also the different communication partners and therefore the composition. One HID models one interaction of a specific set of roles, which together provide a service to the whole system. One role can be part of several interactions. Interfaces between the roles directly result from an interaction. The union of all interfaces allows a role to encapsulate its internal functions and states, as they do not have any connection to the environment outside the role, respective to the agent that has this role.

6   Modeling HERAKLIT Modules with MULAN Concepts

The communication of agents is very flexible. Our agents have no limit on the number of partners with which they may communicate. In principle, every agent can communicate with any other agent. However, in a good system design the communication is well-defined. This is covered by the definition of the possible interactions within a system.

We model the communication flow with our interaction diagrams, the AIPs Sect. 2.3, but these do not restrict how agents may interact. Any kind of interaction protocol can be designed. For a given system, the PAOSE approach defines all possible triggers of interactions as starting points. Assuming a static structure for the Petri-net based agents, where the set of possible interactions for each agent does not change, the cardinality of each agent's interface is limited. A clear definition of the allowed interaction and the interface of each agent exists.

However, MULAN agents, and thus our entities in PAOSE, are based on reference nets (see Sect. 2). By hierarchizing nets within nets and using synchronous channels, they can generate new objects and structures and thereby learn new interaction patterns. While constraints on communication channels and partners are useful for modeling software systems with unique method calls as a form of communication, general-purpose agents require a dynamic interface.

To some extent, this goes beyond the approach of HERAKLIT, where models aim for a static system interface. To remedy this, we propose to extend HERAKLIT modules to HERAKLIT AGENTS in the spirit of MULAN agents from PAOSE. In order to achieve this, we need to adapt the mechanisms built into our PAOSE multi-agent systems: Besides introducing some kind of knowledge base and decision component, a mechanism for starting new net instances is necessary. The HIDs can cover this if they are modeled by reference nets, for example. In addition, agents need to provide a factory-like net component that can instantiate nets based on received tokens (e.g. messages from other parts of the system). However, these dynamic options are beyond the scope of this paper and will be explained in further publications.

Therefore, in this paper, we restrict the HERAKLIT AGENTS to a static interface. Due to this, HERAKLIT AGENTS have a clear separation into a) inter agent communication represented by HIDs and b) internal functionality and states, modeled by further net models that are encapsulated inside the HERAKLIT AGENTS.

HERAKLIT AGENTS are more specialized than HERAKLIT modules. Both have graphs as a basis, but with HERAKLIT AGENTS we restrict the possible forms of the graph. In particular, we propose to model the internal structure as an agent-oriented Petri net (see Sect. 2). In this contribution, we restrict even every HERAKLIT AGENT to a single role. We also consider every HERAKLIT module to be its own agent. Even considering net atoms of HERAKLIT as models, despite their very limited capabilities, they are nevertheless actors in an environment.

In HERAKLIT, composition of HERAKLIT modules follows the ideas of [41] in an elaborate way. This corresponds with our modeling of the interaction of MULAN agents by AIPs. This is why we can apply our AIP/HID modeling to HERAKLIT AGENTS. In this way, a role aggregates the behaviors as HIDs. For each trigger and its corresponding process, each participating role covers its part of a HID. Since several HIDs/processes exist for a system, roles aggregate those partial interaction processes. Furthermore, roles (and hence HERAKLIT AGENTS) contain the internal functionality and states, that are addressed by the interaction.

In terms of a multi-agent system, we use the roles as structuring elements and the HERAKLIT AGENT as the executing elements. The question now is how to compose whole systems with the HERAKLIT approach based on our way of modeling. One interaction in one HID describes a composition like $A \cdot B \cdot C$ (see Sect. 3). A second interaction can be $E \cdot F \cdot G$. In a special case where A and E belong to the same module in the HERAKLIT modeling paradigm, this can also be covered. However, it is necessary to have unique interface labels. The result is to have AE as the new HERAKLIT module. If B, C, F and G are independent, then the composition can be written as $AE \cdot B \cdot C \cdot F \cdot G$. The ordering of B and C, as well as the ordering of F and G, must be kept, however, the two sequences can be interleaved due to the disjunct labeling that we require in our HID.

Using reference nets for our HERAKLIT AGENTS also provides precise semantics and allows dynamic net structures via synchronous channels and net instances, which are not provided in HERAKLIT itself. This allows us to reuse concepts of our agent approach directly without the need to alter these in drastic ways. An examples is the internal functionality and the state of an agent, usually modeled as a knowledge base and decision component (see [6]). Noteworthy is that functionality and state are encapsulated within a HERAKLIT AGENT as there is no intersection with the interface. Just as in our agent approach, they support the concept that the behavior of an agent is influenced by its internal state. Overall, the internal structure of HERAKLIT AGENT is given by our agent-oriented Petri net, whose features are similar to our MULAN agent.

Our multi-agent system from PAOSE also follows the principle of agents residing in platforms to communicate, and the platform itself is considered an agent as well. This form of hierarchization is similar to Heraklits' concept of abstraction, where the composition of modules is in itself a module. We can also apply this concept to our HERAKLIT AGENTS: The composition of different HERAKLIT AGENTS results in a HERAKLIT AGENT, which can be seen as a platform.

As can be seen, we have two forms of hierarchization: The composition of multiple HERAKLIT AGENTS into one, and the division of a HERAKLIT AGENT into various net instances.

In a more detailed view, the HERAKLIT and PAOSE approaches are easily integrated due to the underlying basis of Petri nets: Heraklit modules can be modeled inside of other modules. The overall behavior is then internalized by their composition. Only the interfaces remain accessible in the case that they are not merged inside the module. This is the central property derived from the associative composition calculus of Reisig [41] for Petri net models. Together, Fettke and Reisig extend this to HERAKLIT models, which can, therefore, also be composed in an associative way.

In PAOSE, agents are composed in a similar way. However, the way how a modeler proceeds is different. The whole system is separated by different perspectives (see Fig. 1). In the practical modeling context, modelers start with the idea to cover the reaction of a triggering event of a system.

One can start with a simple scenario and then add roles or interactions as necessary from the application perspective. Indeed, the HIDs start with a simple trigger (initial marking) to model a scenario that can be covered by a causal net model. Adding parameters to the trigger or taking different internal states of the system into consideration, the model covers alternatives (leading to branching processes as models [13]). The case of simple repetitions is covered by loops inside the models, leading from a branching process to a workflow like net [52]. Such an example of a model can be seen in Fig. 12. This kind of net can be tested in PAOSE by considering the initial marking of a net to be the messages that are sent to an agent. In MULAN, the interface of an agent is modeled by only *two* transitions. These *in* and *out* transitions (called *receive msg* and *send msg* in Fig. 2) can be synchronized with the platform of an agent to handle the communication with other agents. All possible markings coming from the *in* transition are the possible initial markings of the respective agent net.

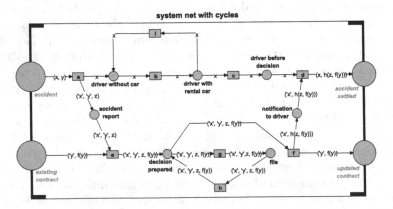

Fig. 12. System net with cycles, from [18, figure 20]

In PAOSE, the internal structure of an agent is fixed. Flexibility comes from the use of net instances that are used inside the agent. All these instances are related in a special structure via synchronous channels. In this way, we can cover a clear separation of the control flow (for which the associative calculus provides the background), the functionality and states.

In the application areas of an organization, workflows are usually used to cover the control flow. The execution of the real work is hidden in so called internal functions. In the context of software architecture, the design of software component interactions is normally modeled by static interfaces. Dynamic aspects are covered e.g. by the contract model of Meyer [33]. In PAOSE this is integrated into the design of the AIPs, which can be considered as contracts. HIDs now transfer this idea to HERAKLIT modules. Restricting the HERAKLIT modules further leads to the notion of HERAKLIT AGENTS.

7   Conclusion

Summary. In this paper we proposed a refinement of our AIPs in the form of HERAKLIT Interaction Diagrams (HIDs). We use these to generate net modules, which will form the basis of the interface components of our HERAKLIT AGENT. They specify the interactions a role is part of. All interactions merged together form the interactions one agent is capable of. To do this, we take the union over the net modules.

We introduced the HERAKLIT AGENTS as specialized HERAKLIT modules. HERAKLIT AGENTS are a fusion of the concepts of HERAKLIT with our agent concepts from PAOSE and MULAN. Internally, it contains all the aspects of an agent, such as a knowledge base and a decision component, and on its interface, it has components corresponding to the possible calls as given by the HID.

Finally, we have shown the usefulness of our approach with modeling examples.

Outlook. Currently, we use a mapping of each agent having one role. In PAOSE, an agent can also take on different roles. Currently, we restricted this for our HERAKLIT AGENTS. In the future, we can imagine to combine different roles into one HERAKLIT AGENT via a union operation and thus cover this aspect of PAOSE as well. The union operation would merge the interface elements and other net structures. We considered using the composition for this, but we would, for example, need to either relabel interface elements or to alter the agents to be merged. This would be necessary, as two agents that are to be merged could accidentally have matching gates due to being automatically generated. This would cause unwanted behavior.

Currently, we more or less use and adapt the architectural dimension of HERAKLIT. It could be interesting to see how the static dimension could fit into our PAOSE approach. Here the intentions of [2] and our PAOSE approach seem to fit very well to address the HERAKLIT approach.

References

1. van der Aalst, W., Moldt, D., Valk, R., Wienberg, F.: Enacting interorganizational workflows using nets in nets. In: Becker, J., Mühlen, M., Rosemann, M. (eds.) Proceedings of the 1999 Workflow Management Conference Workflow-based Applications, Münster, Nov. 9th 1999, pp. 117–136. Working Paper Series of the Department of Information Systems, University of Münster, Department of Information Systems, Steinfurter Str. 109, 48149 Münster (1999), working Paper No. 70

2. Allen, R., Garlan, D.: A formal basis for architectural connection. ACM Trans. Softw. Eng. Methodol. **6**(3), 213–249 (1997). https://doi.org/10.1145/258077.258078

3. Austin, J.L.: How to do things with words. Harvard University Press (1962)

4. Becker, U., Moldt, D.: Objekt-orientierte Konzepte für gefärbte Petrinetze. In: Scheschonk, G., Reisig, W. (eds.) Petri-Netze im Einsatz für Entwurf und Entwicklung von Informationssystemen, pp. 140–151. Gesellschaft für Informatik, Springer-Verlag, Berlin Heidelberg New York, Informatik Aktuell (1993)

5. Cabac, L.: Modeling agent interaction protocols with AUML diagrams and petri nets. Diploma thesis, University of Hamburg, Department of Computer Science, Vogt-Kölln Str. 30, D-22527 Hamburg (2003)

6. Cabac, L.: Modeling petri net-based multi-agent applications. Dissertation, University of Hamburg, Department of Informatics, Vogt-Kölln Str. 30, D-22527 Hamburg (2010). https://ediss.sub.uni-hamburg.de/handle/ediss/3691

7. Cabac, L., Duvigneau, M., Moldt, D., Rölke, H.: Applying multi-agent concepts to dynamic plug-in architectures. In: Müller, J.P., Zambonelli, F. (eds.) AOSE 2005. LNCS, vol. 3950, pp. 190–204. Springer, Heidelberg (2006). https://doi.org/10.1007/11752660_15

8. Cabac, L., Haustermann, M., Mosteller, D.: Renew 2.5 – towards a comprehensive integrated development environment for Petri net-based applications. In: Kordon, F., Moldt, D. (eds.) PETRI NETS 2016. LNCS, vol. 9698, pp. 101–112. Springer, Cham (2016). https://doi.org/10.1007/978-3-319-39086-4_7

9. Cabac, L., Haustermann, M., Mosteller, D.: Software development with Petri nets and agents: approach, frameworks and tool set. Sci. Comput. Program. **157**, 56–70 (2018). https://doi.org/10.1016/j.scico.2017.12.003

10. Cabac, L., Moldt, D.: Formal semantics for AUML agent interaction protocol diagrams. In: Odell, J., Giorgini, P., Müller, J.P. (eds.) AOSE 2004. LNCS, vol. 3382, pp. 47–61. Springer, Heidelberg (2005). https://doi.org/10.1007/978-3-540-30578-1_4

11. Cabac, L., Moldt, D., Rölke, H.: A proposal for structuring petri net-based agent interaction protocols. In: van der Aalst, W.M.P., Best, E. (eds.) ICATPN 2003. LNCS, vol. 2679, pp. 102–120. Springer, Heidelberg (2003). https://doi.org/10.1007/3-540-44919-1_10

12. Christensen, S., Hansen, N.D.: Coloured Petri nets extended with channels for synchronous communication. Tech. Rep. DAIMI PB-390, Aarhus University (1992)

13. Engelfriet, J.: Branching processes of Petri nets. Acta Informatica **28**, 575–591 (1991)

14. Fettke, P., Reisig, W.: Modelling service-oriented systems and cloud services with Heraklit. CoRR abs/2009.14040 (2020). https://arxiv.org/abs/2009.14040

15. Fettke, P., Reisig, W.: Heraklit case study: adder (2020). Heraklit working paper, v1, 5 December 2020. https://www.heraklit.org

16. Fettke, P., Reisig, W.: Heraklit case study: parallel adder (2020). Heraklit working paper, v1, 5 December 2020. https://www.heraklit.org
17. Fettke, P., Reisig, W.: Heraklit case study: retailer (2020). Heraklit working paper, v1, 21 December 2020. https://www.heraklit.org
18. Fettke, P., Reisig, W.: Handbook of Heraklit (2021). Heraklit-working paper, v1.1, 10 September 2021. https://www.heraklit.org
19. Fettke, P., Reisig, W.: Modellieren mit Heraklit. In: Riebisch, M., Tropmann-Frick, M. (eds.) Modellierung 2022, 27. Juni - 01. Juli 2022, Hamburg, Deutschland. LNI, vol. P-324, pp. 77–92. Gesellschaft für Informatik e.V. (2022). https://doi.org/10.18420/modellierung2022-005
20. Fix, J.: Emotionale Agenten: Darstellung der emotionstheoretischen Grundlagen und Entwicklung eines Referenzmodells auf Basis einer petrinetz-basierten Modellierungstechnik. Dissertation, University of Hamburg, Department of Informatics, Vogt-Kölln Str. 30, D-22527 Hamburg (2012). https://ediss.sub.uni-hamburg.de/volltexte/2012/5968/
21. Fix, J., Duvigneau, M., Moldt, D.: Bereitstellung eines Synchronisationsmechanismus für MULAN basierte Agenten. In: Bergenthum, R., Desel, J. (eds.) Algorithmen und Werkzeuge für Petrinetze. 18. Workshop AWPN 2011, pp. 8–14. Hagen, September 2011. Tagungsband (2011)
22. Foundation for Intelligent Physical Agents: FIPA interaction protocol library specification (2000). https://fipa.org/specs/fipa00025/PC00025C.html
23. Gabriel, R.P., Northrop, L.M., Schmidt, D.C., Sullivan, K.J.: Ultra-large-scale systems. In: Tarr, P.L., Cook, W.R. (eds.) Companion to the 21th Annual ACM SIGPLAN Conference on Object-Oriented Programming, Systems, Languages, and Applications, pp. 632–634. OOPSLA 2006, 22–26 October 2006, Portland, Oregon, USA. ACM (2006). https://doi.org/10.1145/1176617.1176645
24. Jennings, N.R.: On agent-based software engineering. Artif. Intell. 117(2), 277–296 (2000)
25. Brauer, W.., Reisig, W.., Rozenberg, G. (eds.).: Petri Nets: Central Models and Their Properties: ACPN 1986, Part I Proceedings of an Advanced Course Bad Honnef, 8–19 September 1986. LNCS, vol. 254. Springer, Heidelberg (1987). https://doi.org/10.1007/978-3-540-47919-2
26. Jensen, K., Kristensen, L.M.: Colored petri nets: a graphical language for formal modeling and validation of concurrent systems. Commun. ACM 58(6), 61–70 (2015)
27. Jessen, E., Valk, R.: Rechensysteme: Grundlagen der Modellbildung. Springer, Heidelberg, Studienreihe Informatik (1987). https://doi.org/10.1007/978-3-642-71120-6
28. Köhler, M., Moldt, D., Rölke, H.: Modelling Mobility and Mobile Agents Using Nets within Nets. In: van der Aalst, W.M.P., Best, E. (eds.) ICATPN 2003. LNCS, vol. 2679, pp. 121–139. Springer, Heidelberg (2003). https://doi.org/10.1007/3-540-44919-1_11
29. Kummer, O.: Referenznetze. Logos Verlag, Berlin (2002). https://www.logos-verlag.de/cgi-bin/engbuchmid?isbn=0035&lng=eng&id=
30. Lehmann, K., Moldt, D.: Modelling and Analysis of Agent Protocols with Petri Nets. In: Lindemann, G., Denzinger, J., Timm, I.J., Unland, R. (eds.) MATES 2004. LNCS (LNAI), vol. 3187, pp. 85–98. Springer, Heidelberg (2004). https://doi.org/10.1007/978-3-540-30082-3_7
31. Maier, C.: Objektorientierte Analyse mit gefärbten Petrinetzen. Diploma thesis, University of Hamburg, Department of Computer Science (1997)

32. Maier, C., Moldt, D.: Object coloured petri nets - a formal technique for object oriented modelling. In: Agha, G.A., De Cindio, F., Rozenberg, G. (eds.) Concurrent Object-Oriented Programming and Petri Nets. LNCS, vol. 2001, pp. 406–427. Springer, Heidelberg (2001). https://doi.org/10.1007/3-540-45397-0_16

33. Meyer, B.: Object-Oriented Software Construction. Prentice Hall, London (1988)

34. Minkowski, H.: Raum und Zeit. Vortrag, gehalten auf der 80 Naturforscherversammlung zu Köln am 21 September 1908. B. G. Teubner (1909)

35. Moldt, D.: Petrinetze als Denkzeug. In: Farwer, B., Moldt, D. (eds.) Object Petri Nets, Processes, and Object Calculi, pp. 51–70. No. FBI-HH-B-265/05 in Report of the Department of Informatics, University of Hamburg, Department of Computer Science, Vogt-Kölln Str. 30, D-22527 Hamburg (2005)

36. Moldt, D.: PAOSE: A way to develop distributed software systems based on Petri nets and agents. In: Barjis, J., Ultes-Nitsche, U., Augusto, J.C. (eds.) Proceedings of The Fourth International Workshop on Modelling, Simulation, Verification and Validation of Enterprise Information Systems (MSVVEIS2006), 23–24 May 2006 - Paphos, Cyprus 2006, pp. 1–2 (2006)

37. Moldt, D., et al.: Renew: Modularized architecture and new features. In: Gomes, L., Lorenz, R. (eds.) Application and Theory of Petri Nets and Concurrency - 44th International Conference, PETRI NETS 2023, Lisboa, Portugal, 26–30 June 2023, Proceedings. Lecture Notes in Computer Science, vol. this volume. Springer (2023)

38. Moldt, D., Wienberg, F.: Multi-agent-systems based on coloured petri nets. In: Azéma, P., Balbo, G. (eds.) ICATPN 1997. LNCS, vol. 1248, pp. 82–101. Springer, Heidelberg (1997). https://doi.org/10.1007/3-540-63139-9_31

39. Mosteller, D., Cabac, L., Haustermann, M.: Providing Petri net-based semantics in model driven-development for the Renew meta-modeling framework. In: PNSE @ Petri Nets. CEUR Workshop Proceedings, vol. 1372, pp. 99–114. CEUR-WS.org (2015)

40. Petri, C.A.: Concurrency theory. In: Brauer, W., Reisig, W., Rozenberg, G. (eds.) ACPN 1986. LNCS, vol. 254, pp. 4–24. Springer, Heidelberg (1987). https://doi.org/10.1007/978-3-540-47919-2_2

41. Reisig, W.: Simple composition of nets. In: Franceschinis, G., Wolf, K. (eds.) PETRI NETS 2009. LNCS, vol. 5606, pp. 23–42. Springer, Heidelberg (2009). https://doi.org/10.1007/978-3-642-02424-5_4

42. Reisig, W.: Understanding petri nets - modeling techniques, analysis methods, case studies. Springer (2013). https://doi.org/10.1007/978-3-642-33278-4

43. Reisig, W.: Associative composition of components with double-sided interfaces. Acta Informatica 56(3), 229–253 (2019). https://doi.org/10.1007/s00236-018-0328-7

44. Reisig, W.: Composition of component models - a key to construct big systems. In: Margaria, T., Steffen, B. (eds.) ISoLA 2020. LNCS, vol. 12477, pp. 171–188. Springer, Cham (2020). https://doi.org/10.1007/978-3-030-61470-6_11

45. Rölke, H.: Modellierung von Agenten und Multiagentensystemen - Grundlagen und Anwendungen, Agent Technology - Theory and Applications, vol. 2. Logos Verlag, Berlin (2004). https://logos-verlag.de/cgi-bin/engbuchmid?isbn=0768&lng=eng&id=

46. Rozenberg, G.: Behaviour of elementary net systems. In: Brauer, W., Reisig, W., Rozenberg, G. (eds.) Petri Nets: Central Models and Their Properties, Advances in Petri Nets 1986, Part I, Proceedings of an Advanced Course, Bad Honnef, Germany, 8–19 September 1986. LNCS, vol. 254, pp. 60–94. Springer, Heidelberg (1986). https://doi.org/10.1007/BFb0046836

47. Searle, J.R.: Speech acts: an essay in the philosophy of language. Cambridge University Press (1969). https://doi.org/10.1017/CBO9781139173438

48. Thalheim, B.: Models: the fourth dimension of computer science. Softw. Syst. Model. **21**(1), 9–18 (2022). Accessed 27 Aug 2022

49. Thiagarajan, P.S.: Elementary net systems. In: Brauer, W., Reisig, W., Rozenberg, G. (eds.) Petri Nets: Central Models and Their Properties, Advances in Petri Nets 1986, Part I, Proceedings of an Advanced Course, Bad Honnef, Germany, 8–19 September 1986. LNCS, vol. 254, pp. 26–59. Springer (1986). https://doi.org/10.1007/BFb0046835

50. Valk, R.: Modelling of task flow in systems of functional units. Tech. Rep. FBI-HH-B-124/87, University of Hamburg, Department of Computer Science, Vogt-Kölln Str. 30, D-22527 Hamburg (1987)

51. Valk, R.: Petri nets as token objects - an introduction to elementary object nets. In: Desel, J., Silva, M. (eds.) 19th International Conference on Application and Theory of Petri nets, Lisbon, Portugal, pp. 1–25. No. 1420 in Lecture Notes in Computer Science, Springer-Verlag, Heidelberg (1998). https://doi.org/10.1007/3-540-69108-1_1

52. Verbeek, H.M.W., Basten, T., van der Aalst, W.M.P.: Diagnosing workflow processes using Woflan. Comput. J. **44**(4), 246–279 (2001)

53. Wagner, T.: Petri net-based combination and integration of agents and workflows, Ph. D. thesis, University of Hamburg, Department of Informatics, Vogt-Kölln Str. 30, D-22527 Hamburg (2018). https://ediss.sub.uni-hamburg.de/volltexte/2018/8995/

54. Wagner, T., Schmitz, D., Moldt, D.: Paffin: implementing an integration of agents and workflows. In: Criado Pacheco, N., Carrascosa, C., Osman, N., Julián Inglada, V. (eds.) EUMAS/AT -2016. LNCS (LNAI), vol. 10207, pp. 67–75. Springer, Cham (2017). https://doi.org/10.1007/978-3-319-59294-7_7

55. Wester-Ebbinghaus, M.: Von Multiagentensystemen zu Multiorganisationssystemen - Modellierung auf Basis von Petrinetzen. Dissertation, University of Hamburg, Department of Informatics, Vogt-Kölln Str. 30, D-22527 Hamburg (2010). https://ediss.sub.uni-hamburg.de/handle/ediss/3920

Author Index

L. Gomes and R. Lorenz (Eds.): PETRI NETS 2023, LNCS 13929, pp. 465–466, 2023.
https://doi.org/10.1007/978-3-031-33620-1

Printed in the United States
by Baker & Taylor Publisher Services